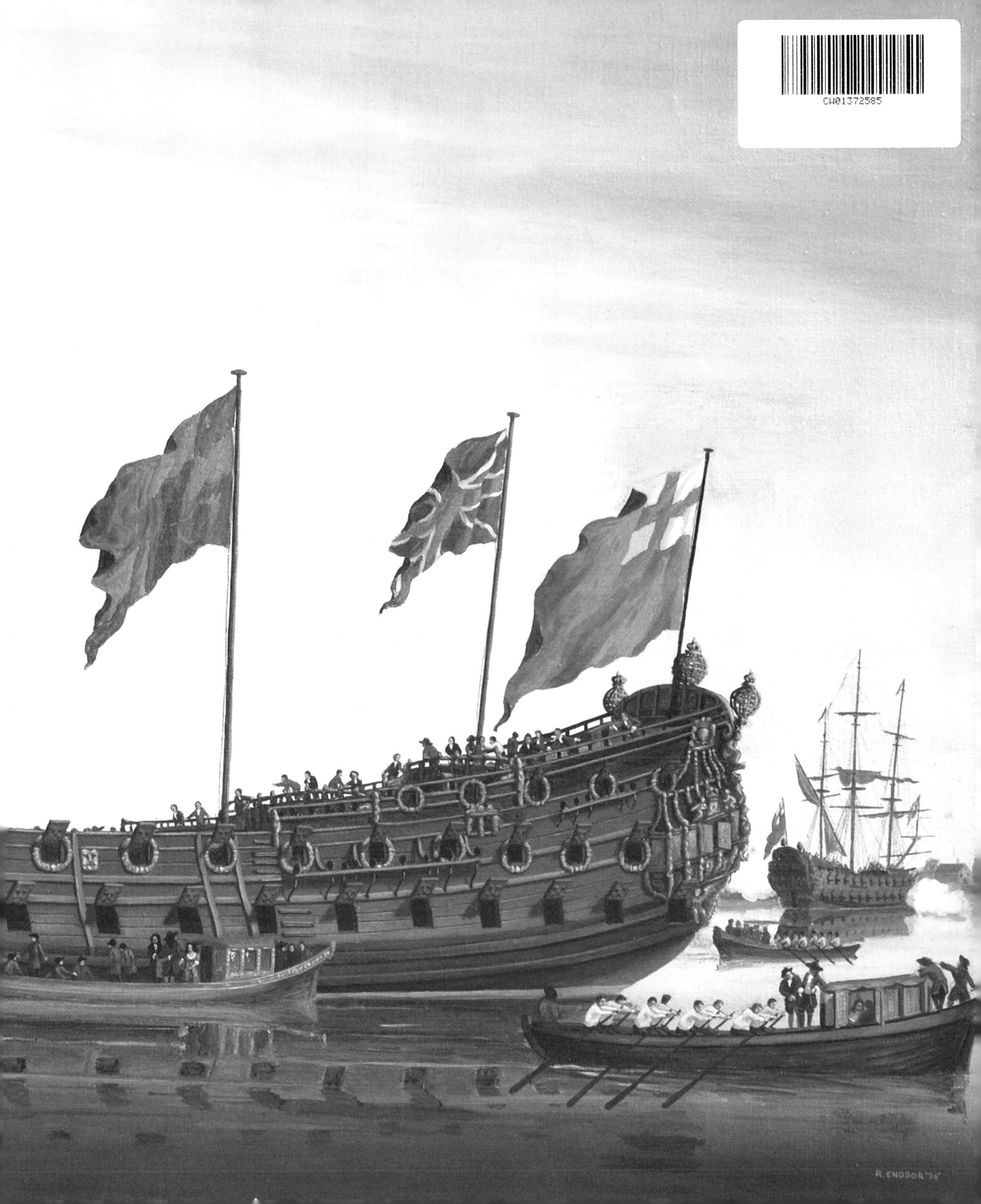
R. ENDSOR '96

THE MASTER
SHIPWRIGHT'S
SECRETS

OSPREY
PUBLISHING

HOW CHARLES II BUILT THE RESTORATION NAVY

THE MASTER SHIPWRIGHT'S SECRETS

RICHARD ENDSOR

OSPREY PUBLISHING
Bloomsbury Publishing Plc
PO Box 883, Oxford, OX1 9PL, UK
1385 Broadway, 5th Floor, New York, NY 10018, USA
E-mail: info@ospreypublishing.com
www.ospreypublishing.com

OSPREY is a trademark of Osprey Publishing Ltd

First published in Great Britain in 2020

© Richard Endsor, 2020

Richard Endsor has asserted his right under the Copyright, Designs and Patents Act, 1988, to be identified as Author of this work.

All rights reserved. No part of this publication may be reproduced or transmitted in any form or by any means, electronic or mechanical, including photocopying, recording, or any information storage or retrieval system, without prior permission in writing from the publishers.

A catalogue record for this book is available from the British Library.

ISBN: HB 978 1 4728 3838 4; eBook 978 1 4728 3839 1; ePDF 978 1 4728 3836 0; XML 978 1 4728 3837 7

20 21 22 23 24   10 9 8 7 6 5 4 3 2 1

Index by Alison Worthington
Originated by PDQ Digital Media Solutions, Bungay, UK
Printed in India through Replika Press Private Ltd.

Front cover: King Charles II visiting the *Tyger*. Author.
Back cover: The fourth-rate *Tyger* as built in 1647. Author.
Unless otherwise indicated all images form part of the author's collection. The Robinson reference numbers in the captions for the van de Velde images held in the collection of the National Maritime Museum refer to Robinson, Michael, *A Catalogue of Drawings in the National Maritime Museum made by the Elder and the Younger Willem van De Velde*, Volume 1, 1958 and Volume 2, 1972, Cambridge: Cambridge University Press.

Osprey Publishing supports the Woodland Trust, the UK's leading woodland conservation charity.

To find out more about our authors and books visit www.ospreypublishing.com. Here you will find extracts, author interviews, details of forthcoming events and the option to sign up for our newsletter.

CONTENTS

Foreword: Charles Berkeley 7

Introduction: The Master Shipwright's Secrets in Relation to the *Tyger* 8

Chapter 1: The Master Shipwright's Considerations 10

Chapter 2: Inventions and Innovations 42

Chapter 3: No Such Thing as the *Tyger* 64

Chapter 4: Planning a New *Tyger* 80

Chapter 5: John Shish's Account of the Dimensions of a Ship 102

Chapter 6: The Draught of the New *Tyger* 122

Chapter 7: Building the New *Tyger* 196

Chapter 8: The New *Tyger* Commissioned 237

Chapter 9: The *Tyger*'s Guns 258

Chapter 10: Contemporary Shipbuilding Contracts Unveiled 271

Appendix 1: The *Medway* Warrant 288

Appendix 2: The *Mordaunt* Survey 289

Endnotes 296

Index 301

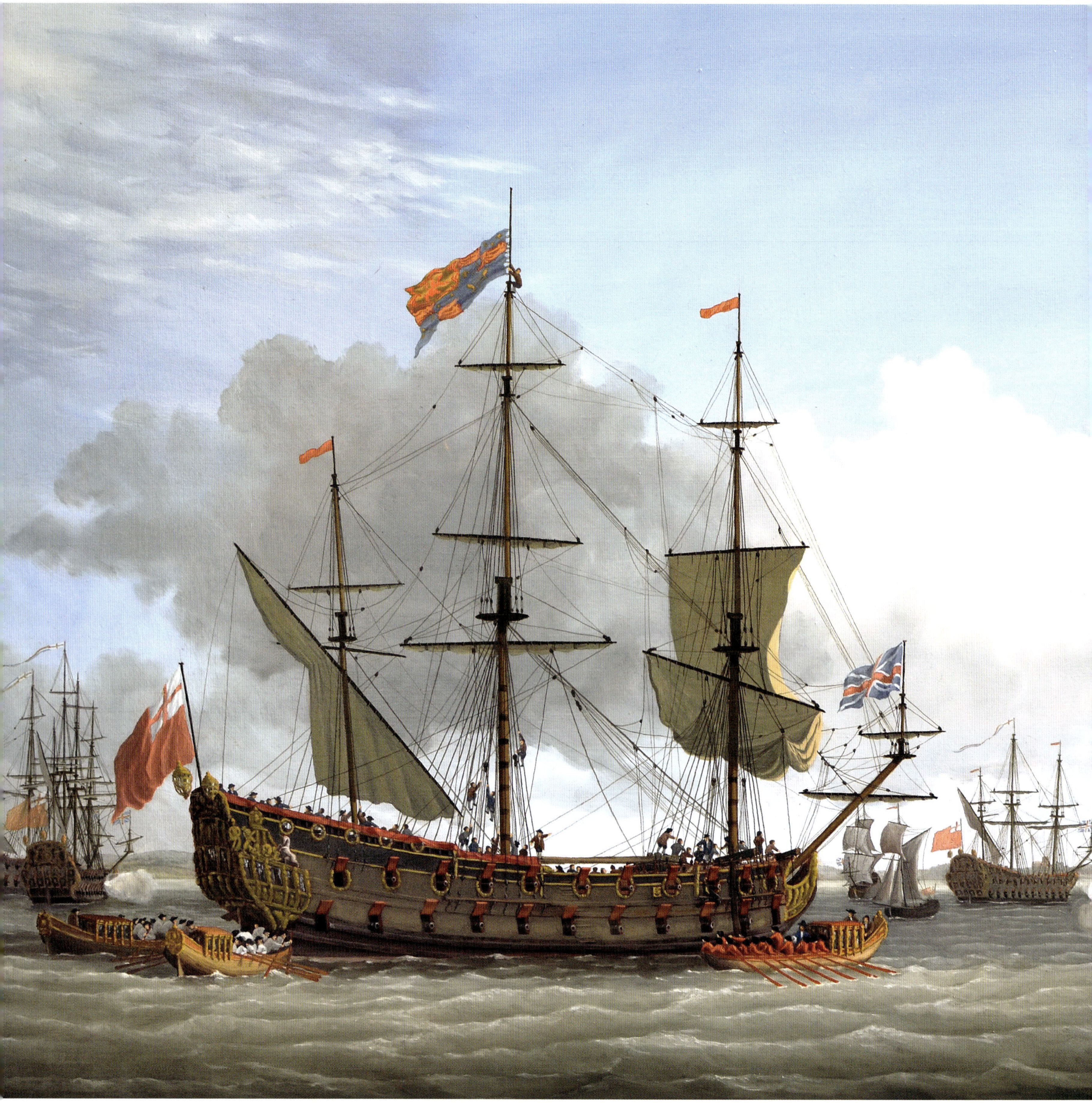

FOREWORD

It was with considerable interest that I heard a book was being written about the warship *Tyger*, built in 1681 for the navy of King Charles II. Charles Berkeley, the 19-year-old 2nd Baron Berkeley of Stratton, an ancestor and namesake of mine was appointed the ship's first captain by the King himself.

This book has taken the author nearly ten years of painstaking research to uncover the extraordinary technical expertise used by the King's master shipwrights. At the heart of the research is a treatise found in the Bodleian Library, Oxford that once belonged to Samuel Pepys the famous diarist who was also Secretary to the Admiralty. The treatise was written by the master shipwright who built the *Tyger* and describes a ship of the same type and size in considerable detail. This, together with all the other research carried out by the author, has resulted in a convincing and accurate reconstruction of the ship.

King Charles II became so closely and personally involved in building the *Tyger* and appointing her captain, that he boarded and dined with her officers during a short cruise. Sadly for my family, Captain Charles Berkeley died aboard the ship less than a year later and was honoured by his family by the commission of a portrait from the studio of Sir Peter Lely and two paintings of the *Tyger* by the famous Dutch maritime artist Willem van de Velde the Elder. Although Charles Berkeley died in the service of his country before he could fulfil his potential, other members of the Berkeley family did rise to become Admirals. Rear Admiral Sir William Berkeley was killed fighting the Dutch in 1665. Charles's younger brother, John, succeeded him as the 3rd Baron Berkeley of Stratton and became a Rear Admiral: one of the ships he captained was the *Charles Galley* featured in this book. James Berkeley, 3rd Earl of Berkeley, was appointed captain in 1701 and enjoyed a long career in the Navy as Admiral and Commander-in-chief of the fleet. A nephew of Charles Berkeley, the Honourable William Berkeley, was first appointed captain in 1727 and by a strange chance of fate died aboard the *Tyger* on 25 March 1733. Some years later in 1766 George Cranfield Berkeley went to sea in 1766 at the age of just 13 and eventually rose to become a Rear Admiral. Visitors to Berkeley Castle today are reminded of this maritime connection by a number of beautiful ship models and the paintings of Charles Berkeley and his ship, the *Tyger*.

This maritime connection at Berkeley Castle extends to the furniture and sea chest of Sir Francis Drake. The beautiful medieval castle is one of the earliest dating from the late 12th century. Romantic as it is, King Edward II was famously murdered here in 1327 which many today believe was carried out in the most barbarous fashion. Later, during the English Civil War, an outside wall was breached and the scars remain to this day. I am grateful that modern research is still able to extend and add to our knowledge of events relating to the Berkeley family and castle, even after hundreds of years.

Charles Berkeley, November 2018
Berkeley Castle

INTRODUCTION

The Master Shipwright's Secrets in Relation to the *Tyger*

One of the great engineers of the Restoration age, Sir Henry Sheeres, Fellow of the Royal Society, wrote to Samuel Pepys concerning shipwrights' secrets and their mysterious lines: '*the rising and narrowing of the breadth, floors, etc are all marked up on moulds and rods which lay up and down among the workmen and marked upon the timbers themselves which marks and measures are the results of those mysterious lines as they are called by which a ship is built*'.

This book is devoted to uncovering the secrets of those mysterious lines used by the master shipwrights – how they were obtained, used in a draught and drawn on the floor of the mould loft. Just as elusive is the way moulds were made and used to mark out the frame timbers. These fundamental skills have almost been forgotten, and considerable research along a neglected and almost forgotten path was required in order to produce an illustrated study of their use.

The master shipwrights of King Charles II did not simply draw their curves at the traditional scale of a quarter of an inch to a foot to be later scaled up on the floor of the mould loft. Instead, actual dimensions were calculated using geometric mathematical formulae to create digitally accurate smooth curves for moulding and placing frames when building a ship. This avoided errors caused by scaling up from the draught and any distortion in the shrinking or expanding of paper.

The ever-inquisitive Samuel Pepys kept a paper written by the shipwright John Shish containing the three-dimensional co-ordinates of the mysterious lines used to create the hull surface of a fourth-rate ship and for placing its frames. By working backwards from his calculations, the formulae used by Shish have been revealed for the first time in 340 years. The ship described bears a great similarity to the *Tyger*, the fourth rate built later by Shish at Deptford in 1681.

King Charles enthusiastically embraced ship development and was interested in fast ships suitable for use in the Mediterranean. He unsuccessfully tried true galleys, then hybrid sailing ships that could be rowed, the galley frigates. After they were criticised, he developed the concept which was to result in the *Tyger*. Under mysterious circumstances following the political turmoil caused by the Popish Plot, Charles rebuilt the old *Tyger* of 1647 which had, in fact, been broken up years earlier. The only person privy to his plan was his master shipwright at Deptford, John Shish. This beautiful ship is fully reconstructed in this book and illustrated with many large-scale drawings. Charles's experiments were a path that led to the evolution of the frigates of the 18th century.

Similar ships to the *Tyger* built along the same stretch of Thames during the same period are also described and illustrated in detail in this book. In addition, other successful and unsuccessful developments of the age and the manner of rebuilds and alterations are examined. The time and expense taken to keep a wooden ship repaired and maintained ready for sea service is also studied. Among the discoveries made along the way is a particular style of ship model-making used at Deptford, which has helped in the identification of surviving models from the period.

A daunting task for the student of 17th-century shipbuilding is reading and understanding contemporary contracts. The difficulty is due to the obscure words used and the barely legible writing. In order to interpret them and make them usable, the five most important surviving contracts for fourth-rate ships have been transcribed, with headings added and indexed with reference to a visual glossary. They span the years from 1649 to 1692 and expose all the changes and developments that took place during that time.

The fourth-rate Tyger flying the royal standard as King Charles II comes aboard. With Grateful Thanks to Berkeley Castle Charitable Trust.

My thanks go to the great institutions, such as The National Archives, the British Library, the Pepys Library and the National Maritime Museum, whose staff make visiting them a pleasure. I have particularly to thank Charles Berkeley and the Friends of Berkeley Castle for their support and access to the castle and its artwork. I have been very fortunate in having the help and support of many friends, particularly Ann Coats, David Davies, Frank Fox, Peter LeFevre, Charles Trollope, Bob Peacock and Jacqueline Stanford. Their diverse and peculiar knowledge of the Restoration Navy was essential to my work. Another friend who gave encouragement was Simon Stephens, Curator of Ship Models at the National Maritime Museum. His practical and enthusiastic help in recording and studying contemporary models in the National Maritime Museum's reserve collection is greatly appreciated. I also thank Richard Wright, my friend of many years' standing, who helped with certain geometric mathematical problems posed by the calculations of John Shish. Another long-standing friend to whom I will always be in debt is Randy Mafit, I thank him for his expertise in finding and often giving collectable source material. I also thank Arnold Kriegstein for his encouraging words and the use of his archives. Finally, I thank my wife, Ilona, who happily gave her support and provided practical help in the IT department when my understanding of such matters failed.

CHAPTER 1

THE MASTER SHIPWRIGHT'S CONSIDERATIONS

King Charles II and Samuel Pepys

During the long years of his reign, King Charles II became extremely competent in understanding all aspects of his navy. He took a particular interest in shipbuilding and was able to discuss and examine all manner of technical details with his master shipwrights. There were four major yards, where the majority of his warships were built – Deptford, Woolwich, Chatham and Portsmouth. The premier yard was Deptford, with its great double dock and the storehouse that supplied the whole navy. Being so close to London, it had the closest relationship with the administration, and particularly with the King, who could and did attend the launch of most ships while at the same time getting to know the Shish family of master shipwrights very well indeed. The master shipwrights' direct superiors were the officers of the Navy Board based in London, but because Chatham and Portsmouth yards were some distance away, a commissioner was appointed to act as a Navy Board officer.

Samuel Pepys

Many extraordinary men rose to prominence during the reign of King Charles II. One of them was Samuel Pepys, who was known only as a naval administrator during his lifetime although today he is more famous for his diary written between 1660 and 1669. He was educated at St Paul's School, London and went on to gain a BA degree at Magdalene College, Cambridge. Shortly after the Restoration of Charles II in 1660 he became Clerk of the Acts to the Navy Board, a position he gained through the patronage of his kinsman, Edward Montague, Earl of Sandwich, who had a leading role in Charles's Restoration after the interregnum.

The Navy Board was the junior administrative body to the Admiralty, and the Clerk of the Acts was the most junior position of the four officers of the Navy Board. Pepys proved to be a man of energy and ability who had a lust for life, qualities admired by the King and his brother, the Duke of York, who also served as Lord High Admiral. Pepys was flexible of mind and effortlessly changed his early Puritan sympathies to become a flamboyant supporter of the court party. At the end of the Second Dutch War, fought between 1665 and 1667, Pepys brilliantly defended the apparent poor performance of the Navy Office at the bar of the House of Commons, inflating his importance, not only in his own eyes but in the eyes of others. In time he was duly rewarded by being appointed Secretary to the Admiralty Commission on 18 June 1673, becoming the most important administrative official in the Navy. As well as a love of regulations and establishments, he set in place a comprehensive archive system to record the Navy's activities. In doing so, he established a lasting legacy by transforming the management of the Navy.

King Charles II, who was renowned for enjoying his pleasures. It is less well known that among those pleasures was a love of ships and shipbuilding. At the time he ordered the repair of the old Tyger, *he attended more Admiralty Board meetings than any other member.*

Samuel Pepys as Secretary to the Admiralty Commission. As well as being a great administrator, he collected material concerning shipbuilding to include in a book about the history of the Navy. It was never completed.

His attitude towards master shipwrights was straightforward. He had his own one and only protégé, the shipwright Anthony Deane, with whom he established a lifelong friendship in the summer of 1662. He shamelessly promoted his abilities above those of all the other shipwrights of the period. At the time, Deane was the assistant master shipwright at Woolwich and was willing to instruct Pepys about all the aspects of shipbuilding as well as about everyday tasks such as measuring timber. He even presented him with a model ship. In return, Deane received unwavering support and praise from Pepys, becoming a man upon whom he could totally depend and one whom he could use.

The Shish Family of Master Shipwrights

By contrast, Pepys regularly slandered Jonas Shish and his sons John and Thomas, throwing many accusations at them. He thought that Jonas could not measure the volume of timber properly[1] and that he depended on his eyes to build ships.[2] Pepys also wrote that Shish wrongly measured ships' blocks by their sheave diameter when the blockmaker said they should be measured by their length.[3]

Perhaps the most famous example of Pepys's bias occurred in 1686 when King James II asked him to make a list of noted shipbuilders who might be considered for a position as a commissioner to repair the fleet, and in the list he drew up he compared them all unfavourably with Sir Anthony Deane. Of Sir John Tippetts he wrote: '*his age and infirmities arising from the gout (keeping him generally within doors, or at least incapable of any great action abroad) would render him wholly unable to go through the fatigue of the work*'; of Sir Phineas Pett, '*in every respect as the first*'; of Mr Lee, the master shipwright at Chatham, '*[he] never built a ship in his life ... also full of the gout and by consequence as little capable as the former of the fatigue before mentioned*'. Of Mr Betts, master shipwright at Portsmouth, he wrote that he '*has built several good ships ... but is illiterate, and not of countenance, method, or authority sufficient for a commissioner of the Navy*'. Mr John Shish, master shipwright at Deptford, was described as being '*old Jonas Shish's son, as illiterate as he ... low-spirited, of little appearance or authority ... little frugality; his father a great drinker, and since killed with it*'. Mr Lawrence, master shipwright at Woolwich, '*has never built a ship in his life but the Little Victory, which he rebuilt at great charge, and when done was found fit for nothing but a fire-ship. A low-spirited, slow and gouty man ... illiterate and supine to the last degree.*'[4] And so the list went on, leaving the reader in disbelief that any of the master shipwrights could ever have built a satisfactory ship. The King, who personally knew them all, must have laughed when he read it. This was Pepys at his best in supporting a friend, and at his worst in abusing others.

Another who compared shipwrights was William Sutherland, author of both *Ship-builders Assistant* (1711) and *Ship-building Unvail'd* (1717), and himself a shipwright. His opinion is likely to be much more reliable than that of Pepys. He thought '*Old Mr Shish and his sons, all eminent builders*'[5] and wrote: '*I could never learn that Sir Anthony was much of a mathematician, or a very great proficient in practice, but had the art of talking well, and gave good encouragement to those men who was well known to be grounded in the practik part of building ships.*'[6]

Pepys looked down on the socially inferior Shish family. The love-of-life attitude of the gregarious Pepys made him very different from Jonas, a God-fearing Puritan who gained prominence during the Commonwealth. Jonas must have known of, and taken a dim view of, Pepys's antics with Mrs Elizabeth Bagwell, his foreman Owen's daughter-in-law. She brazenly exchanged her favours with Pepys for the advancement of her husband, William, and other members of her family.[7] Her efforts eventually helped William achieve the position of master shipwright at Portsmouth.

Old Jonas Shish and his son, John, held the position of master shipwright at Deptford between 1668 and 1686. The family had a long history as shipwrights in the yard and were well liked by members of the local gentry such as John Evelyn, as well as regarding their workmen with sympathy and understanding. On one occasion Jonas received the gift of a shoulder of veal from a woman whose son he had taken on in the yard as an ochum boy. Shish would have none of it, saying, '*Nan, nan (for he spoke thick) take it and roast it and get butter and make soups for the boys, it will fill their bellies bravely.*'[8] Before he became master shipwright, Jonas had already built a number of successful ships, including two fourth rates, the *Foresight* in 1650 and the *Leopard* in 1659. A fifth rate, the *Guernsey*, was built in 1654 and a third rate, the *Cambridge*, in 1666.[9] Although he spoke with a thick accent, '*Mr Shish was very well belov'd by King Charles II who took much delight in the shipwrights art and used to call him his country builder.*'[10]

One would have thought the quality of John Shish's education by his father would have been rather limited. John Evelyn agreed that Jonas was a plain, honest carpenter who could give little account of his art by discourse and was hardly capable of reading. Yet, he continued, he had great ability and was remarkable for bringing up his children well and teaching them to be able shipwrights.[11] As a young man, John Shish would have served seven years as an apprentice, or servant, as they were usually called. He would have learnt practical skills such as hewing, spiling and using staffs with marked-out dimensions to lay out the mould loft and set frame timbers in their correct places. These were skills that every apprenticed shipwright in the yard would have learnt. On 6 November 1665, John Shish was considered sufficiently accomplished to be appointed master carpenter of the *Charles*,[12] a first rate then being built at Deptford. Shortly afterwards, on 22 November, he married Mary Lake and during the next year was blessed with the birth of his first son.

In 1668 Jonas Shish was appointed master shipwright at Deptford and John, at the age of 25, became his assistant. In this elevated position he would have known how to make draughts and would have had an understanding of the hull shape necessary to perform a specific duty. It was an art form more than it was scientific. He also needed the academic ability to calculate the geometry of a ship's rising and narrowing lines to achieve the desired hull shape. These skills were primary among the master shipwrights' secrets. He would have been taught these skills by his father, although it is quite possible that Jonas employed a mathematician to do the endless calculations. If so, then the mathematician would have been entered in the pay books as a shipwright, leaving no trace of his true role. John Shish's education was better than Pepys could have imagined and way beyond an administrator's understanding. John Shish did not depend on his eyes to build ships, for not only did he make draughts, but he digitised his ships' lines by calculation to create perfect curves and eliminate any errors that might have been caused by scaling from drawings.

During the time he spent as assistant master shipwright to his father, John helped build many ships. Among them were the three-decked first rates, the *Charles* of 1668 and the *London* of 1670. Technically, the *London* was rebuilt as she was burnt down almost to the wrongheads at the head of the floor timbers, meaning that everything above was built new.[13] In 1673, the three-year-old yacht *Saudadoes*, built by Anthony Deane, was altered beyond recognition, much to the anger of Deane and Pepys. The work on her was followed in 1674 by the construction of the *Royal Oak*, built as a response to the French *Superbe*, and the largest two-deck third rate yet built, as well as being the first English 74-gun ship. Eight sloops of about 50 tons and two smacks of about 20 tons were also built between 1672 and 1673. One of the smacks was, interestingly enough, named *Young Shish* by the King. The name was probably an acknowledgement of John Shish's first design.

In 1673, during the Third Dutch War, John Shish was appointed master shipwright at Sheerness. The yard had only been established in 1665, as a forward base near deep water so that ships did not always have to navigate the tortuous route to Chatham. It was damaged in 1667 during the Dutch attack on the Medway, but by 1673 was again functional,[14] although no ships had yet been built there. On 1 July 1674, John Shish wrote to Samuel Pepys, the Secretary to the Admiralty, sending him a small treatise entitled *Account of*

Dimensions of a Ship. It reveals the calculated dimensions used to define the rising and narrowing lines of a fourth-rate ship. Three weeks later the *Tyger* arrived at Deptford, and two weeks after that, at Windsor Castle on 2 August 1674, the King approved Jonas's request to exchange places with his son John. John became master shipwright for both Deptford and Woolwich yards and Jonas became master shipwright at Sheerness.[15] A few months later, the King consented that John could also officiate at Sheerness and that Jonas could officiate at Deptford.[16] The close family connections were strengthened even more when Jonas's younger son, Thomas, later become master shipwright at Woolwich.

The Master Shipwright's Duties

Master shipwrights needed a number of vital qualities in addition to being able to design ships. Top of the list was the very different but practical skill of motivating workmen, the majority of whom were shipwrights. Their numbers at each yard varied wildly, between 150 and over 500, depending on the political and monetary situation.

Trades employed in building a ship at Deptford in 1678 by percentage.[17] (Other trades, such as ironmongers and carvers, were provided by outside suppliers.)

Trade	Percentage
Shipwrights	60
Caulkers	3
Joiners	5
Seaman-Riggers	3
Labourers	12
Ocum boys	2
Sawyers	13
Bricklayers Plumbers Scavelmen Teamers and Others	2

Just as important a skill was organizing the supply of vast quantities of wood to keep the workmen busy. Master shipwrights also needed to know how to repair, maintain and rebuild old ships. The duties of all the officers of the Navy administration were formalised at a Privy Council meeting held at Whitehall on 13 June 1673.[18] For the master shipwrights and their assistants, their first duty was to attend all the grounding, graving, docking, repairing and building of new ships. A warrant would be issued by the administration before any of these works could take place. To make sure they maintained an accurate control of cost, the master shipwrights had to keep a counter book, together with the storekeeper, recording the amount of provisions used and their quality. They were also to join with the storekeeper in examining and agreeing the expense of all timber and ironwork issued by warrant to their shipwrights and others working under their supervision, such as carvers and joiners. Even a modest amount of work, such as adding a few extra carvings or making some small repairs, could not be performed before they had received a warrant issued by at least two officers of the Navy Board. When old ships were broken up they had to make sure that all reusable pieces, whether of timber or metal, were saved and not wasted. Care had to be taken that timber was not hewn into chips by the workmen or carried away in the carpenters' private boats. They also had to make sure all the workmen's tools were accounted for after the work was finished.

The master shipwright also had to make sure that the correct amount of carpenters' stores was issued to ships before they went to sea. He also helped the Clerk of the Survey check the amount consumed after its return. Another officer in the yard who assisted them in checking deliveries, primarily oak, were of the correct quantity and quality was the storekeeper. They also worked jointly with the surveyor in making a yearly account of all the ships at the yard, with an estimate of their defects and how long they would remain seaworthy if the defects were repaired.

The care of the men was addressed as well. In addition to making sure they carried out their work properly, the master shipwright also had to see they were properly quartered. In the presence of another principal officer, he rated all the men's daily rate of pay and decided when any of them should be discharged. He had to make sure there was a balance – that is, that there were not too many boys or servants entered onto the yard's pay books. Servants were apprenticed to a skilled shipwright for a seven-year period under the 1563 Statute of Artificers. The shipwright received their wages but was responsible for their welfare as well as their tuition. Another duty of the master shipwright was to observe the time when the men made an appearance in the morning and the time they departed in the evening in order that the Clerk of the Cheque could calculate their wages. Another well-known duty was to be at the dockyard gate when the bell rang in the morning and evening to observe the lawful 'chips' – the offcuts of timber and plank – carried out by the men. He was to make sure that these chips were not made wilfully out of good timber and plank and that they could not be used as fuel for heating the pitch kettle. Those men who offended could be

Whereas I am certainely informed that W.m Wilkins
Robert Maxwell & John Jurden belonging to his Ma.ts
Yard at Chatham were instrumentall & helpefull to one
Mr Bowyer in conveying away out of his Ma.ts sayd Yard
a greate Mast. These are therefore to pray & require yo.u
forthw.th to suspend the sayd Wilkins, Maxwell & Jurden from
working in the Yard & not to make them out Tickets for their
former service w.thout receiveing an Order from the Board
for the same; The like y.ou are Ordered to doe unto Thomas
Eason who conveyed away out of his Ma.ts sayd Yard an
Anchor Stock.
And y.ou are hereby also authorised & required that when
embezelments of any of his Ma.ts Stores in the sayd Yard shall bee
made appeare to have been comitted by any Ordinary
Shipwrights, labourers & the like forthw.th to discharge them
from the Yard & not to make out Tickets onto any of them
without receiveing Order for the doeing thereof. Dated
this 14. July 1668.

Jo: Mennes

To Mr Edw.d Gregory Clarke of
the Cheque of his Ma.ts Yard at Chatham

Left. *This letter signed by the Navy Board officer Sir John Mennes tells how a number of men at Chatham managed to carry a 'great mast' out of the yard. As a main or great mast was in the region of 33 inches in diameter and 95 feet long, and weighed some 7.5 tons, it must have taken some ingenuity for them to take it away. The carpenter's private boat mentioned in the master shipwright's duties comes to mind.*

punished, usually by a deduction of wages, or they could be dismissed for a more serious offence. The master shipwright would have delegated many of the tasks to his assistant but would retain the responsibility for them being carried out.

In addition to all these organizational and technical responsibilities, master shipwrights needed to be aware of the continual improvements and innovations made in shipbuilding. They would have had little difficulty in knowing what they were because the greatest and most enthusiastic promoter of new ideas was King Charles himself. During a conversation with Samuel Pepys he discussed the great improvements in the art of shipbuilding, saying he most truly made it his business to try always for the improving of that matter.[19] Bishop Burnett, who knew the King well, thought that he '*has knowledge in many things, chiefly in all naval affairs; even in the architecture of ships he judges as critically as any of the trade can do, and knows the smallest things belonging to it*'. He added that this interest in a trade was such that '*he knew the architecture of ships so perfectly that in that respect he was exact rather more than became a prince*'.[20] The King's interest in scientific advance had already seen him establish the Royal Society to promote the works of men such as Newton, Wren and Boyle. Members of the Royal Society came up with their own proposals for the design of ships. They must have been confident that their undoubted genius would be of benefit to the likes of the Shish family of Deptford. The first president of the Royal Society, Lord Brouncker, designed a yacht, Robert Hooke proposed a moving keel and Sir William Petty wrote about the principles of ship form.[21] The collective members designed the hull of the second-rate *Royal Katherine* of 1664, but she proved so unstable that she had to go back into dock for alterations. We can almost feel the smug satisfaction of the shipwrights who turned her into a satisfactory ship by widening her with girdling.[22]

New Ships

The King and the Admiralty decided on the need for a new ship or ships and specified what sort of vessels they would be. All three-deck first- and second-rate ships were built in the King's own yards according to instructions issued by an Admiralty warrant. A majority of third-rate ships were ordered in the same way, but more than half of the smaller fourth-rate ships were ordered by contract from well-known and trusted commercial shipbuilders.

The processes involved in ordering ships from these two sources were very different from each other. For ships built in the King's yards, the Admiralty requested a draught for approval together with an estimate of the cost from the master shipwright. If it was decided to proceed, and it usually was, a one- or two-page warrant of about 750 words was issued. It included the important dimensions and usually a general threat or two should the master shipwright take it upon himself to deviate from it. The ordering process did not take too long, and certainly not long enough to allow time for a so-called Admiralty Board model to be built for the instruction of the Admiralty. The Admiralty and its advisory body, the Navy Board, were perfectly capable of understanding drawings and reading specifications without such aids. A typical warrant, for building the fourth-rate *Medway* and dated 17 March 1690, is reproduced in Appendix 1.

When ships were ordered from a private shipbuilder, a comprehensive contract was drawn up giving the precise details and size of just about every piece of timber in the ship. This the Admiralty felt was necessary as builders working for profit would otherwise make the ship as slight as possible. As a result, the contract was five times longer than a warrant. A series of contracts covering the building of fourth-rate ships during the second half of the 17th century is reproduced in Chapter 10. The cost was worked out at an agreed rate per ton and staged payments were made as the work progressed. If the ship was built in the Thames, regular inspections were usually made by members of Shipwrights Hall. If it was built in faraway in places such as Bristol, then a surveyor was appointed by the Admiralty to reside there with authority over the quality of construction and timber used.

Alterations

Many new ships were altered for a variety of interesting reasons. Some were altered to suit the latest aesthetic style, some because they did not meet their intended load-carrying or sailing expectations. Others were altered during building with extra scantling in timber and plank, often by agreement with the King or the Duke of York or by their instruction. In 1674 Anthony Deane received £589 for additional work when building the *Swiftsure* and the *Harwich*.[23]

An interesting example is the *Saudadoes*. In 1670, the master shipwright at Portsmouth, Anthony Deane, built a yacht for Queen Catherine of Braganza which had a keel length of 51ft 6in and was 86 tons burden. The Queen gave her yacht the Portuguese name *Saudadoes*,[24] and it sailed to her native Portugal a number of times. It was, however, a small vessel for such voyages and, on 13 June 1673, the Navy Board received instructions from the Admiralty headed by the King to have her '*lengthened at Deptford about eight feet more of less as the Master Shipwright there shall think fitting and a contrivance made to steer aloft*'.[25] The next day a warrant in pursuance of His Royal Highness's order was sent to Jonas Shish for him to go ahead with the alterations.[26]

Jonas immediately started work, but a month later, on 9 July, he wrote to the Navy Board saying the plank of the upper works was very thin and shaken, and that it required shifting. He also found the body of the ship very full fore and aft, making it impossible for the remaining hull bottom to sail well. To correct her, Jonas said the planks without board needed stripping off down to the keel and the frames moved to answer lines for the quickest way of sailing. Alarmed by this news which implied the ship would be taken apart, the Navy Board replied the same day, advising that its members thought it necessary for him to wait for Anthony Deane to see the ship that he had built only three years before. As arranged, Deane travelled to London and visited Deptford a week later. His jaw must have dropped open in astonishment when '*he enquired what was become of her and found that she was taken into pieces and that her keel was longer and laid on the slip and the new body of a ship in building thereon*'.[27]

At the same time, Samuel Pepys, the newly appointed Secretary to the Admiralty, must have heard from the King or the Duke of their intended improvements to the ship, for he sent a letter to Jonas giving him permission to do anything he liked if he thought it would improve her – or, in his words, '*Not to omit the doing of any work to the Saudadoes now repairing that in his judgement will better her qualities or otherwise render her more fit for service.*'[28]

The distraught Deane made his feelings known to his patron and friend Samuel Pepys, who then suddenly changed his relaxed view regarding the alterations. On 25 July, Jonas was ordered to visit the Navy Board, the junior body to the Admiralty, to give an account of his alterations and to bring with him any orders he had received. According to the Navy Board's account of the meeting, Jonas gave no satisfactory explanation for the work, and '*Commissioner Deane humbly make known the same to His Majesty and also to the Lords of the Admiralty*'.[29] It seems that Anthony Deane attended the meeting, and quite probably his good friend Samuel Pepys did as well, for he signed the account along with the Navy Board officers.

Jonas appears to have been isolated and in desperate trouble with his superiors. But he had the ear and backing of the one man who really mattered. Just a few days after his difficult meeting, Jonas must have taken great pleasure in informing the Navy Board that he had seen the King, who had commanded the building of *Saudadoes* according to a draught had he shown him. The '*eight feet more or less*' was now more than 22 feet, which would increase the ship from its original 86 tons to 180 tons. This must have caused more outrage among the Navy Board officers, for, according to Anthony Deane, the King had told him he would make the keel no longer than 60 feet in agreement with the '*eight feet more or less*' originally mentioned. Trying to cling on to some sort of credibility, the Navy Board argued that the larger ship would be more expensive to operate as it would need victuals and wages for several men more. The Navy Board passed on Jonas's letter, which gave an account of his meeting with the King and details of the vessel, to the Admiralty, adding the following sour note: '*we have received a letter from Mr Shish which he thinks fit for your perusal and therefore do send same to you*'.[30]

Aware of the ill feeling towards him, Jonas wrote another letter to the Navy Board the next day, in which he stated: '*I do see and perceive it is the desire of a gentleman who now sitteth with you at ye Board* (presumably Deane) *that the keel of His Majesty's ship the Saudadoes should be shortened from the draught which I have formerly shown your Honourables. Sir, I know and am assured that there are two great evils which will infuse it. First, the ship will not sail so well neither be so good condition in the sea. Secondly the King's room and the stateroom will be so much*

Sir Anthony Deane, a master shipwright, and friend and protégé of Samuel Pepys.

THE MASTER SHIPWRIGHT'S CONSIDERATIONS

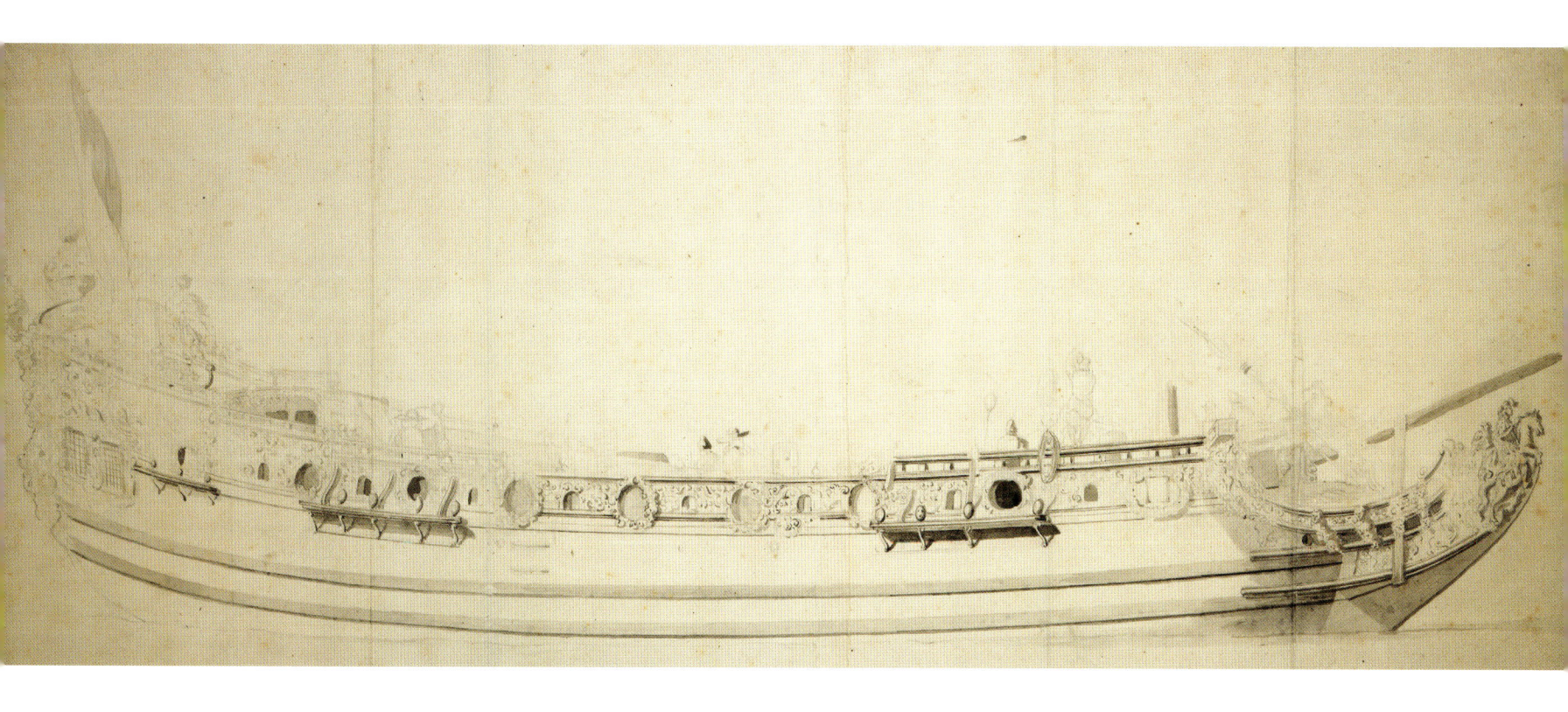

Above. *The* Saudadoes *as rebuilt by Jonas Shish in 1673 (Robinson 1191). Portrait of the 'Saudadoes'. Willem van de Velde, the Younger, c.1676? PAI7303; Photo: © National Maritime Museum, Greenwich, London.*

Below. *Queen Catherine's yacht* Saudadoes *as she was originally built in 1670 by Anthony Deane. Author.* **Bottom.** *The vessel only three years later after she was altered by Jonas Shish. He was supposedly only going to increase her length by eight feet, but King Charles had Shish make a new ship that included very little of the original. Author.*

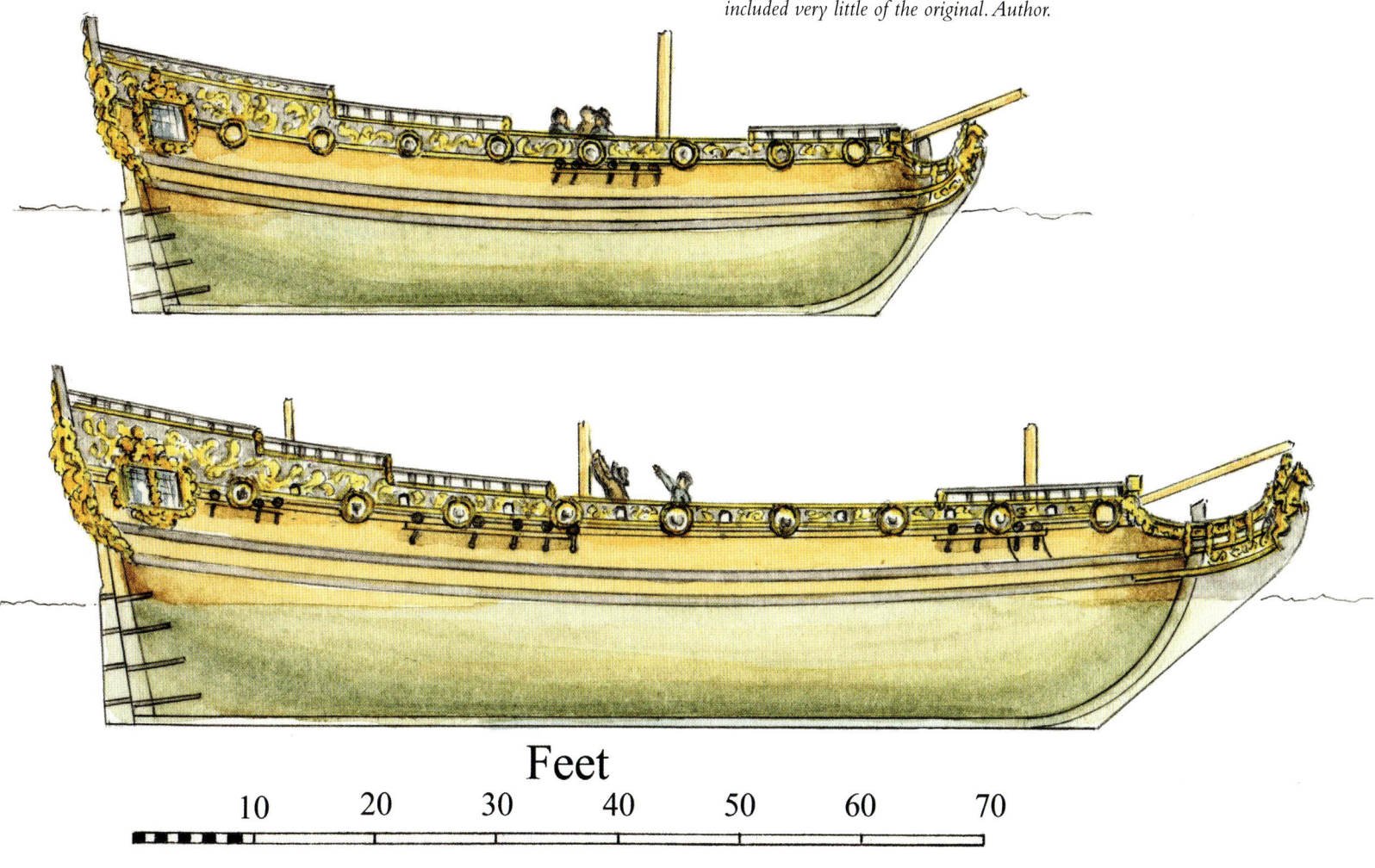

Feet
10 20 30 40 50 60 70

shortened that there will not be sufficient length of room for ye accommodation of the King and Queen. Sir, it is His Majesty's desire that the ship should be the length which is now propounded in the draught and therefore I thought it convenient to acquaint your Honourables that hereafter I may not be blamed. Almighty God doth know that I have no self interest in it but only that His Majesty & the Queen may have a ship which may do them honour & good service which is left for your consideration.'[31]

For the next three months, Shish quietly got on with his work. Then, on 7 October, the Navy Board began to work out the cost '*of ye vessel now in building in His Majesty's yard at Deptford out of the body of the Saudadoes*', even though the final amount could not be estimated as the manner of gilding required by Charles was not yet known. Menacingly, they added that the estimate would include a report touching on, as they put it, the irregularities practised by Mr Shish in his work on the said vessel.[32] The King was out of town at the time, but on his return he gave the Navy Board directions for *Saudadoes* to be furnished with a set of silk pennants and to be gilded like the *Greyhound*, ironically a ship built by Deane the previous year.[33]

The estimate, together with its report regarding Jonas Shish's supposed irregularities, was finished by 3 November. In it, he was charged by the Navy Board of proceeding on the vessel contrary to his original orders of 13 June which mentioned lengthening the ship by about '*eight feet more or less*'. It was passed to the Admiralty for its consideration,[34] but by then its head, the King, had the ship he had always intended to have, while Jonas was left with a slightly bruised superior in the shape of Samuel Pepys, a man who would denigrate him forever after. An anonymous writer in 1717 said with great insight that Shish was King Charles's favourite, '*for which the Commissioners slighted him as all mankind ever was and ever will be for being favourites*'.[35] The episode has clear echoes of the 'gentleman vs. tarpaulin' disputes that raged among the sea officers of the period.[36]

By 1678, toward the end of his career, old Jonas's relationship with the Navy Board was interestingly different and he was writing phrases to them such as '*all the concernments that I have undertaken at sea and ashore I bless God never any miscarry under my hand but have had the love and good wishes of all the masters as I have served*'[37] and '*I have been a servant to this honourable Board 29 years in which time there hath passed through my hands business of great concernments and I bless God I have discharged a good conscience which with great comfort I shall carry on with me to my grave*'.[38]

Girdling and Furring

Very often new ships were found deficient when they went to sea, being tender under sail or having their gunports too near the water when loaded for service. This was not the result of a fundamental design error in 17th-century ships, it was simply that they were loaded with as many heavy guns as possible. For instance, the biggest third rates were the successful third-rate 70-gun ships of the 1677 programme. They were exactly the same length, breadth and depth as the sixth-rate frigate *Trincomalee* that was built in 1817 and that can still be seen at Hartlepool today. Her main armament was 28 eighteen pounders among her 46 guns, while the 70-gun ships of 1677 carried 32 pounders on their lower decks, the same size as the biggest guns on Nelson's *Victory*. Not only did the 17th-century warships carry much heavier guns, but the beams and structure had to be equally massive to carry them. The reason so much of these ships' carrying capacity was devoted to guns was that it was likely any battle would be fought close to home against the Dutch or French, often only a day's sailing away. When this commitment to heavy guns caused inevitable problems, it never seemed to have occurred to anyone that perhaps a few guns should be removed. Instead, the obvious answer was to make the ship wider and bigger underwater so that it would carry the anticipated weight.

There were two ways of increasing breadth and volume underwater. The least drastic of these was girdling, carried out by fixing thick stuff directly onto the existing plank fore and aft. The second, less-used option, was furring, generally employed when even greater breadth was required. Furring involved removing the existing planking and adding timber to the frames to make them wider. The widened frames were then replanked.

After the *Prince* was launched in 1670, it became apparent she required more breadth, and this became the subject of some debate between her builder, Phineas Pett II, the Admiralty and, of course, King Charles. It was decided that she should be girdled with fir timber, as had been done years previously on the second rates *Henry* and *Katherine*. It turned them both from very tender ships with their gunports too near the water into very highly regarded ships that carried their ordnance very well and bore a good sail.[39] Before they were girdled, Pepys described them in less nautical terms as '*useless*'.[40] By 1674, the *Henry* had carried her girdling for 18 years, but it remained in good condition and was very serviceable.[41] The *Prince*'s girdling consisted of 6½-inch-thick good hard yellow fir timber of very great length. She fought through the Third Dutch War and, after its conclusion in 1674, was repaired by Pett in the dock at Chatham. The old fir above water, up to the lower edge of the main wale, was badly damaged by shot received in battle. It was stripped off and replaced with good new oak up to the lower edge of the gunports. Underwater, Pett repaired the old fir to

THE MASTER SHIPWRIGHT'S CONSIDERATIONS

Left. *The midship section of the fourth-rate* Hampshire *according to a contemporary drawing*[44] *showing the framing, planking and eight-inch-thick maximum girdling applied shortly after her rebuilding. Author.*
Right. *The same ship showing how the extra breadth may have been achieved by furring the frames. Author.*

make it very serviceable by firmly caulking the seams to help support the badly split oak plank beneath.[42] King Charles contemplated the work done but did not agree that Pett should leave the old fir in place and approved a letter saying it was his pleasure that all the fir be removed and replaced with an oak girdling of at least seven inches thick.

Pett, relying on his experience and knowing the King would listen to reasoned argument, respectfully replied in a paper in which he said that as the work was of so great a charge he most humbly thought it his duty to let the King know his thoughts. He noted that the existing fir was within half an inch thickness of the proposed oak and being half the weight would be more buoyant. As result, he calculated that there would be a saving of 50 tons and that the *Prince* would carry her ordnance four inches higher. He also noted that the fir was of great length and only spiked in place, whereas oak, coming in much shorter lengths, would need iron bolts. Keeping the fir would therefore also save a great deal of money. The paper was examined by Charles and the Admiralty, who asked the Navy Board, and in particular the Surveyor, Sir John Tippetts, for their opinions. They broadly agreed with Pett and informed the Admiralty with Charles in attendance. The result was that '*His Majesty is pleased to approve of what you therein represent and accordingly we do hereby direct that you give order for the forebearing the doing anything further to the said ships girdling than is present done*'.[43]

The *Prince* remained at Chatham for the next six years with the oak girdling above water and fir below. Then, in 1680, it was found that all the girdling and some of the wale pieces above water were rotten and required shifting. After their removal, it was discovered that the plank underneath needed caulking and that new girdling needed to be fitted.[45] The damage was caused not so much by the use of poor timber but by the unprecedented harsh winters and hot summers which affected all ships during that period. Later, in 1691, the *Prince* was docked at Chatham for repairs and was surveyed by a body of master shipwrights. They found she needed very extensive work which included removing all the plank and all the wales from the top of the sides down to five strakes below the lower wale. The exposed frames, consisting of fully two-thirds of the toptimbers and upper futtocks, were also found to be rotten and required replacing. The new frames would not be put back in the same position as the originals, but instead five inches further outboard. The remaining third of the existing frames, still in place and in good condition, would then be made wider, in and out, by adding five inches of furring to bring them out to the same breadth. Then the *Prince* could be planked in the normal manner without any girdling at all.[46]

Pett was not the only shipwright with the confidence to query his orders for girdling a ship. In 1674 Jonas Shish was asked to girdle a sloop with three-inch-thick fir plank and responded by pointing out the disadvantage of doing so, saying the fixings might split the frame timbers, but if he had to do it he would be very careful.[47] Another example is the *Hampshire*, a typical fourth rate that was built in 1653 by Phineas Pett II at Deptford. She was girdled early in her career, but years later, when she was being repaired, the frame timbers were found to be bored and damaged so much that many were unserviceable and required replacing. She underwent a bewildering number of repairs to the hull during her career.[48]

Accidents

Wooden sailing ships were very vulnerable to accidental damage. In March 1690, the fourth-rate *Advice* hit a rock in bad weather while manoeuvring with her anchors in Plymouth Sound. She made it into the Cattwater, where she was hauled up the graving place for the damage to be inspected.[49] Seven feet of the main and false keel abaft had been beaten off along with five inches of the rising wood above that. The bottom three feet of the sternpost was broken off and missing, while the rudder was broken in pieces to the lower rudder irons.[50] The *Advice* was lucky to have survived, but such a large amount of damage could not be repaired at Plymouth, and she was obliged to sail on to the dry dock at Portsmouth. She arrived at the end of the month, but it took until 15 April before she was unloaded of stores and 91 tons of ballast to make her ready for docking. The repairs themselves only took 16 days and then she was again ready to leave.[51] To have completed such a huge amount of work in such a short time is a testament to the skill and resourcefulness of the dockyard personnel.

The worst kind of accident that could occur was the sudden explosion of the gunpowder magazine. Fortunately, this was not a common event. During the whole of the second half of the 17th century, only about seven warships were lost in accidents of this kind that had no connection with enemy action. Of course, the results were often catastrophic, leaving the ships involved beyond the help of master shipwrights. One such accident occurred in 1669, to the fifth-rate *Oxford* under the command of Henry Morgan. He was leading a buccaneer attack on Hispaniola and, true to the legendary reputation of commanders in that part of the world, held a drunken council of war during which the ship blew up, killing about 250 people.[52] The worst accident occurred to the third-rate *Breda* in 1690 at anchor in Cork harbour, when she suddenly went up in great sheets of fire, killing all but about seven of her crew of over 400 men.[53] Another terrible accident happened to the second-rate *London*. There were about 325 people aboard as she reached the Buoy of the Nore in the Thames on 7 March 1665. She was not yet a fighting ship. Her commander, Sir John Lawson, and the commissioned officers were still ashore and many stores were still being brought out to her. The most senior officers aboard were the warrant officers, who were not from the social upper classes but were men appointed for their ability and experience. One of them, Richard Hodges, the gunner, may have been loading some of the guns for the expected battle to come. As the ship was near land but out of harm's way in the river, life aboard would have been relaxed and noisy. Many men would have been stowing the stores below decks. Some men would have played music, probably the tunes of popular songs, on their own instruments, while others played board games. In quieter corners, many women, for they were allowed aboard during the fitting out, would have been saying their goodbyes to husbands and lovers, hoping for their safe return.

Suddenly, the ship exploded in searing heat and flame, within minutes ending up on the bottom in two distinct parts. Death would have been mercifully sudden for most on board. The dreadful catastrophe must have been caused by an accident involving the tons of gunpowder stored below decks in the magazine. We will never know exactly what happened, but Samuel Pepys recorded the event. '*This morning is brought me to the office the sad news of the London in which Sir J Lawsons men were all bringing her from Chatham to the Hope, and thence he was to go to sea in her but a little a this side the buoy of the Nore, she suddenly blew up. About 24 and a woman that were in the roundhouse and coach* (cabins high up in the stern) *saved; the rest, being above 300, drowned the ship breaking all in pieces with 80 pieces of brass ordnance. She lies sunk with her roundhouse above water. Sir J Lawson hath a great loss in this of so many good chosen men and many relations among them.*'[54]

THE MASTER SHIPWRIGHT'S CONSIDERATIONS

Above. *The damage suffered by the fourth-rate* Advice *after she hit rocks in Plymouth Sound while trying to manoeuvre into the Cattwater. Author.*

Right. *Rudder irons and gudgeon fitted to the stern of the ship to take the pintles of the rudder. From Thomas Blanckley,* A Naval Expositor, *1750.*

Below. *Pintles. From Thomas Blanckley,* A Naval Expositor, *1750.*

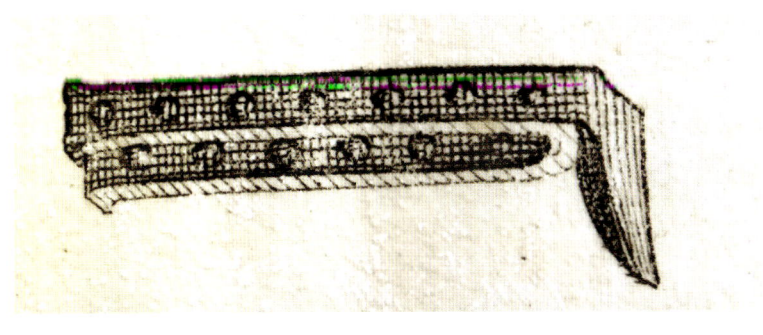

A powder explosion was described graphically after the old *Anne* of 1654 was lost at Sheerness on 2 December 1673. The first account came from John Shish, the master shipwright. The accident happened at about the time he was writing a treatise about shipbuilding at the request of Samuel Pepys. The *Anne* had just come out of the dock and was being made ready for sea and having her rigging set up. Except for some bandoliers and powder horns, no gunpowder was supposed to be aboard, as it should all have been removed before docking. The ship was quietly anchored in 16 fathoms of water and positioned a little within Sheerness point near the fort. Moored alongside her was the *Hestor* fireship. The *Anne* was not yet a fighting ship, and the men would have been amusing themselves or saying their goodbyes to wives, children and sweethearts. At midday she was mustered and found to have 177 men aboard. It was later estimated that during the next couple of hours about 37 had left in the long boat or pinnace, leaving the lieutenant and the purser as the only officers aboard. At the time, a party of four carpenters and three joiners were working in the ship helping fit her out, probably setting up partitions

21

and bulkheads. They were thought to be working in the empty powder room, situated in the bows below water level. Suddenly, at about 2 o'clock in the afternoon, the ship exploded. Her bows were blown apart and the foremast, bowsprit and the head of the ship all blown away. Further aft, the bulkhead of the steerage and coach were destroyed. '*It shattered the greatest part of her almost to pieces,*' wrote John Shish.

Shish believed that the broken ruin remained afloat for about seven minutes before sinking, taking the *Hestor* down with her. Once the wreck had settled, the top of the main mast was all that remained above water. Of all those who had been aboard, only 20 souls survived, and many were so badly hurt that five of them died during the course of the next day. Among them was the purser. As well as all the men, the wives and children of the purser, the master, the carpenter and several other men were also killed. One lucky survivor was the lieutenant, who, having

A ship being careened. The stores and guns were taken out of the ship, enabling her to be hauled down aboard the hulk on her port side so that the bottom could be cleaned and repaired. The hulk was usually an old ship that had been cut down to her gundeck. From A Naval Expoisitor, *Thomas Blanckley, 1750.*

been in the stern of the ship, narrowly escaped.[55]

It was immediately apparent that the explosion had been caused by a large quantity of powder that should not have been aboard. Embezzlement was suspected, and when the gunner turned up two hours after the explosion he was secured in the fort. An enquiry found that when the ship had come into port and the powder taken out, the gunner had taken the opportunity to embezzle some and hide it aboard. He had been tempted by the high value of the powder, for a 100 lb barrel was worth £3 and it would have taken at least 12 barrels to cause such an explosion. Its value was as much as the gunner would earn in a year. Gunpowder was dangerous enough in the secure area of the purpose-built powder room where it was under lock and key. That area also had protected lighting and no iron nails that might have caused a spark. The embezzled powder, however, had been hidden in an unprotected hiding place, and it was assumed that the innocent carpenters working under the light of unprotected candles had accidently set off the explosion.[56] After the accident the remains of the ship caused an obstruction and it was decided to blow them to pieces to clear them away.[57]

Another accident, which occurred aboard the *Eagle* in 1691 showed just how dangerous a small amount of loose powder could be. Thomas Marsh, an apprentice, was working in the powder room, which had been swept with all possible care and afterwards duly swabbed. But inside the room were two chests, one of which contained some paper and parchment cartridges which were thought to be safe as they had been emptied of their powder. Marsh dropped a candle which '*put all into flame and disorder about him, the poor fellow is singed to some purpose*'. The poor singed Marsh was lucky to survive, and lucky too that there was not enough powder to set the ship on fire.[58]

A similar disaster to that suffered by the *Anne* befell another ship. In 1691, the third-rate *Exeter* was alongside the hulk lying in the Hamoaze at Plymouth in order to be careened. All her stores, guns, sails and rigging were taken ashore, when she suddenly took fire and the forecastle blew up. The fire quickly spread aft and blew up the quarterdeck, killing or wounding about 100 men. The hulk managed to cut herself free, while the *Elizabeth* and the *Foresight* anchored nearby cut their cables and hoisted their topsails to get out of the way. The *Exeter* burnt down to the water and sank. A thorough investigation found the cause of the explosion. Before she had been careened, all of the *Exeter*'s 300 barrels of powder should have been taken ashore, but the gunner had embezzled about 20 barrels and hidden them in several places about the ship.[59] To put an end to such '*hellish practice*' in the future, orders were issued that when ships were refitted or laid up they were to be strictly searched by the officers of the yard for any powder in order to prevent accidents or embezzlement.[60]

Estimates were made for rebuilding the *Exeter*. This would involve taking off some of the surviving outside plank in order for some new frame timbers to reach down and overlap those in the lower hull. The biggest problem was finding sufficient timber to carry out the work, as it was extremely scarce at a time when a series of new ships were already being built.[61] As a result, it was decided instead to turn the *Exeter* into a hulk. The estimate for cutting her down to the sill of the gundeck ports and repairing the bows, which had been blown out by the explosion, and then taking her to Chatham was estimated at £2,176 14s 10½d.[62]

Leaks

Sudden leaks were an unpredictable problem and the cause of much consternation to administrators such as Samuel Pepys when his carefully laid plans were thrown into disarray by a ship's unexpected disability. In the summer of 1677 the fourth-rate *Phoenix* of 1647 returned home early from Jamaica, her officers complaining of leaks, much to the scepticism of Pepys. He thought coming home '*upon pretence of the ships disabilities*' should be investigated and ordered a strict survey of the ship. The ship's commissioned and warrant officers wrote a report saying, on oath, that they had done everything they could to find its cause. They said the leak sprang on 2 April, when the ship began to make as much water in eight hours as she normally did in a week. To find out where it was coming from, they brought her by the head, then by the stern, and cut through some of the internal planking every two feet, all to no avail. After that they took up the limber boards next to the keelson but still couldn't find the source of the leak. Nor could they see the outside of the hull as it had been sheathed with lead. They reckoned the probable cause was that the oakum between the planks was going rotten as it was six or seven years old. The leak got worse, and by 20 June they found as much water coming in during an hour as it had formerly in eight.[63] Their theory of rotting oakum seems to have been correct as estimates were made later in the year that included taking off the lead sheathing, making good the plank under it and recaulking the seams.[64]

A similar misfortune befell another fourth rate, the *Reserve*. She started making ten inches of water every four-hour watch, which her crew put down to a leak in the sternpost. Sir Richard Rooth, Commander-in-Chief at the Downs, ordered the carpenters of the *Monmouth*, the *Royal Oak* and the *Mary Rose* to examine her. They found a leak in the sternpost which stopped by itself, but then saw water coming up from under the rider in the breadroom aft, about eight feet below water level. They cut the ceiling on both sides of it but thought it may be coming past rusted iron bolts which went right through to the outside of the hull. Although the leak eased to five inches of water per watch, they found it couldn't be stopped without the ship being docked or laid ashore.[65] A general survey of the *Reserve* later concluded that she was weak in her quarters and that extra standards and a substantial beam were required in that area.[66]

A completely tight ship that didn't let in water at all was unusual and brought its own problems. This was the case with the *Royal Sovereign* in 1691, when the officers made plans to bore a hole to let water in and '*sweeten*' the ship by washing out the dirt and filth through the pumps.[67]

Ice

An unusual but serious problem was caused by climate change in the form of the so-called Little Ice Age. It was particularly severe during the winter of 1683–4, when those aboard the *Eagle* guard ship at Sheerness wrote of a '*miraculous escape we have had of losing His Majesty's ship Eagle and with little hopes of saving ourselves*'. Their ordeal started on Thursday 3 January when the weight of ice broke the swivel bolt of the hawsers, even though it was hauled up above water to save the bridles. Cast adrift, the *Eagle* was taken by the tide from her moorings and ended up halfway to Blackstakes, where she managed to moor with her sheet and best bower anchors. There they remained until Sunday, when a huge piece of ice fell across the hawse and broke the solidly frozen cable of one anchor at the bitts and the other in the shank. Once more cast adrift, they were carried out into the Thames estuary in a sea full of ice with only the smallest anchor remaining, the 3 cwt kedge. With little wind, but helped by the flood tide, the poor, frightened and cold men towed the ship with their boats and anchored with the kedge when the tide ebbed. They managed to keep near the shore, '*and about 4 a Monday morning anchored below Minster Cliff where we stopped by ye kedge in 3 fathoms until seven and then ye wind came at E b N which it pleased God to freshen and with the flood we got with much ado into Sheerness*'.[68]

Another ship in difficulty at Sheerness was the fourth-rate *Diamond*. She was lightened so that she could be brought ashore out of the ice. This was done by those on board together with the shipwrights and caulkers from the yard and soldiers borrowed from a Captain Crawford. It took four days of struggle in terrible conditions to bring both her and the *Arms of Horne* to safety. During the operation, the *Diamond* had her fore yard and the fore part of the knee of the head broken as well as 300 feet of sheathing toward the bow cut through to the plank.[69]

Other ships had similar difficulties, but at Woolwich the ships were all well secured on the beach and shored upright.[70] Taking heed of this, the Navy Board issued warrants for similar precautions to be taken everywhere else the next winter. Ships were to be brought ashore, and to prevent damage caused by the stress of being unsupported by the water, they were to be shored in the hold to strengthen them internally, as well as shored externally to prevent movement. To provide enough shores, a great many masts of six hands had to be found.[71] The damage caused by the harsh conditions was catastrophic to the condition of the ships. The seams and the wood grain were opened up, becoming a source of rot during the hot summers that followed.

Moorings with the detail of the swivel, from A Naval Expositor *by Thomas Blanckley, 1750.*

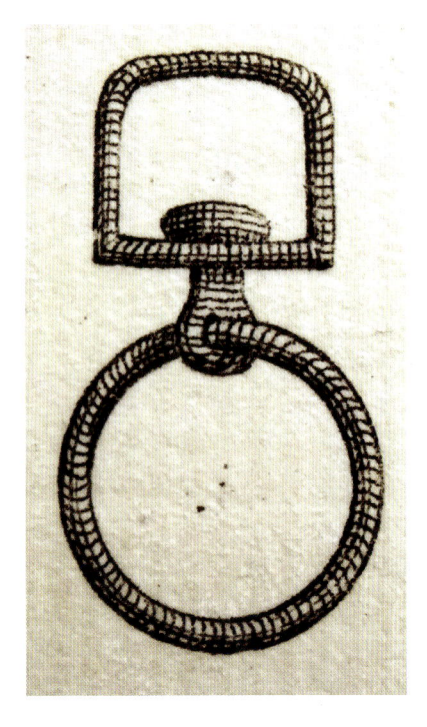

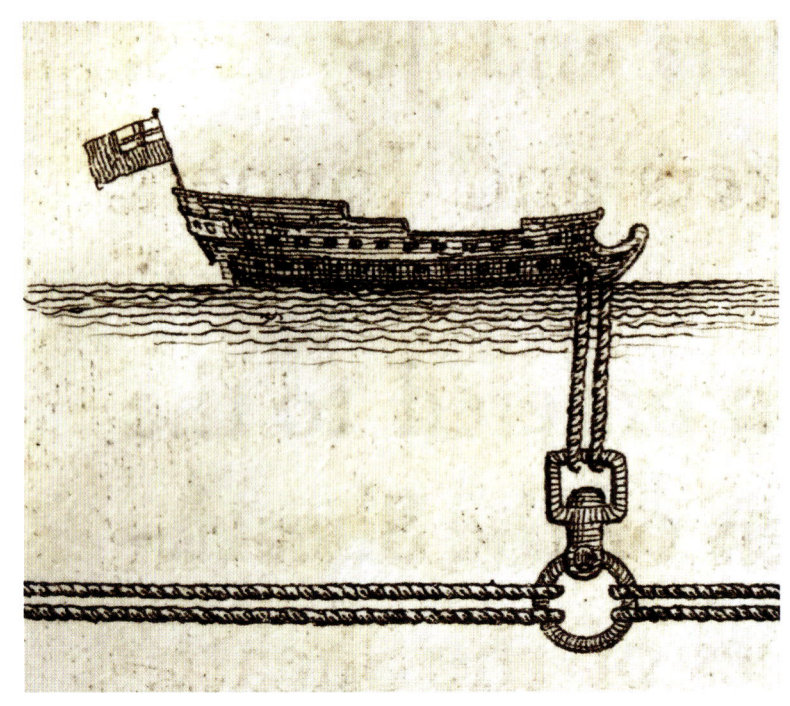

Maintenance

Wood does not generally rot when it is always wet or always dry. However, this state was never possible with wooden ships as they were always exposed to the weather and were partly submerged in water. Above water, they were more susceptible to rot as fresh rainwater is more damaging than seawater. They were therefore in a perpetual state of deterioration and required regular maintenance to keep them in good condition. In service, this normally consisted of a week or so during winter when the ship could be docked to repair and grave the hull. While vessels were laid up in ordinary, a great deal of damage and rot was caused by rainwater dripping in the upper works and to some of the joiners, carvers, plumbers and glass work.[72] Routine maintenance, such as airing during hot weather and closing hatches and ports in wet conditions, was the responsibility of the ship's five permanent warrant officers and their servants. Repairs beyond their capabilities were made on the orders of the Navy Board, usually following the advice of the yard's master shipwright and surveyor, as and when finances permitted. Decay and rot often started before ships were even launched as they were usually built from recently felled green timber. Seasoning, the drying out of the timber, occurred while the ship was afloat, and a good deal of expertise, diligence and good weather was needed for this process to have a successful conclusion.[73]

Very soon after the Restoration of King Charles II it became apparent at faraway Portsmouth that the cost of ship repairs could escalate beyond any reasonable expectation if left entirely in the hands of the master shipwrights. As a result, the Admiralty issued an order in May 1661 that master shipwrights were not to perform any repairs until they had made an estimate of the work, sent it to the Admiralty for approval and received orders to carry out the work.[74] Of course, an estimate made on what is externally visible can prove radically inaccurate when the ship's planking is removed and the internal structure is revealed for examination. This was a problem when work was carried out in the King's yards, but it was even worse when repairs were carried out privately under contract. In one case, a merchant builder, Mr Henry Johnson, was repairing the *Diamond* and the *Dragon* according to the findings of a survey, but when laid open, the ships proved much worse than expected in some places and better in others. The Surveyor of the Navy, Sir John Tippetts, sensibly suggested that another survey was carried out and a new agreement made to prevent any dispute that might otherwise have occurred.[75]

When timber and plank were roughly removed from ships during repairs, many pieces that might have otherwise been fit for further use were split and rendered useless. Instructions were issued, more than once, for shipwrights to be careful in this work and save what they could.[76] Another wasteful practice occurred when ships came into port and were laid up waiting for repair. The storerooms, cabins and other partitions were taken down, but instead of being carefully stored for future use, they were often embezzled. As a result, orders were issued that rooms, bulkheads, cabins, doors, scuttles, shot lockers, locks, bolts and other such materials were to be delivered into the custody of the boatswain, who was to ident for them in the same way he did for other stores.[77]

The decks, gratings and ladders of the King's ships had long been kept '*clean and sweet by washing and swabbing*', a practice also carried out on merchant and Dutch ships. In 1695, the master shipwright at Chatham complained of a new practice of scraping decks. Some ships returning to the dockyard were found to be in such a miserable condition that they could only be made good at great cost. One ship, the *Royal William*, had only been in service for two months and on returning was in as bad condition as if she had been at sea for many years. The Navy Board wrote to the Admiralty asking for the old method to be reinstated, but appeared to receive little sympathy from the sea officers.[78]

In 1697, at the end of William III's war with France, many ships were in poor condition and required attention. A general order was issued which attempted to cover everything related to rebuilding, repair and maintenance. Before carrying out repairs, master shipwrights were to carefully survey the hull, rigging and stores of ships and, as in the past, not start work until a warrant had been received. If extra work was found necessary after they had started, they were not to proceed until they had received an order to do so. The first task was to remove standing cabins, those cabins made from wood rather than canvas, standards on the decks and lyme linings to allow clear access to the decks for caulking and for air to circulate. After the repairs had been made but before the ship returned to service, the standards were to be put back in place but fixed with spikes only at their head and heels. Similarly, to prevent the carved works rotting, they were to be set up away from the plank they were nailed upon to allow air to circulate beneath. On the lower deck, only the first four seams of the planks from the waterways at the sides were to be caulked, in order to leave the rest open to the air. Some long whole-deck planks were not to be treenailed down onto their beams for the same reason. The lower masts, the anchor linings and any other similar items were to be made to fit but to hang loose. To keep the topmasts and main yards out of the weather, they were to be stored ashore in the mast houses, while the smaller yards were to be kept aboard and hung up between decks.

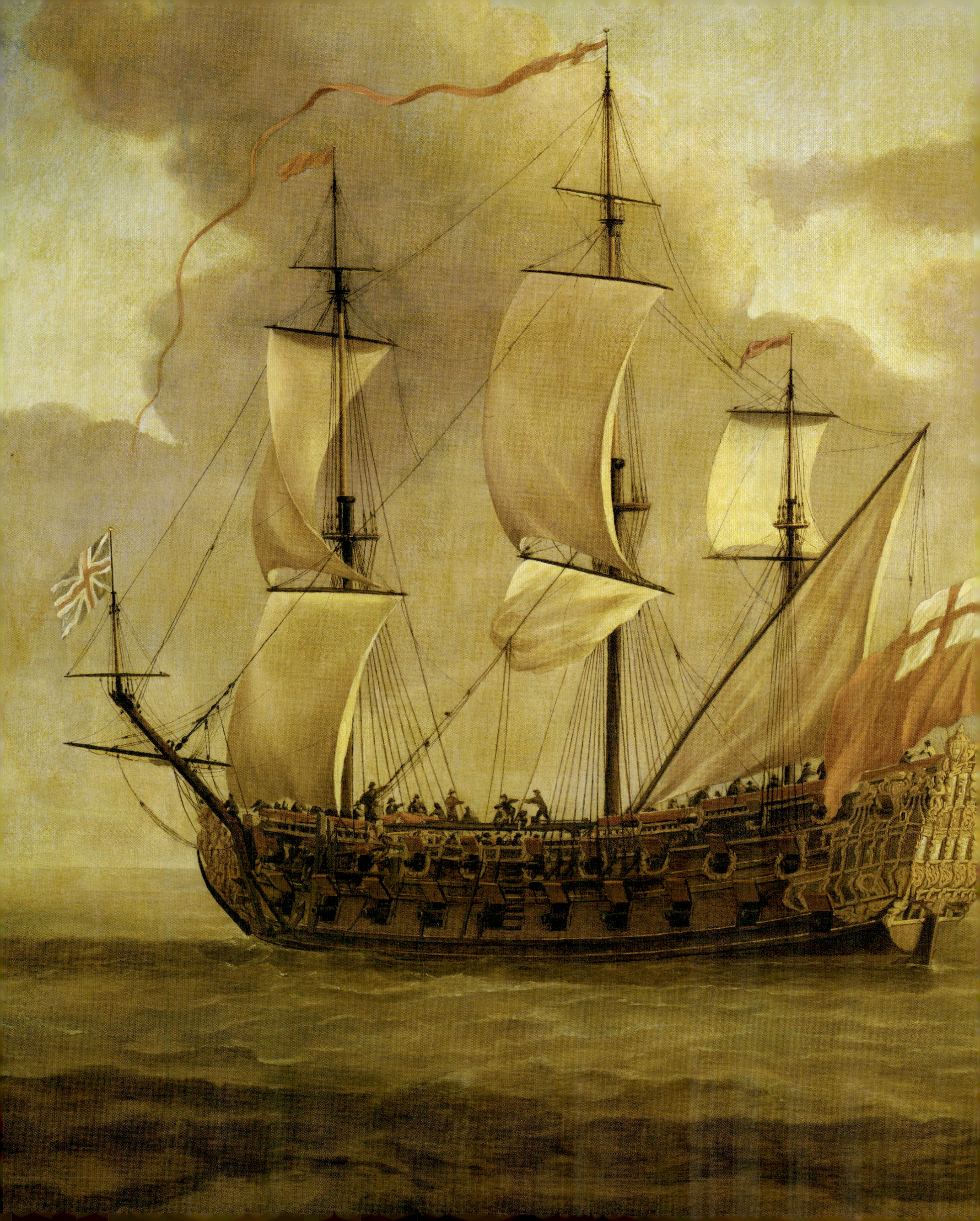

THE MASTER SHIPWRIGHT'S CONSIDERATIONS

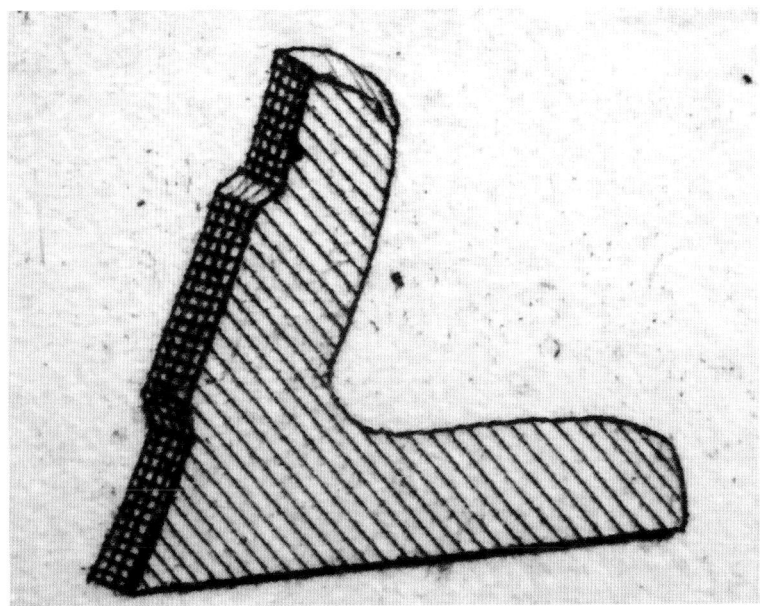

Standards were used to support bulkheads in new ships and were fitted over beams between the deck and the sides. They were also added to strengthen weakening old ships. From Thomas Blanckley, A Naval Expositor, *1750.*

Once repaired and afloat, ships in ordinary were to be ballasted deep enough for the water to support each end to help prevent hogging. Because of their exposure to the weather, ships were to be annually trimmed, caulked and payed, and to be docked at least once every three years for graving. To air the holds in good weather, the hatches, gratings and ports were to be left open, and they were to be shut up in bad. During summer and dry weather, ships' sides were to be watered morning and evening, as had formerly been done to keep them cool, and to stop them drying out too much and cracking. The carpenter warrant officers were to frequently inspect their ships, and any drips were to be stopped by them without delay. The quickwork – that is, the light external covering near the top of the sides that was often clinker laid and payed with tar and blacking – was in future to be painted with oil that was black and timber coloured. While in ordinary, the three standing lower masts and the bowsprit were left standing and were to be payed with rozin and tallow once a year. Once each summer the sails in store were to be well aired, surveyed and mended.

The fourth-rate Woolwich, *painted by Willem van de Velde the Elder. King Charles instructed Phineas Pett III of the Woolwich dockyard to commission the painting from van de Velde and to hang it in the Charlotte yacht[80]. For this reason, it was painted on wood board rather than on canvas. The painting shows a square tuck, a feature not shown on drawings of the ship and only introduced in fourth-rate ships in about 1690. The surviving records of her repairs and refits are the most complete for any fourth rate of the period. The* Woolwich *before a light breeze. Willem van de Velde, the Elder, c.1677. BHC3732; Photo: © National Maritime Museum, Greenwich, London.*

Once a week the master shipwright at the yard was to meet the ships' carpenters to examine and correct any faults. In former times the great cabin and the officers' cabins had been made use of by the warrant officers for their lodgings, and as a result a great deal of damage was done to the carving and paintwork. Now they were to lodge in their own cabins, or in some other cabins between decks. Each quarter year the boatswains were allowed two buckets, 12 brooms, a shovel, a basket and two pounds of candles to keep the ship in order. They were always to have a jack and ensign aboard to be flown on Sundays and holidays. They were also to check the moorings and not let any merchant ships use them. Most of the new comprehensive and sensible instructions were the result of years of experience and reaffirmed previous issues and practices.[79] But as worthwhile and sensible as they were, these instructions would only be effective as long as they were enforced and as long as the men were encouraged to carry them out by such means as keeping them in regular pay.

The Repairs and Refits during the Life of a Fourth-Rate Ship

Wooden ships required regular repairs, refits and the occasional rebuilding, whose frequency and extent were difficult to regulate as there were so many factors involved. These included the quality of the timber and the workmanship during the original build. Ships that were laid up were particularly prone to rot, the severity of the rot depending on their maintenance and the weather. The extremes of very hot summers and winter ice were particularly damaging. Ships sent to sea did not suffer so much from rot, but their fastenings and structural integrity became weaker as a result of the storms they encountered. The variety of factors involved meant that work was often carried out on ships at the point that it became necessary. To gain more of an understanding of the nature of this work, a comprehensive search was conducted to find the most complete surviving records for the repairs and refits of a fourth-rate ship. Those for the large fourth-rate *Woolwich*, between her construction at Woolwich in 1675 and the end of her career in 1698, are recorded here as they were found to be remarkably complete and the most revealing.

The finishing works on the new *Woolwich* were being completed during April 1675. This included fitting the rails, the gunwales, the planksheers, the capstans, the gratings, the ladders, the bulkheads, the standards, the false post, the rudder, the tiller, the chainwales and the gun ports. Other work was being carried out by joiners, painters, plumbers, glaziers, smiths and carvers, and by the caulkers, who were making the planking watertight.

The cost of all this finishing work was estimated at £1,270,[81] out of the total cost for building the hull of £6,147.[82] The *Woolwich* was adorned with an abundance of elaborate carvings[83] produced by John Leadman and Joseph Helby, the approved carvers for the Deptford and Woolwich yards. They cost £285 6s 4d,[84] which was double the usual amount for a ship of her size and far in excess of the £160 paid for the carvings on the much larger third-rate *Lenox* of 1678.[85]

After her launch, the new *Woolwich* sailed to Chatham between 20 March and 10 April 1676.[86] A survey carried out on her in July of that year found there was still some work to be done by shipwrights, carvers, joiners, painters, glaziers, plumbers and masons, as well as the regular caulking and graving. In all, this work was valued at £200.[87] She was fitted out for a voyage to the Mediterranean, departing in May 1677 and returning to Portsmouth in April 1678. She had been lead sheathed for the voyage and this needed repair, as did some copper sheathing in the wake of the anchors near the bow. Copper sheathing must have been used in the vulnerable area near the bow as the soft lead would have been subject to additional wear and damage. By the time she returned home, and in common with many

The amount of time spent in harbour and at sea by the Woolwich, *from her construction in 1675 up until the end of 1682. Author.*

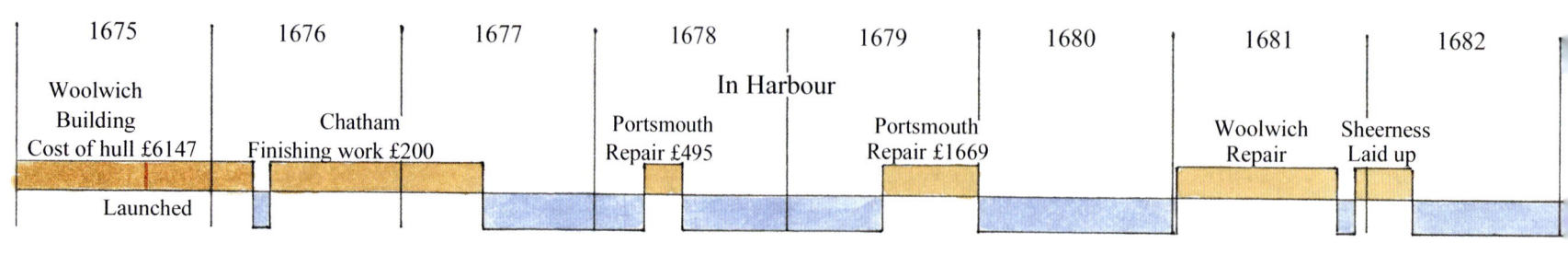

1675	1676	1677	1678	1679	1680	1681	1682
Woolwich Building Cost of hull £6147	Chatham Finishing work £200		Portsmouth Repair £495	In Harbour	Portsmouth Repair £1669	Woolwich Repair	Sheerness Laid up
Launched							
			At Sea				

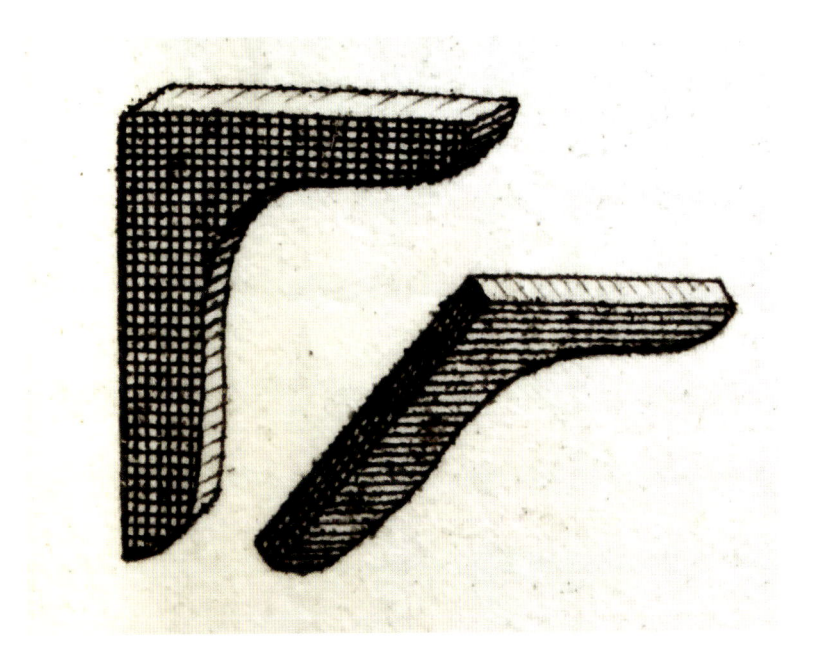

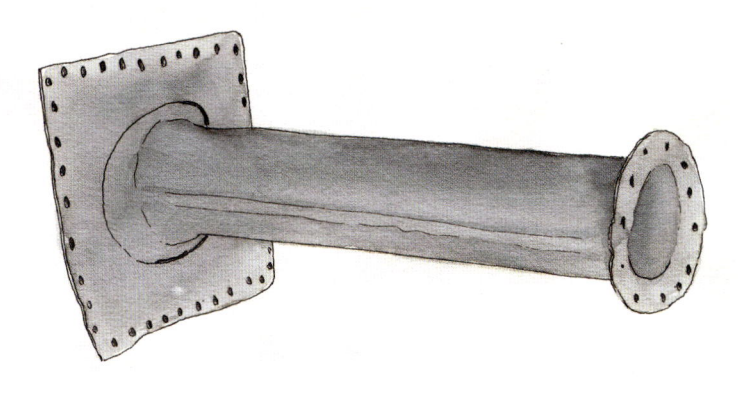

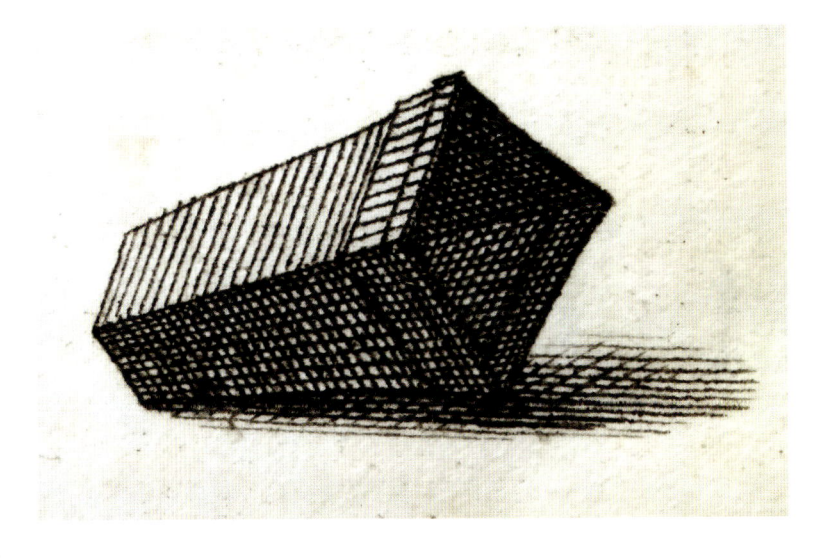

Top. *Most beams had two knees at each end, one of them a hanging or up-and-down knee, and the other a lodging knee, lying horizontally. From Thomas Blanckley,* A Naval Expositor, *1750.*

Centre. *A lead scupper from the wreck of the* Stirling Castle. *The smaller end took water from the waterways at the edge of the deck to discharge it outside the ship. Author.*

Bottom. *A leather scupper. It was nailed to the outside of the lead scupper and freely allowed water to run out of the ship, but collapsed and stopped water entering when underwater. From Thomas Blanckley,* A Naval Expositor, *1750.*

other ships sheathed with lead, some of the rudder irons had become badly eroded by electrolytic action and needed replacing. Other defects included replacing all the lead scuppers on the upper deck which had been broken by the working of the ship. The bricks of the hearth that held the furnaces, or cauldrons, as we would call them, had also been worked loose and needed taking down and setting up anew. Other work included fitting half gunports, which hinged in halves at the side, rather than single gunport lids that were hinged at the top. The powder room forward on in the bows needed alteration and another powder room, or some powder chests, needed to be provided further aft. The main and fore yards needed strengthening with fishes (long, shaped strengthening pieces to fit a mast or yard, spiked and roped in place) and a new studding sail boom was required. The original pinnace was lost and a replacement had to be found. Some work by joiners, carvers, painters and glaziers had to be repaired, and, finally, the hull had to be caulked within board and tallowed without. Tallow was a protective coating consisting mainly of animal fat. This improved the ship's sailing performance, but it did not last long and quickly became foul.[88] Its use seems surprising as the underwater part of the ship was sheathed with lead. All this work was estimated to cost £495.[89]

After her refit at Portsmouth, that took two months, the *Woolwich* set sail in late June 1678 for the Newfoundland and Streights station before returning to Portsmouth a year later, in July 1679.[90] She was now nearly four years old and requiring more serious repair work. A survey dated 21 August 1679 was made by the shipwrights Daniel Furzer and Roger Eastwood. They found some part of the upper deck had to be renewed, as did three pairs of standards on the gundeck. Several hanging knees underneath the same deck also needed replacing. The working of the ship had loosened the cross pillars in the hold and they needed new bolting. Cross pillars appear to have been fitted to ships the size of the *Woolwich* and larger but are not mentioned in contracts for building fourth-rate ships of the period. The ship's movement had also loosened the brick hearth and range, as it had done previously, and once again they had to be made good. Similarly, most of the lead scuppers on both decks required replacing. On the outside, the lead sheathing needed repair and one of the wale pieces had to be renewed. To do this, the fore channel covering it had to be taken off. In much the same way, the lining of the hawse had to be shifted to carry out repairs beneath. At the other end of

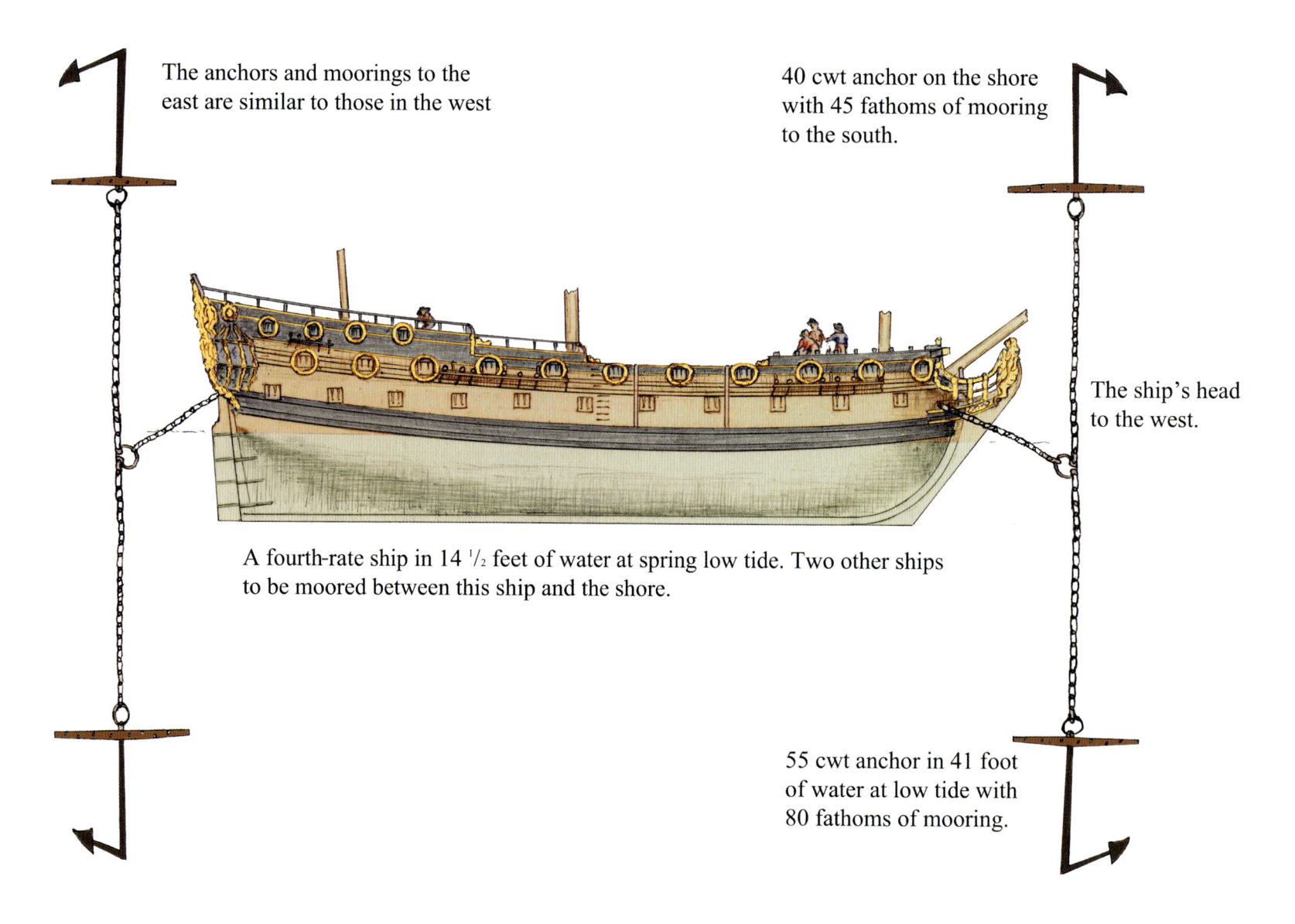

THE MASTER SHIPWRIGHT'S SECRETS

The anchors and moorings to the east are similar to those in the west

40 cwt anchor on the shore with 45 fathoms of mooring to the south.

The ship's head to the west.

A fourth-rate ship in 14 ½ feet of water at spring low tide. Two other ships to be moored between this ship and the shore.

55 cwt anchor in 41 foot of water at low tide with 80 fathoms of mooring.

The method proposed in 1698 for mooring ships at Woolwich by using two anchors at each end. It would prevent ships swinging with the tide and being damaged by passing merchant ships. Author, after Edmund Dummer.

the ship the rudder needed a repair. The main and fore yards needed strengthening and the tops repaired. In addition, the usual work by joiners, carvers, painters and glaziers had to be performed and the ship had to be caulked within and without. The cost of all this would be £1,669 – that is, more than a quarter of her original building cost.[91]

The *Woolwich* remained at Portsmouth until December 1679, when she left to convoy the herring fleet. She was away for a year before returning in January 1681 to Woolwich, where her crew were paid off.[92] At the end of the month Thomas Shish, the master shipwright, reported that the *Woolwich* needed a considerable repair and that it would take some time to complete.[93] After Admiralty approval for the repairs was received, she was docked and the lead sheathing taken off in several places in order to inspect the plank seams and butt joints beneath. Shish found the exposed seams very open and the caulking so rotten it would need clearing out and renewing.[94] All the lead was removed, but as the caulking proceeded, Shish found the plank very rotten in many places, and this would all need replacing. To do this, he needed a large amount of four-inch-thick plank and asked if this

could be supplied from Deptford or elsewhere. He was also short of 2,500 treenails of two feet in length for fixing it to the frames.[95] The rudder irons, which were themselves replacements for those fitted in 1678, were '*so much eaten with ye rust that we were forced to unhang ye rudder and new hang it*'.[96] Lead sheathing had proved a failure due to the damage it caused to ironwork in its near vicinity, and it would not be fitted to the ship again. The *Woolwich* was finally ready to launch from the dock in May 1681.[97] Her works were not '*wholly completed*' until mid-September, after taking more than seven months.

The *Woolwich* would not be sent to sea for the rest of the year, but was instead laid up in ordinary. The Thames at Woolwich was not a good anchorage as '*the King's ships are exposed to damage from Merchant ships passing up and down and also in danger of being put on shore by ye ice in case of a hard winter*'.

It was decided that as soon as they could find enough men they would sail her to Sheerness before winter set in.[98] In spite of the urgency, it was not until early November 1681 that she was rigged and her anchors, cables and sails taken aboard. Although the hold contained 170 tons of ballast, she still needed another 50 tons.[99] Inevitably, just as the *Woolwich* was ready to leave, a merchantman, the *Johns Imployment* of Ipswich, ran foul of her, breaking the jackstaff, knocking the head off the figurehead and breaking one of the rails with all its carved work. Thomas Shish estimated that the damage could be repaired very cheaply, for only £3.[100] The problem of keeping moored warships out of the '*tradeway*' of the river was addressed in 1698, when it was proposed to moor fourth-rate ships with two anchors at each end to prevent them swinging with the tides or the wind.[101]

In April 1682, the *Woolwich* was manned and sent out to escort an Italian convoy before returning home in early March 1683.[102] On her arrival in the Downs, the following estimate was made for refitting her hull, masts, yards, rigging, ground tackle and sails and for furnishing her with six months' worth of boatswain's and carpenter's sea stores:

Refitting work and providing 6 months stores	£
For timber, plank, boards, & masts	150
Ocham, pitch, tarr, rozin, oyle & brimstone	65
Reed, broom, thrums & other petty provisions	30
Perfecting ye carvers, painters, plumbers, glaziers, & bricklayers work	30
Completing the sails to two suites	200
Cables & other cordage and blocks	300
Anchors & other ironwork & lead	20
Kerzey, cotton, colours & hamaccoes	46
Compleating ye Boatswains & Carpenters stores	300
Total	1141

A note of caution was added to the effect that as the *Woolwich* was still at sea, her condition could not be accurately determined.[103] A short time later, on 8 March, she arrived at Sheerness and was safely moored.[104] No major repair work was found necessary, but she was given a thorough refit, especially to her masts and rigging.[105]

A squadron that included the refitted *Woolwich* was to be sent to destroy the fortifications at Tangier under the command of Lord Dartmouth in the *Grafton*. Samuel Pepys also went along, acting as Dartmouth's secretary. By June 1683 she was in the lower Thames, and by the end of the year she was at Tangier. On 14 April 1684 the *Woolwich* arrived back at Sheerness, where her sails were taken ashore[106] and the men paid off. She was then laid up[107] and a survey made of her '*defects*',[108] which resulted in orders being given that she should be repaired and breemed and graved with black stuff.[109] Breeming involved burning off the weed and scraping away all the old protective coating, tallow or tar. Approval was also needed for work to begin on repairing the standing rigging and sails and stropping the blocks.[110] Joseph Lawrence, the master shipwright at Sheerness, then wrote an account of other works to be performed. The first thing he mentioned was the damage to the figurehead, which appears to have been inadequately repaired by the £3 worth of work carried out by Thomas Shish two years before: '*The head being broke off in the middle requires to have a piece scarphed on to take the knee. A new lyon, new wales, timbers, planksheers & gunnels. Some planks to be shifted on the upper deck & forecastle. The gratings, planksheers & gunnels to be repaired. The partners decayed to be new. The bowsprit to be taken out & secured with a fish. The fore mast to have a good paunch* (a mat fitted to the fore side of the lower masts to stop them from rubbing) *brought on afore. The main yard to be well fished. Some waterways defective on the gundeck to be shifted & several lead scuppers that are split. The cross pillars in hold wrought loose to be new kneed and bolted. The channels & wales in the wake of them defective also one of her lower harpins requires to be shifted. Some part of her buttock planks decayed requires shifting. The rudder defective to be secured. Carver's works, joiner's & glazier's to be repaired. The ship to be caulked all over & breemed & graved. The ship to be painted. The ship to be fitted with six months sea stores which charge may amount to the sum of five hundred & fifty pounds or thereabouts. The time the said works may be performed in, if timely supplied with materials, here demanded & sixteen shipwrights – will be three months.*'[111] At the time a great survey was made of all the King's ships as part of a proposal by Samuel Pepys to repair the whole fleet. It included the *Woolwich* and added a couple of additional defects. Once again, the brickwork of the furnaces and the fire hearth was found to be decayed and had to be renewed. Some of the planks under the forecastle and all the gratings also required repair.[112] At the time the reports were written, two shipwrights and five caulkers were working on the upper deck.[113]

Surveys and reports were one thing, carrying out the work was quite another. On 12 August 1684 it was intended '*God willing… if weather permits*' to dock the *Woolwich* with the help of about 20 seamen sent down from Chatham.[114] A month later she was out of the dock but little or none of the repair work had been done, as they reported she had only been cleaned.[115] During September, her standing rigging was refitted, but while strapping the blocks, they found that many needed new sheaves and pins which would have to be sent to Chatham as Sheerness had no blockmaker.[116] Over eight months later, on 28 May 1685, orders were received for fitting out the ship, now ten years old, for sea. None of the repairs had yet been made, and Lawrence pointed out that although the work was considerable, it could be performed at Sheerness, but only if she was taken out of the water and laid on shore on ways.

This was because the scarph, or joint, of the broken head was underwater. Additionally, the rudder needed new irons, and some planking at the water's edge needed replacing, as did some of the rider bolts which went through to the outside of the ship. Lawrence also noticed that another ship at Sheerness, the *Happy Return*, was in a much better state of repair and could be made ready for sea more quickly and cheaply than the *Woolwich*.[117] The Admiralty took Lawrence's advice and refitted the *Happy Return* instead. The *Woolwich* was not overlooked, however, as an estimate was made on 4 June for getting her ready for sea with six months' worth of boatswain's and carpenter's sea stores:[118]

Materials	Repairs £	Sea Stores £
For timber, plank, board & treenails	160	8
Masts, yards, fishes & spars	3	15
Pitch, tarr, rozin, oyle, & brimstone	40	9
Two suits of sails, 2 spare sayles & canvas for stores		436
Cables and other cordage		738
Cotton, kersey, colours & hammacoes		40
Copper, brass, lead, leather & lanthorns	24	96
Ironwork, anchors & grapnels		261
Ironwork, sundry sorts	65	52
Boats, buoys, blocks & oars		80
Other small stores	14	16
Total materials	306	1751
Workmanship & ornament	229	
Total	535	1751
Whereof in store or onboard		1131
Wanting	535	620

The cost of repairs is almost exactly the same as a year before, confirming that little, if anything, had been done to the ship. Indeed, all that was spent on her was for the supply of some tarr and spun yarn.[119]

By 2 July 1685, the *Happy Return* was '*very nigh completed*' and Joseph Lawrence asked the Navy Board if the caulkers from Chatham who had been working on her should be returned or whether they should stay to give the *Woolwich* an '*ordinary*' repair. The ordinary repair involving nothing more than caulking her seams and paying her sides while she remained afloat. Alternatively, he asked, should she be taken on the ways to have her '*full repair*'?[120] A few days later the Admiralty approved a full repair, but wanted it done while the ship remained afloat. Lawrence patiently explained once more the problems and dangers of doing this. He warned that the strong winds could sink her if he removed planking from her sides and some buttock plank near the water's edge. It was also difficult and time consuming fitting stages to the sides for men to work on.[121] He won the debate but after a month's work found that all his store of oak and elm timber was consumed. He requested more, as well as sawn plank, treenails and deals.[122] During September, Lawrence heard that ten loads of large timber were available at Maidstone, but he could not speak to the owners as they only appeared on market day. Each load consisted of 50 cubic feet of hewn timber.[123] The wharfinger, into whose charge this and other timber was entrusted, gave an estimate saying each piece, or tree, contained 80 or 90 cubic feet at a cost of about 55 shillings a load. In addition, there was ordinary-sized oak, containing about 36 or 38 feet in a piece, and costing between 40 and 42 shillings a load, while compass, or curved, timber was 52 shillings a load.[124] The timber was bought and arrived at Sheerness during October.[125] The oak was bought from three partners, Messers Allen, Lee and Spong, and came from Maidstone Quay, high up the River Medway. All the elm came solely from Allen at Otterham Quay near Rainham.[126]

The following table lists the actual timber bought out of a £100 budget given to Joseph Lawrence for repair work to the *Woolwich*. The chart itemises the size of the individual trees in cubic feet:

Large oak timber at 54s a load	Ordinary oak timber at 44s a load	Compass timber at 54s a load	Elm timber at 43s a load
91	64	99	82
76	53	62	31
80½	48	100	34
70	25½		48½
76	49½		60
57	14		54
60	57		50
	19½		44
	45		47½
	33		47
	50		

THE MASTER SHIPWRIGHT'S CONSIDERATIONS

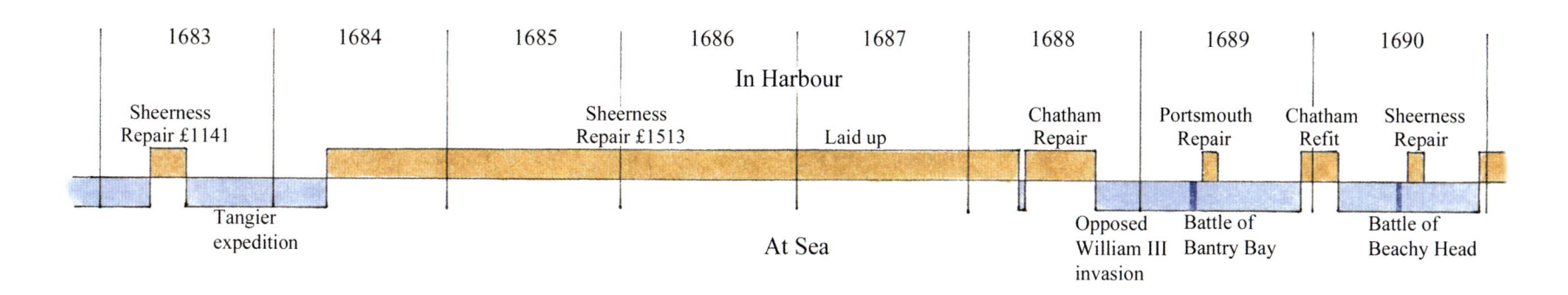

The amount of time spent in harbour and at sea by the Woolwich *from 1683 up until the end of 1690. Author.*

In early November 1685 Lawrence was able to report:

'The lower parts of each gallery has been taken off, the planks being all defective in the wake of them are now shifted & the galleries made up again. We have also shifted about 200 feet of 4 inch plank under the lower ports upon the gundeck. We have shifted several pieces of plank over set 3 standards & new fayed & bolted them again, also the gundeck caulked all over & 2 planks shifted betwixt the bresthooks on the gundeck. Some part of the gundeck ports unhung & new lined and hung again, also one hanging knee fayed under a gundeck beam in hold-aft.

The heaviest of our works are yet to perform. One gundeck beam to shift. The cross pillars to be kneed & fastened. Two new standards to be fayed & several knees. The cookroom to be taken down & new plank to be shifted underneath.

One lower harpin piece (a forward curved piece of wale) & several plank of 6 inch stuff betwixt & under the wales, also some buttock planks to be shifted. A new rudder, the knee of the head to be scarphed with new rails, timbers & carved works. The bowsprit & main yard to be fished & several other works. In regard our number of shipwrights is but five, & seldom but one or the other of them are disabled by one hurt or other… I do reasonably hope the works may be completed in 3 months' time.'[127]

The repairs were completed in early 1686, after which the ship was laid up in harbour. Pepys, in his book *Memories Relating to the State of the Navy*,[128] noted that the estimate of £525 ended up being £1513. The *Woolwich* was moved to Chatham in early 1688, when she was fitted out for sea. On 21 April work started on her main topmast, her fore topmast and her fore yard. Work proceeded slowly; on 28 July, 22 half gunports and six shot lockers were constructed. By August the gratings were repaired, the bucklers made and the shot lockers put in place. Six months after the work started, the *Woolwich*'s top gallants were finished. She had to be docked for graving and to have the iron bolts which passed through to the outside of the hull capped with lead.[129]

The *Woolwich* was now ready for action and formed part of King James II's fleet which was supposed to oppose the Glorious Revolution led by William and Mary. While the rest of the fleet achieved practically nothing, she succeeded in capturing two Dutch ships.[130] Her efforts were in vain, however, as the momentous events ended when James fled to France. The *Woolwich* then became part of a fleet under William and Mary assembled to oppose a French-led invasion on behalf of James's attempt to regain the throne. On 1 May 1689 she took part in the indecisive Battle of Bantry Bay, where *'she was so much disabled in our sails, masts and rigging with several guns dismounted and three split'*. The human cost was 5 killed and 16 wounded.[131] Between mid-May and mid-June, she and many other ships were sent to Portsmouth for repairs to their battle damage. Among other things, the *Woolwich* needed a new main mast, five new guns and several new carriages. By the end of the year the ship was in need of further repairs. She was taken to Chatham and on 5 December entered the dock, where she remained until 28 January 1690. Although she was afloat, it wasn't until 10 February that she was fully repaired and taking in her guns.[132] She joined the main fleet as part of the Red Squadron, but lost her head in an accident at night just before the Battle of Beachy Head on 30 June. As one of

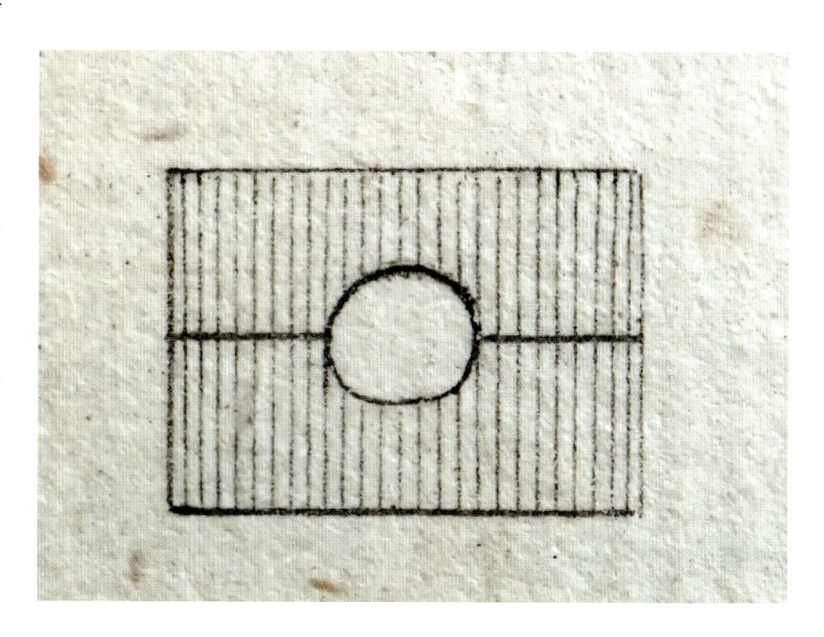

A buckler, used to prevent water washing in at the hawse holes. From Thomas Blanckley, A Naval Expositor, *1750.*

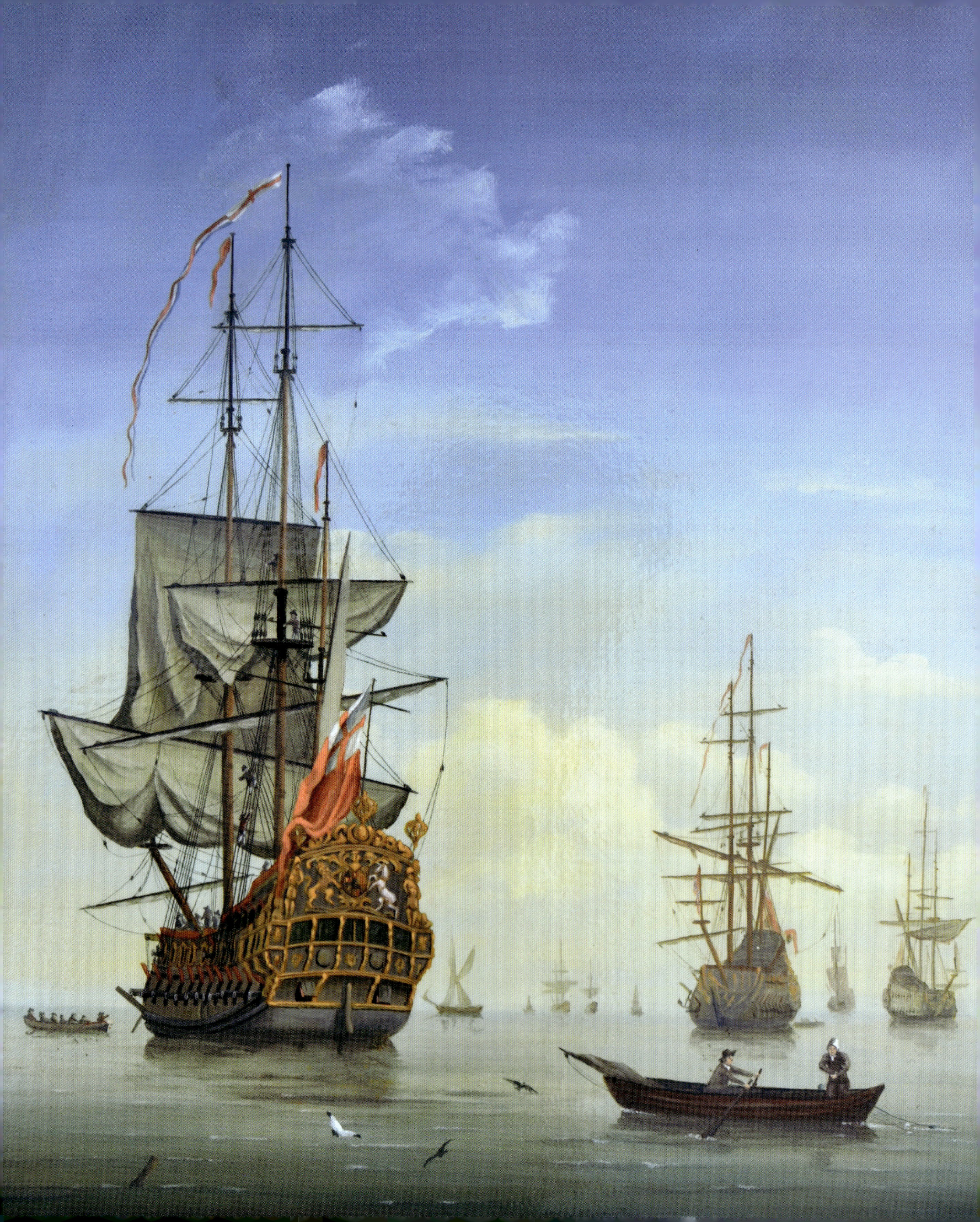

THE MASTER SHIPWRIGHT'S CONSIDERATIONS

Opposite. *The* Lyon, *a ship the same size as the* Woolwich, *which fought next to her at the Battle of Beachy Head. She reputedly dated from Tudor times, and if so, must have had been repaired and rebuilt countless times. She had been rated as a third rate but was reduced to a fourth in February 1690.*[133] *Author.*

the smallest ships in the fleet, she suffered badly during the battle and was very disabled in her masts and yards, especially the main mast, which had to be temporarily repaired with a strengthening '*fish*'. She survived the battle and the following day was hauled over to one side to expose several shot holes below the waterline.

The worst of the *Woolwich*'s damage was repaired near Sheerness between 10 July and 17 August,[134] after which she was sent cruising to the west. She was in a poor condition and was ordered to the Buoy of the Nore late in November.[135] On arrival, she was surveyed by Robert Lee, the master shipwright at Chatham, Daniel Furzer, the master shipwright at Sheerness, and William Bagwell, the assistant master shipwright at Chatham. The shipwrights found that '*the main stern post is shot off about a foot under water in the late engagement with the enemy and the false post within the same rotten in the wake of the bolts. The carpenter informs us she never had any futtock riders & complains the ship is weak, which we attribute in some measure to the want of the same and a pair of top riders in the quarter. One of the gundeck clamps in the luffs* (the curve at the bow) *is sprung and some footwaling in the powder room to be shifted, being rotten, which works cannot be performed without a dry dock and without which we humbly give our opinion she cannot be fit for a foreign voyage.*'[136]

The Navy Board received the report on 12 December and sent a copy to the Admiralty, adding that the ship should be sent up the River Medway to Chatham and taken to the head of the double dock for repair and refitting.[137] On hearing the news, the officers at Chatham pointed out some problems with this arrangement. As the shattered sternpost had to be removed in the dry dock, the *Woolwich* would use up valuable space and time when many ships needed docking for cleaning and graving. Also, while the *Woolwich* remained in the dock, other ships the size of third rates would have to have their bowsprits removed in order to fit in behind her. Moreover, all the hands at Chatham would be busy fitting out the fleet during the winter so that they would be ready for the summer campaign. Three days later, the Navy Board wrote to the Admiralty in agreement with the officers and suggested that the ship should instead be sent up the Thames for repair.[138] She arrived on 31 December and was seen by the master shipwright at Woolwich, who thought the repairs would be greater than he had expected. The problem was resolved the same day when a merchant shipbuilder, Mr Haydon, agreed to take the ship into his dock at Lymehouse for repair and refitting.[139] Mr Haydon was not one of the approved and trusted shipwrights who built ships for the Navy.

A warrant was issued and the work was carried out between 16 January and 19 February 1691. It was overseen for the Navy by Edmund Dummer and Fisher Harding, who wrote that the price for the work was reasonable and the workmanship and materials were good. Unfortunately for Mr Haydon, his bill was not paid until August 1701.[140]

The amount of time spent in harbour and at sea by the Woolwich *from 1691 up until the end of her career in 1698. Author.*

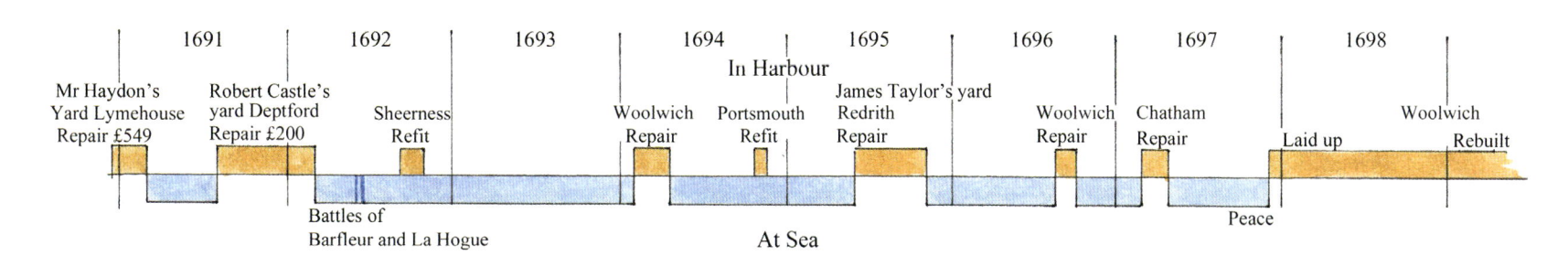

The work carried out on the *Woolwich* at Mr Haydon's dock at Lymehouse in 1691

Work	Cost
Two pieces of false keel of 4 inches thick brought on. Content 56¼ (cubic) feet well fastened with stirrups and staples.	£26 8s
A piece of chainwale, bolts shifted. 15½ (feet) long 6in thick & driving 6 new chain plates in the same.	£ 5
4 inch plank wrought in several places, content 54 feet.	£ 7 4s
Two steps fitted for the sides	10s
3 inch plank on lining for the anchors limber boards, containing 33 feet.	£ 3 6s
A new rail fitted to the wing transom, 23 ½ feet long 4 ½ inch square	£ 1 3s 6d
4 naval hoods brought on abaft to secure the main post & fastening them with 33 bolts	£21
2 pieces fayed between the upper rails in ye head for the fore tacks 5½ foot long each & 9 foot long	£ 5
6 feet of spruce deal without for wash boards under the cheeks of ye head.	7s
3 new brackets in the galleries & repairing the lower part of one of the galleries	£ 2 10s
2 inch plank without in several places 95 feet	£ 5 18s 9d
Taking up 1 old standard on ye gundeck & faying & bolting 3 new standards with bolts in each	£13 10s
Shifting 28 feet of gundeck spirketting 4½ inches thick	£ 4 4s
Workmanship in putting out 2 lead scuppers	5s
2 pieces fayed between the bresthooks in ye powder room 14 foot long each & well bolted	£14
One bresthook fayed upon those pieces 11 foot long and well bolted	£12
Putting in 12 gundeck ledges 4½ foot long each	£ 1 16s
The bulkhead of ye powder room repaired & fitting lockers	£ 1 15s
Taking out & putting in one new crotch within squares for ye cross pillars & bolting the same	£ 3 10s
Faying & bolting 2 new floor riders 22½ foot long each with 14 bolts in each rider	£36
8 new futtock riders new fayed 12½ feet long each with nine bolts in each rider	£96
3 futtock riders more fayed to give scarph to the other riders, 12 foot long each with 9 bolts in each rider	£21
Faying 6 new top riders running through the gundeck 12½ foot long each & fastened with 10 bolts in each rider	£45
2 new standards fayed in ye steerage & bolting them	£ 6
Putting in several pieces after the caulkers without board & on the decks & tearing out some rotten treenails	£ 4
For 120 caps of lead nailed over rider bolts under waters	£ 2 10s
Breeming the said ship, caulking of her from ye strake above the upper wale down to ye keel, graving her with black stuff & tallowing her from the lower wale to the light water draught	£86
For ransacking & caulking all ye upper works to the top of the sides and all the inside works including the gundeck, upper deck, quarterdeck, forecastle & poop	£74
Docking of said ship & shoring her	£30
The use of a storeroom for the Boatswain, another for ye Carpenter & Gunner & a third for the Purser	£ 6
Joyner's work in new ceiling part of ye bread room with slit deal, new bulkheads in ye round house cabins forward, 2 bed sides, the refitting up the state room, bulkhead in ye great cabin & repairing some other works there	£12
To ye carver for carving 3 brackets	£11 1s
Total	**£549 8s 3d**

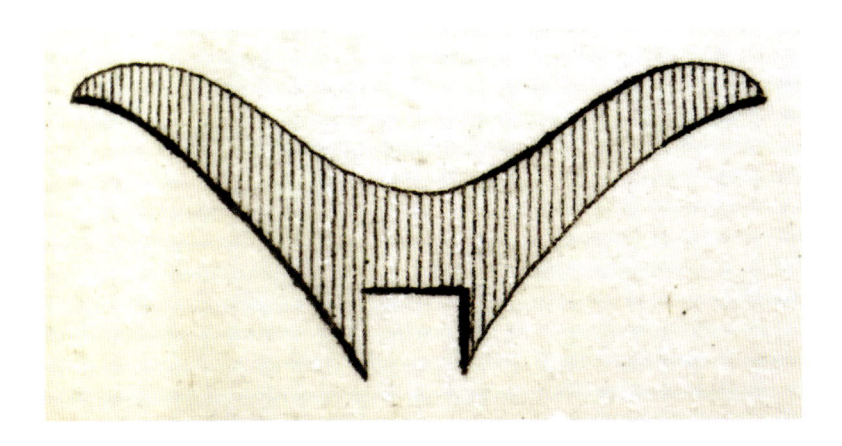

A crotch, mentioned in the list of work to be carried out by Mr Haydon. It fitted at the bottom of the inside of the hull near the stern and helped bind the two halves of the hull together in the wake of the half timbers. The cut-out is made to fit over the keelson. Floor riders, also mentioned in the repairs, were similar and were fitted at the widest part of the ship in midships. They were longer and not so steeply inclined. Also mention are bresthooks, which performed the same function as the crotch at the bows of the ship. Some were fitted down in the hold and others higher up between decks. From Thomas Blanckley, A Naval Expositor, *1750.*

The list of work shows that the shattered sternpost was not replaced but repaired with naval hoods, which were bolsters that strengthened the area where the plank met the post. After the repairs, the *Woolwich* was manned for service and by the beginning of March was in the Thames at Longreach. The repairs were not entirely successful; before the month was out, she had sprung a leak and was surveyed to find the cause.[141] She spent the summer of 1691, from May till August, cruising in home waters. It was only six months since her lengthy repair had been carried out in Mr Haydon's yard, but her officers reported that she was still suffering from serious leaks. As a result, she was paid off and ordered to Woolwich to be refitted. According to the Navy Board, '*Their Majesties Ship the Woolwich has been a crazy ship a great while having had no considerable repairs since she was built and the Master Shipwright… at Woolwich acquaints us that her condition is such that he doubts she cannot be made fit for another voyage without rebuilding.*'[142]

Because it was expected that the repairs would take a considerable time during the coming winter, when the King's yards would be very busy, she was sent into one of Robert Castle's private docks at Deptford.[143] There, she was surveyed by respected shipwrights, including Fisher Harding, Edmund Dummer and Robert Castle. They searched for the cause of the leaks and found one hole in the manger under one of the cheeks of the head and another abaft the mast caused by a swinging anchor. An even worse leak was found in the caulking of a forward plank under the lower main wale, which was found to contain no ochum. They recommended that the lining of the hawse and the cheeks of the head be taken off so that the plank beneath could be firmly caulked. Other work they thought necessary included fitting long, four-inch-thick strings in the waist and making good the rails, gunwales and planksheers. They did not find her in such a poor condition as had been reported by the ship's officers, and estimated the cost of repairs would be about £200.[144] Shortly afterwards, on 12 October 1691, a second survey found that several bolts in the transom knees and bresthooks were badly corroded and needed replacing. The corrosion had probably been caused by the use of lead caps fitted earlier over the bolt heads. Other work included replacing an up-and-down knee to a beam forward on, filling in between the wales with four-inch plank, repairing, yet again, the long-suffering lion figurehead and providing a piece for the knee of the head. At the other end of the ship, the rudder had to be taken off as the main and false sternposts needed replacing. This would involve taking out the buttocks planks aft. A painting of the *Woolwich* by the famous marine Dutch artist van de Velde shows a square tuck, a feature not shown on drawings of her but known to have been introduced on new fourth-rate ships at about this time. It may be that a square tuck was fitted to the ship when her sternpost was replaced and that the van de Velde painting was altered to suit. There was other work to be done, including carving, joiners' work, painting, caulking and graving. Even when all this work was completed it was thought the ship would be fit for only another three years' service if she had the customary ordinary repairs.[145]

The repairs were complete by mid-March 1692, and with the masts set up and ballast put in the hold she set sail from Deptford. It took another month before all her provisions were aboard and she was ready for service.[146] She joined the Blue Squadron of the main fleet and fought at the battles of Barfleur and La Hogue between 19 and 24 May. The outcome was a resounding victory that more than made up for the defeat at Beachy Head. The French fleet would not be able to challenge a fleet action for the rest of the war, but remained a threat through commerce raiding. Afterwards the *Woolwich*'s crew made good her damage, stopping the leaks and repairing the rigging.

On 29 August preparations were made for another refit. The *Woolwich* was taken to Sheerness, where the guns, stores and ballast were taken out so that she could be docked to expose the hull bottom for cleaning.[147] She went into the dock for just one day, on 17 September. Taking the opportunity of being ashore, 50 men, about a third of her crew, deserted the ship and ran off to London. When she was afloat once more, the ballast and stores were put back and the masts and rigging set up. All the work was complete by 16 October and she was able to leave Sheerness.[148]

An engraving of the Woolwich *from a now-lost van de Velde drawing.*

After spending most of the early months of 1693 in the Downs or at Spithead, the *Woolwich* was then sent out cruising in home waters, often on escort duty.[149] A large convoy was assembled to go to the Mediterranean and she was assigned to become part of the escorting squadron. They were attacked by a larger French fleet, and while defending the merchantmen, the *Woolwich* became fully engaged in the ensuing fight.[150] She had the distinction of fighting in every major action of the war. She spent the rest of the year without any major repairs, although the rigging was surveyed on 13 September and she was heeled and cleaned on 6 October.[151]

At the beginning of 1694, the condition of the *Woolwich* was once again causing concern and she was ordered to the Buoy of the Nore near the mouth of the River Medway. She was leaking badly, and a survey was carried out by the master shipwright and master attendant from Sheerness dockyard. They found she needed new decks, and that several broken knees had to be replaced as well as three pairs of futtock riders. On top of that, there were many other unspecified repairs to carry out which could only be done in a dry dock.[152] She was ordered up the Thames to Woolwich dockyard and cleared of ballast, guns, stores and upper masts. She was then taken into the stern of the double dock behind the *James Galley* on 17 February. The work took until the 2 April, and then the *Woolwich* came back out into the river. She was able to set sail three weeks later after her ballast was replaced, her masts set up and her stores put back.[153]

Once more in service, she was ordered to Orkney but suffered some damage, which was reported on her return to the Buoy of the Nore.[154] She was surveyed at Spithead on 17 October and sent into Portsmouth harbour to prepare for a refit and for cleaning in the dock. To prevent a repeat of the mass desertions that had happened in 1692, some of her men were sent on board the *London Merchant* hospital ship rather than given leave ashore. She went into dock on 24 October for a day to be breemed and tallowed on both sides. Before her ballast was put back, care was taken to clear out the limbers, the passages next to the keel that allowed water to run to the pumps. She sailed out of harbour and anchored at Spithead on 7 November.[155]

The *Woolwich* then spent some time at sea on convoy duty, but her condition was again causing concern and on 15 May 1695 she was surveyed at the Buoy of the Nore by the officers from Sheerness. They reported: '*[We] do understand she is a very weak ship about the water's edge, her gundeck in general very defective, the ends of several of those beams rotten, as also that she is treenail sick* (rotten), *and that a perfect account cannot be obtained of her condition until she is brought into dock. We humbly advise her being ordered up to Woolwich, to be disposed of afterwards as upon a second examination as shall be judged best for the service.*'[156]

The *Woolwich* was beginning to reach the stage when repairs were becoming so extensive and frequent that it would be more economic for her to be taken apart and rebuilt or sold off for disposal. As recommended, she was sent up to Woolwich and her men paid off on 25 June. Rather than take her apart, an agreement was made for her to be repaired in the private dock of James Taylor at Redrith near Deptford.[157] The work lasted some five months, and it was not until November that she was rigged and a captain appointed. By the end of year, she was in Longreach taking in her guns.

The first months of 1696 were spent in local waters, but serious leaks began to appear in the bows.[158] The Navy Board ordered her home and, knowing she would need considerable repair work to make her seaworthy, sent her up to Woolwich dockyard to be refitted there or in a nearby merchant dock.[159]

By 6 August her men had been turned over into other ships and the guns and stores sent ashore. She went into Woolwich dock on 18 August and the bowsprit was taken out to find the leaks in the stem. The *Woolwich* remained in the dock for a month and it wasn't until the end of October that she was taking in her guns and made ready for the sea. She was sent to Shetland and the Isles of Orkney late in November, but on arrival she started leaking again, this time caused by loose bolts in the orlop beam knees.[160] She had been at sea for less than three months by the time she was back at the Nore for a further refit on 24 January 1697. There it was reported that she was very weak and in need of rebuilding very soon.[161] Her men were turned over into other ships and all her stores and guns taken out for yet more repairs. These were carried out at Chatham,[162] but it wasn't until 9 August that her guns were brought aboard and enough men found to man her. She went to join the Dunkirk Squadron and was then ordered to Sheerness just as peace was declared in mid-September. The worn-out ship had just managed to survive the conflict. She remained inactive near the Nore until 12 November, when she was ordered to be laid up and her men paid off. Before the month was out she was taken to her birthplace at Woolwich dockyard to be taken apart and rebuilt. Many of the ships in the fleet were in a similar condition and were also rebuilt during the following few years. On 12 March 1698 the *Woolwich* was listed as being at Woolwich dockyard and as being the first scheduled to be repaired as soon as funds allowed.[163] The rebuilt ship was ready for service in February 1702. The new *Woolwich* was almost the same size as the old, but 9 inches longer on the gundeck, and 6 inches wider and 1 inch deeper in the hold.[164] During her 22-year career the *Woolwich* had spent 45% of her time in harbour, and although it is difficult to estimate, the cost of her repairs and refits was probably twice the cost of the original building. As the ship led a typical life, it's reasonable to assume from this that the master shipwrights spent about twice as long repairing ships as they did building new ones.

Rebuilds

A number of contracts exist which describe the rebuilding of ships. It is reasonable to assume this involved stripping off the planking then taking the futtocks out from the frames and renewing them where necessary. Then, as the ship was built up again, good parts from the old, perhaps frame timbers, beams and knees, would be reused. This was not always the case, for a number of ships, such as the fourth-rate *Portsmouth* of 1650, were not taken apart but extensively repaired without any frame timbers being moved at all.[165]

One old fourth-rate ship that was rebuilt by having her frame timbers taken down was the *Assistance*, originally built at Deptford in 1650 by the private shipbuilder Henry Johnson, who was paid £3,386 10s for constructing the hull.[166] A quarter of a century later, in November 1676, she arrived at Deptford, where Henry Teonge, her chaplain, described her as '*the rottenest frigate that ever came to England*'.[167] In June 1684, she was again at Deptford being surveyed by shipwrights. They

THE MASTER SHIPWRIGHT'S SECRETS

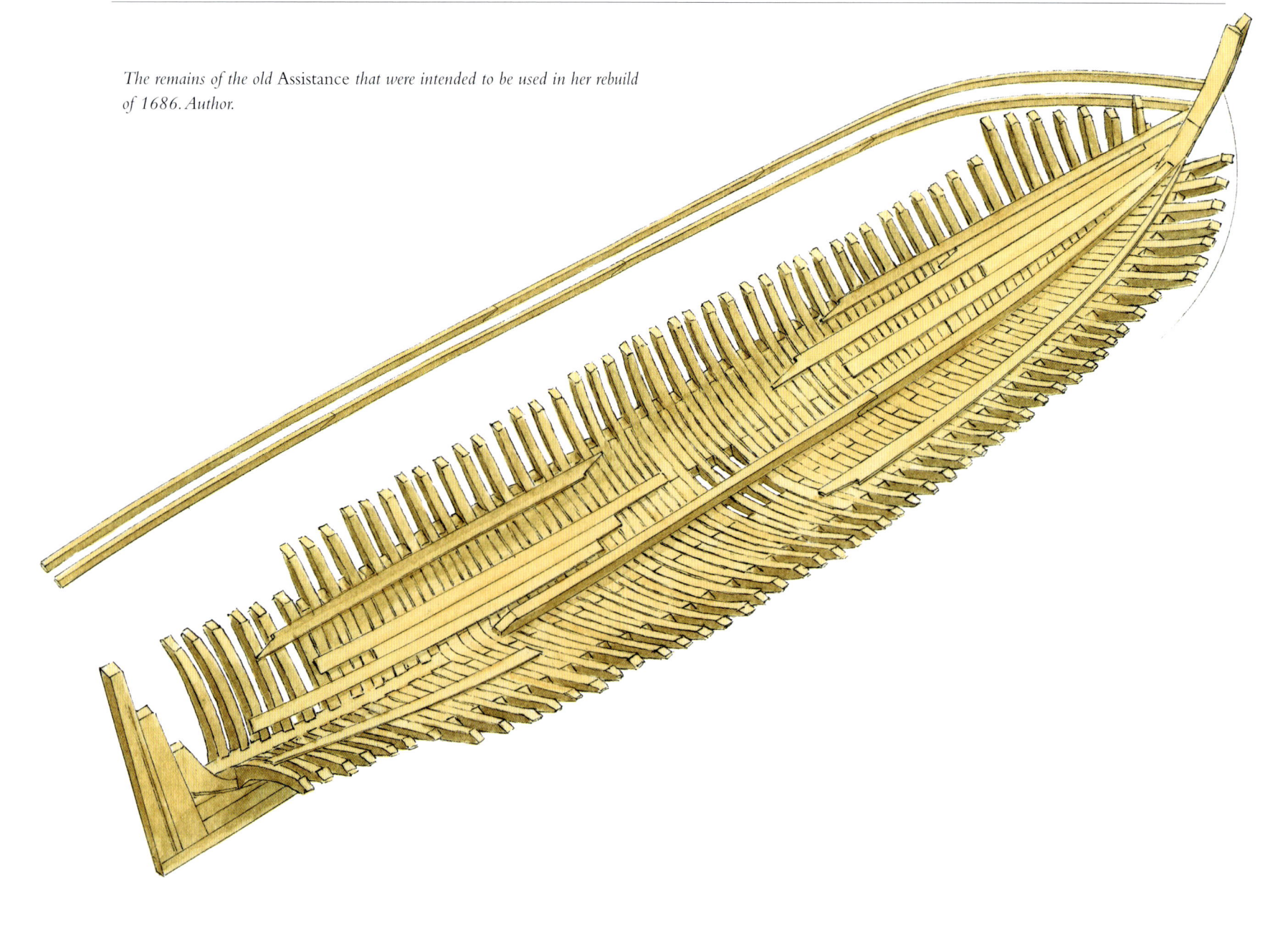

The remains of the old Assistance *that were intended to be used in her rebuild of 1686. Author.*

listed her many defects and estimated it would cost £1,600 to repair them.[168] Little was done at the time, but in 1686 it was decided to have her rebuilt in Mr Castle's nearby private yard. A contract was made out to cover the work. and a rough draught of it survives.[169] Another fourth rate, the *Ruby*, was rebuilt, with almost exactly the same amount of work being performed on her as on the *Assistance*.[170]

The contract for the work on the *Assistance* read as follows:

Mr Castle for the Assistance *to do the work by Sept 29th 86 for £2,900.*

To rebuild the said ship within board & without board. To find all manner of materials & workmanship for the complete finishing of the shipwrights work, smiths work, caulking, graving, joining work, carving, painting, glazing, plasterers and plumbers work fit for the sea, excepting the keel, stern post, stem, floor timbers, lower futtocks, part of ye keelson, & part of the footwaling at the floor timber heads. Plank & wales without board at this time remaining upon the said ship, but if the wales or old work before mentioned upon search shall prove defective it shall be done & repaired at my own charge. I do also further oblige myself that all ye timber, plank, iron work, ornament & other materials to be used upon the said ship, shall be of the like number, dimensions, quality & goodness as is done & to be done to the new 4th rate ship now building by Mr John Shish in his Majesty's yard at Deptford, (the St Albans described in chapter 6) all the works be done to the said ship shall be to the satisfaction & approval of the Commissioners for managing the affairs of his Majesty's Navy for the time being or whom they shall please to appoint on his Majesty's behalf, from time to time to survey the same.

I do also oblige myself to build one water boat 27 foot long, 8 foot broad & 3 foot 5inches deep and one pinnace 27 foot long, 5 foot 10 inches broad & 2 foot 6 inches deep with good oak board & wainscot for his Majesty's service.

The rebuilt *Assistance* left Deptford in August 1687 and would last until the end of the 17th century, when she was rebuilt once again.

The Assistance *as she appeared after her rebuild in 1687. Courtesy Arnold Kriegstein Collection.*

The old fourth-rate *Greenwich* of 1666 was in a poor state by the time she was 25 years old in spite of regular repairs. Her head was weak, with broken rails and gratings having been washed away. The upper deck needed replacing as it was worn so thin it would not hold the oakum caulking. The standards placed on it had worked so loose they needed re-bolting. The gundeck was very decayed near the sides and its supporting beams so loose they worked several inches from side to side. One of the beams was rotten and the standards so loose they lifted at their ends. Several of the hanging and lodging knees were decayed and the bolts loose. The supporting cross pillars in the hold and their knees were split and the bolts worked out. Some of the orlop beams were also decayed and the transom knees split. The bitt pins and cross pieces were worn through, as were the hawse pieces. Her treenails were rotten, making the ship leaky and weak.[171] In spite of her condition she would last a few more years before she was rebuilt. Another ship, the 45-year-old *Constant Warwick*, was so old her upper deck beams rounded downward instead of upward, so that water lay upon the deck without draining.[172] The ship was almost beyond use, but before she could be broken up or heavily rebuilt she was taken by the French.

Old Ships' Demise

One ship that proved beyond repair was the *Victory*, originally built at Deptford in 1620. By 1690 she had been repaired so often it was decided to have her broken up at Woolwich.[173] A contract was made out giving instructions for her decks, knees, internal planking, external planking and frame timbers to be taken apart as far down as the first futtocks. All the parts fit for future use were to be carefully preserved and not split or spoiled during the process. Any middle futtocks in good condition were to be left in place. This would leave only the bottom of the hull partially intact. It was estimated that the work would take two months to complete, at which point the remains would be inspected and a decision made as to whether they should be broken up as well. In the end, it was decided to take the whole ship apart, at a total cost of £270.[174] This was a rare occurrence. In his time, King Charles regarded ships with affection and would always repair them rather than break them up.

CHAPTER 2

INVENTIONS AND INNOVATIONS

So many inventions and innovations were introduced during the Restoration period that it is not possible to give an account of them all – not least because many must have been tried and then, having failed, were totally forgotten about. Some of those adopted were an obvious improvement over previous practice and spread so quickly among the shipbuilding community by word of mouth or private letters that no trace of their introduction has survived. Conversely, some less practical ideas were subject to tests and experiments which required organization and cost money. These have left much more evidence of their brief existence.

The gloomy story of failure that was the fate of so many inventions of various kinds was not always due to them being fundamentally flawed. The apparent conservatism among the naval administration and officers put a stop to some that should have been pursued further. Some, such as lead sheathing and paddle wheels, had to wait for the development of new materials or technology to bring them to fruition. Others, such as iron fire hearths, were later adopted with success. In such matters, King Charles was an eternal optimist in his pursuit of improvement, and during his reign many successful innovations were introduced.

The First Frigate

In 1689, the *Tyger* and the *Nonsuch* captured the French 36-gun *Les Jeux*, a ship newly built at Dunkirk. It was a significant success as Dunkirk frigates were famously fast sailors and had previously always managed to elude the heavily armed ships of the Royal Navy. According to Samuel Pepys, King Charles told him that years before a St Malo Frenchman had invented the 30-gun frigate based on the shape of the local shallops. The Frenchman offered it to Sir George Carteret, who recommended it to Charles I. Charles I's naval administrators, so the story goes, were indignant that a Frenchman should pretend to build better ships than they and sent him packing. The Frenchman then took his design to Dunkirk and the idea was developed into the famous Dunkirk frigate. In 1645, the master shipwright Peter Pett built the fourth-rate-sized privateer *Constant Warwick* based on its design after having the opportunity of seeing one of these frigates after it was brought into the Thames. The early fourth-rate ships that followed the frigate principle of 1645 were regularly repaired over the years and spoiled by alterations and additions such as forecastles and extra guns. The success of the frigate principle did lead to weight as well as strength being considered during shipbuilding. The term 'frigate built' became a term often used and could apply to ships of all sizes.[175]

The nature of Dunkirk frigates was not at all clear at the time and has been something of a mystery ever since.[176] The capture of *Les Jeux*, renamed the *Play Prize*, may hold the answer as she was a direct descendant of the earlier Dunkirk frigates of the 1630s and '40s and probably retained the arrangement of the original successful design.[177] On 28 January 1690 the ship was taken to the Woolwich dockyard, where she was surveyed by the master shipwright, Joseph Lawrence, before he made her ready for service in the English Navy. Not used to the radical design, he considered her ill contrived and recommended a number of changes. He would cut out two additional gunports aft and move the tiller from the hold into the existing captain's cabin on the gundeck, a modification that would require a rudder that was longer by three feet. A quarterdeck would be added under which he would build a new captain's cabin, and which would provide space for two more gunports. A forecastle

INVENTIONS AND INNOVATIONS

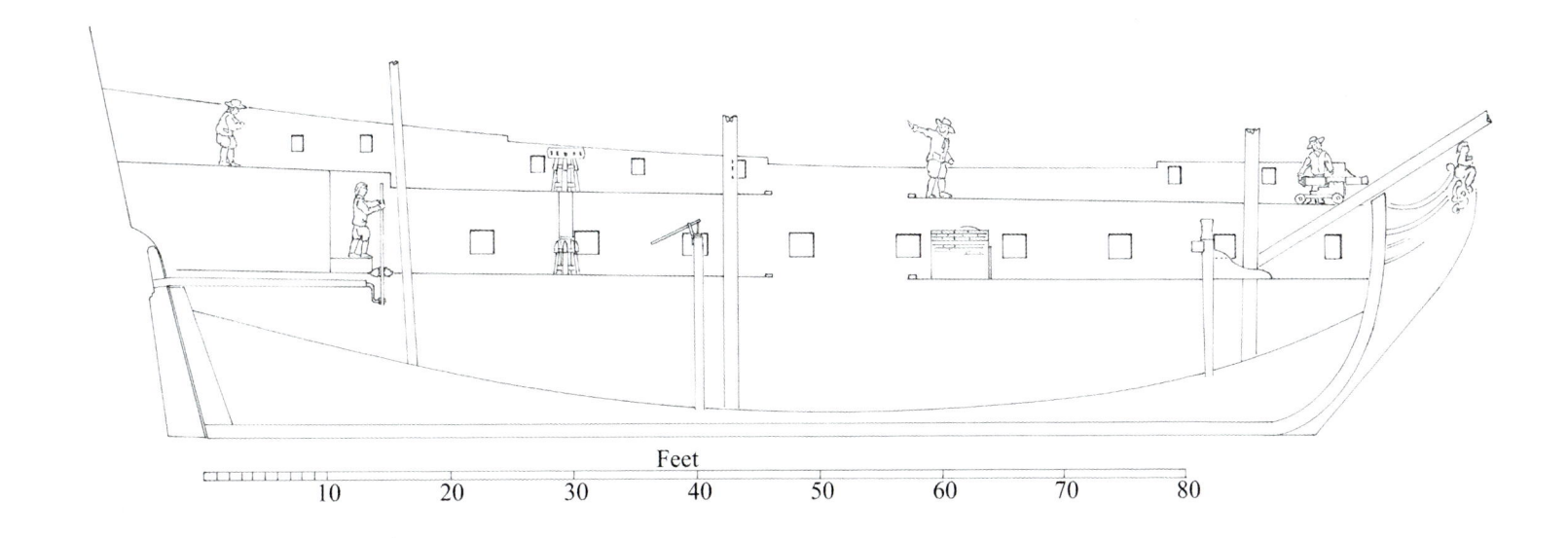

The Dunkirk frigate Les Jeux, *captured in 1689 and renamed* Play Prize. *The arrangement of the ship was described by the master shipwright Joseph Lawrence when she was refitted for Royal Navy service. It fits perfectly with known van de Velde images of the ship. Author and Frank Fox, who helped interpret the Lawrence description.*

The Play Prize *at the time of her capture from the French (Robinson 677). Portrait of the 'Play Prize'. Willem van de Velde, the Younger, 1689? PAH1898; Photo: © National Maritime Museum, Greenwich, London.*

would also be added and the cookroom moved up into it from its position between decks. The concept of a light, fast frigate seemed sadly lost on Lawrence, whose proposed alterations would all add weight and windage above water. Other more prosaic changes included adding chain pumps, moving the bitts further aft, fitting a new capstan and removing the one foot rising aft on the upper deck.[178]

The King's Yachts

King Charles II was famous for his pleasures, and one of them was his love of yachts. They were introduced to England when the Dutch East India Company gave Charles the *Mary* at his accession in 1660. In October 1661, John Evelyn witnessed him enjoying a yacht race against his brother from Greenwich to Gravesend and back, during which Charles sometimes steered himself. During the return voyage, John Evelyn joined the King to eat and talk.[179] But they were not just pleasure boats for Charles. Once their sailing ability was realised, they were produced in some numbers, being used as fast sailing attendants to squadrons, for transporting important passengers and as dispatch vessels. When he decided to build a new yacht in 1677, Charles made it a splendid bargain by donating £500 towards the estimated construction cost of £2,973. This yacht, the *Charlotte*, was lavishly equipped beyond the estimate, with, among other things, three paintings by the famous Dutch marine artists the van de Veldes. The paintings are all of ships built by the *Charlotte*'s builder, Phineas Pett III of Woolwich, and they survive today in the National Maritime Museum.[180] As the yacht neared completion, the matter of the shipwrights' '*extraordinary lavishness*' was discussed at the Admiralty Commission. James, Duke of York expressed '*apprehension of the chargeable consequences of permitting shipwrights to exercise that liberty which they are found daily to usurp of adorning their ships with carving, painting and gilding without any limitation but at their own pleasure*'. King Charles and the lords agreed, adding their own disapproval; however, Mr Pepys pointed out that, when questioned, the master shipwrights *do generally pretend His Majesties verbal command*. With commendable honesty, Charles owned up and agreed that he had given verbal orders for such extravagance, but for the future agreed to a standing rule that they were not to be carried out without written approval from the Navy Board or the Admiralty.[181] It was a promise he found impossible to keep.

The King's resolve to follow his own rules was tested a year later when another of his yachts, the *Charles* yacht, was cast away on the coast of Holland on 27 November 1678.

Above. *The beautiful Louise de Kérouaille, Duchess of Portsmouth, the most important mistress to Charles during the latter part of his reign. Charles named the yacht* Fubbs *after the reputed chubbiness of her naked body. He also honoured their illegitimate son, Charles Lennox, by naming the third-rate ship* Lenox *after him. An even bigger ship, the second-rate* Duchess, *was also certainly named after Louise, although Charles would probably change his mind about that, depending on which duchess he was talking to.*

Right. *John Evelyn, a very close friend of Samuel Pepys who lived at Sayes Court, next to Deptford dockyard. He knew and liked the Shish family of shipwrights, saying many kind things about them in his diary.*

INVENTIONS AND INNOVATIONS

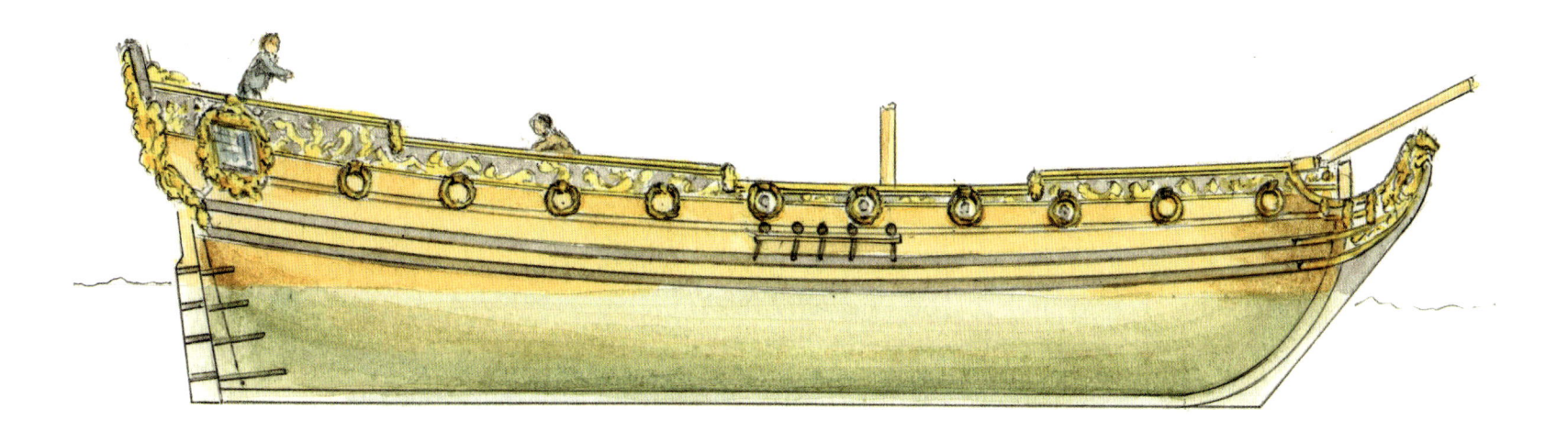

The Charlotte *yacht of 1677. Author.*

The crew's petition for payment of their wages and loss of clothes was discussed by the Admiralty on 1 March 1679. It was agreed to follow proper procedure and examine the loss at a court martial before any payments were made. It was merely a matter of procedure, however, for His Majesty and the lords were already satisfied, in their private opinions, that the loss could not be blamed on the master or the yacht's company.[182] The thought of losing his yacht seems to have stimulated Charles into action. He had Pepys write the following plea to the Navy Board two days later: *'I find His Majesty so very earnest for ye immediate going in hand with his new yacht, he calling upon me every hour almost about it, that I must entreat your board though it is but by word of mouth this day at your seeing of Mr (Thomas) Shish to give him some instructions for ye going in hand with it that so at your seeing him this day you may be able to give him an account of your having so done...'*[183] When his passions were rising, Charles had no qualms in ignoring his own standing order that shipwrights shouldn't act before receiving written orders. Within three weeks, estimates had been agreed and the Admiralty was writing to the Navy Board instructing them to proceed *'as fast as you shall be enabled'*, adding that the King had directed the cost of her ornamentations, gilding and carving not to exceed the costs of previous yachts.[184] Charles named the yacht *Henrietta* at the launch on 23 December the same year and ordered she be rigged and fitted for service as soon as she was launched.[185]

The Double-Bottomed Vessel

The double-bottomed vessel, or catamaran as it is now known, was one of the greatest innovations that managed to avoid being adopted by the Navy. Starting in 1663, Sir William Petty designed a series of successful vessels that culminated in the *Experiment*. She was much larger than the earlier vessels, being 60 feet long and carrying 16 guns. Although she was owned by private investors, the King and the Duke of York attended the launch. The *Experiment* was described as having a shallow draft with the two hulls held together with huge timbers that could take a monstrous broad sail. In trials, she outperformed conventional fast-sailing single-hull boats. She could sail close to the wind but seemed to have difficulty turning, to the extent of braking her rudder on one occasion. The Royal Society set up a committee to examine her. They reported that she sailed well but that when going swiftly into a head sea she

Sir William Petty.

The double-bottom ship design of 1664, based on drawings and model by Sir William Petty. Author.

could plough into waves and be submerged. When she was lost in the Bay of Biscay during a storm, doubts set in and official interest declined, even though other conventional ships were lost at the same time.[186]

Petty remained undeterred, however, and in 1683 held a meeting at his house in Piccadilly attended by Samuel Pepys, Sir Anthony Deane and Henry Sheeres, the engineer. Petty showed them models of an even bigger vessel, about which Deane and Sheeres expressed reservations. Sheeres said he would put his down in writing, while Deane gave an account of the advantages of a single hull.[187] Shortly afterwards Petty wrote to Pepys outlining his further development which took in Deane's suggestions: '*I have in honour to Sir Anthony Deane contrived a single body which in many particulars doth answer the double body: But must tell you, that when this single body is well considered, it is but a double body disguised.*' He also reminded Pepys that Sheeres had promised to give him his objections but had not yet done so. Knowing his design was regarded as a failure, he added: '*Pray let me hear from you in what manner and measure I am ridiculed at court concerning this matter and not sparing to speak plainly.*'[188]

Unfortunately, this project failed and the principles of the design were forgotten. Although it's difficult to be certain, the surviving drawings and model suggest that the two hulls with guns were too heavy to be joined with timber beams. It's a shame that Petty preferred larger vessels, because if he had persevered with his earlier and smaller designs he may have had a much greater chance of success.

Decoration

For a long time before the reign of Charles II, the decoration style of two-deck ships remained much the same, with a single row of windows across the stern. They were situated at the upper-deck level for the benefit of the captain's great cabin. At the bow a lion, generally called a beast, with a whorl above its head featured as the figurehead. This changed after Charles had seen the French ship *Superbe* at Spithead during the Third Dutch War. She had had another row of stern windows added in the cuddy at the aft end of the quarterdeck for the convenience of the master and the lieutenant. This style was adopted even though it left a smaller space on the stern for the royal arms. At the same time, the whorl on top of the beast's head was changed to a crown.[189] The design of the carvings does not appear to have been included on ships' draughts as there are many instances of master shipwrights asking the Navy Board for the manner of the decoration. It may be thought that the cost of such lavishly decorated ships would have been enormous. In fact, however, the art of carving was so common it cost only £160 out of the £12,000 total cost of building the hull,[190] or a little over one per cent.

Sometimes master shipwrights took it upon themselves to adorn ships with carvings and were known to gild boats to excess. To prevent this happening, the Admiralty issued orders that any extra work done without its warrant must be paid for by the person who directed it.[191] Such penalties seem never to have been enforced, for when offenders were interrogated they always had a ready defence, invariably saying they had received verbal orders from the King.

Cookroom

By the early years of Charles II's reign the cookroom layout had become standardised and fulfilled most requirements. In fourth-rate ships, it was situated in the forecastle, where the crew's meat was boiled in one of the two copper furnaces.[193]

INVENTIONS AND INNOVATIONS

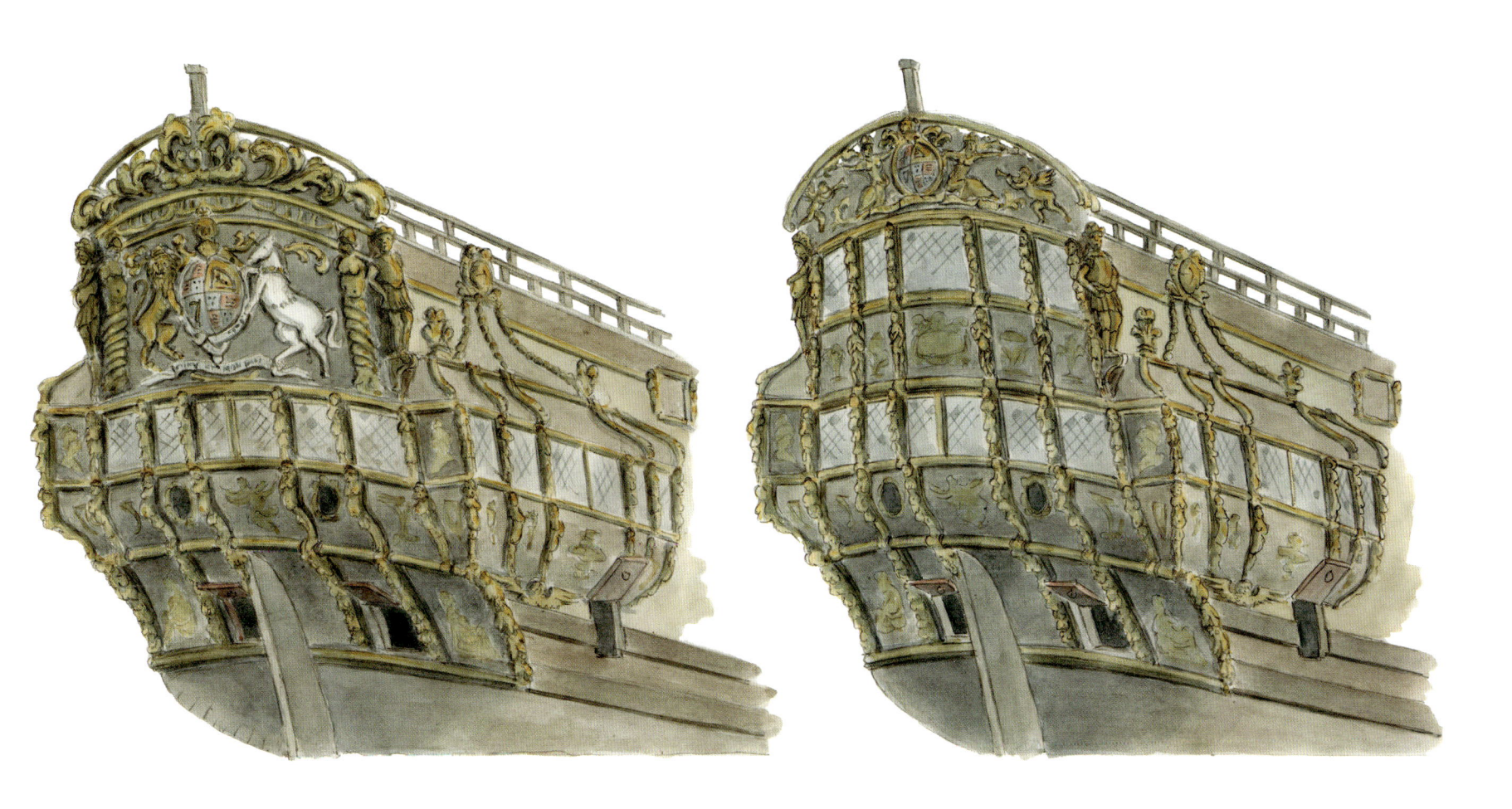

The stern alterations to the fourth-rate Hampshire, *showing the old style to the left and the new-style alterations, made after her rebuild in 1677, to the right. Author.*

Each of these had a brass cock soldered at the forward end to drain the residue away. The furnaces were enclosed in a brick hearth fuelled by wood loaded through a fire door. At the aft end of this was a range for roasting. It was covered by a copper hood and chimney which took the smoke up to vent above the forecastle deck.[194] Furnaces were often covered with lead sheet, presumably to seal the awkward and complicated gap around the copper hood.[195] Due to the working and motion of the ships, brick hearths required frequent repair and were often taken apart and rebuilt. They were also very heavy, requiring extra, or closer-spaced, beams in the deck beneath.

Iron Range

In January 1676, King Charles met Mr Pett, the master shipwright who had come to town to talk about shipbuilding. As was his pleasure, the King entered into a discourse about a flyboat Pett was building for him at Woodbridge. Charles required that '*her furnaces be placed in an iron range between decks according to ye late new invention in case* (i.e. unless) *yourselves upon survey shall not judge it hazardous to ye safety of ye said ship*'. Its advantage over a brick range was that it would be much lighter and require less maintenance.[196] During the next three months, warrants were issued for the construction of two lightly built galley frigates for use in the Mediterranean – just the type of ship that would benefit from having an iron range. These ranges were invented and patented by a Colonel Ewbank, who had sold some for use in merchant ships.[197] Just what they looked like is unknown, but they probably followed the shape and style of brick hearths. Charles, ever interested in such inventions, met Colonel Ewbank himself and gave verbal orders for him to provide an iron range for each of the two recently launched galley frigates, the *Charles Galley* and the *James Galley*. Charles doesn't appear to have told anyone else about his order, for their delivery seems to have come as a surprise to the Navy Board and the Admiralty.[198]

Perhaps irritated at Charles's flouting of all the rules, the Admiralty soon found fault with the new iron ranges, writing that '*we also understand that some inconveniences do already appear*

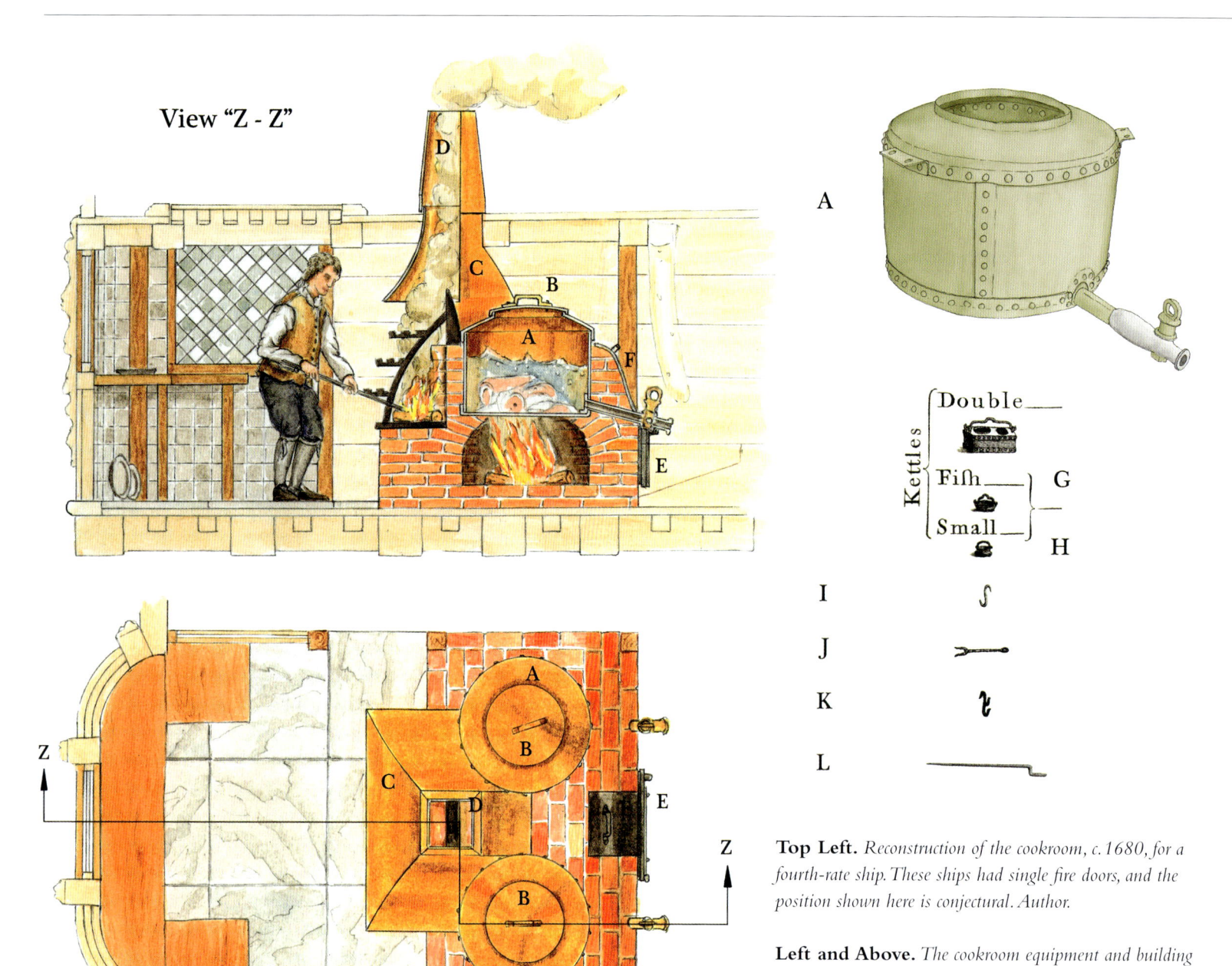

THE MASTER SHIPWRIGHT'S SECRETS

View "Z - Z"

Top Left. *Reconstruction of the cookroom, c.1680, for a fourth-rate ship. These ships had single fire doors, and the position shown here is conjectural. Author.*

Left and Above. *The cookroom equipment and building materials. Author. A. Two copper furnaces; B. Two copper furnace covers; C. One copper hood; D. One copper chimney; E. One fire door; F. One oven lid; G. One fish kettle; H. One small kettle; I. One esse hook for kettle; J. Two flesh hooks; K. One kettle hook; L. Two spits; two racks; one fire fork; one fire shovel; one pair of fire tongs; one iron fender; two fish hooks; two gaming hooks; 1,200 bricks; paving tiles; plain tiles.*[192]

in that of the James by the fires being placed to near the foremast besides some other suggested from the want of proper conveyance for the smoke'. In order to fully inform the King about the usefulness or otherwise of the new invention, it ordered the Navy Board to have a report made in respect of safety, service and good husbandry in comparison with the old brick hearths.[199] The very same day, a number of prominent master shipwrights, including Phineas Pett of Woolwich and John Shish of Deptford, reported that the forecastle bulkhead of the *James Galley* should be moved four feet further aft to provide more space between the hearth and the foremast. If this modification was made,

they wrote that they '*conceive that the range will do very well and will boil meat enough for the ships company*'. Shish had the materials ready to perform the work, and a warrant for carrying it out was issued.[200] Two weeks later, Captain Canning of the *James Galley* complained that the lack of space in the forecastle precluded a baking place in his hearth, a feature said to be commonly allowed in His Majesty's ships. Charles himself replied, indicating that a separate convenience would be supplied for baking, of the kind provided on board other ships.[201] Colonel Ewbank seems to have anticipated the problem, delivering an apparatus to Deptford made of iron which could bake and stew. It was inspected by John Shish and Phineas Pett, who judged that it would perform well.[202] The baking facility's primary function was probably for making bread[203] and for baking the captain's private provisions. It may have been placed at the ship's side, as indicated in a van de Velde drawing of the *Charles Galley*.[204] The ship also needed a modification to her iron range involving the addition of some iron plates between the new hearth and the anchor cables.[205]

During December, the new-fashioned fire hearths supplied by Colonel Ewbank were discussed by the King, the Admiralty and the Navy Board officers. They expressed their approval of the invention, conditional on future experience with its use.[206] They then turned to the consideration of its cost, which Colonel Ewbank calculated as follows:

October 2nd 1676 Delivered on board the *Charles Galley* Frigate

	Cwt Qtr Lb		£ s d
Iron hearth	13　2　14	at £3 per Cwt	40　17　6
Double copper furnace with copper funnel and 2 covers	2　1　26¼	at 20p per lb	23　03　8
			64　01　2

November 20th Delivered on board the *James Galley* Frigate[207]

	Cwt Qtr Lb		£ s d
Iron hearth	12　3　8	at £3 per Cwt	38　09　4
Double copper furnace with copper funnel and 2 covers	2　1　20¾	at 20p per lb	22　18　8½
			61　08　0½
		Total £125	09s　2½d

By way of comparison, a brick hearth for a fourth rate, including materials and labour, was quoted in April 1680 as £8, making the new iron hearth five times more expensive.[208] The copper furnaces, funnels and cover should have been about the same price. The Navy Board investigated the high cost of £3, or 60 shillings per hundredweight, and discovered that before the patent was issued to Ewbank, the iron master who made them, probably Robert Foley, supplied them to merchantmen at 40 shillings per cwt. Taking advantage of having had a patent granted, Ewbank increased the cost to 56 shillings per cwt. The Navy Board concluded that Ewbank was overcharging, and that 45 shillings per cwt was reasonable.[209] Using this figure, John Shish did his own estimate for the iron fire hearth of the *James Galley* and reckoned it should cost £29 10s 4d.[210] During further debate, Colonel Ewbank claimed the prices were the same as those he had charged for the iron hearths he supplied to merchantmen,[211] and as a trump card, mentioned that he had been personally promised ready money by King Charles for those supplied to the galley frigates.[212] Ewbank was duly paid in ready money as promised, but for the future he would be paid as usual by course along with the Navy's other creditors. Four years after its installation, the *Charles Galley*'s fire hearth needed some repair with new plates.[213]

At least four ships were supplied with iron hearths made by the ironmonger Robert Foley during 1680[214] and another, the *Kingfisher*, was given one in 1685.[215] After the death of Charles in 1685 interest in iron ranges ceased, and even 50 years later small ships of 20 guns still had brick hearths.[216] It was not until the late 1750s that iron hearths were eventually reintroduced and fully accepted.[217]

Steering Wheel

One of the greatest innovations in the development of the sailing warship was the replacement of the whipstaff by the ship's steering wheel. No written evidence has yet been found to determine who invented it or when it was introduced. From the evidence supplied by models, it was relatively simple in its early form and easily made by shipwrights, or even by the ship's carpenter. In 2002, an archaeological survey was made of the third-rate *Stirling Castle*, wrecked on the Goodwin Sands in the Great Storm of 1703. On one of the last dives of the season, Bob Peacock of Seadive Organisation found the fundamental double block necessary for the steering wheel to work. It would appear to be of such an obvious benefit that it would soon replace the whipstaff. In fact, however, the whipstaff remained in use as a redundant back-up steering system until at least 1717.[218]

THE MASTER SHIPWRIGHT'S SECRETS

Inches

Above. *The fixed double block, as recovered from the wreck of the Stirling Castle, lost in the Great Storm of 1703. It is now on display at the Shipwreck Museum, Hastings. Author, courtesy* Mariner's Mirror, *February 2004.*

Above Right. *The early steering wheel arrangement of the type found in a model of about 1702 and as described by William Sutherland in* Ship-building Unvail'd, *of 1717. Author, courtesy* Mariner's Mirror, *February 2004.*

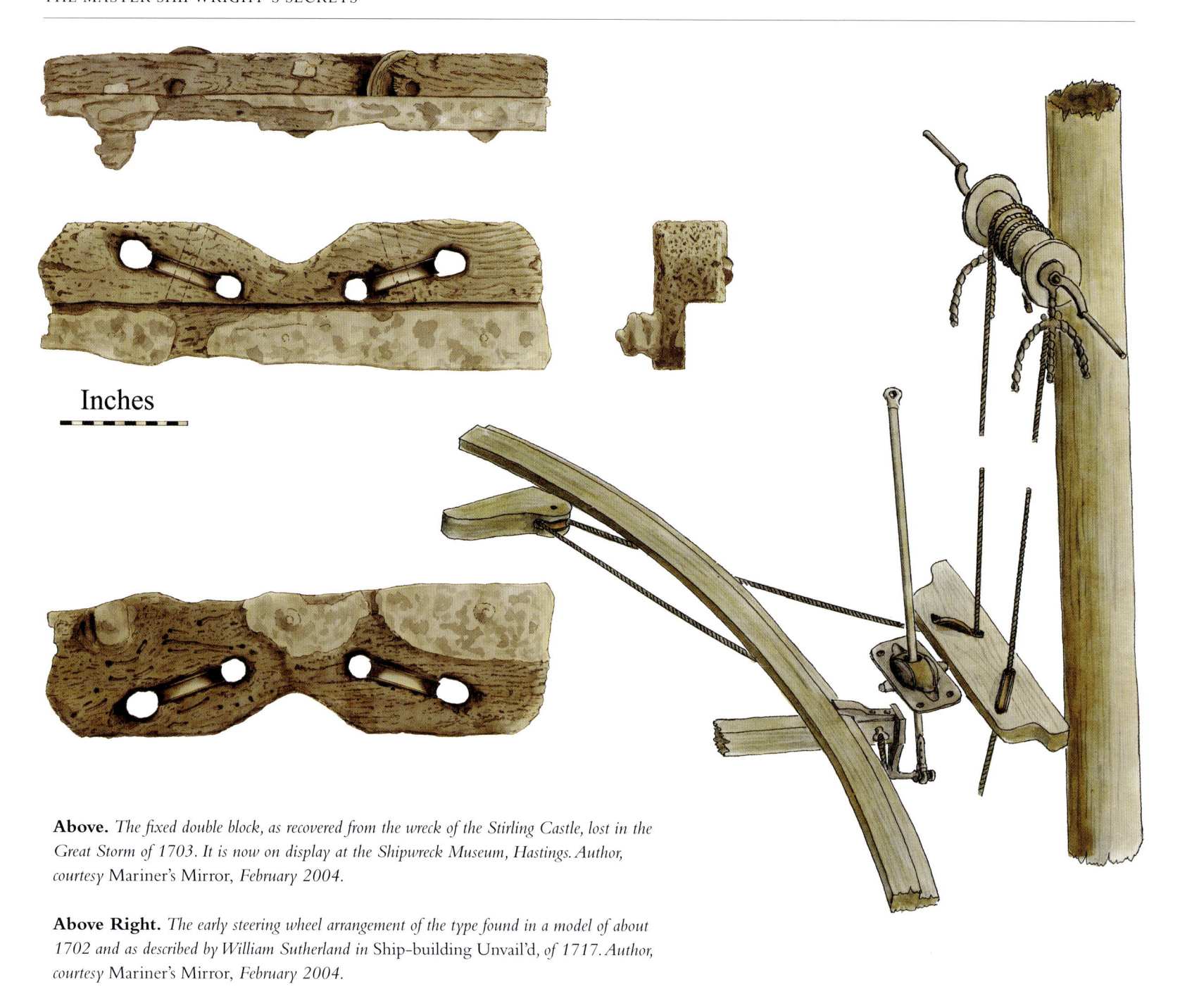

Drumhead Capstan

After the introduction of the capstan in the middle of the 16th century there was only one fundamental change to its design over the next 300 years. This was the replacement of the crab capstan with the drumhead. Instead of having four long bars passing through the spindle at necessarily different heights, a large-diameter drumhead was fitted with slots for 12 half-bars all at the same convenient level. Contemporary models suggest that the change occurred during the 1670s, but using models for evidence is unreliable as such small details as detachable capstans would have been easily lost and might perhaps have been replaced with a later version on these models. In 1686,

Sir Samuel Moreland claimed he had invented it some years back,[219] but the earliest reliable evidence dates from September 1676, when the old fourth-rate *Diamond* was modified by having new whelps and drumheads fitted to both her main capstan and her jeer capstan.[220] The way a drumhead was fitted to an old-style capstan is explained in an estimate of work to be carried out on the *Warspite* in June 1680. This simply involved cutting the head off the capstan and putting a drumhead in its place.[221] The change seems to have been made general at the time, for a similar modification was carried out to the main capstan of the *Reserve* during the same month.[222]

INVENTIONS AND INNOVATIONS

Left. *The original main capstan of the fourth-rate* Diamond *of 1651. Author.*
Right. *The same capstan with the head sawn off and fitted with a new drumhead and whelps, as modified in September 1676. Author.*

Above. *The upper end of the Roman chain pump found at Gresham Street, London. Based on a reconstruction by the Museum of London. Author.*

Pumps

Mechanical pumps that lift water vertically by some sort of iron chain have been used since ancient times. The remains of a Roman pump were found in Gresham Street, London and excavated by the Museum of London in 2001. Although the driving wheel at the top was not found, parts of the chain survived in good condition. This gave enough evidence of its workings for the museum to build a credible working reconstruction.[223] In operation, the looped chain drove wooden boxes, or buckets, filled with water from the bottom of the well to the top, where their contents were tipped out as they passed round the drive wheel. All knowledge of chain pumps appears to have been lost after the Romans, but it resurfaced in Tudor times, when Raleigh wrote about '*the chain pump which takes up twice as much water as the ordinary did*'. Three ships had chain pumps installed in 1577, but what sort of pump the 'ordinary' was is not known.[224] Perhaps the ordinary pumps were similar to the Roman chain and buckets, but more probably they used the principle of a simple valve called a 'burr', or a common suction pump.[225] The new chain pumps referred to by Raleigh may well have been similar to the example found in the Great Hall undercroft at Hampton Court Palace in 1923. This pump consisted of a fabricated square-section wooden tube through which a chain drove square valves. A description of this type of pump was given in 1556 by a Portuguese priest, Gaspar da Cruz, who mentioned that '*in the midst of this piece of wood is a square little board almost a hands breadth...*'[226]

The Hampton Court pump would appear to be a direct ancestor of those used during Charles II's reign. Although the design principle was the same, it had evolved by then, with many improvements. The square fabricated duct was replaced by a round bored-out elm tube, and the wheels at top and bottom were more specialised. In 1679, William Purser delivered to Deptford the ironwork for four pumps, each of which was described as having an iron rowl, two cheeks, a bottom plate, about 25 lengths between each leather to leather of chain, the iron chain, ten forks for the wheel, two round plates, two square plates, an axletree, a winch and nails to fix the plates.[227] Iron rowls were not universally used: a ship built at Deptford the year before had rowls made of brass.[228]

Pumps were vital for the safety of ships and constantly under development to be more efficient and reliable. One of most prominent inventors of the day was Sir Samuel Moreland, an inveterate inventor of naval engines, one-time spy, adventurer and Fellow of Magdalene College, Cambridge. In September 1673, he developed an 'engine' that could remove water efficiently from the dock at Chatham using energy from a windmill, or a horse should the wind fail.[230] The mill pump

THE MASTER SHIPWRIGHT'S SECRETS

Opposite. *The late-17th-century chain pump. A. Axletree; B. Winch; C. Dale; D. Wheel; E. Case; F. Pump; G. Cistern; H. Sprockets or forks.*[229] *Author.*

Above. *The upper end of the Tudor-period Hampton Court pump. It has been suggested that it was used to remove sewage, but its size and ingenious compact design may indicate a naval connection. The lower (idle) wheel is similar to the upper but not driven. Author.*

was ready for installation by December,[231] but nearly three years later it was still not being used. As it was situated inconveniently near the dock, it was decided to either put it in working order or have it removed.[232] Soon afterwards, Moreland went to Chatham with Pepys and King Charles for a meeting[233] at which improvements were recommended and suggestions made for a final trial.[234] This must have been unsuccessful, for by mid-1679 the mill pump lay abandoned in the innermost plank yard, from where one of the labourers, or some such other person, had stolen some of its parts.[235]

In August 1677 Moreland came up with another idea for ship pumps which Charles wanted to see tried out before he went to Newmarket. This pump harked back to the Roman design of buckets and a chain. At the time Moreland invented, or reinvented, the design, he was living at Vauxhall, only two miles away from where the Roman original was lying undiscovered in Gresham Street. (Undiscovered, that is, if we eliminate the possibility that part of it had been revealed after the Great Fire in 1666 and that Moreland had in fact seen it.) Orders were sent to the fourth-rate *Jersey* moored at Woolwich for one of her conventional chain pumps to be replaced by the new design for a trial before the King.[236] The pump was not adopted, but Moreland remained undeterred and persevered with his designs. King Charles was not always such a pushover for pump designers, however. When Thomas Smith petitioned for an allowance to '*make proof of a pretended invention of his of*

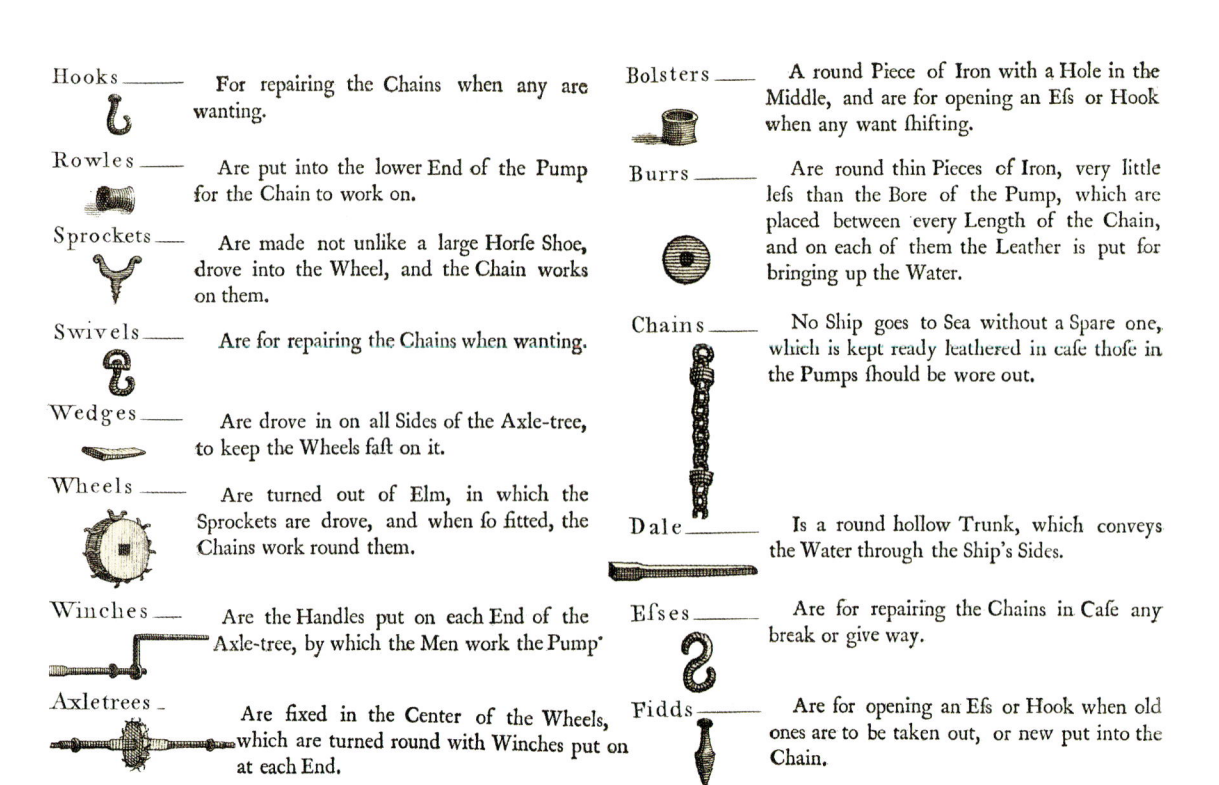

Hooks	For repairing the Chains when any are wanting.
Rowles	Are put into the lower End of the Pump for the Chain to work on.
Sprockets	Are made not unlike a large Horſe Shoe, drove into the Wheel, and the Chain works on them.
Swivels	Are for repairing the Chains when wanting.
Wedges	Are drove in on all Sides of the Axle-tree, to keep the Wheels faſt on it.
Wheels	Are turned out of Elm, in which the Sprockets are drove, and when ſo fitted, the Chains work round them.
Winches	Are the Handles put on each End of the Axle-tree, by which the Men work the Pump.
Axletrees	Are fixed in the Center of the Wheels, which are turned round with Winches put on at each End.
Bolsters	A round Piece of Iron with a Hole in the Middle, and are for opening an Eſs or Hook when any want ſhifting.
Burrs	Are round thin Pieces of Iron, very little leſs than the Bore of the Pump, which are placed between every Length of the Chain, and on each of them the Leather is put for bringing up the Water.
Chains	No Ship goes to Sea without a Spare one, which is kept ready leathered in caſe thoſe in the Pumps ſhould be wore out.
Dale	Is a round hollow Trunk, which conveys the Water through the Ship's Sides.
Eſses	Are for repairing the Chains in Caſe any break or give way.
Fidds	Are for opening an Eſs or Hook when old ones are to be taken out, or new put into the Chain.

Left *The parts of a chain pump, from Thomas Blanckley,* A Naval Expositor, *1750.*

52

INVENTIONS AND INNOVATIONS

THE MASTER SHIPWRIGHT'S SECRETS

Lead ——— Some Ships have one fixed, whose Pipe goes down the Knee of the Head, and is there placed for washing the Decks.

Hand ——— Hangs over the Side, and lashed there for washing the Decks, and sometimes are put down into the Well, for freeing the Ship when she makes more Water than the Chain Pumps can throw out.

Above. *In addition to chain pumps, fourth-rate ships were fitted with a number of hand pumps, as shown by Thomas Blanckley in* A Naval Expositor, *1750. Sutherland, in* Ship-building Unvail'd *(1717), gives the size of the pump in the well as being 7 inches diameter with a bore of 2.83 inches.*

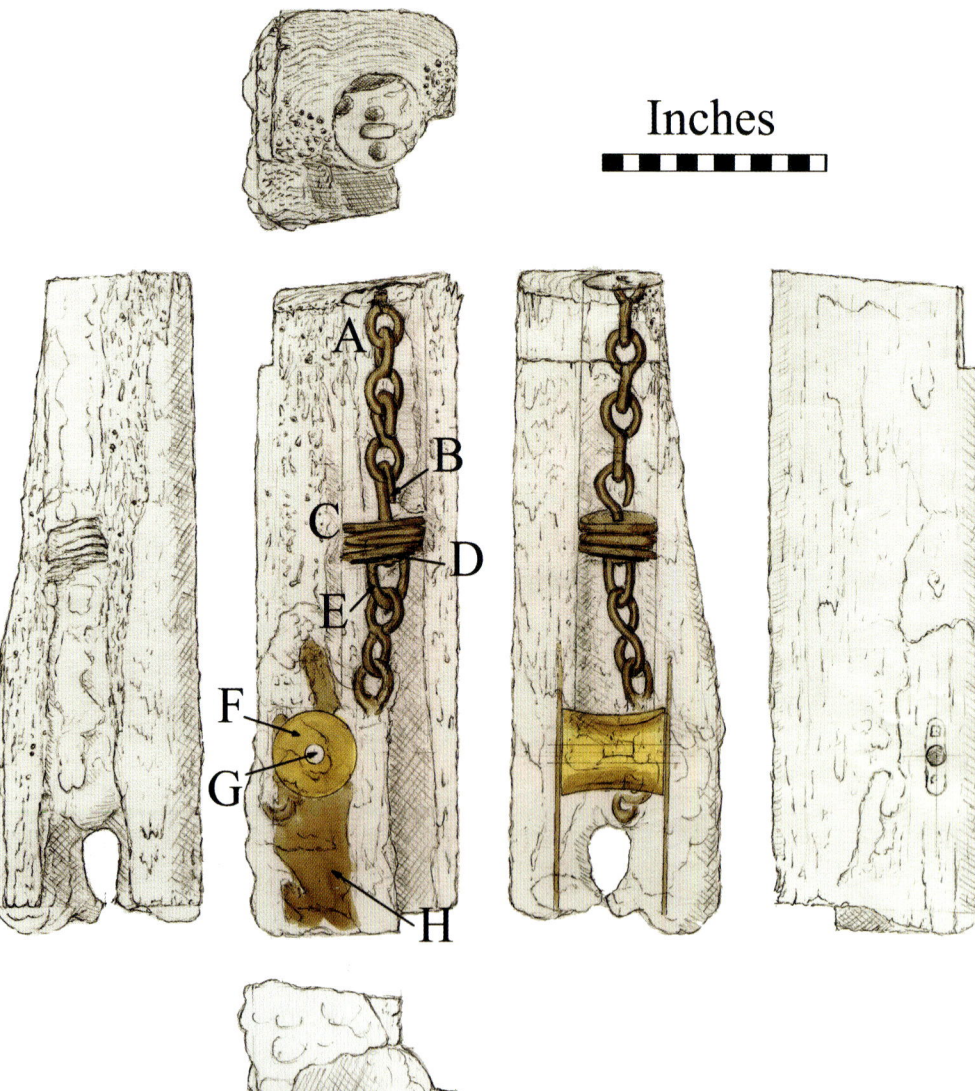

Inches

Left. *The bottom end of the pump of the Northumberland, lost in the Great Storm of 1703. The coloured internal arrangement is shown as revealed by CT scan carried out by the University of Southampton. The metal work is generally of wrought iron, with a lead bronze alloy rowl. The burrs were situated 30 inches apart. A. Esse hooks; B. Hooks, the lower ends are riveted through the swivel; C. Leathers, three layers; D. Burr, iron plate to support the leathers; E. Swivel; F. Brass rowl; G. Pump bolt; H. Bolt plate, to secure the rowl and pump bolt. Terminology from the* Lenox's Building List *of 1678 and from Thomas Blanckley,* A Naval Expositor, *1750. Courtesy of Robert Peacock and Daniel Pascoe of Seadive Organisation.*

Left. *Sir Samuel Moreland, one of the great minds of his age. As well as being an inventor of pumps, he was a mathematician, a linguist, a cryptographer and a diplomat, among other things. Private collection, via author.*

an extraordinary pump, His Majesty who seemed already informed and satisfied in the folley of the said invention was pleased to reject ye said petition'.[237]

Charles died in 1685, but his brother succeeded him as James II and shared his interest in the Navy. In June 1686 Moreland demonstrated another sea pump to the great satisfaction of James, who had one of the *Charles*'s chain pumps replaced by the new sea-pump engine, for which Moreland received £120.[238] Two months later, on 17 August, at seven in the morning, Sir Robert Gourdon, another inventor of pumps, demonstrated his engine in St James's Park in front of King James.[239] It was successful enough for the Navy Board officers to discuss its further progress.[240] They arranged to meet at Deptford during September regarding the development of the pump,[241] but a new trial before James was not arranged until December.[242] Yet another inventor of pumps, a Robert Lodgingham, put forward his proposal and James decided that he should also have the opportunity of proving his design along with Moreland and Gourdon.[243]

The great trial of the three pumps was arranged to take place at Deptford. Lodgingham, however, failed to make an appearance. Sir Robert Gourdon's pump was satisfactorily stowed and secured aboard the *Swallow*, while Sir Samuel Moreland's was allocated to the *Portsmouth*.[244] The master shipwright, Fisher Harding, looked after Moreland, but after studying the ship for an hour to find a place to fit his engine, and having brought no materials with him, Moreland eventually said he would give the Navy Board an account before he proceeded further. Then, changing his mind, he decided that a one-foot-square hole would have to be cut in the gundeck at the side near the waterways, and another hole measuring eight by ten inches cut right through the side of the ship, one foot above the deck. He also wanted one side of the well taken out, two stanchions put up, and one pump taken out. The Deptford officers were rather shaken by Moreland's impulsive demands and sensibly declined to carry out such drastic modification unless they received direct orders from the Navy Board.[245]

There was good reason for Moreland's prevarications. Like many men of immense intellectual ability, he seemed to lack ordinary common sense. His third wife had died in 1680, and he was in debt with creditors who were threatening him with debtor's prison. A solution was presented to him by a friend whom he had helped in the past with '*a thousand kindnesses*' and who said he wished to return the favour. He generously found Moreland a new wife, who was not only '*very virtuous, pious and sweet dispositioned*' but also an heiress, with an income of £500 a year from land she had inherited and £4,000 in ready money. She also received £300 a year from a mortgage and owned plate and jewels. Moreland instantly fell in love with her and married her just as his pump was being readied for the trial. Shortly after, he wrote to Samuel Pepys about his marriage, saying he had been '*led as a fool to the stocks, and married a coachman's daughter, not worth a shilling, and one who about 9 months since was brought to bed of a bastard. And thus I am become both absolutely ruined in my fortune and reputation and must become a derision to all the world.*'[246] In 1688 Moreland heard that his wife had been cohabiting with Sir Gilbert Gerrard '*and besides had ye pox*'. He went to law and a '*sentence of divorce was solemnly pronounced in open court against that strumpet for living in adultery*'. His success was somewhat spoilt, though, as he was ordered to pay all the debts she had accumulated during the marriage.[247] The derision he received at his misfortune would seem as justified now as it was then.

In the face of such modest opposition from Moreland and Lodgingham, it was decided that the pumps used by Sir Robert Gourdon in the St James's Park trial would be tested at sea. Two ships, the third-rate *Anne* and the fourth-rate *Sedgemoor*, were duly fitted out, receiving one pump each.[248] The pump of the *Anne*, as described by her captain, Cloudesley Shovell, seems to have been a type of suction or 'force' pump, for it had two staves, a plug, two boxes, two valves and two small iron rods no bigger than curtain rods. Unfortunately, the rods were apt to bend and break. Shovell and the ship's carpenters suggested that the rods be made bigger, to measure ¾ inch thick and 7½ feet long. A major difficulty was that neither they nor Gourdon's men could make it work in less than two feet of water.[249] Gourdon perceived even more serious problems, telling King James '*of some sinister practices used by persons employed on behalf of the King of France for the discovery of the secrets of his new invention of sea pumps*'.[250]

Gourdon's elevated opinion of his supposed internationally famous pump was groundless, for Captain Lloyd of the *Sedgemoor*, which was equipped with the second pump, wrote that the '*new engine pump proves unserviceable as it was so tiresome to the men as the* (whole) *ships company cannot free so much water in one hour as may be done with ten men* (on the remaining chain pump).'[251] After considering this, King James signed an order on 19 May 1688 for a conventional chain pump to be sent out to him as a replacement.[252] Later the same year, the captain of the *Swallow* wrote that Sir Robert Gourdon's '*projection*', as he described it, placed in his ship as an experiment, had proved useless and had had to be replaced by a chain pump.[253] By 1694 the Navy began to tire of hearing proposals for new suction or 'force' pumps. When William Tindell and Thomas Crew proposed replacing chain pumps with their force pump, the Navy Board considered the proposal and reported to the Admiralty that '*Sir Samuel Moreland and Sir Robert Gordon both great Mechanics have often essayed to improve the art of pumping in ships beyond that of the chain pump and yet after diverse, chargeable experiments which they made in the Reign of King Charles ye 2nd it does not appear they brought anything to perfection but on the contrary. That the conveniences in general which are to be found in the chain pump have been thought preferable to whatever they could propose by new improvements and therefore the custom and use of them is still continued in their Majesties Navy.*'[254]

In 1691 Robert Lodgingham made another attempt at improving chain pumps. He offered a model and papers relating to a modified version with an extra drive wheel situated on the orlop deck. His chain was also different in using round rather than the usual 'S' links. The Navy Board was doubtful the design would be an improvement and also criticised the loss of valuable space on the orlop deck.[255] Lodgingham persevered, and in 1695 petitioned for a patent for another new invention for improving chain and hand pumps, for which he made a model.[256] It was intended to try the pumps out in the third-rate ships *Hampton Court* and *York*, but they were not ready in time.[257] The Navy always remained open to new ideas, including an "*ingenious invention*" by a Colonel Richards in 1696 which the Navy Board thought worthy of a trial.[258] However, despite repeated attempts to radically improve the performance of pumps, the chain pump remained in use until the end of the days of sail. Indeed, the list of pump parts used in the *Lenox* of 1677[259] appears the same as those listed by Thomas Blanckley in 1750.[260] Many minor changes were made to the design of the chain pump over the years, improving its performance and reliability, but most of them were never recorded.[261]

Hawse Rollers

Sir Samuel Moreland did not confine himself to the unsuccessful improvement of ship pumps. He pursued an idea to ease the weighing of anchors by preventing cables chafing as they passed through the hawse hole. Unwisely, the ship chosen for his invention was the largest and most prestigious ship in the Navy, the *Royal Sovereign*. His new idea consisted of brass and forged iron rolls running on iron spindles, all of which were turned before being fitted to an iron frame. The device must have been placed inboard in the wake of the manger. The structure had to be very strong as the forces involved were enormous; the frame for the four hawses weighed 17 cwt, the brass rolls weighed 3½ cwt and the iron rolls 5¾ cwt, and altogether they cost £189 18s 2d. The sea officers judged it too great a weight and not fit to be used.

Sir Samuel, it was recorded, '*notwithstanding did not so discourage ... that he still pursued ye invention and brought it to such perfection*'.[262] Surprisingly, Charles believed Moreland's claim about his invention but prudently decided to have it installed on a yacht rather than on the biggest first rate. The yacht, the *Charlotte*,[263] was supplied with brass frames and rolls[264] for her hawse at a cost of £46 10s.[265] No other vessels appear to have been fitted with the invention.

Iron Knees

One of the best inventions to narrowly miss becoming adopted in the 17th century was the iron knee. Oak knees were the most expensive and difficult-to-find form of timber used in shipbuilding. At least as early as 1690, small iron knees were used on boats such as 30-foot pinnaces, where 12 bracket-shaped iron knees were used to bind the thwarts and the transom to the sides.[267] In boats, the thwarts acted in the way beams do in big ships. The use of iron was gradually adopted, and by 1750 it was used on the beams of yachts and as deck standards in warships to replace oak upper futtock riders.[268] Toward the end of the 18th century a cheap and cost-effective method using a wrought-iron frame was widely adopted. The iron frame overlapped the joint between the beam and a simple straight-grained wedge beneath. In the early 19th century, the iron frame was replaced by an iron knee that resembled the brackets of the earlier design.[269]

This long period of evolution was nearly broken in May 1670 by the master shipwright at Portsmouth, Anthony Deane. At the time, he was building a prestigious first-rate ship that was to become the *Royal James*. Pepys wrote to him after hearing that he '*presumed to lay aside the old secure practice of fastening your beams in your new ships with standards and knees, and in the room thereof taken upon you to do it with iron*'. It seems Deane had done this without informing anyone about the monumental change in shipbuilding practice that this constituted. News of Deane's innovation soon went up the command chain to reach the ears of King Charles and the Duke of York, gathering complaints from conservative naval officers as it went. At Pepys's suggestion, Deane wrote a letter in response which Pepys read to the King, the Duke and

INVENTIONS AND INNOVATIONS

Above. *The bracket-shaped iron knee used early in the 18th century in place of standards and riders.[266] It is probable that Anthony Deane used this type of knee fitted underneath the beam and against a simple straight-grained oak wedge at the side of the ship. This design allowed a common-shaped iron knee to be used in conjunction with a shaped wooden wedge to accommodate any difference in angle between the beam and the side of the ship. It is probable the knees were made of wrought iron rather than cast. Cast iron would have been cheaper but more brittle. Author.*

Right. *The iron knee used towards the end of the 18th century. It allowed easily obtainable straight oak to be used as a wedge directly beneath the beam rather than expensive knee-shaped oak bolted to its side. Author.*

members of the Navy Board. Although others argued against it, Charles expressed his interest in the innovation, saying that Deane's method must be stronger, and adding '*if he had any doubt about it, it was that it was rather too strong, and would not leave the ship so loose as perhaps might be requisite for her to be; as also that this would probably be somewhat more chargeable than the old*'. Charles then took the letter in hand and perused it, '*and defended it so well against those that would oppose it, that they were silenced*'. Charles would have enjoyed taking pains to explain the new method to others attending the meeting who had not fully understood it as he had.[270] Launched in March 1671, the *Royal James* was burnt and lost a year later in the Battle of Solebay. Her iron knees went down with her and seem to have been forgotten except for one other intriguing reference. The Navy Board minutes for 2 January 1674 recorded that '*Mr Shish's proposals of making knees appears good, if made of good iron; consult ye surveyor for his proposal*'.[271]

Lead Sheathing

Lead sheathing has often been commented on as probably the most unfortunate of 17th-century innovations. The corrosion resistance of the metal had always been known, and in the late 1660s a method of rolling it into thin sheets called 'milled lead' was developed by Sir Phillip Howard. This came to the attention of Anthony Deane, the master shipwright at Portsmouth, who thought it seemed like a good solution to the age-old problem of organic wooden hulls being eaten by gribble or teredo worms and being overburdened by the build-up of weed and other growths, which reduced sailing performance.

In 1671, it was decided to sheath Deane's new fifth-rate *Phoenix* with 5½ tons of lead and to use 6¾ cwt of nails to attach it.[272] After two voyages to the Mediterranean, in 1673 the *Phoenix* was careened at Sheerness, where she was inspected by King Charles himself. The hull was in such good condition that the Admiralty gave the following order to the Navy Board concerning all new ships: '*in regard of the many good proofs and usefulness … they shall for the time to come be sheathed in no other manner*.[273] This soon resulted in a number of other ships being similarly sheathed. In May 1675, it was noted that the *Phoenix* had suffered damage as the sheathing was not high enough. Consequently, when the third-rate *Harwich* was also ordered to the Mediterranean, she was sheathed higher to prevent similar damage occurring.[274] Unfortunately, just a month later, the first signs of trouble began to appear when a letter bringing worrying news arrived from Sir John Narborough in the Mediterranean. When the *Henrietta* was being careened he found the ship in good condition but the iron on the sternpost, rudder, keel stirrup and bolt heads quite eaten to pieces to an extent he had never seen before, even though the iron was only two years old. He doubted it was the lead or copper nails causing the problem as other ironwork completely covered by lead was hardly damaged at all. He also found that the washing by the sea had removed some of the soft lead, leaving the plank beneath bare in several places and infested with worm. Unfortunately, the infestation couldn't be killed with fire as the heat would have melted the remaining lead. However, where the lead was intact, the plank beneath was preserved and had not become foul.[275]

As the reason for the corroding ironwork was not understood, sheathing with lead continued. Nevertheless, in November 1676, the *James Galley*'s sheathing was fixed with copper nails, and in areas of high wear, such as the back of the sternpost, the bearding of the rudder, the foreside of the gripe and in the wake of the anchors, copper sheathing was fitted instead. Her lower wale was sheathed with wooden board laid with tar and hair, and the operation altogether cost £317 15s.[276] Similar procedures were carried out on the *Foresight*[277] and the *Lyon*.[278] Regrettably though, the problems kept occurring, with the rudder irons of the *Rose* needing replacing and her lead needing to be repaired,[279] while the *Bristol*'s lead needed repair where it had rubbed off in the wake of the bolts, the false keel, the rudder irons and the anchors.[280] King Charles, Anthony Deane and Sir John Tippetts, the Surveyor of the Navy, researched the problem and found it had also been encountered in the distant past. When applying lead, their forefathers had always payed over the heads of bolts and other ironwork with 'stuff' to prevent rusting. Before condemning the otherwise advantageous use of lead sheathing, they ordered the revival of this old practice for ships going to the Mediterranean.[281] At about this time during a visit to Deptford, the King '*observed they were sheathing one of his ships with fir board of ¾ of an inch thick. The King taking notice says, Why was not this ship covered with lead? The Managers replied, She was a weak ship and required to be strengthened. Lord have mercy upon the poor men (says his Majesty) that depend on that sheathing.*'[282]

By early 1678, however, it was clear that lead sheathing was doing more harm than good. Charles agreed to stop the practice until an understanding of the cause of the corrosion was known. It was a very perplexing problem that affected some ships more than others and seemed to have no logical cause. Anthony Deane, a supporter of lead sheathing, disagreed with the decision,[283] but now that *Henrietta* had returned to Chatham, the opportunity was taken for her to be thoroughly inspected. The damage proved to be as bad as Narbrough had said. To prove the point, some corroded floor and futtock rider bolts were sent up to London; they had '*been eaten almost to pieces and some bolts were so much decayed as they were only converted to rust and thrust out with sticks*'.[284] Although not understanding that the problem was caused by electrolytic action, the Navy Board concluded in October 1682 that the practice of lead sheathing needed to be discontinued and ordered that it be removed from ships and new ironwork supplied where defects were found.[285] Charles must have been disappointed, but, always open to new ideas, he remained undeterred in the pursuit of protecting his ships from worms. In late 1677, Charles instructed Samuel Pepys to write a letter to the Navy Board. In it, Pepys could hardly restrain his scepticism. '*This comes by His Majesties special command to signify his pleasure to you that a barrel of tarr be forthwith sent hither for ye making of an experiment by a stranger of mixing ye same with such ingredients as shall effectually render ye ships side that shall be payed with that stuff proof against ye worm. But ye ingredients ye proposer keeps as a secret from all but His Majesty to whom (as I am told) he is willing to discover them without any reward expected till His Majesty be convinced of their efficiency. So as ye trial being likely to cost His Majesty nothing of substance but only employing of one barrel of tarr you may please to order a barrel thereof to be sent hither to Derby House so as to be here some time tomorrow where ye proposer is to have a room to himself to make his preparation in to be afterward shown to His Majesty and then sent back to you to be proved in its practice.*'[286]

A similar claim was made in 1685 by Henry Aldred, a gentleman who reportedly had discovered the art of preserving ship plank from being eaten by worm. The Navy Board thought his method deserved a trial, although one does not appear to have taken place.[287] In 1695, a Charles Ardisoif claimed that by carrying out a great study in the theory of paying and graving he had invented "*a composition far beyond anything of that nature yet known against the worm, durableness, cheapness and cleanliness for sailing*".[288] He persuaded the Navy to carry out a trial, which resulted in the *Tyger* having a plank payed with his stuff. Two years later, at Cadiz, the ship was careened and the captain and his carpenter inspected the plank and reported '*the aforesaid plank as much eaten with worm as any other part of the ship*'.[289] The problem was eventually solved in 1768 by the discovery of vast quantities of copper at the Parys mine in Wales. From then on, the sheathing and all other associated metal work could be made from copper or its alloys.

Engine Propulsion

Today, the term 'engine power' implies a fuel-burning heat-driven machine, but in the 17th century anything powered by wind or muscle that drove some sort of mechanism was called an engine. In 1678, a twin-hulled horse-powered paddle-wheel boat was built by a Mr Beane for towing warships down the River Medway from Chatham against the wind.[290] It was reported to be successful in trials, and was almost identical to designs widely used in the 19th century. However, in 1694 the Navy Board wrote that it was never found useful in practice.[291]

During the 1670s, many fourth-rate ships were employed in the Mediterranean against Barbary pirates or to escort merchantmen. To enable them to manoeuvre in little or no wind, these ships had provision for using oars between the gunports. Some ships were built with small oar scuttles, but most had them cut out retrospectively during a refit. The idea of using an engine to power these ships exercised the minds of at least two individuals. One was Edmund Bushnell, whose book, *The Compleat Ship-Wright*, contained in its 1678 edition a final additional chapter, 'Concerning Rowing of Ships when they are becalmed'. He reasoned that the circular motion of the capstan was more efficient than the use of oars and devised a method of transferring the power to a crank that turned the oars in a circular motion. Among the many problems shown in his design is what seems to be a lack of support at the aft end of the engine. No evidence can be found of the arrangement ever having been tested.

As Edmund Bushnell describes it, the two lines *CD* and *CD*, represent the sides of the ship. The line *CC* represents the breadth of 16 feet. The four bitt-pins are represented by *d, d, d, d*, with the aft cross piece being *ef*. Between the forward bitt pins is a rowl, or windlass, driven by the capstan by means of the viol rope, normally used in helping to bring in the anchor cable. Two winches, *a* and *a*, are fitted into each side of the rowl that runs in hollows at the head of the bitt pins. The two frame timbers for the oars are represented by *bb* and *bb*. Two other timbers, *nL* and *mL*, are fastened to the frame at *L* and *L* by a swivel bolt, and attached at *n* and *m* to the handles of the cranked winch by a hole. The oars are fixed each side of the frame at points *1, 2, 3, 4, 5, 6* and *7* by a mortise and tenon on the oar with clearance to provide movement. Along the ship's side are placed thoul pins for each oar, as in boats. Bushnell estimated that two, three or four leagues could be rowed in a watch.[292]

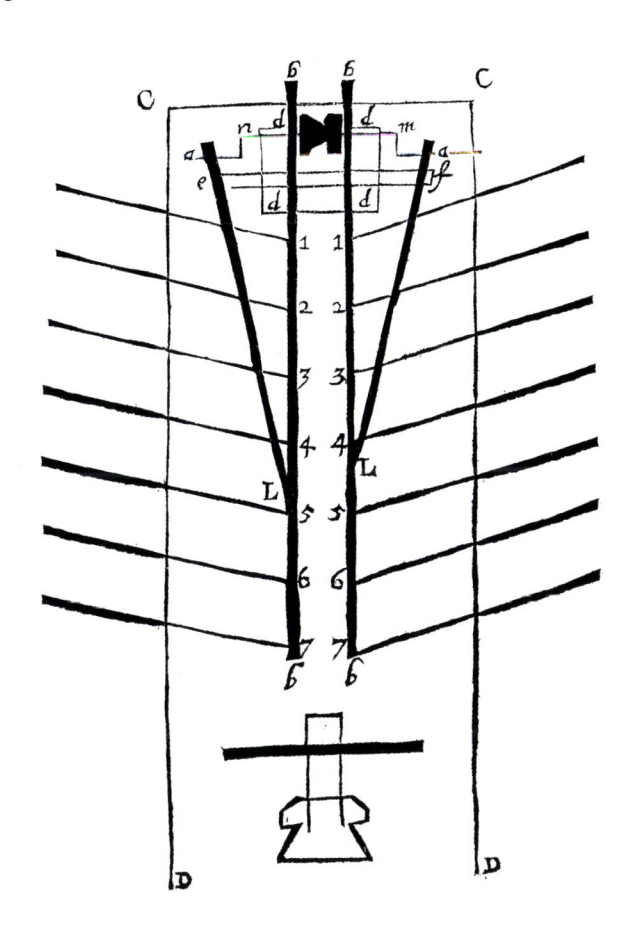

The layout of Edmund Bushnell's arrangement for rowing using the main capstan. Private collection, via author.

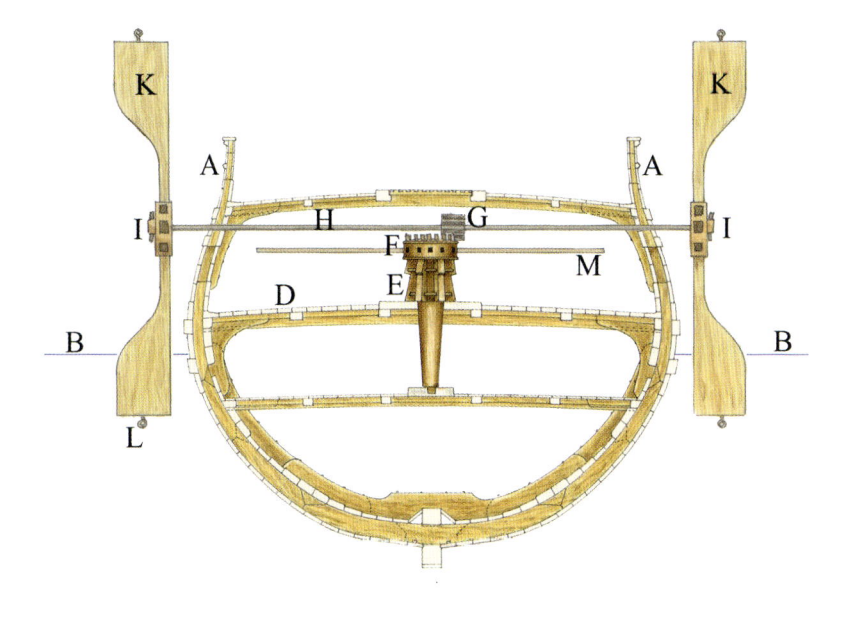

Thomas Savery's 1694 design for a fourth-rate-size ship of 32 feet beam with detachable paddle wheels driven by the capstan. A. Section through the ship; B. Waterline; D. Deck; E. Capstan; F. Gear wheel fitted to the top of the capstan; G. Trundel or pinion wheel; H. Spindle going through the pinion and ship's side; I. Drumheads similar to the capstan drumhead, recessed to take the detachable paddle blades; K. Paddles; L. Iron eyebolts to hoist the paddles, and the location for ropes to strengthen and equally space the paddles around their circumference; M. Capstan bars. Author.

A simpler and more practical proposal was made some years later by a Thomas Savery, who also used the capstan for providing power, but instead of rotating oars, it turned paddle wheels.

The Navy Board studied Savery's proposal but was extremely sceptical, remembering the failure of the horse-powered Chatham paddle-wheel boat and imagining all sorts of accidents likely to occur at sea. Its report ended by saying that '*in our opinion can never by any knowing sailor or other intelligent man be thought practical at sea*'.[293] In 1698, Savery wrote a book on the subject, *Navigation Improv'd*. The design was revived a century later in 1799, when the 38-gun *Active* was fitted with the same arrangement.[294]

Desalination Engine

Although it was obviously beneficial to turn seawater into drinking water, it was extremely difficult to achieve this economically. In March 1680, John Shish and an engineer named Mr Walters attended an Admiralty meeting to discuss making fresh water from salt.[295] Unfortunately, nothing else has been found concerning the project, but five years later a letter was written from Barbados by Jonathon Martin, master of the *Elizabeth*, saying that he was satisfied with some patentees' engine for making seawater fresh. He reportedly found the water very wholesome and beneficial and the engine worked as well in foul weather as it did in fair.[296] In April 1686, the inventors met the Admiralty and persuaded them to proof test the engine aboard a Navy vessel.[297] No desalination engines appear to have been supplied as a result of the experiments.

Fire Buckets

In 1680, Commissioner Godwin, based at Chatham, wrote to the Admiralty suggesting that leather fire buckets be hung aboard His Majesty's ships in the steerage or the great cabin. They agreed to such a sensible idea for preventing accidents, and orders were issued immediately for their introduction to all ships in harbour.[298] From then on, fire buckets were issued to all ships of the Navy for the rest of the days of sail.

Central Bowsprit

Up until about 1675, the bowsprit was offset to starboard in order to miss the stem and foremast, which were both situated on the centreline. In a reasonably sized warship, the bowsprit ran some six feet beyond the foremast.[299] Subtle changes in design allowed it to be moved onto the centreline directly over the top of the stem. The inner end was rebated into a large partner situated between the beams of the main and upper decks. There seems no written account of the change, and the best evidence of it happening is from surviving models of the period.[300]

Bumpkins

The bumpkin was generally thought to have been introduced in about 1710,[301] but Sir Phineas Pett told Samuel Pepys that King Charles himself invented and introduced the bumpkin, which was found to be very useful.[302] The bumpkin was a slightly curved spar pointing forward and outward over the headrails. At the outer ends were blocks for the lower corner of the foresail to help keep the sail open to the wind. Perhaps Charles was referring to two square-rigged sloops having bumpkins when he wrote in 1673 that they '*have my invention in them, they will outsail any of the French sloops*'.[303] They were definitely introduced before August 1680, when Thomas Shish, master shipwright at Woolwich, fitted the fourth-rate *Newcastle* '*with 2 bumpkins for hauling down ye foretack*'.[304] They were a valuable contribution that would remain in constant use for square-rigged ships after their invention. There were many other minor alterations made to the rigging and sails by unknown seamen during the late 17th century, too numerous to mention.

Keel Length

On 16 January 1668, Samuel Pepys wrote in his diary, '*My work this night with my clerks till midnight at the office was to examine my list of ships I am making for myself, and their dimensions, and to see how it agrees or differs from other lists; and I do find so great a difference between them all, that I am at a loss which to take.*'[305] Twenty-four years later, in 1692, little had

INVENTIONS AND INNOVATIONS

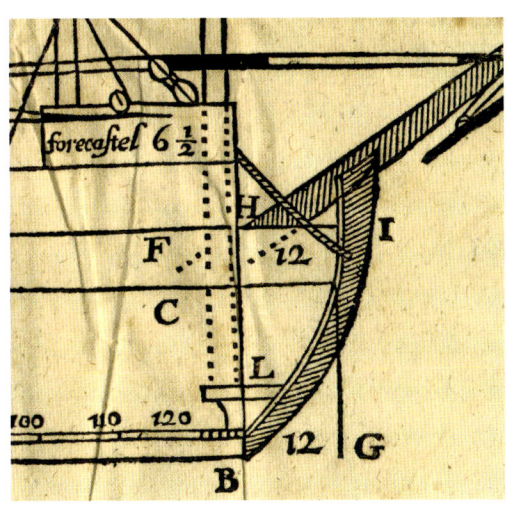

Top Right. *A leather fire bucket, from Thomas Blanckley,* A Naval Expositor, *1750.*

Above Right. *Offset bowsprit. The head of a ship, dated 1664, from* The Compleat Modellist *by Thomas Miller. Miller shows the offset bowsprit placed at H,F,C running some six feet beyond the foremast.*

Above. *Bumpkins. The lower corners of the mainsail and foresail, the clews (A), were controlled by the sheet (B), tack (C) and clew garnet (D). The sheets hauled the clews downward and aft while the clew garnets hauled them up to the yard. The tack was a single rope that held down the clew forward. Usually, and for most of the period, the tack ran to the foretack dead block (E), going through the mouth of a carved face mask. The bumpkin (F), introduced by King Charles II, had greater reach and helped keep the foresail open to the wind. It may owe its origins to a method used in earlier times. Author.*

changed, with Edmund Dummer, Surveyor of the Navy, writing in a similar vein, '*Having obtained with some labour several catalogues or lists of the ships of their Majesties Navy containing the dimensions and burdens of them as they have been at different times and by different hands taken and calculated; and being compared one with another it seems altogether doubtful from the vast disagreement in one and the same ship what the real & true dimensions & burden of any of them is.*'[306] Many of the lists Pepys and Dummer considered still exist, and studying them today, it is remarkable to see how much in agreement the breadths, depths in hold and gundeck lengths are, considering the difficulties involved. Usually they vary by only a few inches, which is quite a testament to the care taken. For example, imagine how hard it would have been to measure the length of the gundeck from the rabbet of the stem to the rabbet of the sternpost with all the obstacles such as capstans, bitts, pumps, masts and bulkhead in the way – especially as the measurements were taken with just a carpenter's rule.[307] The only measurement Pepys and Dummer had cause to question was the length of the keel, where it is common to find recorded differences of between five and ten feet.

Many contemporary ship lists describe exactly what is being measured – for instance, '*breadth to outside of plank*'.[308] Unfortunately for keels, '*length by the keel*' is the only description ever given for them, even though it could have at least three different meanings – namely, the *touch*, the *tread* or the *calculated length*. There was therefore always confusion regarding exactly what the length of keel was. There was no programme for

THE MASTER SHIPWRIGHT'S SECRETS

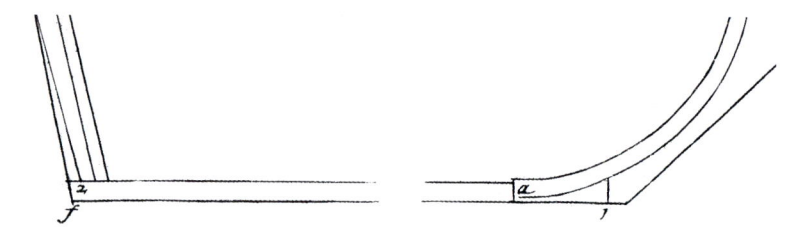

Top. *William Sutherland's drawing showing the touch and tread of the keel. a. 2. Length of the keel from the back of the main post to the Touch of the stem. f. 1. Length of the keel from the back of the false post to the extreme forepart of the keel or Tread. Extracted from William Sutherland's Shipbuilding Unvail'd, 1717, p.26.*

Bottom. *The head of a ship with a circular stem, extracted from William Sutherland's Ship-builders Assistant, 1711, with additions. a. Stem; b. Knee of the head (its foremost part is barbed away and is called the cut-water); c. Forefoot; e. Lyon; d. Cheeks; f. Rails of the head; g. Brackets; h. Gripe; l. Rabbet; m. False keel.*

systematically measuring the ships in the fleet and no procedure for describing how the keel should be measured. It appears to have been measured whenever it was convenient to do so, by diverse shipwrights, at the time of building or during repair, when the ship in question was in dry dock.[309]

The primary keel length was the touch, described by William Sutherland in 1711 as follows: '*When the keel is put in order, set off the exact length forward and afterward, from the observation of the rising of the keel by shipwrights called the Touch, or place where the keel's upper part ends to be straight.*'[310] This was the only keel length of interest to master shipwrights when they created the geometry for draughts. In 1664, in *The Compleat Modellist*, Thomas Miller described how the keel length is bound by two perpendicular lines, one at the back of the sternpost, and the other at the rising of the stem. The stem was then drawn to follow on from the forward perpendicular line.[311] In 1670, Anthony Deane wrote his *Doctrine* describing the same process.[312] In 1678, the same method was described by Edmund Bushnell in his book *The Compleat Ship-Wright*,[313] and even earlier, in about 1625, it is similarly termed in another manuscript.[314] Interestingly, none of these authors found it necessary to call it the touch, describing it as simply as the keel length. Apart from the small difference described in the later period by Sutherland in his 1717 print, where he measures from the back of the false sternpost rather than the main post, it seems the touch of the keel is very clear.

The tread is shown by Sutherland's 1717 print to extend from the back to the fore end of the keel, including the forefoot but not the gripe. Although dating from between 1733 and 1737 and in French, a description by Blaise Ollivier confirms the tread as being '*measured in a straight line from the angle of the forefoot to the angle of the keel with the sternpost*'.[315] The forefoot is the foremost keel piece which curved upward at the forward end to receive the lower piece of stem. In 1692, the tread was also mentioned by Edmund Dummer in an instruction for surveying and measuring two new ships then being built, which included the following line: '*a c. Is the tread of the keel*'.[316] Unfortunately the original enclosed sketch showing '*a c*' is now missing. R. C. Anderson obtained the sketch in 1917 and published it in 1919.[317] However, it is unclear in the 1919 print whether the keel is measured to the forefoot or to the gripe, as they appear to be in about the same place. In 1679 when Dummer measured the *Grafton*, he appears to have measured the keel to the fore end of the gripe and called it the tread. It is little wonder that this has continued to cause confusion.

Upright Stem

At an Admiralty Board meeting on 7 July 1677 a discussion took place about the King's idea for building some new ships with a more upright stem. An upright stem would probably increase buoyancy at the bows and help to reduce pitching and avoid hogging.[318] Phineas Pett III, the master shipwright at Woolwich, was chosen to build one of the ships, and he made journeys specifically to show Charles and the Duke of York draughts of this ship.[319] When it was completed, the *Captain*, rather than being laid up with the other new ships, was ordered to sea as it was important to have '*proof made of ye success of ye practice lately introduced by his Majesties special direction of building with upright stems for ye discovering of any inconvenience yet unforeseen that may arise from ye same before ye said method shall have taken too much place in ships of so great value*'.[320] The *Captain* was under the command of Sir John Holmes, who reported her working and sailing to the King. Pepys duly recorded Charles '*taking great pleasure in recounting it as often as any come in his way (that) can be thought in anywise to understand it... I have not any time known the King more gratified in anything of this kind in my whole life than in this account of the success of his art in this ship.*'[321]

Since 1628 the keel length to the touch was generally used in a rather crude and inaccurate formula to obtain the burden tonnage of ships.[322] The tonnage formula evolved over the years, and by the 1670s the general rule was '*multiply keel by main outside breadth and the product by the half breadth and divide the result by 94*'. By having a more upright stem, the tonnage formula would make the ships appear bigger than they really were as the traditional circular stem assumed that more of the ship was in front of the touch of the keel.

In 1677, Parliament had provided funds to build ships of a certain burden, and the error which resulted from using the touch of the keel with an upright stem had to be addressed. The Admiralty discussed '*Whether it be not fit, since the measure of the tonnage of the ships depends upon the length of the keel and breadth of the ship and that of late some ships have their keels much longer in respect to the length of their gundecks than was formerly in use and approved of by His Majesty and thereby measure more than otherwise they would. That the reputed length of the keel be ascertained by supplemental bill to be taken from the length of the gundeck between the rabbets and the stem and stern posts, thus, viz, abating thence 3/5 of the main breadth of the ship and a fourth of the perpendicular height of the stern post to the gundeck, the remainder to be accounted the length of the keel.*'[323] This new method was the third way of defining the keel length, and was known as the calculated keel length. It may be that this keel length was sometimes given in the lists of a ship's dimensions, but it was only a theoretical length based on the length of the gundeck.

Needless to say, the new formula was subject to change. One example of the calculated keel length was described by Sutherland in 1711 as follows: '*Some say, the general method which has been pitched upon by the greater number of shipwrights and others and may be termed Shipwrights Hall Rule, is to take the length of the keel measured from the back of the main post to the fore side of the stem, at the upper edge of the lower harping by a perpendicular made from thence to the upper or lower edge of the keel only 3/5 of the main breadth, from the outside of the plank of one side to the outside of the plank of the other side, at the broadest place of the ship, being set backward or aftward from the right angle made by such a perpendicular and base. Observing also that as several ships and vessels have no false post, in such a case there ought to be allowed 1/3 of the main post from the after part of such a stern post.*'[324] The added complication of a false sternpost is evident in the 1717 Sutherland print, where it is shown as stopping at the upper surface of the keel. A model dating from about 1670 shows a similar false post, except it reaches to the bottom of the keel.[325]

The proliferation of ways used to define the keel length makes them virtually useless as a means of determining the actual dimensions of a keel. It may be based on the length of the gundeck, or it may have been calculated on the touch or the tread. All we do know is that the touch would have been shorter than the tread. It is tempting to try and identify keel lengths, but after spending a considerable amount of time studying this issue, the only conclusion is that Pepys's and Dummer's words of despair were well founded.

The forefoot. This was the tabled scarph at the front of the keel with a complex shape that received the matching lower piece of stem. Author.

CHAPTER 3

NO SUCH THING AS THE TYGER

The *Tyger* arrives at Deptford

At the end of the Third Dutch War an old and worn-out battle-damaged ship arrived at Deptford. It was Tuesday 21 July 1674, a cloudy day with light winds, as the 38-gun fourth-rate ship *Tyger* moored in the river near the old dockyard founded by Henry VIII. She was returning to the place of her building 27 years before by Peter Pett II. In 1646, three successful frigate-built ships were followed by four more in 1647, one of which was to be the *Tyger*. This was during the time of the English Civil War and the new ship formed part of the Parliament's fleet. In 1649 she managed to capture the Irish ship the *Mary Antrim*. Then, as part of Admiral Blake's squadron during the pursuit of Prince Rupert, she took a Portuguese 36-gun ship in single combat. This success was followed by the capture of a Dutch ship, the *Morganstar*, without losing any of her men. During the First Dutch War she fought at the Battle of Portland in February 1653, at the Battle of the Gabbard in June and at the Battle of Scheveningen at the end of July.

Early in the Second Dutch War the *Tyger* was put under the command of Phineas Pett, son of Sir Phineas Pett I, the famous shipwright. She fell in with a Zealand 40-gun privateer and Pett was killed by the privateer's first broadside. The lieutenant continued the fight for a further six hours, by which time the masts and rigging of the *Tyger* were so badly damaged she could not close with the enemy. Later, in July 1666, she fought at the Battle of Orfordness. A month later she became the flagship of Sir Robert Holmes for an audacious attack by small warships on the Dutch merchant fleet in the Vlie, an action that became known as Holmes's Bonfire. The number of merchantmen destroyed was put between 150 and 170 and the best conservative estimate of the ships' cargos was twelve million guilders, or a little over a million pounds.[326] To give an idea of what a million pounds was worth at that time, Parliament raised £600,000 in 1677 to pay for 30 large warships consisting of 10 three-deckers and 20 large third rates with all their guns and stores. A million pounds would have bought a fleet of 50 such ships. This was the size of the immense damage caused to the Dutch economy.

In 1672, during the Third Dutch War, the *Tyger* was again in action at the Battle of Solebay. She then distinguished herself under Captain Harman when attacked by eight large Dutch privateers while convoying a fleet of colliers to London. She defended them so well against the overwhelming odds that not one collier was lost. The *Tyger* was then sent to the English base at Tangier before receiving orders to return home in January 1674.[327] On 22 February she reached Cadiz harbour soon after a Dutch ship, the *Schakerloo*, had entered under Captain De Witte. De Witte decided to fight the *Tyger* in single combat, but to make sure of victory, Admiral Evertson, who was already at Cadiz, sent him two lieutenants, 70 soldiers and 60 seamen from other ships. De Witte then had 270 men against Harman's 184. Coming out the following morning, the two ships fought in one of the greatest ship duels, closing to half pistol-shot and exchanging broadsides. The *Tyger*'s accurate and relentless shot did the most damage, killing or wounding 80 of the *Schakerloo*'s men and disabling her topsail yard. Harman then closed and boarded at the bow. After half an hour of desperate and bloody fighting, the miserably torn and shattered *Schakerloo* surrendered, with 140 men killed and 86 wounded. Harman was also wounded by a musket shot which struck under his left eye and came out between his ear and jawbone.[328] After the action the *Tyger* continued on her way home to Deptford. She had certainly done her duty, having few rivals for the number of actions fought and economic damage inflicted on the King's enemies.

The *Tyger* was moored fast against the fifth-rate *Falcon* of 1666[329] and her crew were paid off and discharged.[330] Because of the limited space, the depth of the water and the difficulty in getting big warships up the Thames, only the smaller-rated ships were moored there. The largest were fourth rates the size of the *Tyger*, while the larger first, second and third rates were kept at Chatham or Portsmouth. Although the crews were paid off, the five warrant officers and their servants remained with the ship. In addition to the ships laid up in ordinary there were between about four and eight other ships awaiting repair. There were also several smaller vessels that remained in sea pay on the Thames in the vicinity of Greenwich and Deptford, most notably the royal yachts.

Survey of the *Tyger*

The variety and number of vessels belonging to the King at Deptford were recorded on 6 July 1674, just before the arrival of the *Tyger*:[331]

Type	Name	Built	Burden	Guns
Fourth rate	*Portsmouth*	1640	463	40–46
	Dragon	1647	470	40–46
	Assurance	1646	340	36–42
	Foresight	1650	522	42–48
	Stavoreene	taken 1672	544	42–48
Fifth rate	*Guernsey*	1654	245	28–30
	Speedwell	1656	233	28
Sixth rate	*Francis Frigate*	1666	140	16
	Fanfan	1666	33	4
Dogger	*Dover Dogger*	Taken 1672	73	6
Sloops	*Hunter*	1673	46	4
	Hound	1673	50	4
	Vulture	1673	68	4
	Spy	1666	28	4
	Chatham	1673	50	4
	Chatham Double	1673	50	4
	Invention	1673	28	4
	Fox	?		
	Prevention	1672	46	4
	Bonetta	1673	57	4
	Emsworth	1667	39	4
Smack	*Bridget Smack*	1672	21	–
Fireship	*Castle*	Bought 1672	329	8
	Young Spragge	Bought 1673	79	10
	Eagle	1654	299	32

Immediately on her arrival, a survey was made of the *Tyger* by old Jonas Shish and his two sons, John and Thomas,[332] after advice from the ship's warrant officer carpenter. Only three months before they carried it out, Jonas told the Navy Board exactly how he conducted such surveys: '*My constant custom hath been that so soon as any of his Majesties ships arrive here to be repaired to go onboard each ship and receive an account off the Carpenter of her defects. After that I make it my whole business carefully to search her and inform myself of every particular work (both to what I am informed) and likewise what doth otherwise appear to me to be defective and taketh a survey of the same with an estimate of the charge and send the same to the Surveyor of his Majesties Navy.*

And after your Honourables order for the repairing of each ship I immediately proceed in the performing of every particular work mentioned in our survey and (it) is my constant practice daily to see the works well and securely done and commonly in the diligent searching of the ship further defects appear to us that was not discovered in our first survey, yet notwithstanding that survey we order and direct such works to be carefully and securely performed and am daily overseeing the works well wrought and fastened. I shall for the future upon any discovery made by me give the Surveyor an account of what further defects we find (not mentioned in our first survey) all (of) which is left to consideration.' [333]

After the Shish family completed their survey it was sent to the Navy Board, along with an estimate of the work to be done and its cost. It makes gloomy reading:

The £2,016 estimate for the work compares with the original building cost in 1647 of £3,003[334] and reveals that only about a third of the old ship was expected to survive the rebuild. During the next eight years, of the fifteen fourth rates and dozens of smaller ships for which estimates of repairs were made at Deptford, only the *Assistance* would rival *Tyger* for the amount and cost of the work.[335]

Survey of the *Tyger* dated 29 July 1674

	£ Cost
To drive out all the bolts out of ye floor and futtock riders being much taken with rust, to repair and make good all ye platforms, storerooms, cabins, well etc, shot locker both afore and abaft and to shift 300 feet of ceiling in hold being rotten with one new breast hook, per estimate:	95.
To shift 4 gundeck beams & 8 knees being rotten. To new lay ye gundeck wholly with 3 inch plank, new waterways, spirketting & ceiling between ye ports and likewise most part of ye carling & ledges to be new. To take up six pair of standards, to new fay & bolt them. To have new fore partners, new hatches, head ledges & coamings. Also taking down ye furnaces & to set them up again. To have one new breast hook between decks about ye hawse, new bitts and cross pieces with a wing transom & knees to ye same ye old ones being defective. All to be well fastened, per estimate:	417.
To have new upper deck clamps, 14 upper deck beams, 20 knees and most part of ye carling & ledges to be new being much defective, all to be well fastened and bolted. To new lay ye upper deck wholly with spruce deal and 2 inch plank, likewise with new waterways and spirkett rising, head ledges, coamings and gratings. To take down the bulkheads of the steerage & fore peake to be new set up again, also to new lay ye quarter deck with spruce deal & waterways with ye spirkett rising to be shifted and to take down ye roundhouse bulkhead & to set it up again, per estimate:	297.
To strip ye ship down two strakes below ye lower wales, to bring on a new tier of toptimbers on each side fore and aft ye old ones being shot to pieces & new hawse pieces & to birth all up again with timber for wales. Plank with new rails, gunwales & planksheers fore & aft. Likewise to make good all kevills, ranges, sheet blocks with her channels, per estimate:	750.
To build a new head with rails, timbers, brackets, cheeks & trail board. To have a new upper piece of stem & a new stern post (being rotten & much defective) To unhang ye rother, to have new irons & to hang him again. To take off ye sheathing in ye wake of the new work to make it all good again, with painters, joiners, carvers, glaziers & plumbers etc works, per estimate:	276.
To have a new main mast & main yard & some other small masts and yards, per estimate:	55.
To caulk her fore and aft within board & without & to grave her with white stuff, per estimate:	95.
To fit ye ship with 2 boats, per estimate:	31.
Total is:	2016.

Money Shortage

The end of the Third (and last) Dutch War brought considerable changes to the Navy as it adjusted to peacetime. The crews and many dockyard workers needed paying off as soon as possible and the many battered and worn-out ships required repair. Although a peacetime navy was much cheaper to run than one in wartime, a sudden and considerable sum of money was required to pay off the crews and unnecessary dockyard workers. Such money was not readily available, even though the Admiralty expressed genuine concerned over the welfare of the '*poor seamen*'. It proposed paying the ships in cash up to Michaelmas 1672 and giving them an IOU ticket for the remainder. This would have the unfortunate consequence of forcing many seamen to sell their tickets to speculators, often landladies, at a discount. To ease the pain and make them more valuable for trade, an interest of 6% would be paid when the tickets were eventually cashed in.[336] Care would also be taken to pay off as many men with the least outlay, such as those from ships that had been in pay for the shortest time. Also requiring attention were the orphans and widows, and the suppliers of goods to the Navy. Those workmen who were discharged from the King's yards were not given the same consideration as the seamen because, it was reasoned, they could more readily find work outside in merchant building yards.[337] Not only did they have better work opportunities, but a shipwright's pay of 2s 1d[338] a day was over double that of an able seaman's 24s a month.[339]

As the Navy struggled to pay off the seamen, on 12 October 1674 Commissioner Sir Anthony Deane prepared a plan for repairing ships, many of which still had battle damage received at the battle of Texel fought the previous year in August.[340] He gloomily compared the English fleet with those of France and Holland and raised the idea of building 20 new large third-rate ships to bring it up to parity. He also identified the old ships that required major repair and listed them according to their condition. His plan was as follows:

To be new built:

Type	Name	Built	Burden	Guns
First rate	*Sovereign*	1637	1605	100
Second rate	*Old James*	1634	890	70
	Rainbow	1617	868	64
	Unicorn	1634	823	64
Third rate	*Defiance*	1666	863	64
	Plymouth	1653	742	60
	Revenge	1654	766	62
Fourth rate	*Tyger*	1647	453	44
Fifth rate	*Richmond*	1656	233	28
	Eagle	1654	299	32

As well as those needing rebuilding, Deane listed a further 11 ships that required a '*very great repair*' and 56 needing a '*least repair*'.[341] He also estimated that it would take two years to complete the ships needing rebuilding or great repairs, provided the money and men were available. On the other hand, he pointed out that the 56 ships which required the least repair could be made ready for service more quickly and should therefore be taken in hand first. It appeared the *Tyger* could be in for a long wait before work started on her. The report was discussed on 4 November by the Admiralty Commission and circulated to the lords and Mr Pepys.[342]

To keep the records up to date, the Navy Board organised further surveys of the ships, to be carried out by the master shipwrights of all the King's yards and to specify the amount of repair work needed and its cost. These reports, which included Shish's *Tyger* report, were carefully catalogued in an indexed journal.[343]

Deane's paper and the master shipwrights' surveys were considered by the Admiralty Commission on 28 November, but no decisions were made due to the lack of money.[344] In early December, the only progress being made was the preparation of a paper giving a general representation of the work for the King's consideration.[345] As the plans slowly developed, Pepys wrote to Deane and the Surveyor of the Navy, Sir John Tippetts, on 11 May 1675 for estimates of men and materials necessary to build new ships and repair the old ones.[346] These estimates for repairs reached £144,000, which was far too much for the Treasury to supply. In the meantime,

Thomas Osborne, Earl of Danby and Lord Treasurer. Many ships required repair after the Dutch wars, and Danby was determined to keep the cost of repairing them under control.

the King considered it important '*that no time be lost in forwarding the repairs as far as the state of his treasure and condition of his stores will admit*' and to make more preparations by updating the surveys.[347] The cost of building a huge number of new ships was far more than the King had available through his ordinary revenue and would require Parliament to provide what were termed 'extraordinary funds' by raising tax.

Ships requiring repair were continually taken in and out of the Deptford double dock. Among them were two small vessels, the *Holmes* (of 220 tons and captured from the French) and the *Eagle* (an old fifth rate of 299 tons, built in 1654). Both were to be repaired and converted into fireships.[348] On 31 May, just after the *Holmes* was completed, Jonas wrote to the Navy Board suggesting that he launch her, then dock the *Eagle* and some other ship they wished to have repaired. Sir John Tippetts wrote in response saying that when the dock was clear, the *Tyger* should go in the head of the dock, with the *Constant Warwick* and the *Falcon* astern.[349] Ships had to be made as light as possible before entering dock, and thus 100 tons of decaying ballast from the *Constant Warwick* and the *Tyger* were removed into the *Constant Mary Lighter* and put ashore at the dockyard.[350] The *Tyger* and the *Constant Warwick* were docked by the beginning of July but the *Falcon* was left in the river. It was feared three ships in such close proximity would be a danger should there be an accidental fire when weeds and filth were burned off the hulls.[351] With the *Tyger* safely in the dry dock, John Shish was able to remove some of the plank that normally lay underwater and inspect the frames hidden beneath. On 10 September, he wrote to the Navy Board, enclosing a copy of the *Tyger* survey made the previous year in July and adding:

> *In obedience to your commands we have cut out a plank between the wales and another at ye wrongheads of his Majesty's ship ye Tyger, likewise taken up her limber boards and surveyed ye timbers and bottom of ye said ship and do find both her timbers and plank underwater to be very sound and good. We have also considered that ye charge specified in ye foregoing survey will make her a good serviceable ship & fit for His Majesty's use and finding ye bottom so well conditioned we judge ye King may save in ye hull of ye ship in rebuilding eight or nine hundred pounds.*[352]

In other words, in return for spending the survey's estimated cost of £2,016, the Navy would have a ship to the value of between £2,816 and £2,916.

The *Tyger* had been at Deptford for well over a year without any repair work having started on her as the underfunded dockyards tried to repair the least damaged and decayed ships. The Admiralty discussed the backwardness of the repairs which Parliament expected to be paid out of the revenue already allowed the King for that purpose, and it was reluctant to raise a separate tax to make up any deficit. The Lord Treasurer, the Earl of Danby, complained that repairs were not being kept strictly to estimate and that work had been carried out without the approval of the Admiralty Board. In defence, Pepys pointed out that all the repairs were fully debated, usually with the King being present.[353] He may have added that the cost of repairs could easily escalate as unseen defects were uncovered during the work. In order to update and accurately estimate the cost of repairing the ships, a new State of the Navy survey was commissioned on 25 September.[354] In an accountant's way of thinking, Danby, the Lord Treasurer remained determined to control the repair costs, and with the King present, suggested that the '*ancient method*' of the Navy be practised. This laborious method involved the estimate of cost

Opposite. *The fourth-rate* Tyger *as built in 1647. Although the portrayal of the ship is made as accurate as possible, the scene is fictitious and was painted for the book* Gentleman Captain *by J. D. Davies, 2009. She is flying Commonwealth colours. Oil on canvas. Author.*

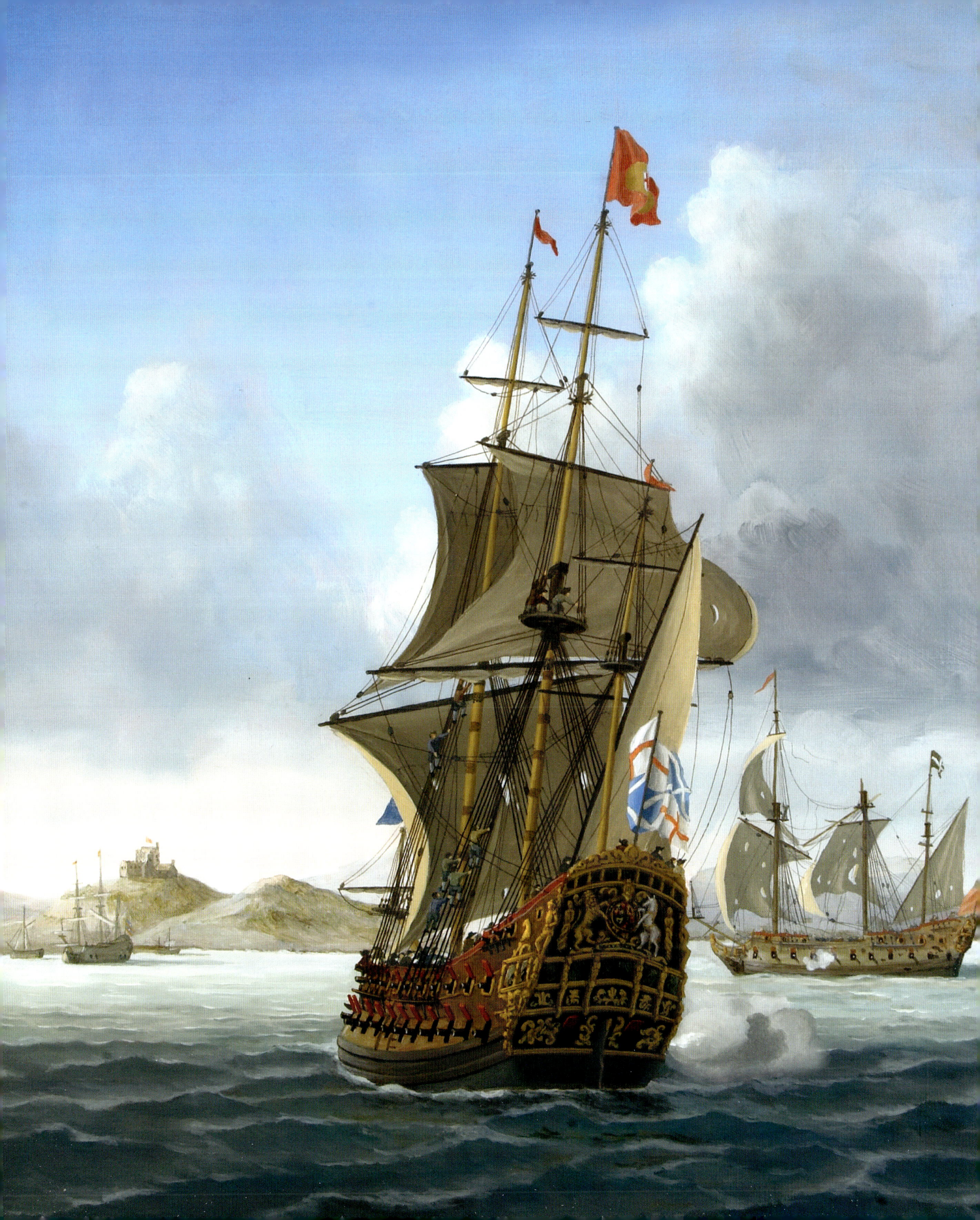

THE MASTER SHIPWRIGHT'S SECRETS

being signed by the officers of the Navy, then by the Lord High Admiral, followed by approval from His Majesty in Council; only then would a privy seal be granted to the Lord Treasurer to issue the money.[355]

State finance was raised by a number of taxes, the oldest and best regarded being the customs duty applied to shipping, which was updated under the Tonnage and Poundage Act established at the start of Charles II's reign. Another was the excise applied mainly to barrels of beer and ale. These indirect taxes were seen as fairer to the poor as, of course, nobody was forced to pay. Less popular was the hearth or chimney tax, which could be a cause of conflict with an unhappy householder at the point of collection. Also available were direct taxes, such as a poll or land tax, which were regarded as exceptional, or as taxes of last resort for emergencies or large projects.

The Navy Board was deeply worried about the state of a financial settlement between the King and Parliament for the repair of ships, partly because the customs duty failed to bring in as much as had been expected. It wrote to the Admiralty on 2 February 1676 complaining that if the money was not forthcoming, it would have to stop the repairs and discharge the workmen.[356] Two weeks later, and accompanied by Samuel Pepys, the Navy Board met Danby, the Lord Treasurer, who said that he had studied the board's estimates for the Navy and that £2,300 a week, or thereabouts, would be supplied until the end of December and £5,000 a fortnight to be continued from the customs duty.[357] This *'seasonable proportion'* of money for wear and tear of the fleet could be employed with liberty for repairing ships.[358]

The matter was discussed again on 4 March at an Admiralty Board meeting at which the survey of ships was presented

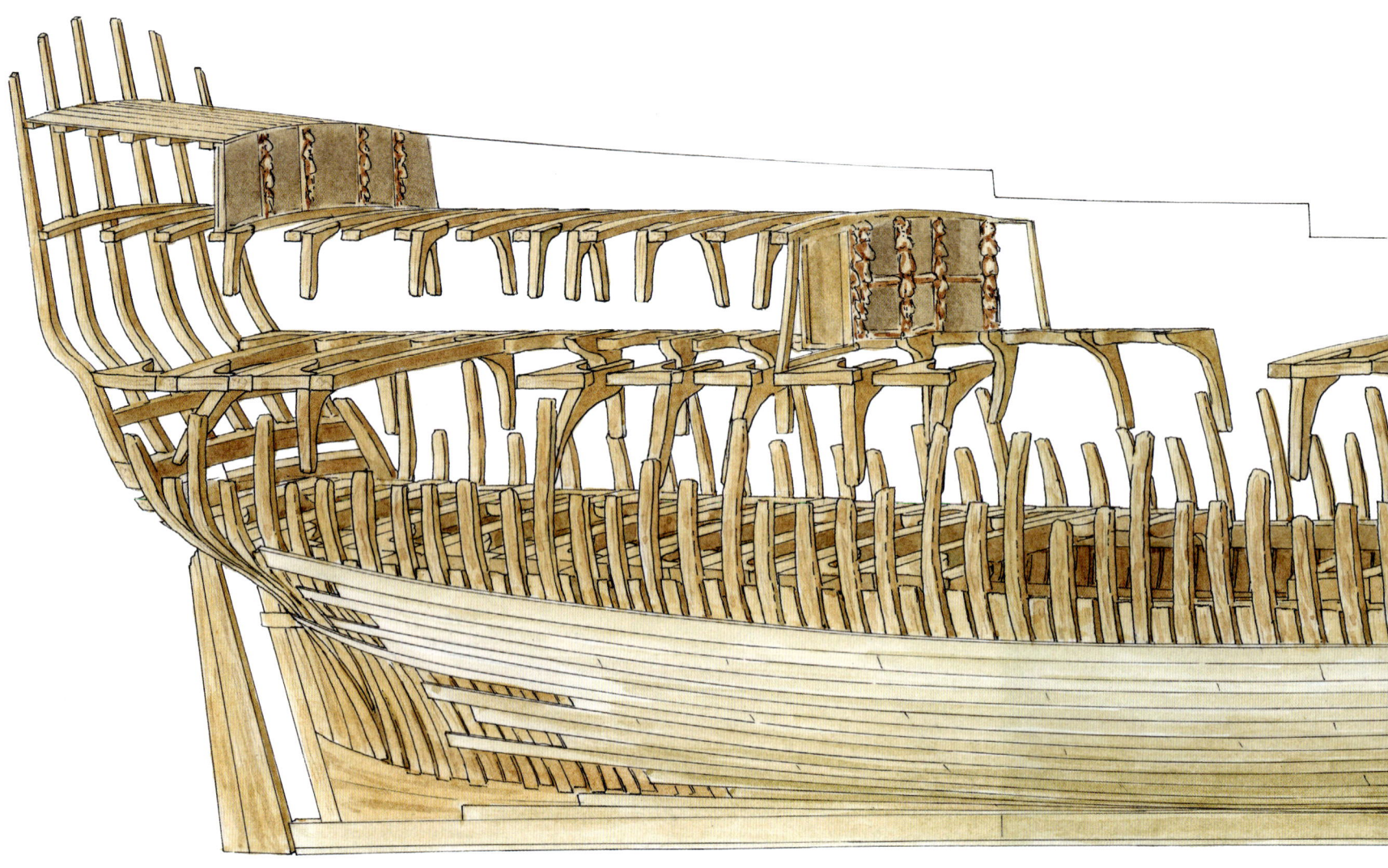

Opposite. *King Charles and the Lords of the Admiralty discussing the design of ships. Model at the Science Museum. Author's photograph.*

Below. *The expected surviving structure of the* Tyger *after rebuilding according to the survey of 29 July 1674 by Jonas Shish and his sons John and Thomas. Author.*

NO SUCH THING AS THE TYGER

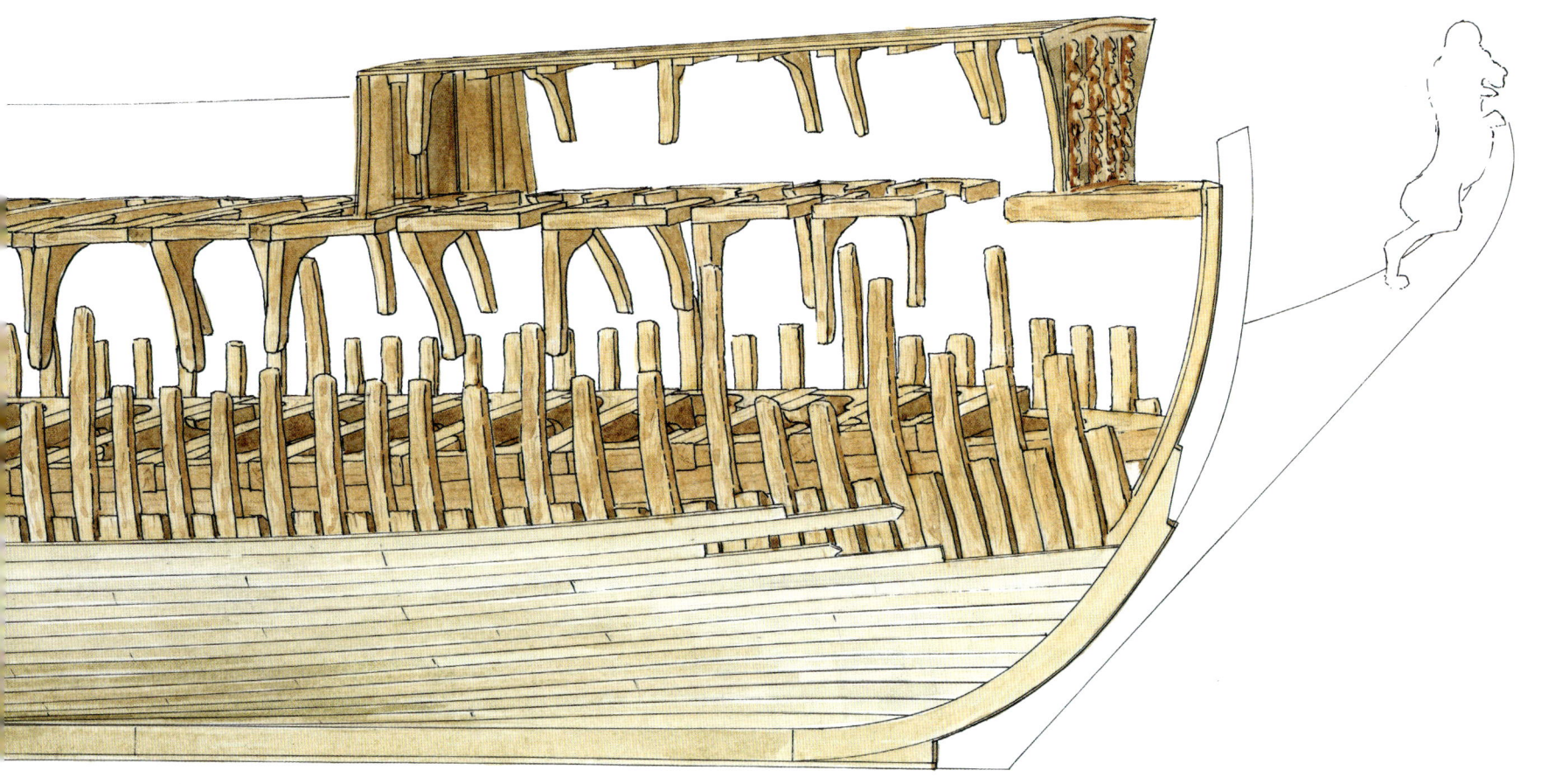

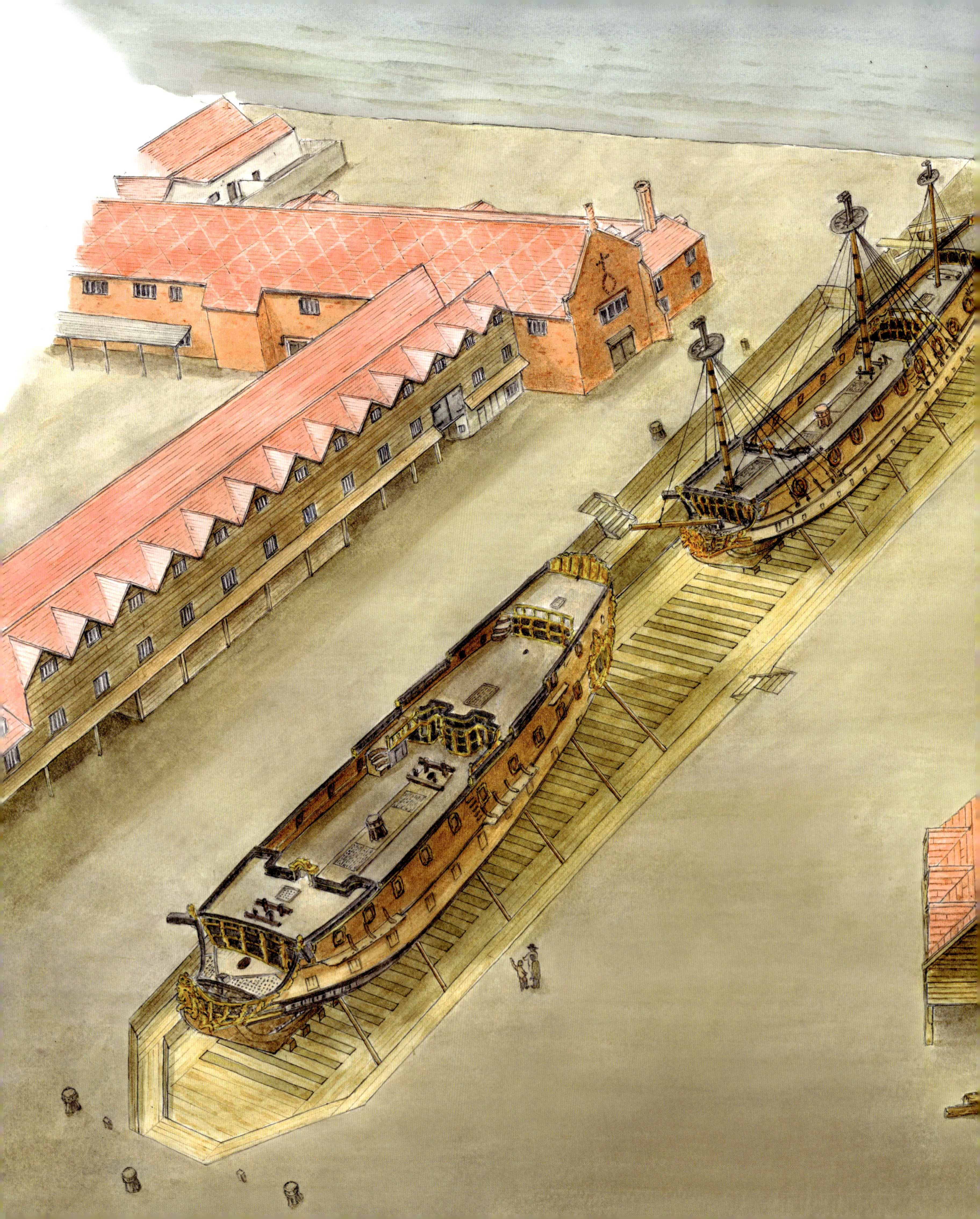

NO SUCH THING AS THE TYGER

and the King was reminded that he should pay for the repair of ships from his usual revenue.[359] Five months later, on 2 August, the ships at Deptford were surveyed again, but the new report for the *Tyger* was the same as the old, except the wording was shuffled around to make it appear fresh and the costs increased from £2,016 to £2,206.[360] By the end of September, the Lord Treasurer informed the Navy that £50,000 would be available by 1 January of the coming year for the extraordinary repairs of ships.[361]

The funding situation remained desperate, and by 25 October 1676, the Navy Board was complaining that it was five weeks in arrears of the £2,300 a week it had been promised and, worse, £70,000 in arrears of the £5,000 a fortnight it had been assured of getting. Although it was informed that substantial sums, including £20,000 in customs bonds, would be provided, it was nevertheless worried that its credit would soon be at an end.[362]

The *Tyger* in the Double Dock

Work proceeded intermittently, or barely at all, on the *Tyger* as other ships requiring fewer repairs entered and left the stern of the dry dock. On 20 October 1676, the *James Galley* was ready to launch and Shish asked whether the *Guernsey* or the *Roebuck* should be docked next. The Surveyor of the Navy, Sir John Tippetts, thought both could fit in, but if not, then it would be acceptable to dock either one of them. It was decided to dock only one of the ships, allowing some resources for work to proceed on the *Tyger*.[363] She had been in the double dock for over a year and a half before work commenced. The outside planking on both sides was removed down to the lower end of the toptimbers, the area near the waterline at the widest part of the ship. This revealed many of the upper futtocks which fitted between them and which reached as high as the lower edge of the gunports. Concerning them, John Shish wrote, '*In the midships on both sides the upper futtocks appears very bad and split, that requires to be shifted, which hath been occasioned by girdling the ship twice in bolting the said girdling on. Her upper futtocks afore and abaft indifferent good and may serve again.*' Shish suggested that in order to replace the defective upper futtocks it would be best if he took off all the footwaling on the inside of the ship as some of it had already been removed. That would leave the planking on the outside undisturbed, and it could remain in place as it was in

The old Tyger *in the head of the Deptford double dock and the* James Galley *in the stern. The* James Galley *had been newly built in the private yard of Henry Johnson and was being fitted out for service in the Mediterranean against Barbary pirates. Author.*

73

An approximation of the surviving remains of the old Tyger *that were used in the new ship. Author.*

very good condition. Tippetts agreed, and responded by instructing him to '*take up ye footwaling and ceiling and tell us how you find the timbers under*'.[364] Less than a week later Shish had taken off more planking down below the toptimbers.[365]

At precisely this period, in February 1677, Parliament voted to raise £600,000 through a land tax for the building of 30 new ships following a dramatic debate led by Samuel Pepys, who, as well as being the Secretary to the Admiralty, was also an MP. The day drew to a close before any discussion of money to repair the existing ships took place, but nonetheless, building 30 new ships was a massive programme that would have dramatic consequences for dockyard resources.[366] To help the MPs' deliberations concerning the repair work, a copy of the Navy Board reports was sent to the Speaker.[367] A month later, on 12 March, the King's financial worries eased when Parliament agreed to continue an additional duty of excise on small beer, strong beer and ale for three more years.[368] Wanting to keep a tight control on the shipbuilding programme, Danby requested more information, including the order in which the first batch of 33 ships were to be repaired, where the repair work was to be carried out, the type and quantity of materials needed and the cost.[369] One ship not on the list was the *Tyger*[370] – as the 33 more easily repaired ships were to be completed first, her long-term work was of a lower priority. It was estimated that a further 50 shipwrights would be needed at Deptford to work on the repairs.[371]

On 6 April 1677, John Shish made a new evaluation of the amount of work required to rebuild the *Tyger*. He had stripped another five or six planks down on the outside of the hull, leaving only about half the outside planking below the wales in place. Following Tippetts' advice, he had also removed 1,000 feet of footwaling and clamps in the hold, some of which he thought may be usable again. He exposed more of the upper futtocks that he had previously supposed were only bad and split in midships. Unexpectedly, he found that out of a total of about 100 upper futtocks, 80 were rotten and broken. Able to inspect the lower futtocks for the first time, he estimated that 40 of those were in the same poor condition. As a result, the cost of rebuilding increased by £610 over the previous estimate and now stood at £2,816.[372]

Shish wrote to the Navy Board on 18 May saying he was '*in hand to plank down the Tygar and rebuild her, we will undertake to perform so much work to her where she is as to launch her out of the dry dock in eight or ten weeks time and finish what works left undone in the wet dock which I judge may be more advantageous to His Majesty than to have her up the launch. I humbly desire your Honourables to signify your pleasure herein for the ship will be worth rebuilding and that plank that is left on may serve again which will not if taken off, her floor timbers and all the after & forward lower futtocks are good and serviceable.*'[373]

Shish's suggestion for rebuilding the *Tyger* would probably have been carried out had Parliament not attached conditions to the bill for building the 30 new ships, one of them stipulating that they must all be built within two years from 24 June.[374] The *Tyger* had resided in the head of the double dock for nearly two years and now the space she occupied was urgently needed to build one of the new ships. Two officers of the Navy Board, Sir John Tippetts and Sir Richard Haddock, visited Deptford about a week later and gave John Shish directions to take down all the *Tyger*'s structure to a level below the gundeck and launch her out of the dock the next spring tide. He was then to heave her up the launch where the *Saudadoes* was rebuilt in 1673 and continue to work on her there. Such verbal directions were not enough in themselves, and Shish had to write to the Navy Board for confirmation in the form of a written warrant.[375]

The *Tyger* Sunk

The rebuilding of the *Tyger* had so far been a very gloomy affair that had slowly reduced her to what must have looked like a low, flat barge. Things were to get even worse. Following the receipt of a warrant to launch the remains of the *Tyger*, she was slid down the sloping dock to the water's edge on 6 June 1677. At half flood, she began lifting in midships, but with less buoyancy at the stern, she retched so much that there was a danger her back would be broken. Desperate measures were taken to prevent this happening, with holes cut through the bottom to sink her where she lay. Shish wrote to the Navy Board the same day saying that if they were to get her out of the dock the next tide, he would need two lighters to lift the stern and casks to lift her afore. He was also concerned that even if they managed to float her out, it would be difficult to heave her up the launch as the remains of her hull were weaker than expected. As a further option, he suggested taking her to pieces where she lay.[376] It was this alternative that was to be her fate. For the moment, she was left where she was, but at least the head of the double dock was now clear for John Shish to start building one of the 30 new ships. He immediately started preparing the dock floor and a week later received approval of his draught for building a new ship.[377]

In spite of the terrible condition of the *Tyger*'s remains, the Navy Board asked for a new estimate for rebuilding her, which was ready by 20 August 1677.[378]

Estimate for rebuilding the *Tyger*

According to your Honourables order we have considered of ye charge of rebuilding his Majesty's ship ye Tygar, we have estimated the same with an estimate of what may be saved in ye old timbers that is serviceable. Viz: To rebuild ye said ship from ye floor quite upwards to have all new futtocks & toptimbers with 10 new rising timbers & to have some new footwaling & to be new planked with new wales to have new decks, quarter deck & forecastle to be wholey new excepting some knees that may be serviceable, to fit her with ports, rails, gunwales, planksheers, head, stern, galleries all joiners, painters, plumbers, glaziers, carvers, smiths etc works to make her a complete ship. I judge per estimate may be worth:	£3122 10s plus £277 10s
We judge in ye old bottom there may be saved in that of ye old timbers & knees that may be serviceable ye value of	£200
John Shish Thomas Shish	£3600

Note: There seems to be a mistake in the cost calculations.
The £200 should have been deducted from, not added to, the total, which should be £3,200.

Work started on removing timbers from the dock floor that were not good enough for use in a rebuild. On 5 October it was found that the keel was good and sound, as were 37 floor timbers which were probably the rising timbers referred to in the last estimate. One piece of keelson and all the rising wood were also found to be good and serviceable to rebuild upon. However, they found more '*bad*' than in the August estimate, and as a result the charge rose from £3,122 10s to £3,272 10s, for which sum they expected to make her a good and serviceable ship.[379] It really should have been enough money for such a transformation, for only a fraction of the old ship would survive the rebuild. Some in authority may have considered the *Tyger* still existed purely as a way of building a new ship without having to navigate the naval and financial bureaucratic channels. However, William Fownes, the Clerk of the Cheque at Deptford, had little interest in such ruses, for two weeks later he wrote, '*Since there is no such thing as His Majesty's ship Tiger I thought it my duty to desire your Honourables directions from what day the standing officers must be discharged.*'[380]

The Tyger Taken Apart

Work on the new ship in the head of the double dock proceeded rapidly throughout the winter of 1677/1678 while the remains of the *Tyger* behind her were slowly taken apart and removed piece by piece. On 28 January 1678, it was recorded that part of her was still in the stern of the dock,[381] but by 12 April she was completely cleared out for the new third-rate ship to slide down the ways and be launched.[382] A week later a report addressed the repair and rebuilding of the 22 old ships that were in the worst condition, that would take place after the more easily repaired 33 (now 28) were completed. Among them was the *Tyger*,[383] whose remaining broken-up fragments were taken and stored for later use.[384] Following the sad trend of ever-diminishing quantities of usable timber being found, '*only some part of her old keel, stem and some few of her (frame) timbers were made use of in the rebuilding of her*'.[385] King Charles seems to have considered ships with a great deal of sentimentality, and thought it was far better to heal or repair them than sign their death warrant. He prevaricated over the fate of the 48-year-old *Old James* and was several times on the point of ordering her uneconomic rebuilding before finally ending his misery in 1682, when she was disposed of. Apart from her, he only disposed of one English fourth rate and one third rate in his entire 25-year reign.[386]

Warrant Officers

Although the *Tyger* was now just a few pieces of disjointed timbers, her permanently appointed warrant officers remained in pay with no duties to perform. William Fownes, who was responsible for calculating their wages at Deptford, waited in vain for instructions to discharge the purser, the boatswain, the gunner, the carpenter, the cook and their four servants. The warrant officers received £8 2s 6d a quarter year, except for the cook, who received half, and the servants, or apprentices, who received around £2 8s.[387] This compares very favourably

The remains of the Tyger, *with her warrant officers and servants, at Deptford dockyard in early 1678. The men, maimed or worthy in other ways, together with the ship's remnants, were kept on the books as part of King Charles's benevolent policy towards their subsistence. Author.*

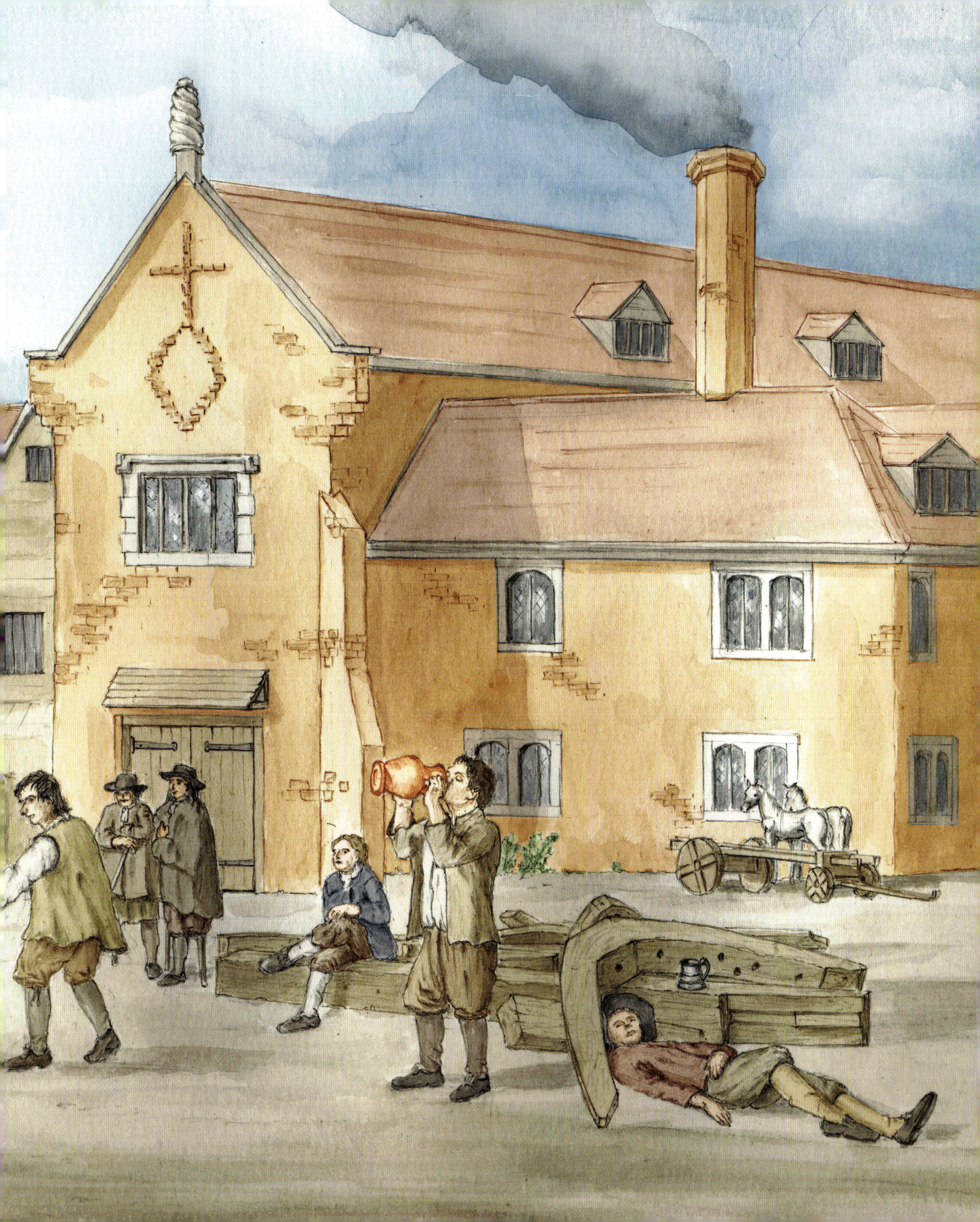

THE MASTER SHIPWRIGHT'S SECRETS

Part of the side of a broken-up ship at Deptford dockyard as revealed by recent excavations. Parts of ships were often used in dockyard buildings and docks Author.

with the mean quarterly income per household of around £1 15s 6d,[388] and with the £6 5s a quarter for a hard-working skilled shipwright.[389] In July 1678, Robert Gross, the gunner, was removed from the non-existent *Tyger* into the *Hampton Court* as it neared completion on the launch at Deptford. The warrant for his appointment was signed, as most were during this period, by Samuel Pepys and the King.[390] In May, another warrant was issued to move the carpenter, James Kelly, from the *Tyger* into the *Yorke*, while Samuel Newman was moved to the *Tyger* from the *Dartmouth* to replace him.[391] In February 1679, Edward Solby was appointed purser in the place of Daniel Whitfield, who had surrendered his position.[392] An explanation for some of these curious appointments may lie with the strict instructions that were issued about warrant officers keeping watch aboard their ships. On reading them, some of the dockyard officers guessed the reason *'we having*

Deptford as it was in the early 18th century. The dockyard storehouse and the master shipwright's house are the brick buildings to the left. The church next to them is St Nicholas, where a memorial to the Shish family survives. The Baroque church to the right is St Paul's, built between 1712 and 1730.

several bourn here (we suppose for their subsistence) from whom little or no duty is required'.[393]

Good evidence for this dates from April 1678, when Abraham Hamm, the boatswain of the *Tyger*, was removed into the *Revenge* and replaced by Cuthbert Sparkes,[394] who only had one leg. In the circumstances, a one-legged boatswain of a few ship's timbers was every bit as good as a two-legged boatswain. Sparks performed his peaceful duties for three years and three months, right up until the time the newly rebuilt *Tyger* was ready for sea. Sparkes then petitioned the Admiralty Commission with a reminder that he had lost a leg, which would be a great hindrance to him if he ever went to sea. He organised a solution to the problem himself by persuading Henry Morgan, the boatswain of *Grafton*, which was laid up in ordinary, to exchange places with him. The sympathetic Admiralty Commission agreed to the arrangement, but only for as long as the *Tyger* was at sea. As soon as she was paid off and laid up, Sparkes would resume his tranquil duties aboard her.[395] There are other examples of disabled men being looked after by the Navy. Jacob Little, the cook of the *Captain*, was blind, and when the ship was ordered to sea, he exchanged places with the cook of the *Kent*, which was still laid up. As an added benefit, he was allowed a deputy, in reality a carer, to perform his duties.[396] In September 1681, permission was granted for Robert Bird, the cook of the *French Ruby*, to be excused from doing any duties aboard her as he had lost both his hands in an engagement against the Dutch in 1672.[397]

The benevolent policy towards the disabled men of the Dutch Wars is noticeable after 1674. Although the Chatham Chest pension fund did its best to provide for disabled seamen, Charles and his administration considered it their responsibility to provide subsistence by keeping them in pay. They were generally employed as cooks, but as there

Deptford dockyard as it is today showing a proposal to build replica of the 17th-century warship Lenox *built by John Shish. Beyond her can be seen the surviving brick built master shipwright's house and the* Cutty Sark. *Author.*

were not enough such vacancies for them all, they were also the first to be given places in hospitals or almshouses. To further help them with their duties, it was agreed in September 1676 that those appointed to a ship should be allowed a servant.[398] The matter was discussed again at an Admiralty Board meeting on 9 November 1676, when Samuel Pepys opened by telling King Charles of *'the case of several poor seamen* (the estimate exceeded 40)*, who by having lost two of their limbs are not capable of enjoying the benefit of His Majesty's favour to maimed seamen in giving them employment of Cooks on ships by reason of their being unable to execute the same and propounding therefore that in compensation of the extraordinariness of the charity due to the poor wretches.'* Charles agreed to take them into his charity during their lifetime by allowing them a pension equal to the wages they had enjoyed at the time they were maimed. The only thing to be decided was a process of payment by an establishment at the Council Table. The subject was discussed again two days later, and it was resolved to continue to pay the seamen and provide them with a servant to do their work, or else to provide them with a pension.[399] The only naval personnel who were kept in continuous pay were the warrant officers, as seamen were paid off at the end of a ship's commission along with the senior officers. Making the maimed seamen warrant officers ensured that they would remain in pay as long as the ship they served still existed. This may have been part of the reason why the *Tyger* was not allowed to die in the stern of the Deptford double dock.

CHAPTER 4

PLANNING A NEW TYGER

War with the Barbary States

For the time being, the dismantled remains of the old *Tyger* lay somewhere near the Deptford wet dock, slowly decaying as time passed. There was little threat of war against the major maritime powers of France and Holland, but a problem that had existed for many years previously and that would remain for many years afterwards concerned the attacks by corsairs from the Barbary Coast on merchant ships trading in the Mediterranean. There were four principal states involved: furthest to the east was Tripoli, and to the west, Tunis, Algiers and Sallee. It should not be thought these states were backward, for in many ways, especially in shipbuilding, they were very innovative. With scant resources from their own lands, the lure of heavily laden and valuable merchant ships from an alien culture often proved too much of a temptation. They gained wealth not only from the ships they captured, but also from ransoming the crews, who were often used as slaves. To escape this fate, some captives changed their allegiance and became renegade pirate corsairs themselves. Through the influence of captured ships and renegade seamen and shipwrights, the traditional corsair galleys developed into stouter sailing ships that could range far afield.

The type of ship the new *Tyger* became was determined by the nature of the conflict with the Barbary States. Since the beginning of the 17th century there had been a monotonous cycle of merchant ships being taken, warships being sent as a response, a peace treaty being made, the warships returning home and the peace treaty inevitably being broken. Sometimes the corsairs were severely punished, most notably in 1655, when Blake burned nine Tunisian galleys at Porto Farina, and again in 1671, when Spragge destroyed ten Algerine ships in Bugia Bay. Just after the old *Tyger* arrived home for repairs at Deptford, a squadron of some six ships under Sir John Narborough was ordered to Algiers to redeem slaves and then to proceed to Tunis and Tripoli to ratify peace treaties.[400] When two merchant ships, the *Hunter* and the *Martin*, were seized and taken into Tripoli, Narborough was sent there with all his ships to insist upon a restitution of the vessels.[401] The harbour blockade lasted throughout 1675[402] until midnight on 14 January 1676, when Narbrough sent in boats under the castle walls. Within an hour and without loss to themselves they captured the guard boat and burnt four warships, the *White Crown Eagle* of 50 guns, the *Looking Glass* of 34 guns, the *St Chiara* of 24 guns and a French petach of 20 guns. The action resulted in Tripoli agreeing to the restitution of the *Hunter* and the *Martin* and settling another peace treaty.[403] After serving on station there for two years, Narbrough returned home in September,[404] leaving behind a small number of ships. This number was added to early in 1677 when the two new galley frigates, the *Charles Galley* and the *James Galley*, together with the *Assurance*, were sent to the English base at Tangier.[405] However, the reduced presence soon resulted in the disappointing news that merchant ships had been taken[406] and violence offered by the Algerines to the King's subjects. Plans were made for Narbrough's return[407] with sums of money to pay for stores and redeem captives.[408] He set sail aboard the third-rate *Plymouth* in July 1677 with a squadron of several warships. His arrival brought the number of ships in his fleet up to 12, consisting mostly of fourth rates but also including numerous smaller support vessels.[409]

Opposite. *The Barbary Coast, showing the states of Tripoli, Tunis, Algiers and Sallee. Tangier belonged to King Charles II from 1662 to 1684. From a print by H. Moll, 1720.*

Their presence subdued the Barbary Coast, especially when the *Charles Galley* and the *James Galley* took the Algerine *Golden Marigold*, a large warship of 500 tons burden. The ship was good enough to be taken into the Navy, but was wrecked in a storm at Tangier in 1679. This success was followed by the *Mary* and the *Rupert* capturing the *Tiger Prize* of 48 guns and 475 tons.[410] Narborough returned home in mid-1679, leaving the remaining ships under the command of Arthur Herbert.[411] Although the fleet's numbers were reduced, there were still seven or eight fourth rates based at Tangier in mid-August 1680 when the new *Tyger* was being made ready to join them.[412]

The patrols by a permanent squadron of English ships based at Tangier seem to have had unforeseen consequences. The Algerines found a weakness in the King's strategy, as it meant that the English Channel and the Soundings were now patrolled with far fewer ships than the Mediterranean.

On 5 July 1679, it was reported that two corsair ships from Algiers chased some small vessels into Weymouth Road.[413] Worse followed a week later when the corsairs came into the King's Channel at the entrance to the Thames, where they took several ships before leaving to cruise between Ushant and Land's End.[414] Worrying intelligence was then received in a letter from a James Pevans in Algiers, who reported several ships were being made ready to sail for the Channel. Pevans was the master of a merchant ship which had recently been in a fight with an Algerine man-of-war near Land's End. He fought for three hours before his ship blew up and he was taken to Algiers as a slave. The *London Gazette* reported the incident and that a fishing boat and a ship off the coast of Ireland had also been taken.[415] That ship, the *Angel Gabriel*, was left a wreck and adrift, with all her men and furniture gone, 25 or 30 leagues west by south off Cape Clear.[416]

To counter the growing Algerine threat so close to England, in July 1680 Herbert was given orders to concentrate his efforts against the Algerians.[417] Less than a month later the East India Company sent a letter giving an account of four Algerian ships in the Soundings, leading to the Admiralty issuing orders for the *Constant Warwick* and the *Dartmouth* at Plymouth and the *Oxford* in the Downs to sail against them.[418] The *Constant Warwick*, a fast sailing ship based on the Dunkirk frigates, saw two of the corsairs at six in the morning of 8 August in a fresh gale southwest of the Scillies and immediately gave chase. She fetched up to them, but at 11 her main topmast went by the board, forcing her to give up. It was three days before the weather eased enough to allow her crew to get up the spare topmast. Then, on the 12th, at five in the morning, they spotted the Algerine corsairs once again about two miles to leeward sailing with only their main courses set. The corsairs quickly made sail, but by nine the *Constant Warwick* was within half a mile, only for the recently replaced main topmast to be carried away once more. She kept up the chase, but the wind eased, allowing the corsairs to escape.[419] Adding to the woes of the English was worrying intelligence from a released captive that the corsairs intended to keep up their campaign and send three more ships to cruise between Scilly and Cape Clear.[420]

The war with Algiers was causing serious disruption to the Mediterranean trade, for not only were ships being lost but there were long delays while the English waited for convoys to assemble. Some outward-bound ships carried goods such as lead, tin, fish, calicoes (cotton cloth), goods from the plantations, while others picked up freight from Genoa, Livorno, Venice and other ports as the opportunity offered. They returned home from the same places and many others, including Constantinople, Smyrna, Alicante, Malaga and Gallipoli, bringing back such goods as wines, raisins, oils and other perishables at different times according to the seasons. They often had to wait three or four months in these ports for safe passage, during which time their cargos deteriorated, while the crews still needed feeding and paying. In August 1680, the Levant Company merchants petitioned the Admiralty, saying they could not possibly carry on their trade without '*many mischiefs and inconveniences*'. They suggested a regular convoy system carried out by six frigates, and this was broadly and successfully implemented by the Admiralty.[421] Forced to take the Algerians seriously, the Navy kept a strong presence off their shores and began to have success.

In 1681, a heroic action was fought by the *Kingfisher*, whose captain was killed defending a convoy against seven Algerines.[422] During the same period came very agreeable news '*to ye King as well as the merchants*' that the *Golden Horse*, a large warship of 46 guns and 722 tons, had been taken by the *Adventure*.[423] Admiral Herbert was directed to give the colours of the *Golden Horse* to Captain Booth of the *Adventure* as a deserved mark of honour. Then, showing that it was not only the Barbary States that traded in slaves, he was directed to sell those taken from the *Golden Horse*, or to have them sold through the consuls of Cadiz, Alicante or Malaga and give one third of the proceeds to Captain Booth and the officers and seamen of *Adventure* for their services.[424] After these deductions were made, the proceeds were to be used to pay for the redemption of English captives held in Algiers.[425] The selling as slaves of men taken from the Algerian warships to reward the men who captured them and to pay ransoms for English captives became a regular practice. The anguish felt by some of the families of English captives is recorded in their petitions to the King. In August 1681, Leah Pye wrote about her husband, John, master of the *Trevola Merchant* of London, and her son, Francis. They were both taken on 2 December 1677 and carried into Algiers, where they remained enslaved. She had four small children to look after and had no means to raise the ransom without the favour of the King.[426] Similarly, Mary Richardson's father was a slave in Algiers and she requested that the payment from a bill of freight for over £36 would go towards his ransom.[427] This war was very different in many ways from the recent conflicts against the Dutch and required rather different qualities from the ships involved. These qualities would greatly affect the design of the new *Tyger*.

Corsair Ships

A shipwright, Jasper Goodman, was held at Algiers for three years during the mid-1670s, and on being repatriated he wrote down some notes on the shipbuilding techniques he had observed. He noted that the ships' frames were not marked out full size in chalk upon a platform or mould loft as was the English practice, but were built by eye without a draught. Only the lowest frames that crossed the keel, the floor timbers, where strength was really important, were made of oak. Above that they used ash to about the water level and the toptimbers above were made of pine. He also observed that the frames were very open, or spaced far apart, whereas we '*stuff our ships and clog them with timber*'. The frames were so small they could not use large-diameter treenails to secure them, but only small iron spikes. The fore and aft planking used for strengthening on the inside of the frames (the ceiling), that extended to the top of the sides in English ships, only reached up to cover the overlap between the oak floors and ash futtocks on the Algerian ships. The planking was taken from prizes, which meant they had to fill up the old fixing holes to seal them. He further noted that although the frame timbers were small, they were also strong as they would only use those that had the correct compass (grain flow) to follow the shape of the frame. The Algerians were also aware that the position of the masts had a great effect on sailing performance and changed the masts, yards, ballast and everything else until they found the best trim, and afterwards did not change it for as long as the ship lasted. To maintain this trim in service, they filled empty freshwater casks with salt water to keep the same ballast level. Once they were satisfied and the ship was in use, they maintained her good sailing performance by cleaning the bottom every five or six weeks.

The Algerians followed good engineering practice in making their performance-related structures as light as possible but within the limits of failure. Goodman noted the drawbacks with this technique, however, saying that under sail the whole hull trembled, especially the whipstaff in the helmsman's hand, to the point that he could not hold it steady. The structure was so fragile they could not take the ship ashore for graving and it was old and worn out in seven or eight years.[428] Although he does not say so, Goodman was probably referring to smaller ships of about 200 tons and under. During the same period, further details of their construction was provided by Sir John Narborough, who brought the *Londra*, a vessel from Tripoli, '*for information for His Majesty of their manner of build*'.[429] During a visit to

A near contemporary view of Algiers. Private collection, via author.

THE MASTER SHIPWRIGHT'S SECRETS

Deptford in 1683, Sir Anthony Deane pointed out to Pepys a Turk prize in the wet dock, apparently built from fragments of other broken ships and roughly put together.[430] It was to light ships such as this, designed to operate in the usually benign Mediterranean, that King Charles had to counter a response.

The *Margaret Galley*

King Charles's possession of Tangier and the construction of a mole allowed specialist English Mediterranean craft to have a permanent base there. Perhaps the most specialised and ancient type of vessel was the galley, which could be very useful against becalmed sailing ships, as it had two or three heavy guns firing forward over the bow. In 1671 two galleys, a small one from Genoa and a larger one from Livorno, were constructed for the Navy. With no English expertise in the building of galleys, Sir Jean Baptiste Duteil, a Frenchman who had served in the French galley fleet and who was assumed to be an expert, was appointed as overseer[432] even though he could not speak English. He was made captain of the *Fountain* by the Duke of York in 1665 and of the *Jersey* in 1666. He supposedly fired on his own flagship and was turned out by Prince Rupert and the Duke of Albemarle.[433] At the end of 1673, the hull of the galley at Genoa was built, but the vessel could not be completed in time for the following summer's service. The other at Livorno was rigged and could be ready by springtime but was short of 70 bonovogly (paid crew members), 48 officers, 48 marines and 35 slaves to form a complement of 324.[434] In January 1674, it was reported to the Navy that 40 Moorish slaves would be sold as '*merchandise*', of whom about 25 or 30 were fit for the oars, but they could only be purchased with ready money as the nearby states were getting their galleys ready and would be competing customers.[435] By July the shortage of oarsmen was so desperate that soldiers from Tangier were sent to make up the numbers if Captain Duteil could not recruit some locally.[436] The lack of oar slaves resulted in the smaller galley at Genoa being abandoned while the larger, named the *Margaret Galley* after the Grand Duchess of Tuscany, eventually sailed for Tangier at the end of 1674. On her arrival there, the lack of slaves kept her in harbour during the summer of 1675.[437] As a war was being waged with Tripoli, orders were sent to the Mediterranean squadron that any Turks or Moors taken from its vessels should

The squadron of warships sent to the Mediterranean in 1687 to enforce peace treaties with the Barbary States. They are shown here in the Grand Harbour, Malta. The large ship in the foreground is the third-rate Anne *of 1678. The squadron was a balance of ships containing fire power and speed. Oil on canvas. Author.*

be sent to Tangier to serve as galley slaves.[438] A Tripoli ship manned by 24 volunteer Greek seamen was taken in July 1675 and found to contain a cargo of 98 negro men, women and children. Orders were given for the negroes to be sold at Malta and for the Greek citizens of Tripoli to be taken to Tangier to make up the number of galley slaves.[439]

Even though the galley lay idle, she still required provisions of fresh beef and mutton as agreed with the supply contractor. Unfortunately, the contractor found it impossible to find fresh meat in those parts, but managed to provide salt beef and pork instead. This in turn created another problem as the *Margaret Galley*'s Moors would only eat the beef as their religion forbade them eating pork.[440] As well as food, clothes and other necessaries had to be purchased for them as well.[441] The shortage of slave oarsmen and the unwillingness to buy more ensured that the galley stayed in harbour at Tangier throughout the following summer.[442]

The high cost of maintaining a galley which never went to sea was considered by the Admiralty in February 1676. With few options available, the decision was made to lay it up safely at the west end of the mole. It was hove up within three feet of high water, where it would be protected against the weather under the battery of York Castle. The crew would be paid off and the '*strangers*' (probably French and Italian galley specialists) taken in the *Quaker Ketch* to Livorno. The slaves, who had remained idle for so long, would at last be found useful employment working on the mole. Their numbers around this time were augmented by a shipment of 178 Algonquian Indians captured in New England by the English colonists during a bloody conflict.[443] To house the diverse assortment of Moors, Greeks and North American Indians in the town, lodgings were rented under the care of the governor of Tangier.[444] The landlords were soon complaining of the considerable damage caused to most of the houses they were lodged in, particularly Mr Nevill's, where they burnt timber boards and spoilt his garden.[445] The sorry tale continued when it was discovered in June 1677 that the bottom of the galley was badly damaged by shipworm,[446] and rather than have it repaired, it was given away to Sir Roger Strickland and George Legge.[447]

The Galley Frigates

The failure of the *Margaret Galley* demonstrated how difficult it was for a northern European navy to change their culture and become Mediterranean slave masters. In 1693, the English consul at Genoa sent a proposal to the Admiralty from a Venetian shipwright concerning building a galleass for the Navy. A galleass was similar to a galley but also had guns mounted on the broadside. The Navy Board, including Edmund Dummer, who had been to Venice, responded by

A reconstruction of the Margaret Galley *based on her recorded dimensions and on contemporary drawings of galleys in the Venetian State Archives. She had 27 benches each side, with six men to each oar.*[431] *Author.*

PLANNING A NEW TYGER

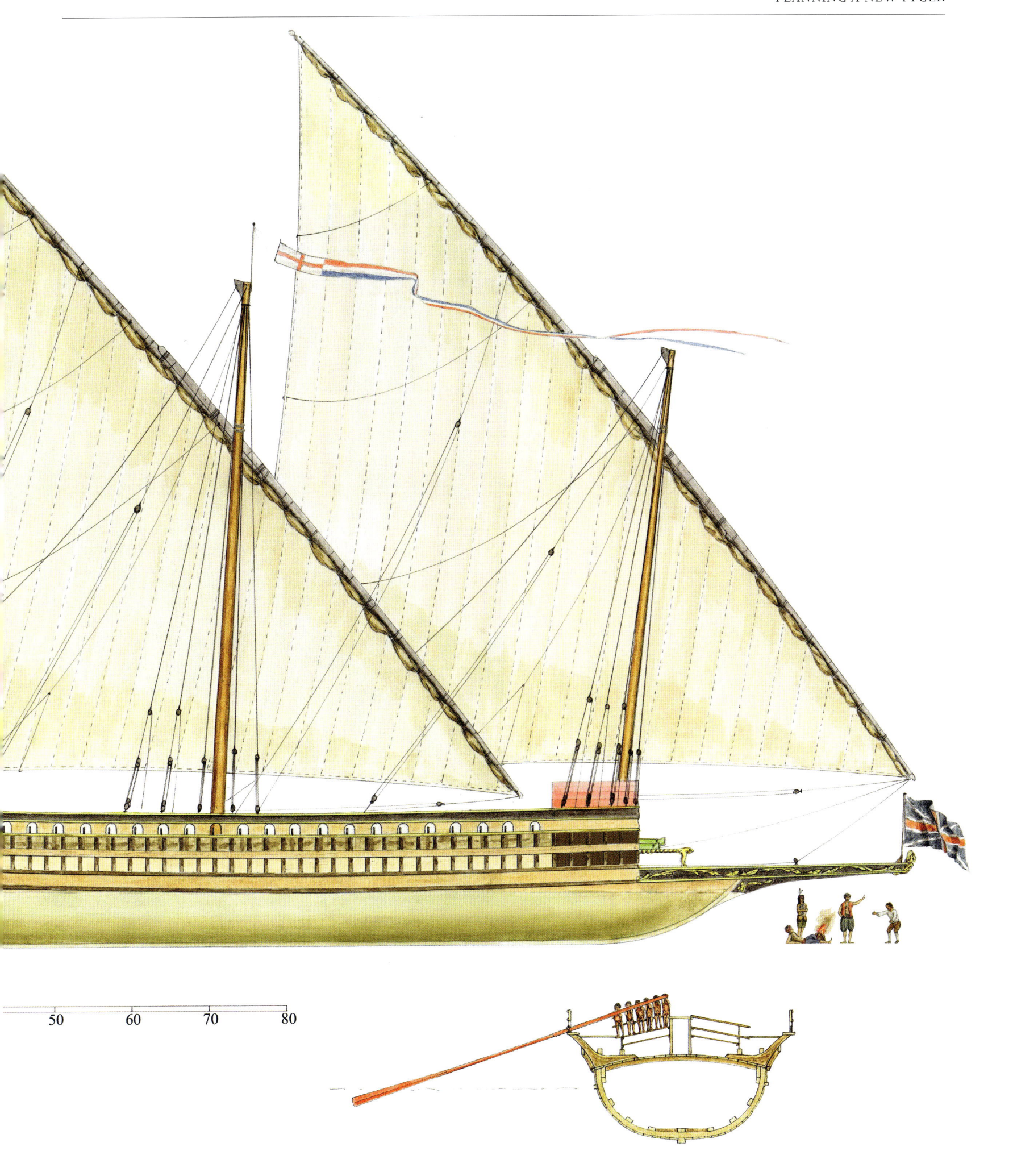

pointing out, among other things, the horrifying reality of the galleass, telling the Admiralty that '*these as well as other galleys are rowed by slaves who are chained and do all their offices of nature at the bank, therefore they have only one deck flush fore and aft over which they spread an awning when in port and are open on service that ye slaves thereby (being 7 and 8 at an oar) may be kept tolerably sweet which otherwise is not very likely to happen… we are unaccustomed to slaves to tug at them*'.[448]

Something else was needed to fulfil the warship role in the Mediterranean, and thoughts turned to a design that was a compromise between a true galley and an English frigate. This was unimaginatively named a 'galley frigate', a ship about fourth-rate size but four or five feet narrower and with the lower-deck guns largely replaced by oars. Not having to bear so much weight, the scantling (the size of the timbers) was also smaller and lighter, which also made it easier to row. There was nothing new about the design principle as 130 years before, in 1545, Henry VIII had built four such vessels, including one called the *Tyger*. A later drawing of a similar ship[450] that was probably never built, dating from the reign of Charles I, shows that the idea never died. Charles II would have been aware of these earlier ships and would have been reminded of them when Captain Thomas Willshaw informed him the French were building several galley frigates at Toulon but could give no clear description of them.[451] Willshaw had been to the Mediterranean during the first six months of 1674 in the *Reserve*[452] and may have seen the first of them, the 300-ton light frigate *Bienaimee* of 24 guns.[453] Wanting to know more, Charles authorised Sir Anthony Deane's son to visit Toulon to view them and take their dimensions. Charles also met the Venetian ambassador, Paolo Sarotti, and discussed the type of vessel with him, hoping he would have had some knowledge of them. The ambassador reported the matter to the Doge and Senate, informing them: '*The oars are so disposed that they can be withdrawn inside the ship, when required, without moving them from their place and without getting in the way of the artillery and its use. His Majesty has decided to have one of this sort built at once; the King himself told me this. The Duke of York remarked to me that he could have wished that the King had waited to hear how one of the (French vessels) turned out. He asked me various questions about*

Above. *Mr Nevill finds the galley slaves burning timber boards at the house rented from him for their lodging. They consisted of Moors, Greeks and North American Indians. The costumes are speculative and probably overstated. Author.*

Left. *A Maltese galley similar to the* Margaret Galley *(Robinson 684). A Maltese galley. Willem van de Velde, the Younger, c.1694? PAH9369; Photo: © National Maritime Museum, Greenwich, London.*

the shape and use of them. Prince Rupert afterwards expressly asked me to get one (drawing) for him from Venice. He said, with the utmost courtesy, that they had not the remotest idea of having galleasses built, which would be of no use in these seas, but to consider the position of the oars and the artillery and other things which might throw a great deal of light upon the ship in question. They only want a drawing upon a sheet of paper.'[454]

At the Admiralty meeting of 6 February 1676 at which it was decided to pay off the *Margaret Galley*, it was also decided to '*estimate at ye same time being presented to my Lords by ye officers of ye navy of ye charge of building a vessel of ye design lately proposed by his Majesty for ye answering the service both of a frigate and a galley as ye occasion shall require it. By which estimate it appeared that such a vessel of 120 foot long and 27 foot broad might be built… which vessel would answer all ye purposes of a galley and yet be capable of doing ye service of a frigate in all seas & weather, besides being maintained at much less charge.*'[456]

Two days later, the King and the Admiralty informed the Navy Board that they were resolved upon building a ship by contract with Anthony Deane Junior, son of the famous shipbuilder Sir Anthony Deane. Deane's proposition, drawn at the King's direction, was for a ship to row with 36 oars and to carry 32 guns. Evaluations continued for a week before the dimensions were finally settled at a keel length of 104 feet and a breadth from outside the plank of 27 feet 8 inches. The building cost of £3,341 14s for the hull was to be made up in a series of staged payments.[457] This cost was calculated at £7 18s per ton using the formula of keel length x breadth x half breadth, divided by 94. On 19 February the estimate was passed by the Admiralty for a privy seal to be issued.[458] When finished, the ship's keel length was measured at 104 feet from the back of the sternpost to the touch afore, and the breadth was 28 feet, which increased her tonnage from 423.4 tons to 433.7 tons. This was calculated

Below. *Speculative layout of the oar-equipped* Tyger *of 1545 based on the Anthony Roll[449] and several other 16th-century sources which are difficult to reconcile with any certainty. Of about 200 tons, she fought in the Armada campaign and pursued the Spanish as far north as Newcastle. Author.*

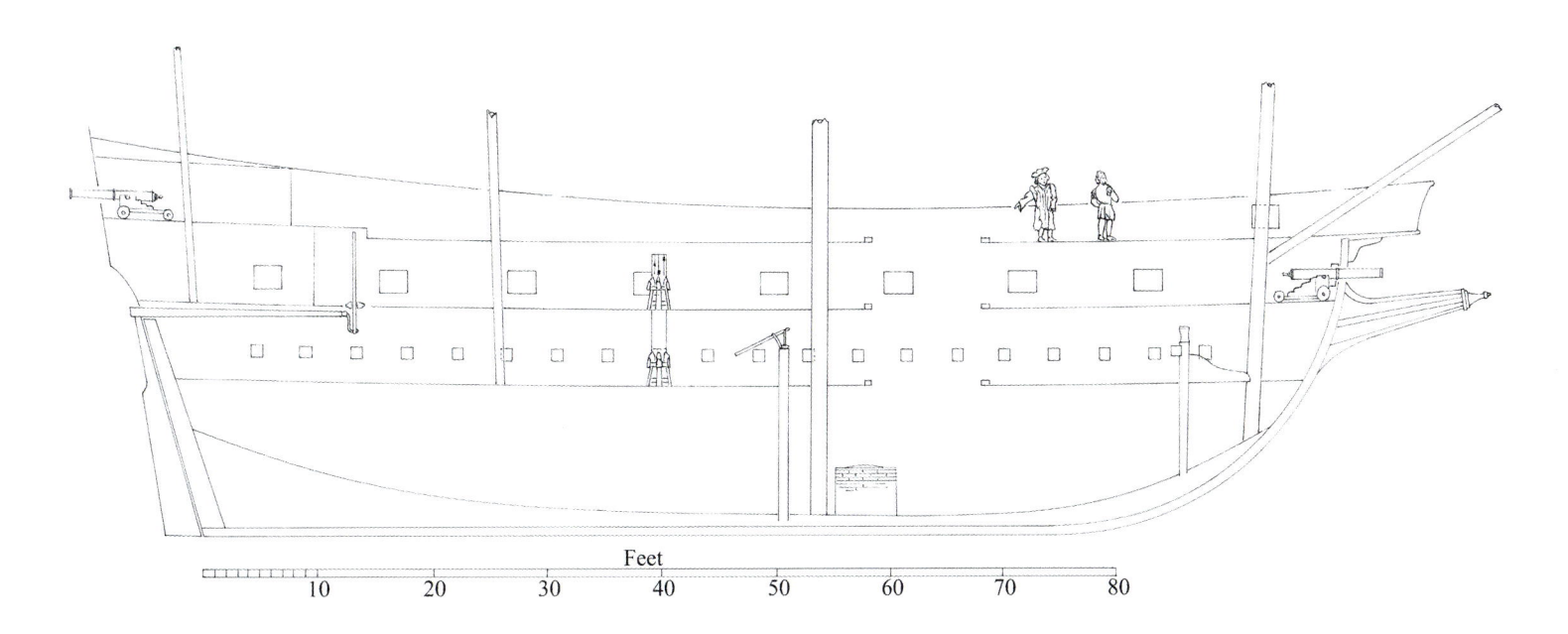

THE MASTER SHIPWRIGHT'S SECRETS

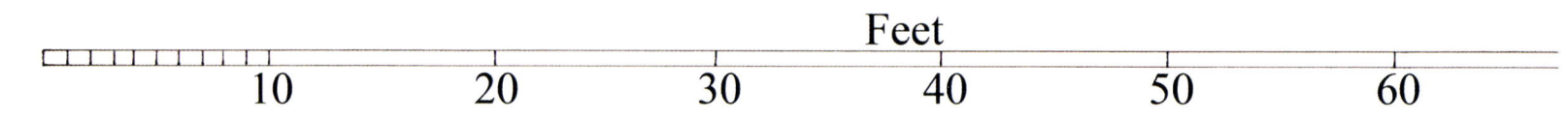

Feet
10   20   30   40   50   60

PLANNING A NEW TYGER

The James Galley *(above) and the* Charles Galley, *at a scale of 1/96 (⅛ foot). These fourth-rate ships were narrow, lightly built and fast. The oars were manned by the crew and not by slaves. In spite of their success, the galley frigates were reduced to fifth rates in 1690, partly because of their small size, but also to reduce the cost of employing sea officers who were paid according to the rate of the ship in which they served.*[455] *The small difference between the length of the keel and the length of the gundeck of the* James Galley *suggests she had an upright stem. Author.*

The launch of an English ship, probably the 'James Galley' (Robinson 542). Willem van de Velde, the Younger, c.1676? PAF6610; Photo: © National Maritime Museum, Greenwich, London.

to increase the cost slightly to £3,424 13s. In addition, Deane was allowed another £40 for adding a forecastle.[459] Her anchors, rigging, sails, oars, boats and six months' stores cost a further £1,670 17s.[460] Intriguingly, Deane junior didn't build the ship himself but subcontracted it to a private shipbuilder, Henry Johnson, to be built in his yard at Blackwall for £6 10s per ton, by an agreement made on 1 March. This was £1 8s less per ton than Deane junior's contract with the Navy, making him a healthy £607 profit for doing virtually nothing other than procure workmen for the project and give a '*present*' of £50 to Mr Johnson's eldest son, while Mr Johnson promised to make a present of £30 to his foreman.[461] The sub-contract was signed by Deane Junior and Sir Anthony Deane, a member of the Navy Board, showing that the deal must have been arranged by Sir Anthony with the blessing of his benefactor, Samuel Pepys. Such obvious conflict of interest was thinly circumvented by awarding Deane's son the contract to build the galley frigate. That being so, it's probably safe to say Sir Anthony Deane designed the ship which Henry Johnson built. Previously, in 1674, Deane had made a similar arrangement with Johnson for building the sixth-rate *Lark*.[462]

Great care had to be taken with the design of such narrow ships as many fourth rates of the period built four feet wider had to be subsequently girdled in order to carry their guns far enough above the water. Nevertheless, Charles was confident in the success of the galley frigate and declared his intention a day later to have a second built in his own yard by Phineas Pett III, the master shipwright at Woolwich. He would not need to find much ready money as it could be built out of timber already lying in the yard and at Deptford.[463] The Navy Board drew up estimates for a larger ship with a keel length of 114 feet, a breadth of 28 feet 6 inches, with 44 oars and of 492 tons. Although the total cost was estimated at £5,435, only £3,670 would have to be ready money, the rest being provided by a little credit.[464] A few days later, on 22 March 1676, orders were issued to build her.[465]

The radically narrow and light ships were inspected as they were being built by the Surveyor of the Navy, Sir John Tippetts. He was alarmed by what he saw being built at Woolwich and arranged for members of Shipwrights' Hall to inspect Pett's ship. They found the frame timbers placed too far apart and some of them insufficient in scantling, as well as lacking sufficient scarph (overlap), being of poor quality, and too small for a frigate of her size.[466] In spite of these concerns, the ship was not altered. The concerns seem to have been well founded, for when she was inspected before major repairs were carried out many years later in December 1692, she was found very weak in the wake of the upper deck, having no toptimber frames sufficient to secure the deck. They were also so few, so far apart and so weak that there was nothing on which to fasten the plank. Not only that, but during building the frames were not moved or spaced in such a way as to miss the gun and oar ports but cut through as and where needed. The numerous oar ports '*cut and mangled*' the frames so much the ship was not fit for sea.[467] In contrast, the ship being built by Anthony Deane in Henry Johnson's yard was found to be '*well and sufficiently performed both in materials and workmanship proper to a vessel of her intended quality and dimensions*'.[468] Pett's ship was launched on Tuesday 12 September 1676 and named the *Charles Galley*,[469] and she was followed a month later by Deane's *James Galley*. Phineas Pett was concerned to know how many of the carvings of the *Charles Galley* the King wanted to gild or have painted. He was instructed to gild the shield, the royal letters, the crowns, some highlighting on the stern and part of the figure of the head, all of which was estimated to cost £35.[470]

The dimensions of the ships have been compiled from several sources.[471] The most detailed were made by Fisher Harding, who recorded the *James Galley*'s measurements along with good explanations of what the dimensions were. 'Between decks' is the height from the gundeck plank to the upper-deck plank in the midships. 'Depth' is depth in hold from gundeck plank to ceiling next to the keel. 'Breadth' is breadth from the outside of the plank on one side to the outside of the plank on the other at the greatest breadth not including the wales. 'Gundeck length' is the length of the gundeck from the rabbet (a rebate to take the outside plank) of the stem to the rabbet of the main post. The gun and manning list was compiled from the proposed 1685 gun establishment.[472] It shows the ships in their original configuration, and although some sources give different figures, those listed below seem the most likely or reliable, with the exception of the between-deck height of the *James Galley*. She is listed as having only a six-foot height between decks, which seems unlikely for a ship with oars. The *Mary Galley*, built in 1687 and '*answerable*' to the *James Galley*, had a seven-foot height between decks.[473]

Dimensions of the *Charles Galley* and the *James Galley*

	Gundeck length	Keel length	Breadth	Depth in hold	Draught	Between decks	Number of oars	Burden
Charles Galley	131' 1"	114' 0"	28' 8"	9' 1½"	12' 0"	7' 0"	42	498
James Galley	112' 0"	104' 0"	28' ½"	9' 2"	12' 0"	6' 0"	38	435

	Men (in war)	Men (in peace)	Guns	6-pounder	Saker (5¼ pounds)	3-pounder
Charles Galley	220	220	32	26		6
James Galley	200	200	30	10	16	4

To save space in the holds and keep the centre of gravity as low as possible, 40 tons of iron ballast was ordered for each ship. This could take the form of ingots, old, unserviceable broken guns or shot.[474] The *James Galley* was launched from Henry Johnson's yard at Blackwall on 2 October 1676[475] and taken into the stern of the great double dock at Deptford for fitting out. Ahead of her was the old and worn-out *Tyger* waiting to be rebuilt. The *James Galley* was to be sheathed with lead, fitted with chain pumps, furnished with a roundhouse built on the quarter deck and have pieces fitted for the oars to lay on.[476] After she was launched from the dry dock, another 47 tons of ballast was put in the hold, carefully supervised by Sir Anthony Deane himself. In service, the ship would take on more ballast to compensate for the reduction in weight as stores were consumed, in the same manner as the Algerian warships. At one time three tons of salt water was put into the casks to stiffen her during a storm.[477]

Rowing

In 1681, there was an oar makers' workshop at Deptford that was in poor condition and that required making good with brick along the back wall.[478] It was a modest affair which in 1685 employed only one oar maker who could not keep up with demand. Rather than having to rely on purchasing them from outside, it was decided to employ three or four men to convert the '*rafters*' into oars as the Navy required.[479] Oars were made in a great variety of sizes ranging from 14 feet long for small boats to over 30 feet long for ships such as the *Tyger*. The inboard section of the oar was made bigger for balance, and although most were square loomed, some were round.[480] All the smaller oars used in boats and barges were made from ash,[481] while large oars of the

A ship's oar capable of being rowed by five men, from A Naval Expositor *by Thomas Blanckley, 1750.*

type used in ships were made from fir.[482] All the oars were protected from rot by being primed three times.[483]

There were many things to consider when producing oars for a lot of different purposes, and sometimes they failed to meet expectations. In 1681 about nine dozen square-loomed oars made of English ash were supplied to Deptford from Portsmouth but found to be very heavy, ill shaped, warped and not fit for service until the oar maker had overhauled them.[484] In 1680, the *Anne* yacht was being fitted out at Portsmouth and was supplied with 20 oars for rowing. Her captain complained that they were such clumsy things they were only fit for lighters and '*so weighty that two men have burden enough to carry one of them*'. When they were shipped in rowing position, he found they were at least two and a half feet too long inboard and very weighty outboard, and unless shortened, they would be unserviceable.[485] In 1686, the fifth rates *Garland* and *Sapphire* were at Portsmouth being made ready for service in the Mediterranean and it was intended to equip them with ship oars. The *Garland* had no oar ports, and although the *Sapphire* had eight a side, there were no ship oars of the necessary length of between 24 and 26 feet. Cutting the ports in the *Garland* was considered less of a problem than getting oars for the *Sapphire* sent down from Deptford.[486]

Oars supplied for the *James Galley* were between 28 and 29 feet long and cost 14 shillings each.[487] They were longer than the inside breadth of the ship by a foot. As the oar ports were 15 inches wide[488] and only a little bigger than the oar blade there must have been a lot of interference when they were shipped, causing much banging and bruising. Being longer than the deck beams above, they could not be stowed breadthways but instead lay awkwardly in the fore and aft direction. The tholes, the vertical pins that retained the oars where they passed through the ship's side, were made from ash in boats[489] but from iron in larger vessels[490] such as the galley frigates. It seems that each oar

Opposite. *Portrait of the 'James Galley' (Robinson 541)? Willem van de Velde, the Younger, 1676? PAH1855; Photo: © National Maritime Museum, Greenwich, London.*

Below. *The* Charles Galley *by van de Velde the Elder (Robinson 538). Note the smoke from the iron baking oven. Portrait of the 'Charles Galley'. Willem van de Velde, the Elder, 1676? PAI7276; Photo: © National Maritime Museum, Greenwich, London.*

was attached to a single iron thole by a withe, or strap. It seems safe to assume the oarsmen in the galley frigates did not sit on benches, as in true galleys where only a short stroke was possible, but remained standing,[491] allowing them to take a much longer stroke of about four steps forward and aft.[492] Each oar was designed to be rowed by at least two men,[493] but sometimes they were manned by three[494] and were probably capable of being manned by four or even five.[495]

Immediately after the launch of the *Charles Galley*, the officers of the Navy Board contemplated the use of oars in the galley frigates without using slaves. They supposed that Thames watermen were generally more expert than other seamen in their use and suggested a warrant be sent to the rulers of the Company of Watermen requesting 80 men be sent to each galley frigate for '*the swifter rowing at sea when there shall be occasion for it*'.[496] Warming to the task, they also decided to have an extra boatswain extraordinary for the governance of the men at the oar[497] and issued a set of instructions for him to follow:[498]

He shall see that every oar and scuttle be marked with either figures or letters so as upon all occasions the oars may be readily put out for service in their proper places.

That the oars shall be always kept clear and ready upon all occasions and that before the ship go out of the River he see that scuttles and oars be fitted with iron bolts, oar handles, withes, or ought else proper for the better rowing of the ship and if anything be wanting he is then to apply himself to the Commander for it if the Carpenter or Boatswain have it not on board.

He is by his Captain or superiors approval to quarter two men at least to each oar expressing the names of the persons to which oar they belong & the make or number of the said oar they are to row at that thereby every man may know his station and avoid confusion in time of service.

When the men are all quartered to their oars he shall exercise them while the ship is in any river an hour or more every day and also at sea in calm or little winds if the scuttles may be opened without prejudice to the ship that thereby the men may be the more expert in that work unless his Commander shall think fit to command ye contrary for some other service of the ship.

If any oars shall be broken or lost he is to acquaint the Commander thereof and also of the reason & by whose neglect the expense of which is to be noted in a journal which he is to keep, not only of that but every days rowing expressing the day of the month, the number of oars used, where and how long they rowed and the occasion whether in the river to exercise the men or at sea in little wind or calms or in chase after ships. His journal at the end of the voyage is to be presented to the Comptroller of His Majesties Navy upon which he shall receive his pay. Lastly upon his performing this duty we humbly conceive he may deserve 40 shillings a month which we leave to consideration.

In obedience to the King's order in council, the rulers of the Company of Watermen and some of the principal officers of the Navy selected 80 young watermen to serve aboard each ship. Generally they were all experienced, with four or five years' service each.[499] They were chosen in a public hall, where about half were prest (conscripted), while the rest volunteered for the Mediterranean voyage. Some may have been influenced by the prospect of adventure and some by the presence of ice in the River Thames that winter. They were all described as fit young men ranging in age from 18 to 27 years old, and although most were servants (apprentices), many were fully qualified watermen. Nearly all had previously served at sea, which may well have been a reason for their being chosen. In spite of the seriousness and formality of the events, Captain Canning of the *James Galley* disregarded the process and when the watermen came aboard he turned many off the ship and replaced them with seaman of his own preference, probably because they had served him before. This soon reached the ears of Samuel Pepys from the masters of some of the dismissed servant watermen who had gone to a great deal of expense and trouble in seeing their servants properly clothed and equipped for the voyage. Pepys had the crew mustered to prove the complaints valid before writing a stiff letter to Captain Canning. He let him know the King was greatly dissatisfied with his proceedings, and because of the great dishonour he had brought upon the service he was strictly ordered to receive back the discharged watermen and to discharge those taken on in their place.[500] To help Canning along, two senior officers were sent from Chatham, and they found the turned-off watermen aboard a yacht at the Nore, counting 14 from the *Charles Galley* and seven from the *James Galley*. They returned the men back to their ships and removed the same number of '*least useful persons*' to bring the ships to their established complements.[501]

The Performance of the Galley Frigates

During December 1676, the *James Galley*[502] was joined by the *Charles Galley* toward the mouth of the Thames in the Hope for their voyage to the English base at Tangier. King Charles desired to test them against the *Katherine* yacht as they sailed down river through the Downs and on to Portsmouth. They proved to be extraordinary sailers even coming up to the *Katherine* yacht when sailing large, giving everyone confidence that they would prove to be the answer against the Turks.[503] At anchor in the Downs, Captain Canning wrote from the *James Galley* that he believed both frigates sailed extremely well but couldn't say which had an advantage until further trial.[504] He also wrote that '*we find our anchors too light for though the ship rides as easy as any in the world, yet every puff brings our best bower (anchor) home*'.[505] During February in Plymouth Sound, Captain Dunbar of the *Charles Galley* reported that they had experienced mighty winds and he was worried the rigging was too small for the masts, '*but ye ship will answer what is expected from her for we find that she sails well and works well*'.[506] Stopping in Cadiz Bay, Captain Canning found '*both our frigates very wholesome and good conditioned and of such equal sailing that neither can justly brag of wronging the other*'.[507] A later captain of the *Charles Galley* reported that she was the best sailor in the fleet.[508] However, in May 1679, while in Tangier, Sir John Narborough in the *Plymouth* ordered the *Charles Galley* to chase him, but after sailing 85 miles and reaching the Spanish town of Fuengirola the latter had gained nothing and gave up.[509] This may have been a worthless comparison as the unknown cleanliness of the hulls would have had such an important influence on such a race.

Oars were sometimes used in the *Charles Galley*. At Cadiz in February 1679 during fair weather she was rowed from Puntal to above the castles, presumably Rotta, a distance of ten miles. Three months later, off Sallee, some suspicious sail was seen to the southeast and with little wind they got out all their oars and rowed up to it. Disappointingly, the suspicious sail turned out to belong to the *Hampshire* and the *Nonsuch*. The next recording of her oars being used was in July during the night. Then, a month later, on the morning of 8 August, a ship was seen three leagues to the east of Cape Tetuan. They sailed after it, but at one in the afternoon the little wind died away and as it turned calm they got out all their oars and rowed after it. By four o'clock they were within shot of what proved to be the Algerine warship the *Great Pearl*, of 34 guns. Rather than fight, the crew of the *Great Pearl* ran their ship ashore, set it on fire and ran off. Captain Hamilton boarded it and tried unsuccessfully to put the fire out but was forced to leave just before it blew up at half past five. The pursuit had been a great success, directly attributable to the use of oars. They proved their worth again on 20 April the following year when the wind died away, leaving the ship becalmed off Cape Spartel. The tide began to take the *Charles Galley* onto dangerous rocks until the crew got out the oars and rowed to safety. After the long commission the ship returned home at the end of July 1680. She sailed up to Woolwich, where the guns were taken out, before the crew rowed the last six miles upriver to Deptford. There, they hauled aboard the hulk, made ready to unrig and towards evening got out all the gunpowder and the gunner's stores.[510] Nine years later, the then captain, Jeremy Roch, mentioned the performance of the *Charles Galley* under oars: '*At half an hour past 5 the Start bore N.N.E one mile off. Now, having little wind, I had a mind to try how the Galley would row, so I ordered the sails to be hauled up and ran out all our oars being 42 and put 3 men to each oar and found we could row 3 miles an hour; upon which my men were so proud that they would row all the way into Plymouth Sound.*'[511] Indeed, his log book confirms they rowed into Plymouth Sound and anchored in seven fathoms at about six in the afternoon. The only other time oars were used during the voyage was at the very end, when they rowed from above Gravesend to Longreach.[512]

Captain Canning in the newly built *James Galley* with 80 watermen aboard made little or no use of the oars on his way to Tangier or while cruising on in the Streights. On 28 September 1677 he, in company with the *Charles Galley*, met the *Golden Marigold*, one of the best Algerian warships, of 500 tons, mounting 30 cannon and with a crew of 400 men. The *James Galley* came up to her first, followed by the *Charles Galley*, and after exchanging fire they managed to board. During the boarding, both galley frigates had their foremasts and heads badly damaged. Their fire cleared the upper deck of the *Golden Marigold*, forcing her crew below to the gundeck. They continued to fight ferociously for another hour and only surrendered after 200 of their men were killed. During the action, Captain Canning and several men in both frigates were killed. Narbrough was so impressed with the *Golden Marigold* that he had her repaired and taken into naval service as a 44-gun fourth rate. He was also impressed with her former crew, who he described as '*resolving to fight it out to the last man… I am doubtful we shall have some of our ships taken from us by these fellows, they are resolute and understand fighting and have good ships.*'[513]

The taking of the *Golden Marigold* by the galley frigates led the King and Admiralty to reflect upon their '*special usefulness*' in the Streights. They asked the Navy Board to plan the building of two more vessels to the same dimensions as the *Charles Galley* by the same hands as before in merchant yards.[514] Their confidence in galley frigates seemed well founded, for the *James Galley* went on to take part in the capture of two other Algerian ships, the 32-gun *Half Moon* and the 34-gun *Flowerpot*. Less than two weeks after the Navy Board received the request, the following estimate of costs for the two new galley frigates was made:[515]

November 15th 1677
An estimate of ye charge of building a *Galley Frigate* for His Majesty of ye dimensions of ye *Charles Galley*. Viz:

	Feet Inches	Tuns
Length by ye keel	120 0	527
Breadth to ye outside of ye plank	28 9	

	Cost £
For ye hull, masts & yards, completely fitted	4216
For ground tackle, rigging, sails, 6 months stores for Boatswain & Carpenter	1976
	6192
For one other of same dimensions	6192
For both ye sum of	12384

After a few days' further consideration, the King decided the proposed two new galley frigates should be six inches wider than the *Charles Galley*. Then, satisfied with the dimensions and estimates, on 19 November he gave approval to the Navy Board for the ships to be built, one in the King's own yards by Phineas Pett III, master shipwright at Woolwich, and the other by commercial contract on the River Thames by Sir Anthony Deane. They were further ordered to build the ships as soon as financial terms were agreed. This was confirmed by the King in council, and a Privy Seal was issued.[516]

The building of the two new galley frigates had almost started when Samuel Pepys received a letter from Henry Sheeres, the prominent military engineer and highly respected Fellow of the Royal Society. He was then at Tangier supervising the building of the mole and offered his opinion about the *Golden Marigold* action: '*In the taking this last Turk (who had escaped the whole fleet had not the galleys by accident shot his main topmast by the board) I observe a remarkable defect in the said galleys which laid him aboard, which you know have no guns upon their gundecks and consequently were incapable of molesting the enemy after they had forced them between decks (to the gundeck) where they were able to maintain the fight without receiving the least damage from us and would always consequently render the taking them very difficult to these sort of ships unless provided by such as could annoy them between decks which these cannot. I therefore humbly offer my advice that they may have some small matter of force mounted between decks which may be no impediment to their rowing and that for the future all vessels of this sort may be contrived with consideration had thereunto.*'[517]

In other words, the lightly armed galley frigates with nearly all their guns on the upper deck were so close when they boarded the *Golden Marigold* that they could not fire down into its gundeck where the crew had taken refuge. To address the fault, Sheeres advised having more guns on the gundeck. On receiving the report in mid-December, Pepys read it out at the next Admiralty Board meeting and had copies sent to the Navy Board for their comments, and for Sir Anthony Deane's in particular.[518] Another event occurred that also required consideration by these official bodies – namely, the recent Act of Parliament granting the King £584,978 to build 30 ships of war, all to be completed within two years. This enormous programme would of course have the effect of increasing the cost of timber and labour. The whole matter was discussed at another Admiralty meeting on 27 December 1677 attended by King Charles and the Navy Board officers. The cost of the new galley frigates would increase from the £7 18s per ton paid for the *James Galley* to £8 5s for the proposed new ships. Finally, the King could not see how more guns could be provided on the gundeck without interruption to the oars, nor thought it very useful even if it could be done. It was therefore decided that the proposals should be laid aside and that the resolution lately taken should be revoked.[519] Years later, in 1686, when the *James Galley* was rebuilt, part of the work included reducing the number of oar ports to 34 to make space for ten guns on the gundeck.[520]

New Admiralty Commission

Although the ships were cancelled, the matter was not forgotten, and it appears that a surviving draught based on the *Charles Galley* was altered to include extra gunports.[521] Discussions continued and contracts were agreed with Deane and Pett for building the two new galley frigates; even when Pett died in February 1678, the Admiralty agreed that his widow should build a ship according to the design and draught made by her husband during his lifetime.[522] Then, unexpectedly, the smooth-functioning naval administration led by King Charles, who attended more Admiralty Board meetings than anyone else, suffered a severe setback as a result of the unlikely allegations of the one-time chaplain of the *Adventure*, the homosexual Titus Oates, a man known for coarseness, thieving, perjury and stupidity.[523] He alleged to have uncovered a popish plot to assassinate King Charles and replace him with his Catholic brother, James, Duke of York. Amid mass hysteria, a political crisis led by the Shaftsbury opposition developed which caused the downfall of the Admiralty Board and the resignation of Samuel Pepys on 21 March 1679. Opposition politicians with no previous experience of naval administration formed a new board.[524] Charles no longer attended the meetings, but instead rather obstructed them by acting alone. With its members lacking experience, as well as facing severe budget cuts by the Treasury and being undermined by Charles, the new board was condemned to failure. As Pepys said, Charles '*put it into hands which he knew were wholly ignorant thereof, sporting himself with their ignorance*'.[525] Sadly, due to the want of money and the distractions caused by the Popish Plot, Mrs Pett never got to build a ship for the King.[526]

Planning a New *Tyger*

In 1678, with the proposal for building two more galley frigates foundering, King Charles started to think about the next evolution in the development of ships for use in the Mediterranean. Although he was persuaded oars were very useful,[527] the true galley had proved a dismal failure. The galley frigates were compromised when fighting at very close quarters or when boarding enemy vessels. In any case, they were seriously under-gunned compared with other warships with a 220-man crew. The *Charles Galley* fired a broadside weighing only 174lb, compared with the similarly manned fourth-rate *Oxford*, which fired a 733lb broadside. Another obstruction was the serious limitation of funds for building new warships. Ever inventive, King Charles knew money was available from the customs revenue for the repair of ships, and that it could be spent on 'repairing' the few pieces of timber that may have come from the old *Tyger*. In reality the ship didn't exist at all, but she was still on the 'ordinary' list as a means of providing a benevolent income to old or maimed seamen incapable of doing any work whatsoever. As long as it was called the *Tyger* and it could be said that some few remnants of the old ship existed somewhere, a new one could be built with the repair revenue without troubling the bureaucratic process.

With his funding problem brilliantly solved, King Charles must then have considered the design of the new ship that the *Tyger* was to become, and must have spoken to the master shipwright at Deptford, John Shish, and his father, Jonas. These meetings probably took place at Deptford when Charles attended the launchings of three of the 30 new ships – on 12 April 1678 for the *Lenox*, on 10 July 1678 for the *Hampton Court* and on 17 May 1679 for the *Duchess*. Charles must have decided to build a fourth-rate-sized ship with two tiers of light guns, taking into account Henry Sheeres' comments. Not carrying such a heavy armament as other fourth rates, it could have fine lines with a narrow floor. The layout would have just enough space to have two oar scuttles in between each of the gundeck ports, giving a total of 32 oars. The concept may not have been completely novel as a contemporary model of an unidentified ship exists with '*Bristol 1666*' painted on the stern, but with two oar scuttles more and a gunport less each side on the gundeck.

The King did not inform the Navy Board or the Admiralty about his plans for the new *Tyger*,[528] which was to be a compromise between the traditional fourth-rate frigate and the galley frigate. Although the ship could not be rowed as effectively as the galley frigates, it could still be rowed and manoeuvred to bear its broadside towards an opponent. With guns on both decks, it could fight at close quarters when the crew were preparing to board an enemy vessel. Even though the weight of broadside would be half that of many fourth rates, the ship's 316lb was nearly double the weight of the *Charles Galley*. It may be coincidence, but the size and the necessary fine lines were within the building tolerance of a treatise made by John Shish for Samuel Pepys a few years previously.[529] Although King Charles would have led the discussions and specified what sort of ship the new *Tyger* would be, he would have left the details and draughts to Shish. Charles and the Shish family had history when it came to ignoring the naval administration: in 1673, they took apart Sir Anthony Deane's almost new *Saudadoes* and built another ship with the same name. More recently, in November 1678, Charles had unmercifully bullied Pepys to get Thomas Shish to start a new yacht to replace the recently lost *Charles* yacht in spite of standing orders for shipwrights not to do any work without written orders from the Admiralty.[530]

Next Pages. *King Charles II attending the launch of the* Lenox *at Deptford on 12 April 1678. It was on this and other similar occasions that he discussed shipbuilding with his shipwrights. He probably discussed the fate of the remains of the old* Tyger, *lately removed from the stern of the double dock visible in the background. Oil on canvas. Author.*

THE MASTER SHIPWRIGHT'S SECRETS

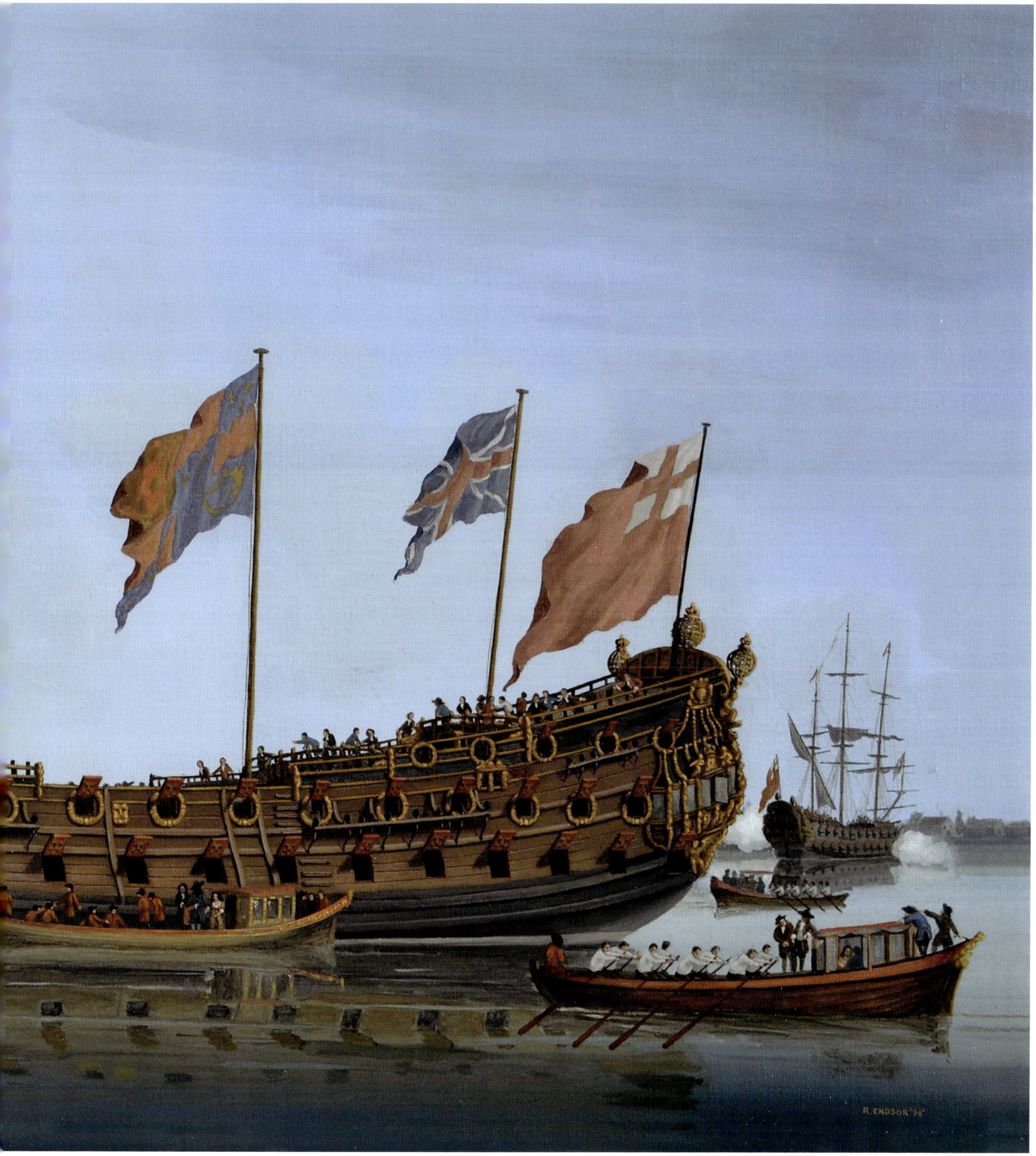
R. ENDSOR '96

CHAPTER 5

JOHN SHISH'S ACCOUNT OF THE DIMENSIONS OF A SHIP

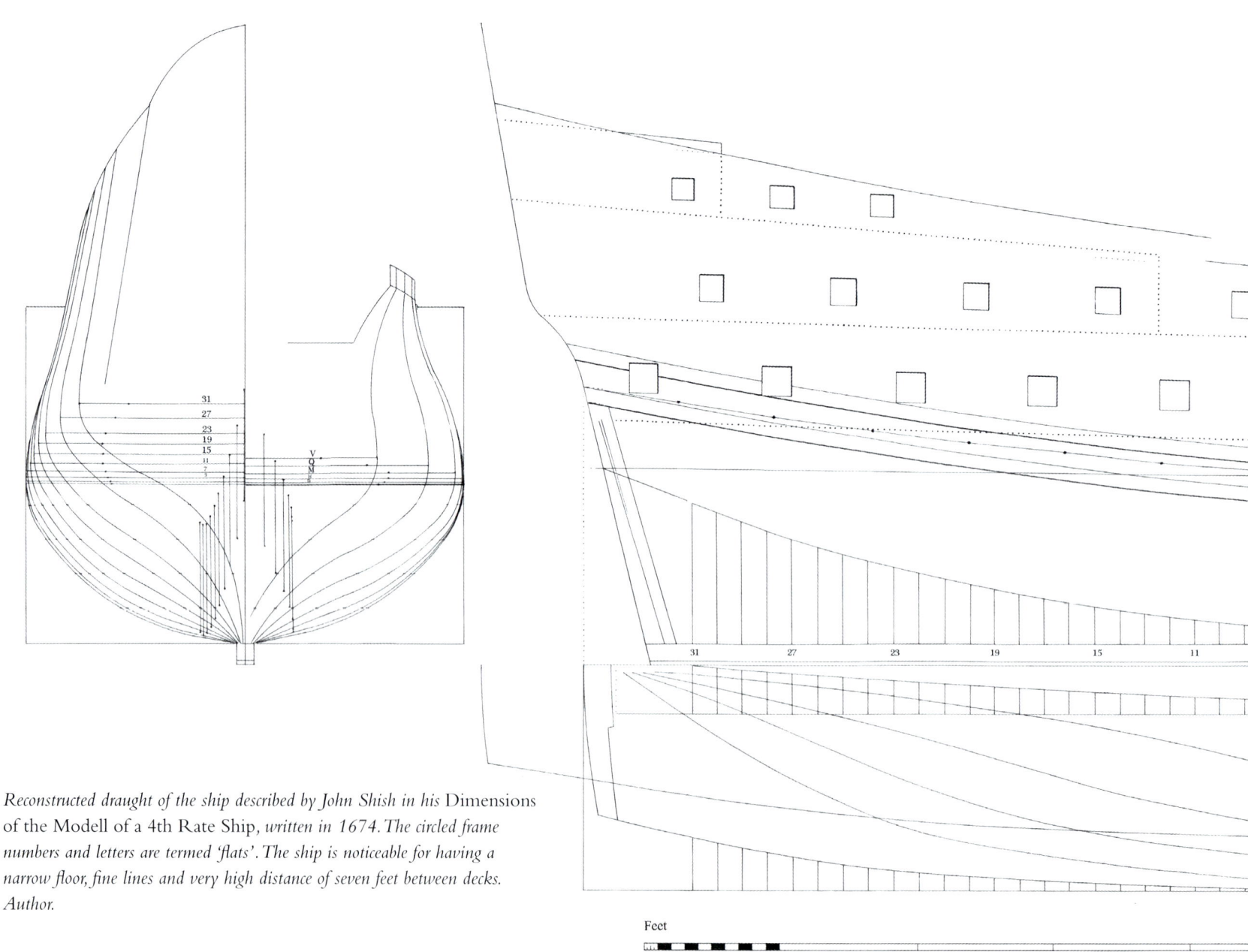

Reconstructed draught of the ship described by John Shish in his Dimensions of the Modell of a 4th Rate Ship, *written in 1674. The circled frame numbers and letters are termed 'flats'. The ship is noticeable for having a narrow floor, fine lines and very high distance of seven feet between decks. Author.*

On Wednesday 1 July 1674, John Shish, the master shipwright at Sheerness, sent a small treatise to the Secretary of the Admiralty, Samuel Pepys.[531] Most of the written communication between them concerned mundane business matters such as the progress made in fitting out a ship and the like. This one was different – it was a personal and private correspondence entitled *The Dimensions of the Modell of a 4th Rate Ship*, and it goes into considerable detail.

Pepys's Collection of Shipbuilding Documents

The events leading up to John Shish's treatise began years earlier, during the summer of 1664. Samuel Pepys's naturally enquiring mind was stimulated when he and the Duke of York's secretary, Sir William Coventry, discussed the writing of a history of the Navy of England.[532] From then on, Pepys made a habit of collecting interesting material relating to the Navy for his personal use. Although he never wrote the work, he carried out a comprehensive research programme and collected a vast amount of material that survives to this day.[533] This material includes many precious and irreplaceable works relating to shipbuilding which vary considerably in their depth and approach to the subject.

The earliest is *Fragments of Ancient English Shipwrightry*, dating from the late Tudor period.[534] Among the works that are contemporary with Pepys is Edward Battine's *Method of Building Ships of War*, written in 1684 and containing lists of the principal dimensions of a ship's structure and the size, or scantling, of all its individual pieces.[535] There is also Mr Dummer's *Draught of the Body of an English Man of War*, a series of drawings of sections through the hull of a first-rate warship.[536] An anonymous undated paper from the same period, *Method of laying down the dimensions of a Ship 78ft long by the Keel*, describes a simplified method of designing a ship. It includes drawings of a 'bend of timbers' and the rising line of the floor aft.[537] There is another anonymous work, *A New Method & Proportions Calculated for the more ready demonstrable & artificial delineation of a Ship's Body humbly presented to his Majesty*.[538] This work contains the dimensions of the rising

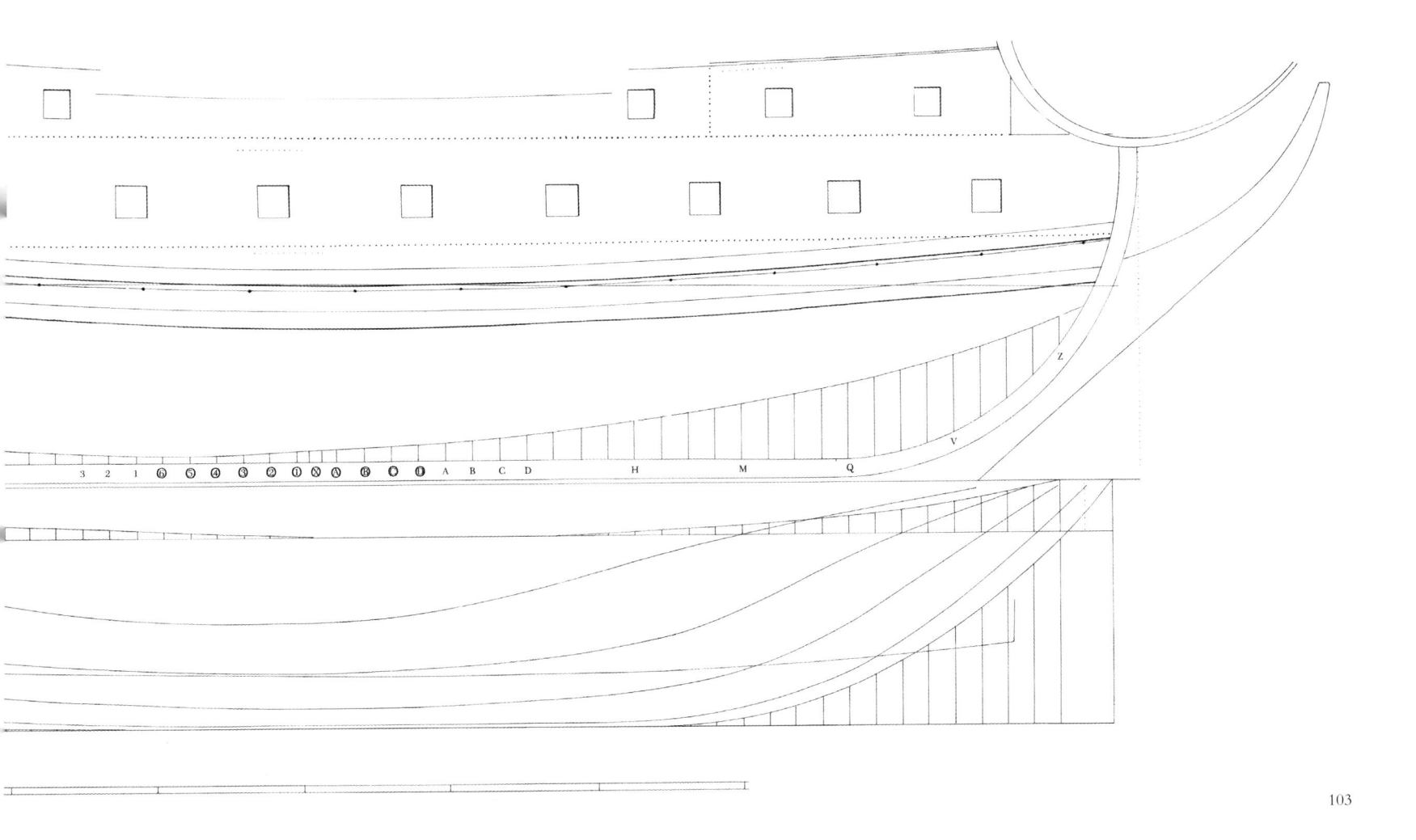

THE MASTER SHIPWRIGHT'S SECRETS

and narrowing lines of a ship's body set out as proportions that could easily be applied to a ship of any size. It would save the shipwright considerable time when calculating his curves from basic principles. Being written anonymously, it provides no clues as to whether it was actually used by a known shipwright or whether it was a hypothetical study by a theoretician. All that can be said with some certainty is that '*his Majesty*' was either Charles II or possibly James II, as these monarchs would have been most interested in such material.

The most significant and useful work collected by Pepys is Sir Anthony Deane's *Doctrine of Naval Architecture* of 1670. It describes in some detail, with text and drawings, how to make the draughts for a third-rate two-deck warship.[539] Being a more comprehensive work, it fills in many of the gaps left by John Shish's treatise. Although the work is a manuscript and the title page says it was written in 1670 '*at the insistence of Samuel Pepys Esq*', there are many references to '*the young artist*', indicating that it may have been intended for publication.[540]

John Shish's Treatise

John Shish called his treatise *The Dimensions of the Modell of a 4th Rate Ship*, and it could be thought that '*model*' means a small-scale wooden ship. But, as William Sutherland wrote in his 1711 book *The Ship-builders Assistant*, '*The draught or model is generally described on large paper*',[541] and in his glossary, he says a draught is '*the model or figure of a ship, or any of her parts, described upon paper*'.[542] A treatise on shipbuilding dating from c.1625 states: '*These lines being all drawn the plot is finished, which ought to be the true model of the ship proposed to be built.*'[543] Thomas Miller, himself a shipwright, wrote a book about shipbuilding called the *Compleat Modellist*[544] in which he made it clear that in Shish's time, the term '*model*' meant a draught or, as in this case, drafting data.

Shish's treatise may well owe its existence to the same request from Pepys that prompted Anthony Deane to write his *Doctrine* – a request for master shipwrights to describe their designs for a ship. The Shish document and Deane's *Doctrine* do not appear to describe any particular ship but are theoretical papers produced for the benefit of Samuel Pepys. Shish's treatise is almost unique in layout and content and is nothing like the usual official description of a ship set out in either contract or draught form. This would indicate that it was almost certainly a document that Pepys requested from Shish to add to his growing personal collection. The surviving document reveals that it is a replacement for an original that was either lost or destroyed. It is extremely unlikely that Pepys would simply lose a document, which makes it probable that the original was destroyed by accident, quite possibly when the Navy Office burnt down on 29 January 1673. The surviving document,[545] dated 1 July 1674, is transcribed below in its entirety.

St Nicholas Church, Deptford. Being close to the river and dockyard it was the shipwrights' church. The Shish family are buried there and, although the church has been rebuilt a number of times, the ancient tower survives.

ST. NICHOLAS CHURCH, DEPTFORD.

Honourable Sir Ye enclosed is ye Account of ye Dimensions of a ship which I promised so long since, for which delay I humbly Crave your pardon, & hope it may tend to your Honourables satisfaction it being exactly according to ye same you had formerly. I take leave & remain Your Honourable's most obedient Servant John Shish, 1st July 1674 Sheerness

The Dimensions of the Modell of a 4th Rate Ship

	Feet	Inches
Length by the Keel	106	0
Breadth from outside Planke to outside Planke	33	0
Depth in hold from Planke to Planke	13	0
Draught of Water Afore	13	0
Abaft	14	2
Rake of the Stem to ye Outside	21	0
Rake of the Post	5	0
Height of the Transome	20	0
Breadth at the Transome	22	0
Length of the Gundeck from the inside of the Post to the inside of the Stem	127	6
Height between Decks	7	0
Hanging of the Gundeck	0	11
Height of the Ports from ye gundeck	2	2
Ports on ye lower Tire in No. 13 being square each Port	2	2
Length of the upper Decke	138	0
Rake of the Counter	4	6
Height	7	6
Rake of the Upright	3	2
Length of the Upright from ye Counter to ye upper part of the Taffarell	19	0
Ports on each side on ye upper Tire in No. 9		
Height of the Ports from the upper decke	1	11
Hanging of the upper Decke	1	1
Length of the quarter Decke	47	8
Height of the Quarter Decke abaft	8	0
Height of Ditto afore	5	8
Ports on each side on ye Quarter Decke in No. 3		
Length of ye Roundhouse	16	0
Height of the Roundhouse Afore	5	4
Abaft	5	6
Length of the forecastle	20	8
Height of the forecastle Afore	6	0
Abaft	5	0
Length of the upper Decke from Bulkhead to Bulkhead	60	4
Height of the waist	3	2
Length of the Knee of ye Head on a straight Line	18	0
Hanging of the Head	2	8
Length of the upper rails	20	6
Hanging	6	1
Height of the upper Edge of the lower Harpin afore	14	2
Height abaft	20	0
Height Midships	11	8

Names of the Timbers	Rising abaft	Narrowing of the floor abaft	Narrowing of the breadth abaft	Names of the Timbers	Rising afore	Narrowing at ye floor afore	Narrowing at ye breadth afore
	Ft. Ins	Ft. Ins	Ft. Ins		Ft. Ins	Ft. Ins	Ft. Ins
X	00.09	..	..	A	00.10½	..	..
1	00.08	00.01	..	B	00.11	..	..
2	00.07	00.02	..	C	01.00	..	..
3	00.06	00.02½	..	D	01.01	..	..
4	00.05	00.03	..	A	01.02	..	..
5	00.05½	00.04	00.00½	B	01.04	..	..
6	00.06	00.05½	..	C	01.06	..	..
1	00.06½	00.06½	00.00¼	D	01.08	..	..
2	00.07	00.07½	00.00½	E	01.11	..	..
3	00.07½	00.08½	00.00¾	F	02.01	00.01	..
4	00.08½	00.09½	00.01	G	02.04	00.02	..
5	00.10	00.10	00.01¼	H	02.07	00.03	..
6	00.11½	00.10½	00.01½	I	02.11	00.04	..
7	01.01	00.11½	00.01¾	K	03.02	00.05	00.01
8	01.03	01.00	00.02	L	03.05	00.06	00.03
9	01.04½	01.01	00.02½	M	03.09	00.07	00.06
10	01.06½	01.01½	00.03	N	04.01	00.09	01.00
11	01.09	01.03½	00.03¾	O	04.05	00.11	01.04
12	01.11½	01.04	00.04½	P	04.10	01.00	02.00
13	02.02½	01.05	00.05	Q	05.01½	01.02	02.09
14	02.05½	01.06½	00.06	R	05.06½	01.04	03.05
15	02.08½	01.07	00.07	S	06.00	01.06½	04.05
16	03.00	01.08½	00.08½	T	06.06	01.10	05.05
17	03.03½	01.10	00.10	V	07.01	02.01	06.07
18	03.08	01.11	00.11	W	07.07	02.04	07.11
19	04.00	02.00	01.00	X	08.02	02.08	09.03
20	04.04	02.01½	01.01½	Y	08.08	03.01	10.08
21	04.08	02.02½	01.03½	Z	09.05	03.06	12.03
22	05.00	02.04	01.05				
23	05.06	02.05	01.07				
24	06.00	02.06	01.10				
25	06.05½	02.08	02.01				
26	07.00	02.09	02.04				
27	07.07	02.10½	02.08				
28	08.01½	03.00	03.00				
29	08.09½	03.01½	03.04				
30	09.06	03.03½	03.09				
31	10.03	03.05½	04.01				

The first part of the document gives the principal dimensions of a fourth-rate ship and is the type of information found embedded in contemporary shipbuilding contracts but more comprehensive, giving many more dimensions than normally found, such as the height between decks, the length of the quarterdeck and the height of the lower wale at three different points. While Shish's quarterdeck is fully dimensioned for length and height both afore and abaft, contracts, by poor contrast, usually say something like '*To make a large quarterdeck and forecastle with two round bulkheads*'.[546] Shish's treatise gives not only its length but also the height, both afore and abaft.

From the dimensions given in the document a draught was recreated at the normal shipwright's scale of quarter of an inch to the foot, or $1/48$. Rather surprisingly, no conflicting dimensions were found, although omissions meant that some amount of interpretation was necessary to complete it. The only dimension that did not immediately fit on the draught was the length given for the upper deck, which, at 138 feet, appears to be about 12 feet too long and clearly a mistake. If this is ignored for a moment and all the other dimensions drawn, it is found that the length from the aft end of the poop to the fore of the beakhead at the stem is 138 feet. It is this dimension that Shish must have intended to write in the document.

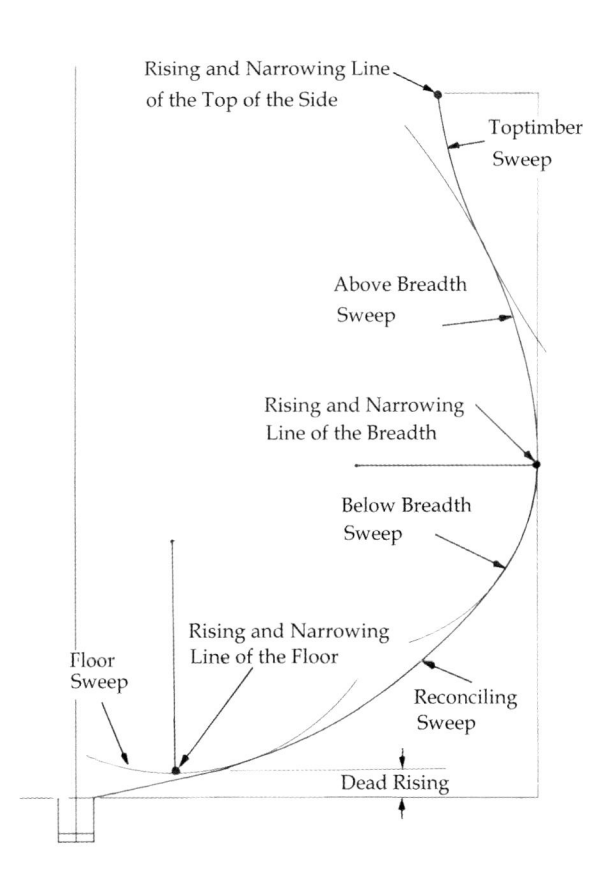

The midship flat, showing the rising and narrowing lines and sweeps (arcs) described in John Shish's treatise. Following contemporary practice, the drawing is made to the outside of the frames and does not include the plank. Author.

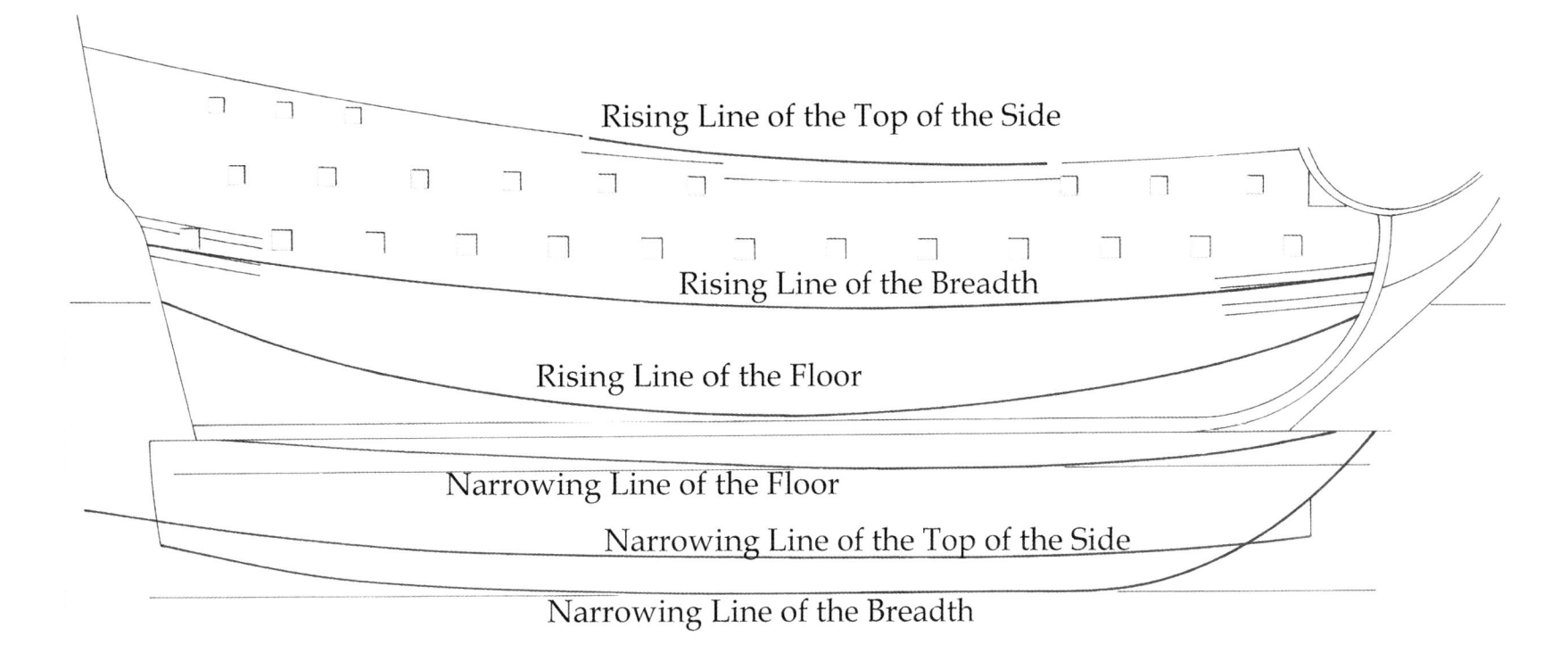

The three rising and narrowing lines that control the shape of the hull. The rising and narrowing at the top of the side line above water was less important and received less attention from the master shipwrights. Author.

In plotting out the draught it becomes apparent that some information is missing. The spacing of the gunports is not given, and it is presumed that the foremost gundeck port was situated immediately aft the toptimber that formed the edge of the beakhead bulkhead. The ports are 2 feet 2 inches square, which is rather small compared with the usual 2 feet 10 inches for the period.

Although Shish recorded the rising and narrowing lines in his treatise, he made no mention of the radii of the mould sweeps. Leaving them out seems an odd omission, but it may be that Shish did so deliberately in order to protect some of his secrets. Fortunately, they are recorded in at least three contemporary sources[547] which are all much in agreement and based upon proportions of the ship's breadth. Anthony Deane's *Doctrine* specifically stated his proportions were based upon the moulded breadth – the distance outside the frames and inside the plank. The below-breadth sweep did not remain constant towards the ship's stern. It reduced in radius to allow a good flow of water to the rudder while allowing a wide stern for guns and the captain's great cabin above. Deane clearly shows the reduction in his draughts and it is seen on all models of the period. William Keltridge, in *His Book*, gives measurements for the reduction.[548] Apart from a few frames at the extreme ends the floor sweep remained a constant arc, and as late as 1754 Mungo Murray wrote, '*It is usual for all the floor sweeps to be of one radius.*'[549] Another sweep worth mentioning is the hollow sweep at the ship's ends that extends from the keel to blend with the floor sweep. Both Deane in his *Doctrine* and Keltridge in *His Book* give this as a true arc; however, all other evidence shows a complex curve that gradually tightens. By using different sections of the same curve all the frames can be described, with the possible exception of those at the extreme ends.

Estimated radii of Shish's sweeps given in his treatise

	Floor sweep	**Below-breadth sweep**	**Reconciling sweep**	**Above-breadth sweep**	**Toptimber sweep**
Typical proportion	¼ main breadth	7/9 floor sweep	20/36 main breadth	17/18 half breadth	17/18 half breadth
Tyger	8' 1½"	6' 4" midships 3' 8" sternbend	18' 1"	15' 4" 11' 3" sternbend	15' 4"

Although the document identifies all the frames in the usual way, with letters forward and numbers aft, the space and room, or the spacing of the frames, is missing. Fortunately, we know where the narrowing lines must start and end, and by carefully testing their positions, the space and room was found to be 22 inches. The midship flat is shown at frame X and there are four flats, lettered A to D, forward of midships and six aft, numbered 1 to 6. The frames forward of midships omit the letters J and U in common with the Latin alphabet.

Shish does not given a name to the ship he describes, which has the very fine lines of a fast warship, similar to the frigates of the 1640s. Another similarity to these early ships is an unarmed waist with a gap where three gunports a side could have been positioned. In spite of this resemblance, it cannot be one of these ships as it is much larger, being 614 tons burden against the early frigates' 350 tons. In size, it is much closer to a number of fourth rates built between 1650 and 1654, although these ships had a complete upper tier of ports. Only the largest of these ships, such as the *Newcastle* of 1653, carried 13 ports a side on the gundeck, the same number given in Shish's treatise. In common with all these fourth rates, Shish gives her a depth in hold of 13 feet. The height between decks is given as 7 feet, which is very high, as contracts for the period, draughts and the lists of Battine and Keltridge all give 6 feet 6 inches or 6 feet 7 inches. The only known fourth rates with 7 feet between decks are the *Mary Galley* of 1687[550] and the *Charles Galley*. With such a height between decks and very fine lines, it may well be that the treatise describes the type of ship the new *Tyger* would become, a type of ship that normally uses sail but is also designed from the outset to use oars. It is possible that Shish proposed the ship described in his treatise to King Charles during their discussions concerning the rebuilding of the *Tyger*.

The last page of Shish's treatise is very rare, if not unique. It contains the three-dimensional co-ordinates for the rising and narrowing lines of the floor and for the narrowing line of the breadth for every frame in the ship. Although they are always called '*lines*' in contemporary works, they are in fact curves. Shish's curves are complex and intended for a real ship. They are not simple arcs used as instructional examples of the type described by Deane in his *Doctrine* and in some other works.

This incorruptible digitally recorded data is far more precise than any draught could ever hope to be. The co-ordinates were very important as they were marked out full size on rising staffs and half-breadth staffs for use in the mould loft during ship construction. The rising of the breadth is not mentioned, probably because it has very little movement. Edward Battine puts the rising of the breadth[551] in a very similar position to that given by Anthony Deane in his *Doctrine*, as starting abaft at the upper edge of the main transom, even with the waterline in midships and finishing at the upper edge of the upper wale at the stem. Battine's work, dating from 1684, differs only a little from that of Deane, who, in 1670, has it finishing a little above the lower edge of the upper wale.[552] The rising and narrowing at the top of the side is also omitted by Shish, although the rising can be easily discerned from the reconstructed draught.

The general shape of the rising and narrowing lines determined whether ships sailed well or not, but the complexities and compromises were such that it could only be determined by experience. Pepys had some understanding of this and, referring to men like those of the Shish family, lamented, '*Is there anything in the whole art of building or guiding of ships that was ever found out by much learning, but all the plainest and most unlearned builders and boatswains? And therefore why should anybody think of there being any great mystery therein, or anything required but trial and experience.*'[553]

Many ordinary shipwrights knew how to build ships using staffs and moulds, but it was the art of producing the rising and narrowing lines for draughts that was the master shipwright's jealously guarded secret. As Edmund Bushnell, author of *The Compleat Ship-Wright* of 1664, put it, '*Yet their knowledge they desire to keep to themselves, or at least among so small a number as they can; for although some have many servants* (apprentices), *and by indenture (I suppose) bound to teach them all alike the same art and mystery that he himself useth; yet it may be he may teach someone, and the rest must be kept ignorant, so that those shipwrights, although bred by such knowing men, yet they are able to teach their servants nothing, more than to hew, or dub, to fay a piece when it is moulded to his place assigned, or the like.*'[554] Also in 1664, Thomas Miller described how to draw a model for rigging and masting and gave helpful advice to keep it secret so that '*it will not be so easie for any one to steal away the use of your model*'. He recommended drawing the yards and rigging with black lead pencil and make little pricks in the paper at the ends of the yards. Then, after taking and writing down the measurements '*with crums of white bread and a clean linnen cloth I rub them out again ... that you may easier draw them again if you have occasion*'.[555] Indeed, after the secrets of the rising and narrowing lines, the positioning and sizing of the masts and yards had the most effect on a ship's sailing ability. In 1680, while the new *Tyger* was under construction, the eminent engineer Henry Sheeres also referred to the secret rules used by the master shipwrights to obtain their rising and narrowing of the breadth and floors and preserve their mysterious lines: '*It is not in the power of shipwrights to make a secret of those rules by which he builds a ship as regard every workman who has the gift of a measure by a rule may be master of the design in as much as the rising and narrowing of the breadth, floors, etc are all marked up on moulds and rods which lay up and down among the workmen and marked upon the timbers themselves which marks and measures are the results of those mysterious lines as they are called by which a ship is built.*'[556]

If the draught were drawn freehand, then scaling it up from the usual scale of $1/48$ would inevitably result in errors and discrepancies, no matter how regular the drawn curve appeared to be. Shish's digitally recorded risings and narrowings, which must have been obtained by calculation, eliminated the problems caused by scaling, drawing accuracy or any shrinkage or damage to the paper. As the anonymous author of a treatise dated c.1625 wrote, '*There are many good artificers that can draw a plot well and build a ship also, that if their work be compared with their plot you will find them very little to agree; and so many times good ships are spoiled in the building whose principal lines were well contrived in the plotting. The chiefest reason is want of skill in arithmetic and geometry to take all things truly off the plot.*'[557] This clearly means that some shipwrights built their ships by scaling off from their draughts, creating many errors. In late Tudor times, it was suggested that if a shipwright was '*unskilful in arithmetic and geometry*', a line should be fastened with a nail and stretched taught enough over the necessary length to produce the bow required, and the measurements taken off at frame positions.[558] Such practical but imprecise solutions had almost certainly ceased when the expensive warships of Charles II's Navy were built.

The necessary accuracy for producing smooth, even curves was obtained by calculation and trigonometry, and Shish's treatise gives convincing results for a real ship. In one respect this is far more advanced than Deane's *Doctrine*, where the curves are simple arcs. In practice, Deane probably used complex trigonometry, but used arcs for simple explanation and possibly to preserve his shipwright secrets. Shish, on the other hand, gives the results of his calculations but does not let us know how the calculations were obtained, thereby preserving his secrets in an entirely different way. The dimensions he gives are the same as those marked out on the staffs for use in the mould loft. A number of works from the period survive that describe some of the different formulae used to obtain the curves.

Methods of Calculating the Rising and Narrowing Lines

Anthony Deane, 1670

In late Tudor times, calculations and the use of curves for the rising of the floor were described in *Fragments of Ancient English Shipwrightry*, written for the most part by the shipwright Matthew Baker.[559] Some 80 years later, Anthony Deane included some good examples for calculating the curves in his *Doctrine*. All his rising and narrowing curves are true arcs, and it may be that he used them in his *Doctrine* only to simplify his explanations. He certainly fails to mention such complexities as the dead rising and changing radii of the lower breadth sweep, although they are shown in his illustrations. In actuality he must have used something more complex when designing real ships. The example he gives for the rising of the floor aft starts with a calculation to find its radius at the midship bend, one third of the way along the 120-foot-long keel. The radius is therefore calculated using the 80-foot distance to the aft end of the keel, at which point Deane required it to be 22 feet high. The method of calculation was discovered by the Greek mathematician Euclid, who worked in about 300 BC. As Deane describes it:

$$\frac{80^2/22 + 22}{2} = \text{Radius of rising line is 156 feet 5½ inches}$$

To simplify things for further calculations Deane rounded the radius down to 156 feet. It was then necessary to calculate the height from the keel to the rising line at every frame station. This he did by using Pythagoras' theorem – that is, the work of another ancient Greek.[560] These rising heights at every frame bend, or more probably at every frame, would be marked out full size on the staffs for use during the construction of the ship moulds and frames. Deane's manuscript was seen by others, including John Evelyn, who thought it an '*extraordinary jewel*',[561] and William Keltridge, a shipwright who copied the calculations into his *Method of Building Ships of War*, written in 1684.[562]

$$156 - \sqrt{(156^2 - 60^2)} = \text{Height of rising line at 60 feet is 12 feet}$$

$$156 - \sqrt{(156^2 - 40^2)} = \text{Height of rising line at 40 feet is 5 feet 2½ inches}$$

All Deane's rising and narrowing lines were calculated using the same method as shown in these examples.

After calculating the radius of the arc to be 156 feet for the rising line of the floor aft, Anthony Deane used Pythagoras' theorem to obtain its height at different frame positions. Author.

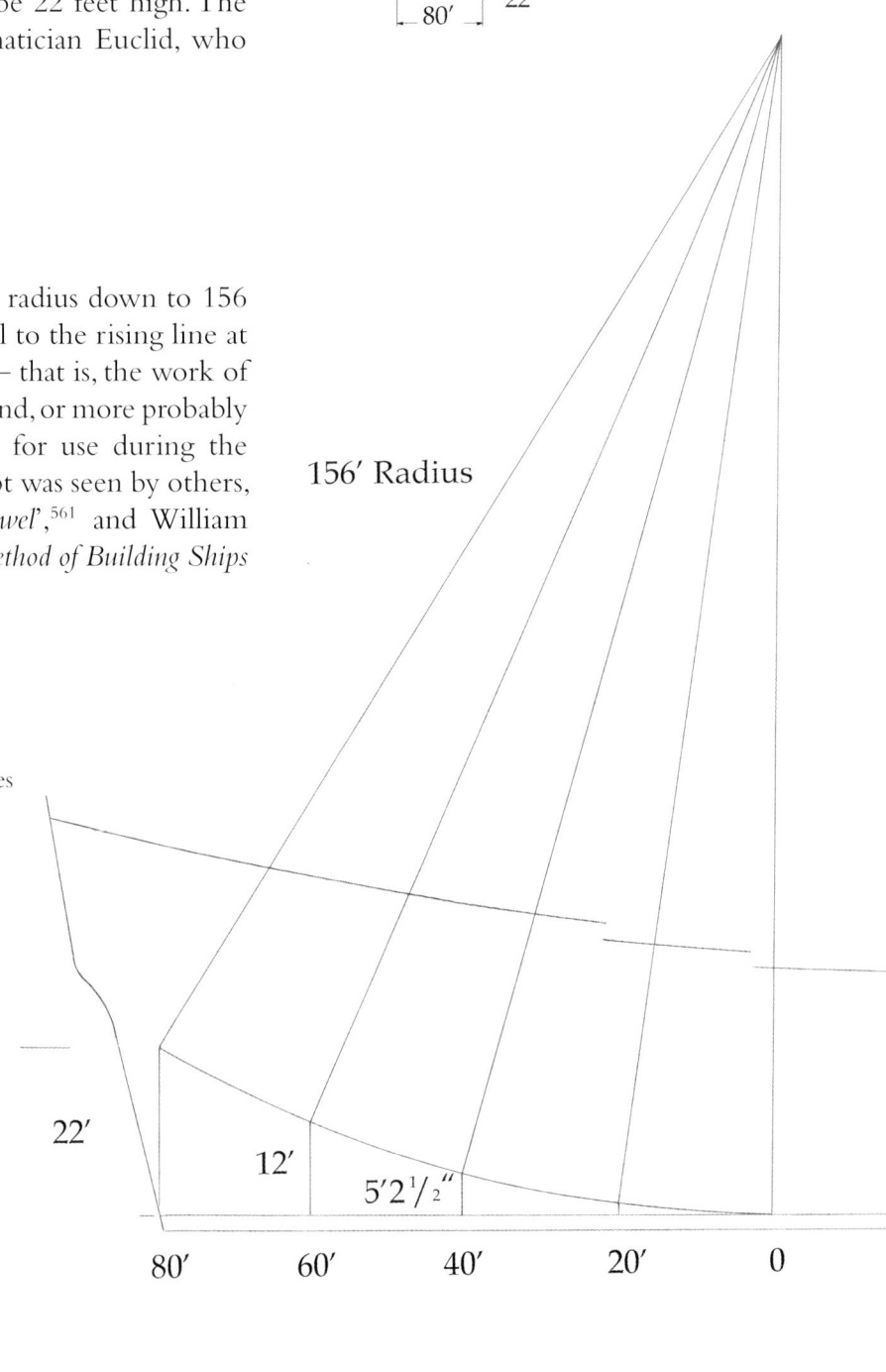

William Sutherland, 1711

Another method of creating the rising and narrowing lines is given by William Sutherland in his book *The Ship-builders Assistant*, published in 1711. He had close links with Deptford, for his uncle was William Bagwell, famous for having his compliant wife seduce Samuel Pepys for her husband's advancement. Sutherland also describes the method used by Old Mr Shish (Jonas) and his sons for calculating the rake of the sternpost.[563]

The method described by Sutherland[564] produces curves that bend more quickly towards the ends, which may have been more useful than Deane's arcs for the rising line of the floor forward. It seems less useful aft, and a draught, illustrated elsewhere in Sutherland's book,[565] can be seen to have been created by another method.

Sutherland acknowledges that his figure 'A' on p.58 of his book would only be suitable for round-, or pink-sterned ships as the narrowing of the breadth aft meets on the centreline, whereas warships have a wide stern.[566] There is no need in Sutherland's method to calculate the radius of an arc as it is the same as the maximum distance covered by of the rising or narrowing lines. Sutherland does not go into the mathematics of finding the heights of the rising line at the frame stations. Unlike Deane's method, which involved measurements of hundreds of feet, Sutherland's geometry was small enough that it could be drawn directly onto the floor of the mould loft. The discerning shipwright who required complete accuracy could always use Pythagoras' theorem as described by Deane. It seems probable that Sutherland, in the same manner as Deane, describes a simplified method that would be more complex when used in practice.

Among some of his unlikely suggestions, Sutherland advocates the use of circular hull sections based on a surface that lay somewhere between a cone and a cylinder. This he calls a '*conide*', but acknowledges that perfect circular bodies have been universally condemned.[567]

Opposite.
Diagram of Euclid's work used by Deane to calculate the radius of the rising floor to fit the 80-foot distance between the midship bend and the aft end of the keel, at which point it is to be 22 feet high. The 290 feet 11 inches and 22 feet dimensions are added together and divided by two to find the radius. This arc is tangential to the arc of the rising line forward. Diagram by Mr Richard Wright.

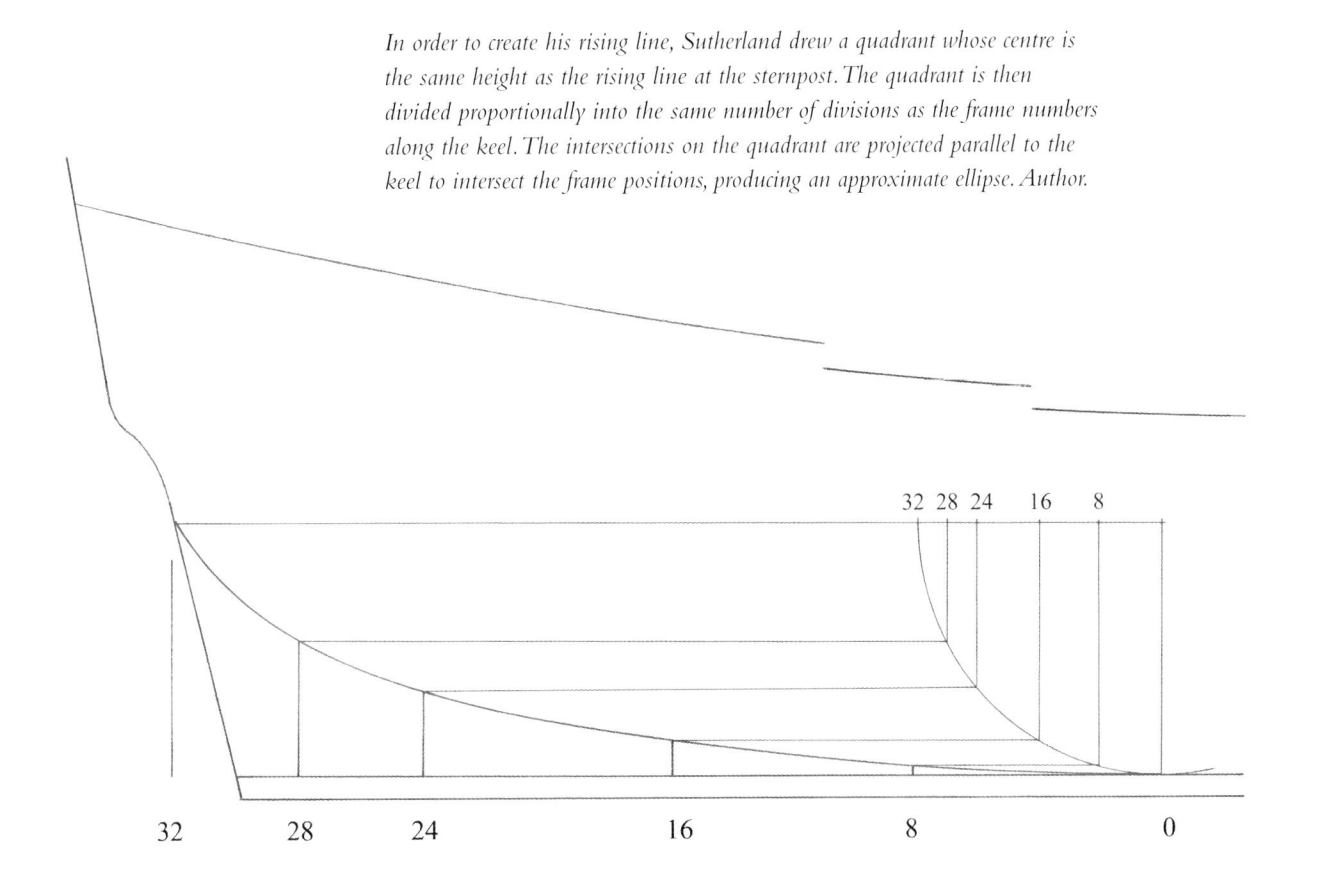

In order to create his rising line, Sutherland drew a quadrant whose centre is the same height as the rising line at the sternpost. The quadrant is then divided proportionally into the same number of divisions as the frame numbers along the keel. The intersections on the quadrant are projected parallel to the keel to intersect the frame positions, producing an approximate ellipse. Author.

THE MASTER SHIPWRIGHT'S SECRETS

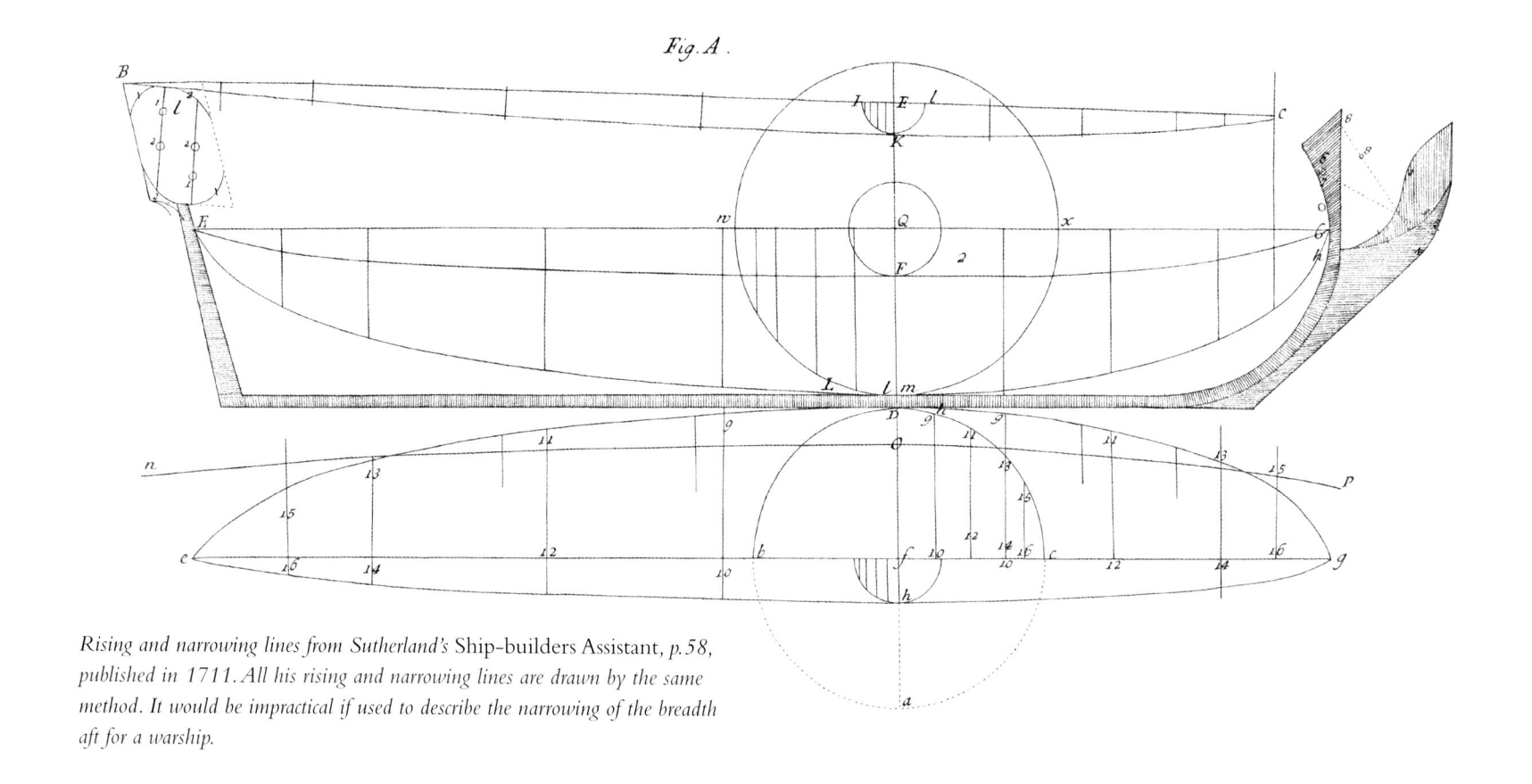

Rising and narrowing lines from Sutherland's Ship-builders Assistant, *p.58, published in 1711. All his rising and narrowing lines are drawn by the same method. It would be impractical if used to describe the narrowing of the breadth aft for a warship.*

Anonymous Treatise c.1625

Although written earlier in the 17th century, an anonymous treatise uses another way of calculating different rising and narrowing lines. It is more sophisticated in that the method can be altered in a number of ways to suit the required curve. Although it is early in date, the author uses decimals rather than fractions, which were used by Deane. The care taken by the author in obtaining precisely the right curves may indicate that his work relates to actual ship practice. The treatise also contains a table of results necessary for making moulds or staffs similar to those contained in John Shish's treatise.

Rising Line of the Floor Aft

Rather using than a formula based on geometry, the anonymous author uses a factor to obtain the curve. The factor is obtained using the number of the aft frame and its known height. The aft frame number to the power of 3 is taken and the result divided by the height in inches.[568]

To find the factor for the rising line of the floor aft:

$$\frac{\text{Frame } 28.33^3}{\text{Height } 124"} = \text{Factor of } 183.37$$

To find the height of the rising line aft at frame 24:

$$\frac{\text{Frame } 24^3}{\text{Factor } 183.37} = 75.39 \text{ inches}$$

And for Frame 20:

$$\frac{\text{Frame } 20^3}{\text{Factor } 183.37} = 43.63 \text{ inches}$$

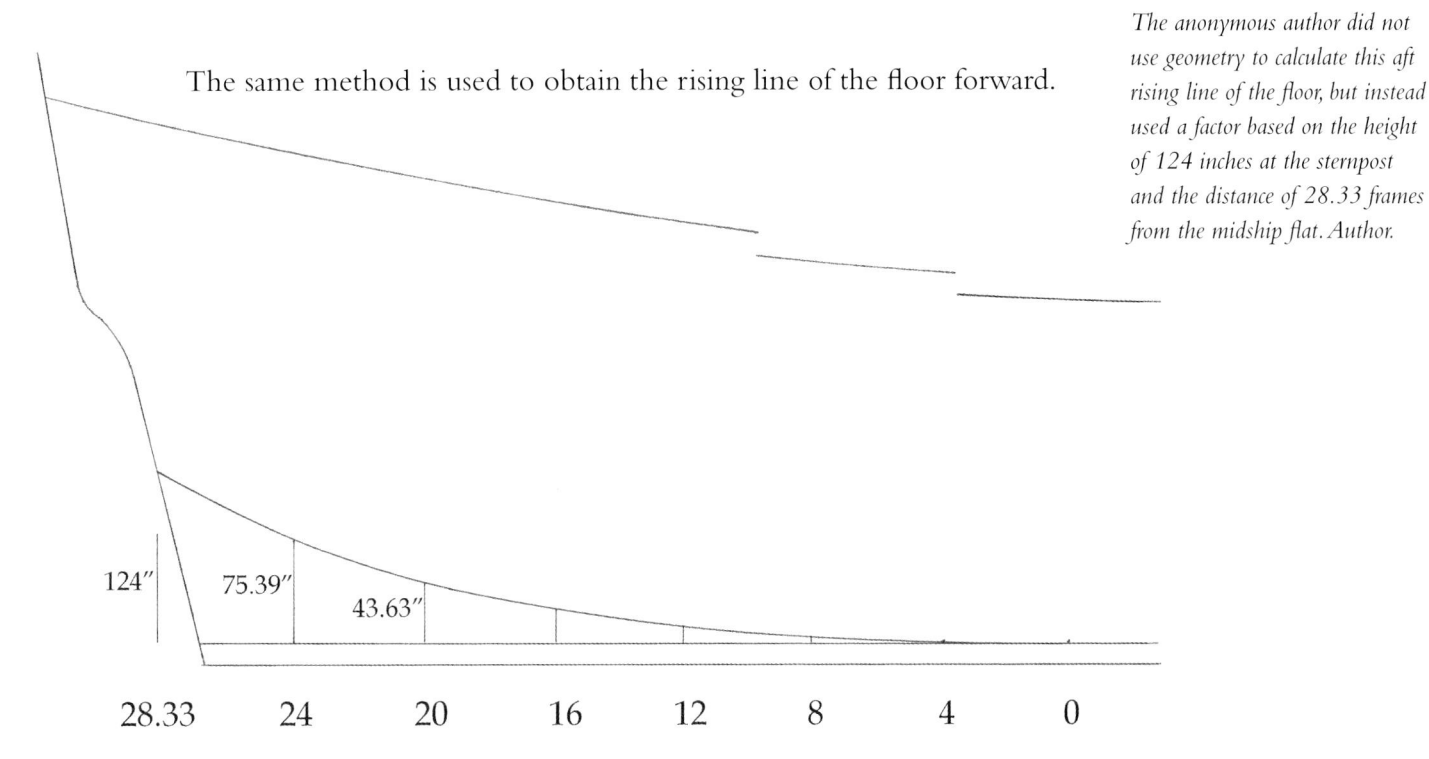

The same method is used to obtain the rising line of the floor forward.

The anonymous author did not use geometry to calculate this aft rising line of the floor, but instead used a factor based on the height of 124 inches at the sternpost and the distance of 28.33 frames from the midship flat. Author.

124"   75.39"   43.63"

28.33   24   20   16   12   8   4   0

Rising line of the breadth

As there is little movement along the length of the rising line of the breadth, the author uses a straight line in midships leading into true arcs at each end. In calculating the radius and risings of the arcs, he uses the same method, based on Euclid and Pythagoras, as Deane.[569]

Narrowing line of the breadth aft

The factor method used for the rising of the floor aft is also used for the narrowing of the breadth aft.[570]

Narrowing line of the floor aft

The curve for the narrowing of the floor aft required particular attention. The author rejected the factor method, probably because it produced too great a movement towards the end. Instead, the anonymous author used a modified stretched-circle method described by Sutherland but with a larger radius, and made use of only the flatter part of the curve, which he considered most useful. The calculations were made using Pythagoras' theorem.[571]

THE MASTER SHIPWRIGHT'S SECRETS

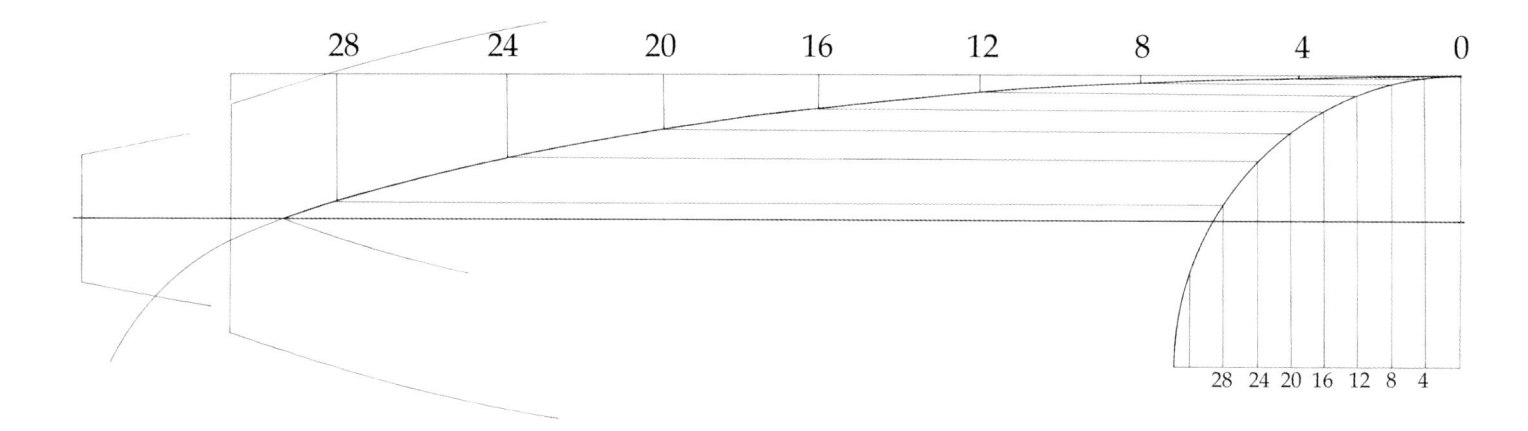

A similar geometric method to that described by Sutherland was used to obtain the narrowing line of the floor aft, but modified so that only part of the curve was used. Author.

Narrowing line of the breadth forward

The ever-inventive anonymous author found yet another solution to produce the curve for the narrowing line of the breadth forward. He used the scalar method similar to that for calculating the rising line of the floor aft, except he used the power of 4 rather than the power of 3. As a further adjustment, he added another foot to the 18-foot distance of the curve.[572]

To find the factor for the narrowing line of the breadth forward:

$$\frac{\text{Frame } 20^4}{\text{Narrowing 228 inches}} = \text{Factor of } 701.75$$

To find the narrowing line of the breadth forward at frame 17:

$$\frac{\text{Frame } 17^4}{\text{Factor } 701.75} = 119 \text{ inches}$$

And for frame 14:

$$\frac{\text{Frame } 14^4}{\text{Factor } 701.75} = 54¾ \text{ inches}$$

The narrowing line of the breadth forward was obtained by another factor method but used a false narrowing of 19 feet, or 228 inches, rather than the true 18 feet. Author.

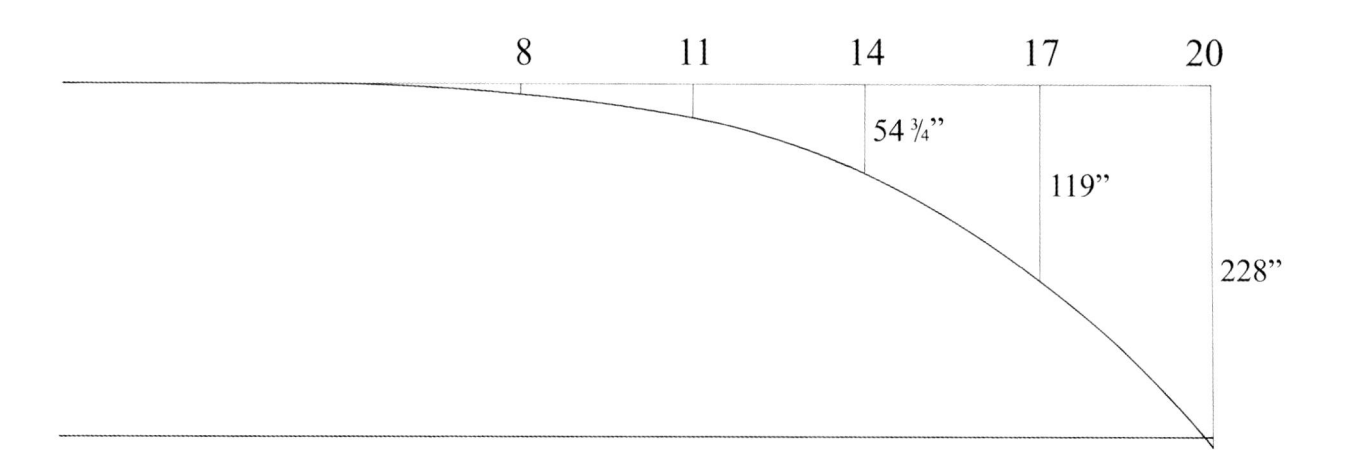

114

Narrowing line of the floor forward

The factor method is again used, but in a more complex manner:[573]

To find the factor for the narrowing of the floor forward:

$$\frac{(3 \times \text{Frame 17})^2}{\text{Narrowing 110 inches}} = \text{Factor of 23.645}$$

Edmund Bushnell, 1664

Edmund Bushnell's book *The Compleat Ship-Wright* went through a number of editions and was copied into another book entitled *Marine Architecture* as late as 1748. He used the same simplistic method of arcs for the rising and narrowing lines as described by Deane but in one case made an interesting modification. He describes 'how to hang a rising line by several sweeps to make it rounder aftward than at the beginning'.[574]

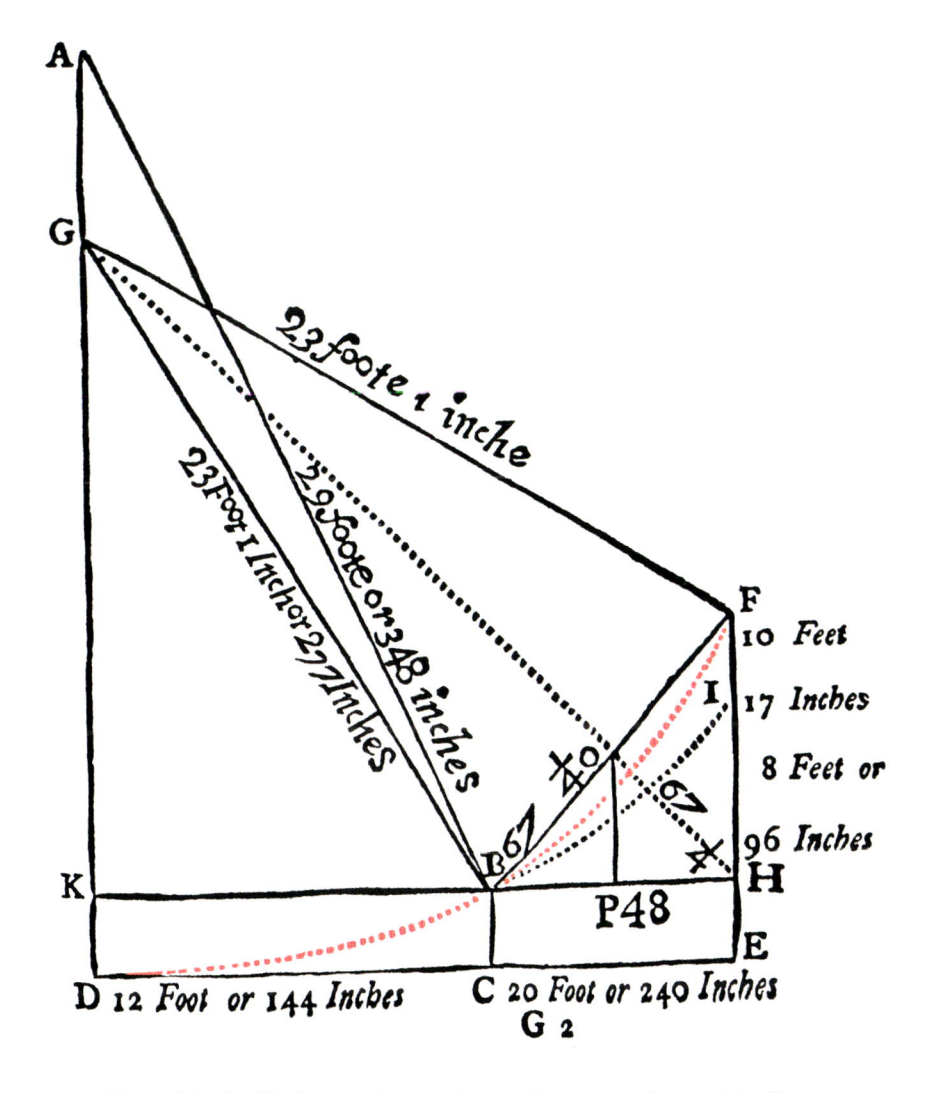

Edmund Bushnell's diagram showing the use of two arcs to form a rising line of the floor, highlighted in red. The arc from D to B has a radius of 348 inches, and from B to F, 277 inches. The arcs are not tangential but give the shipwright a simple and practical opportunity to create the curve of his choice. Private collection, via author.

John Shish's Calculations for His Account of the Dimensions of a Ship

With a broad understanding of the different mathematical ways master shipwrights obtained their rising and narrowing lines, it was attempted to work back from John Shish's co-ordinates given in his treatise to find the methods and formulae he used to obtain them.

Three years after he wrote his treatise, an interesting insight into how Shish worked his calculations was given when he wrote to the Navy Board giving the burden of a small ship hired for service in Virginia.[575] This was the *Adam and Eve,* and he estimated her as being $189^{35}/_{47}$ tons.[576] Burden was an approximate calculation of capacity and was about 25% less than the displacement of a fully loaded ship.[577] The formula used was:

$$\frac{\text{Keel length x Breadth x ½ Breadth}}{94}$$

The letter is written by a clerk, but at the bottom in a different hand are the calculations used to obtain the burden. They reveal how the awkward problem of calculating in feet and inches was resolved and that the mathematician preferred to work in feet and inches rather than converting all into inches. A small error was made by leaving off the 6 inches when adding up the figures to obtain the result of multiplying the keel by the breadth. The small error may also have been the result of sensible rounding up, as the true burden of the ship should be 189.772 tons, as against Shish's calculation of 189.745 tons.

Length of keel multiplied by breadth	Length of keel 66 feet Breadth 23 feet 3 inches
Multiply by 3'	
Multiply by 20'	
Multiply by 3" or 1/4 foot	
	Result of multiplying keel by breadth Half breadth 11 feet 7 1/2 inches
Multiply by 1/2 breadth	
Multiply by 1'	
Multiply by 10'	
Multiply by 6" or 1/2 foot	
Multiply by 1 1/2" or 1/8 foot	
	Result of multiplying the keel by breadth by 1/2 breadth

Part of the calculations made by John Shish to determine the burden of the hired ship Adam and Eve. *It shows his method of multiplying feet and inches with each other. A good example of how he worked with an awkward dimension such as 11 feet 7½ inches is given in the second part of his calculation. He breaks it down into 10 feet, 1 foot, 6 inches (½ foot) and 1½ inches (⅛ foot).*

John Shish himself probably calculated the burden of the *Adam and Eve* and the risings and narrowings for his treatise. We know master shipwrights made these calculations, for Anthony Deane performed them for his *Doctrine* and Phineas Pett was reckoned the most proficient shipwright in mathematics. The writing of the numbers in both Shish's treatise and the *Adam and Eve* letter are similar, which suggests they may be Shish's work. The only other shipwright who may have been privy when Shish was at Sheerness in 1674 and at Deptford in 1677 and capable of such calculations was William Stigant.[578] He was the first shipwright listed after

JOHN SHISH'S ACCOUNT OF THE DIMENSIONS OF A SHIP

Shish on the Sheerness paybook, and by the time he was at Deptford, he had risen to become the master boatbuilder.[579] On 10 January 1682, he was appointed assistant master shipwright at Portsmouth,[580] and he went on to become a master shipwright. He may possibly have been the mathematician responsible for the calculations, but even if he was, he would have been following Shish's method and formulae under his direction.

Trying to discover the formulae used by Shish to produce his lines was probably an impossible task in the 17th century. It is still very difficult today, but computer software used in modern aircraft production was a great help in testing possible solutions against Shish's figures. Patiently running test formulae in a loop to generate point data for comparison against Shish's figures eventually produced very reasonable results.

Comparing the very accurate computer-generated results which calculate down to the sixth decimal place with Shish's figures reveals that he generally worked to within half an inch accuracy for the aft end of his ship and one inch accuracy forward. His rounding up and occasional small errors mean the results from a reconstructed formula can never be a perfect match with the figures given by Shish. An oddity is the midship flat: although placed at the usual two-thirds the length of the keel from its aft end and marked with an X, it is not at the lowest height of the rising line of the floor, but is four frames further aft.

Rising line of the floor aft

The curve for the rising line aft was taken from its lowest point, called the dead rising, which, unusually, falls at frame flat 4 rather than at frame X. The effect of this is to move the section of hull with the greatest area further aft by three frames, possibly in an attempt to lessen water resistance. After testing dozens of possible formulae, a very close match was found using the approximate ellipse method explained in an anonymous treatise of *c*.1625.

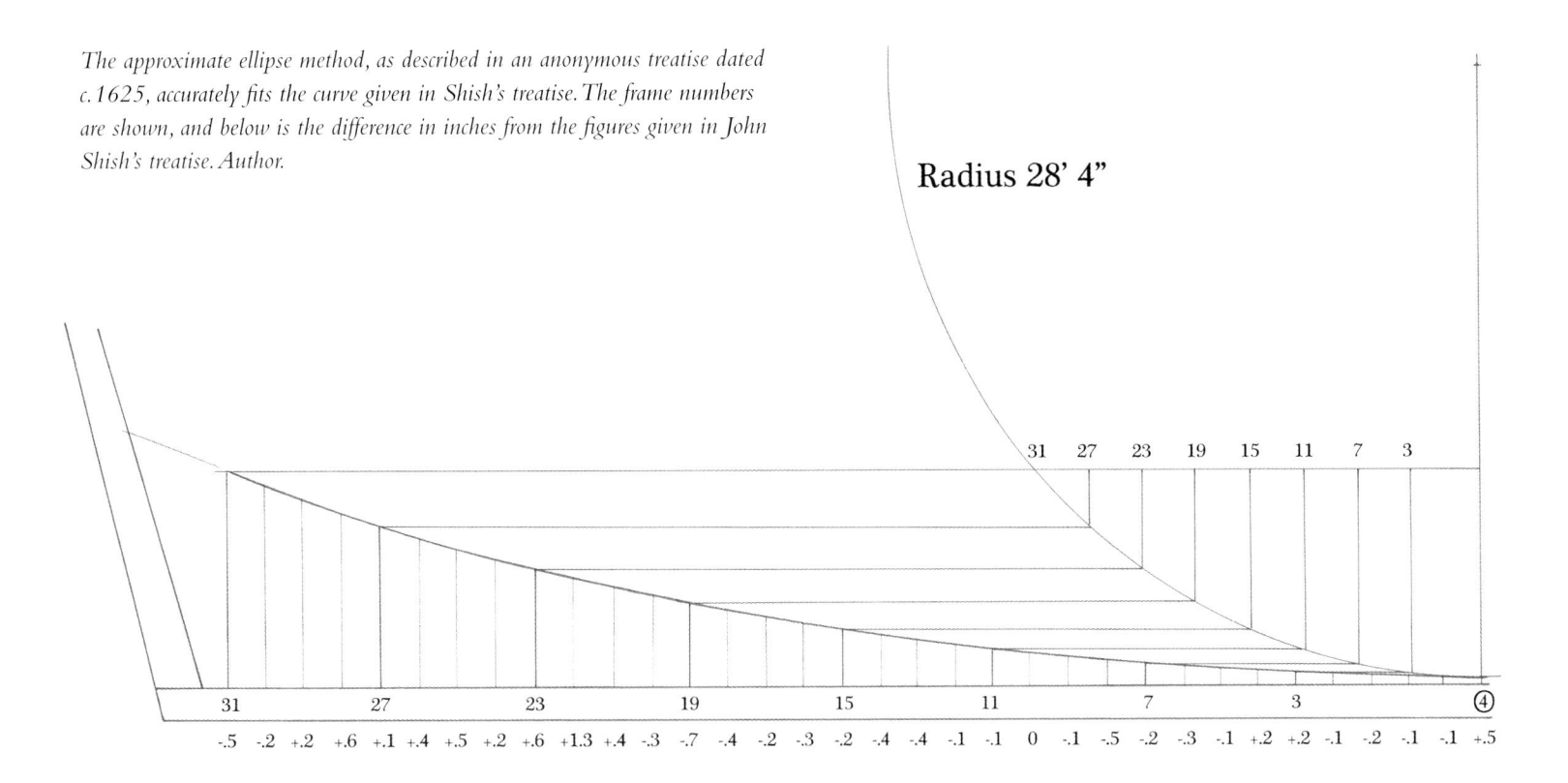

The approximate ellipse method, as described in an anonymous treatise dated c.1625, accurately fits the curve given in Shish's treatise. The frame numbers are shown, and below is the difference in inches from the figures given in John Shish's treatise. Author.

Radius 28' 4"

117

Rising line of the floor forward

A quick look at the progression of the rising line over nine frames, from frame flat 4 to frame A, shows an even rise of 1 inch per frame. This is of course a straight line, except for a peculiar inexplicable deviation of ½ inch at flat frame A. The rest of the rising appears to be made up of two separate arcs which meet at frame S. This solution is made more probable by the fact that the intersection point chosen by Shish is exactly 6 feet high. The method closely follows that described by Edmund Bushnell in his book *The Compleat Ship-Wright* – except, that is, for the straight line, which is surprising to find and difficult to understand, for if Shish had continued the large radius arc it would have conveniently met near the rising line of the floor aft. Perhaps he considered the difference between the arc and a straight line for the first nine frames to be so small that he pragmatically used a simple solution.

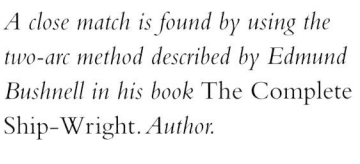

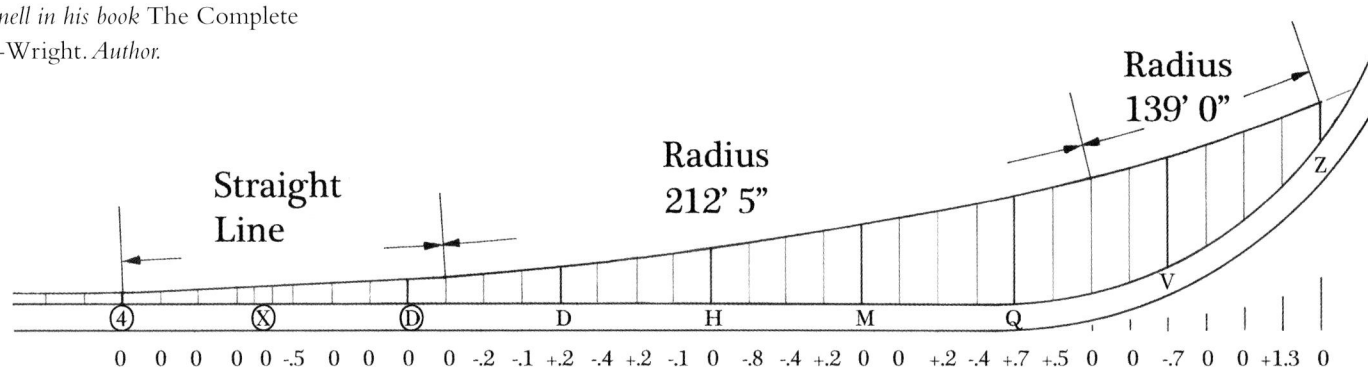

A close match is found by using the two-arc method described by Edmund Bushnell in his book The Complete Ship-Wright. *Author.*

Narrowing line of the floor aft

The figures given in the treatise are very uneven and a two-arc solution does not accurately fit Shish's curve. It is almost a straight line from frame X to frame 19, a distance of 550 inches, over which it bows less than 2 inches. The best result was obtained using a straight line from frame X to frame 14 followed by an arc.

The curve deviates very little over a long distance, and as Shish only worked to the nearest ½ inch, any one of a number of the known methods used by shipwrights could have produced results close to Shish's curve. None of them are totally convincing. Author.

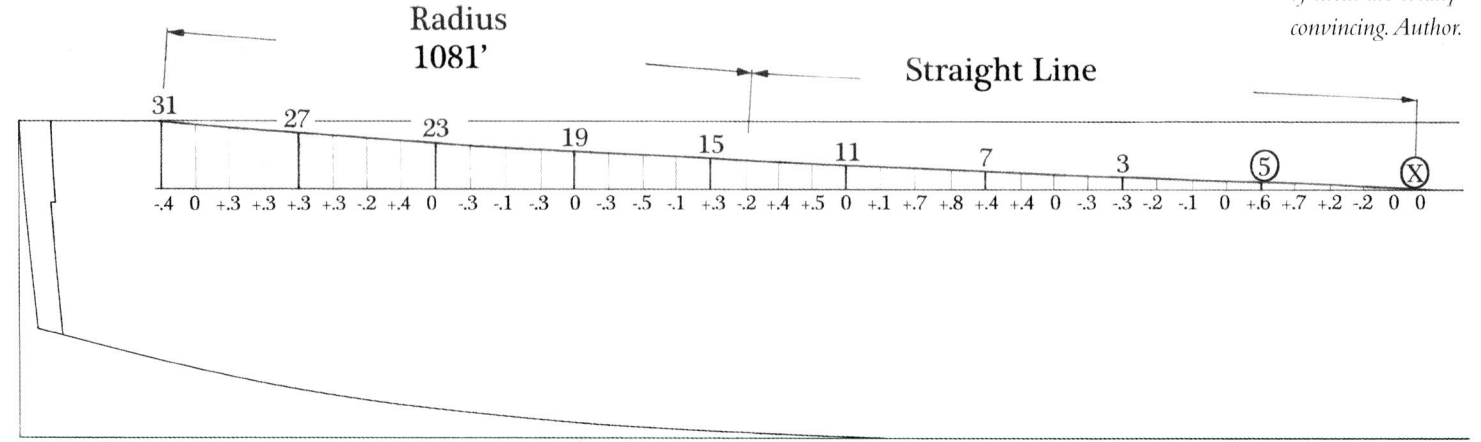

Narrowing line of the floor forward

From a parallel floor narrowing of 3 feet 6 inches, Shish starts his curve at frame E and reduces it by one inch per frame over the next several frames. This straight line then curves, accurately following the approximate ellipse method, turned at the same angle as the straight line. Possibly Shish calculated the approximate ellipse method in the normal view then accumulatively added one inch per frame to produce the curve. A small deviation is evident in that Shish gives the half breadth of the floor forward as 3 feet 6 inches but as 3 feet 5½ inches abaft.

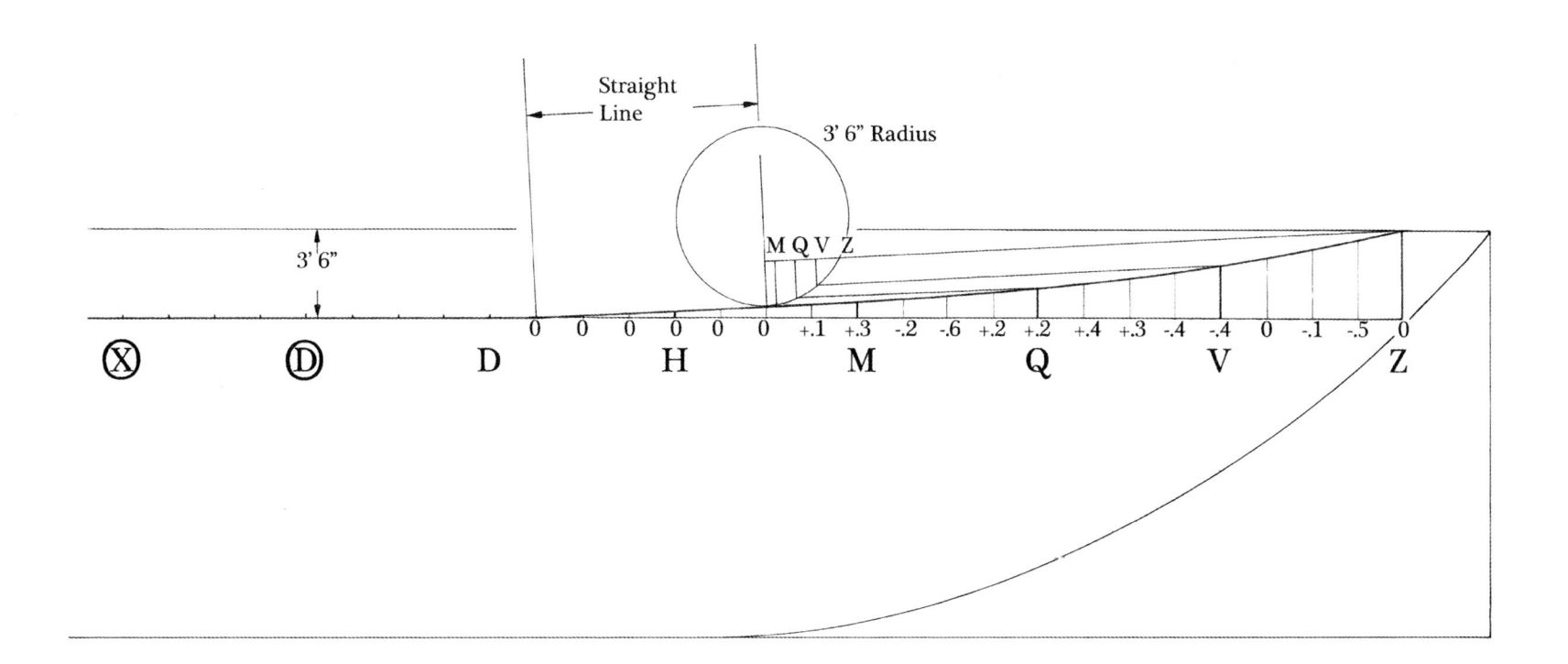

A version of the approximate ellipse method credibly produces a curve very close to that in Shish's treatise. Author.

THE MASTER SHIPWRIGHT'S SECRETS

Narrowing line of the breadth aft

The sides of the ship retain their full breadth parallel to the keel from flat frame 6 aft to I forward, a total of some 19 frames – apart from, that is, a peculiar anomaly of ½ inch at flat frame 5. The narrowing then follows a straight line and two arcs, with results very close to Shish's figures. The intersection point of the arcs at frame 19 is exactly one foot. The last frame is at station 31, where the narrowing is 4 feet 1 inch. Beyond that is the assembly of the stern frame, of which the widest part is the wing transom. Shish gives it as being 22 feet wide; taking this from the breadth of the ship, and allowing for the 3-inch plank, leaves a narrowing of 5 feet 3 inches.

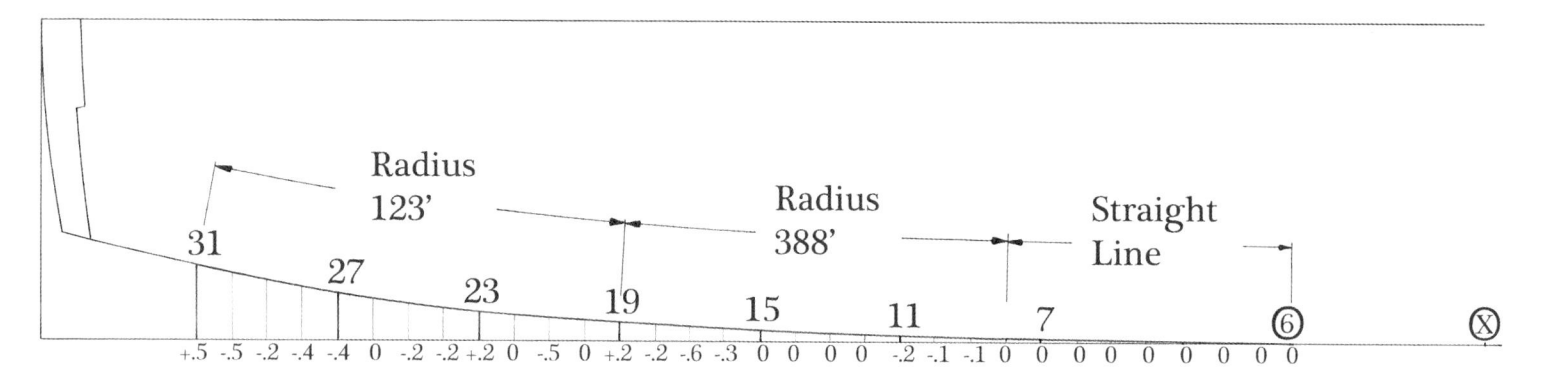

A straight line and two circles closely follow the figures given in Shish's treatise. Author.

Narrowing line of the breadth forward

The factor method as described in an anonymous treatise of *c.*1625 was used to calculate the narrowing of the breadth forward. The alternative use of two arcs produced similar results and highlighted the same irregularities or unevenness in Shish's figures, indicating that the original calculations contained some errors or severe rounding up. It is noticeable that Shish calculated only to the nearest inch. The factor was obtained using a power of 2 rather than a power of 3.

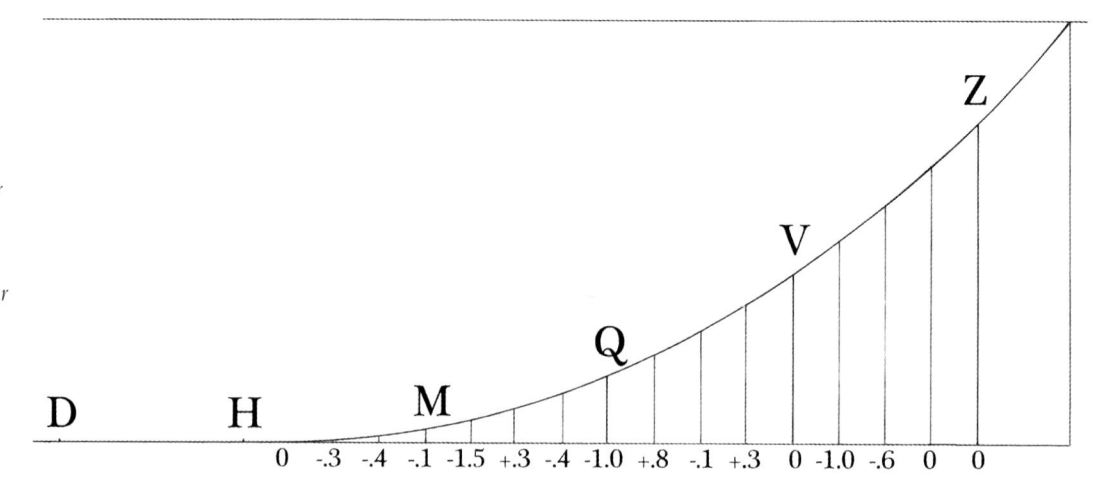

The factor method used in the c.1625 treatise for the narrowing of the breadth forward can be applied to produce similar results given in Shish's treatise. Author.

It is not possible to say with absolute certainty that John Shish used the methods of calculation described here, but because they fairly accurately reproduce his results, they can be used to recreate ships he designed. It can also be said with some confidence that even if Shish didn't use precisely these methods in his draught for a new *Tyger*, he must have used something very similar. Indeed, it is also reasonable to conclude that Shish did not make the numerous calculations necessary for every ship he built or that would have been needed to produce his treatise, but instead used work done earlier for another ship and scaled the figures to fit his work. He also appears to have been pragmatic and knew when it was acceptable to use a straight line to make things fit. This pragmatism may explain some of the straight lines in his treatise. Pepys certainly didn't think the master shipwrights spent much time making endless calculations, for he noted '*How little mathematic is required or to be found among our best shipwrights*'[581] – not that too much notice should be taken of his opinion regarding them. Knowing the boundaries and tolerances to work in is the mark of a good engineer and shipwright.

Another field of calculation that some shipwrights may have indulged in was calculating the volume of water displaced by the hull. They knew full well that the weight of the displaced water was equal to the weight of the ship. Deane describes in his *Doctrine*[582] how he made an approximation of the hull volume, which meant little if he did not also calculate the weight of the ship fully loaded. This exercise was demonstrated by William Keltridge in *His Book* of 1675, where he worked out the weight of water displaced by the hull of the *Royal James* when she drew 20 feet 6 inches abaft and 19 feet afore at just over 2,423 tons, the same as the weight of the hull, guns, ballast, masts, men and all her stores.[583] Her burden, the figure usually obtained by the crude shipwright's formulae, was 1,422 tons.

The description and calculations used to describe a similar sized ship to that in Shish's account. The Shish family are mentioned in the print. William Sutherland, The Ship-Builders Assistant, *1711, p76.*

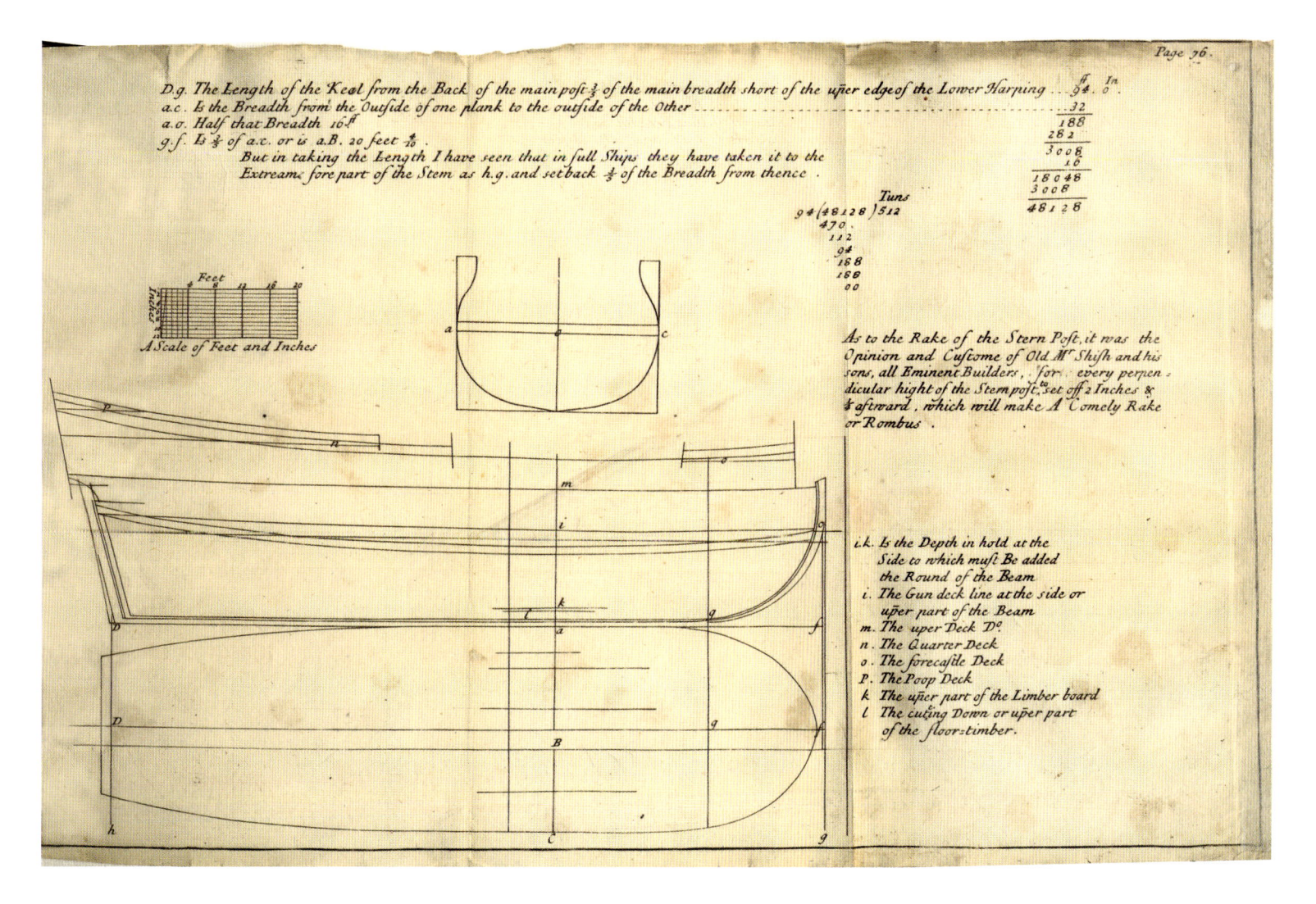

CHAPTER 6

THE DRAUGHT OF THE NEW TYGER

There are hardly any authoritative draughts surviving from the 17th century, and there is certainly not one for the new *Tyger*. To help understand some of the popular design features of the time, two other ships closely associated with her were examined in detail. One is the Trinity House model of *St Albans*[584] of 1687, a fourth rate designed by John Shish shortly after he finished the *Tyger*. The other is the *Mordaunt* of 1681, a ship built at the same time as the *Tyger* by the respected merchant builder Captain William Castle in his private yard at Deptford. As well as being her contemporaries, these two ships are relevant to the *Tyger* as they come from Deptford and are both nearly identical in size, being within 20 tons or so of her. In any case, their stories are worth the telling.

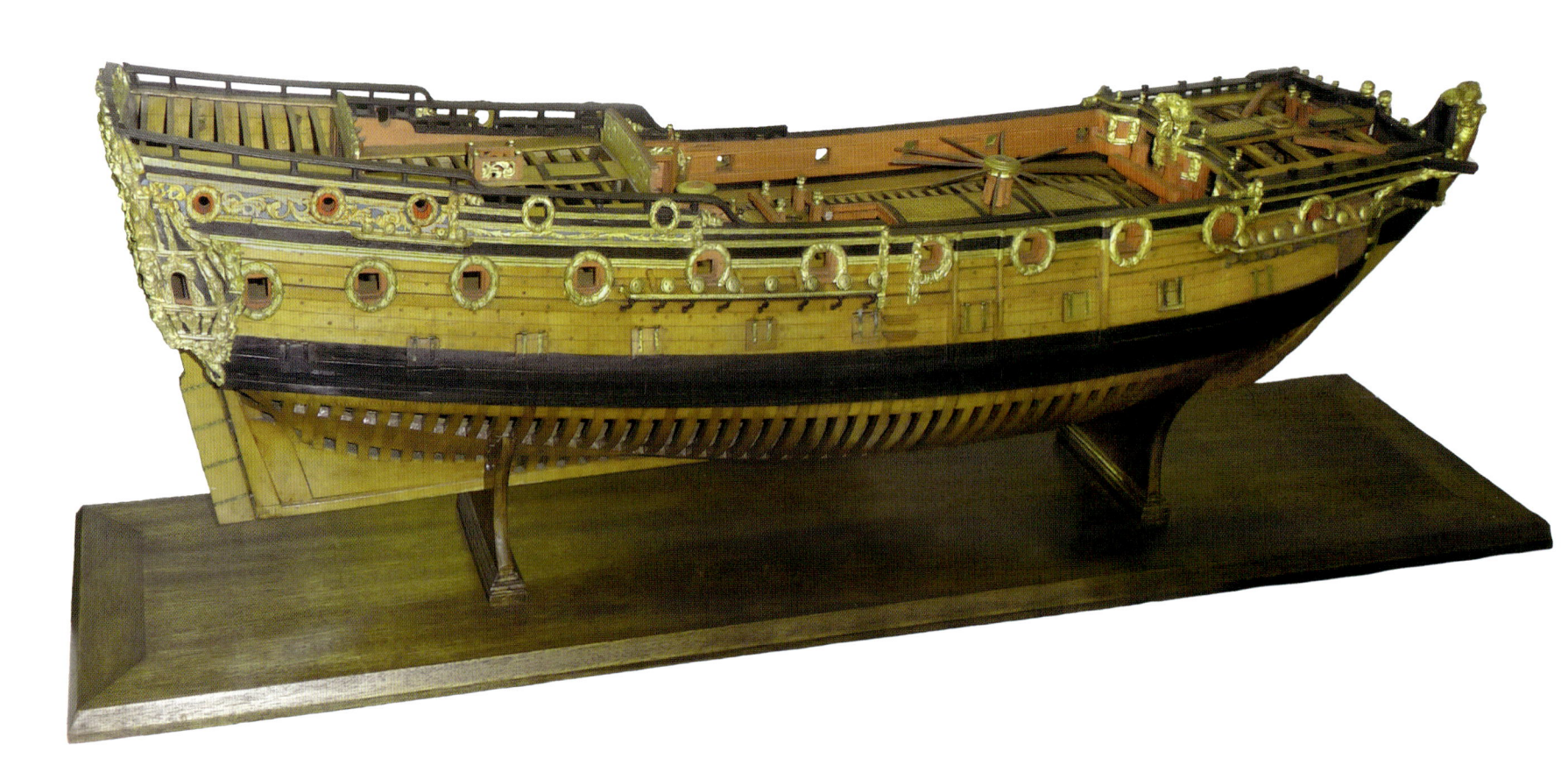

The Purchase of the *Mordaunt*

The *Mordaunt* was not built for the Navy but was a private venture for a syndicate led by Lord Mordaunt. The Spanish ambassador complained about the ship, supposing she was to be part of a fleet led by the Elector of Brandenburg to collect by force a disputed debt from Spain. Lord Mordaunt was examined by the High Court of the Admiralty on 30 June 1681, at which the Admiralty Marshall described how formidable the ship was, saying she was as good as the new *Tyger*, which had been launched the week before. In his defence, Lord Mordaunt said that even though his ship was powerful enough to sail independently of a convoy, he was willing to fix her armament and crew at a figure agreeable to the court.[585] Apparently unconvinced, the court issued a warrant for the arrest of the ship and ruled that Lord Mordaunt had to give bail for the ship's good behaviour.[586] After his ship had lain idle for months in the Thames, Lord Mordaunt gave up the venture and began negotiations for her sale to the King. On 9 February 1682, the Navy Board was ordered to estimate the value of the hull, the masts, the yards and the furniture,[587] but its officers lacked any real interest and nothing transpired. Seven months later, on 19 September, King Charles II[588] ordered another survey to assess her dimensions and worth as the ship would have deteriorated during the delay and it was feared that some of her stores may have disappeared.[589] The survey was carried out by two of the elder brethren of Trinity House and two members of the Company of Shipwrights – John Shish and the master attendant of the Deptford yard.[590] Their survey was delivered to the Admiralty, which resolved to acquaint the King and know his pleasure in the matter.[591] Early in October it was agreed to buy the ship according to the survey estimate of £5,842 8s 6d,[592] but again things dragged on, and in February 1683 the authorities who had previously surveyed the ship were ordered to do so again.[593] They found that all the stores were according to their original inventory and that the ship had been caulked, tarred and well kept.[594] At the same time, and in anticipation of a successful conclusion, the Navy appointed the ship's complement of five warrant officers.[595] However, unfortunately for Lord Mordaunt, the Admiralty became worried about outstanding wage claims by the seamen he had employed on the ship. This caused yet further delay,[596] and it was not until 1 October 1683 that the ship was finally bought into the Navy.[597] The survey of September 1682 contains a great deal of information, giving such details such as the lengths of all the decks, the rakes fore and aft, the height between decks, the lengths and diameters of the masts and yards, and the size and height of the gunports from the decks. The full survey is given in Appendix 2.

Reconstructing the *Mordaunt*

In addition to the survey, a contemporary quarter-of-an-inch-to-the-foot ($^1/_{48}$) Navy Board model of the ship exists in the stores of the National Maritime Museum.[598] It is one of the best and most beautiful in the collection. The first published identification of the model was made by R. C. Anderson in 1912, when it was in the museum of the training ship *Mercury*. Anderson identified it by the arms of Lord Mordaunt carved onto the roundhouse bulkhead, by its likeness to the van de Velde drawings, and by the dimensions, which closely match those recorded in Pepys's *Register of Ships*. It is difficult to be sure whether the model was made before or after she was bought into the Navy as the uncrowned lion figurehead and Mordaunt's arms suggest before, but the Admiralty badge and King's arms at the stern suggest after.

As a safety feature, all known two-deck ships built for the Navy had two pairs of bitts fitted on the gundeck for securing the anchor cables. The model has only one pair, and for a Navy ship, these are unusually placed on the upper deck, just aft the forecastle bulkhead in the manner normally found on single-deck ships.[599] Associated with this, the hawse holes are placed higher than usual, with long, sloping troughs leading the cables to the bitts on the upper deck. The main capstan is also on the upper deck rather than the gundeck in order to pull the cables to the main hatch. Being led upward and along the upper deck, the cables would have been drier than if they went along gundeck before being coiled in the hold. The arrangement would also have the benefit of clearing space on the gundeck for stores, as would be found on a merchant ship. Another oddity is the absence of the fore jeer bitts normally located just abaft the foremast on the forecastle and used for raising the fore yard. Most probably the *Mordaunt*'s jeers ran through the deck to blocks beneath in the forecastle and then aft to the jeer capstan. There are scuttles through the forecastle bulkhead and a single bitt inside the forecastle on the larboard side[600] which may be associated with this. In addition to the survey and the model of

Opposite. *The contemporary model of the* Mordaunt, *built by Captain William Castle in his private yard at Deptford, has several features very similar to the* Tyger. *Both ships were built in 1681. Author, by kind permission of the National Maritime Museum.*

The Mordaunt *viewed from the starboard quarter, showing strong similarity to the model (Robinson 597). Portrait of the 'Mordaunt', Fourth Rate, built 1681, 46 guns. Willem van de Velde, the Elder, c.1683. PAI7281; Photo: © National Maritime Museum, Greenwich, London.*

the *Mordaunt*, there are at least four van de Velde drawings that are invaluable in understanding the ship's layout. A reconstruction of her draught was made using the dimensions from the Navy Board survey and the lines from the model. The method used to record the lines was inspired by the printed squared paper method of Edmund Dummer, used to take the end views (later called the body plan) of the new ships of 1677.[601] A modern version was devised to record them without actually touching the valuable model. The end views were recorded at intervals of four feet (one inch on the model) using a simple straight-line laser from a DIY store and a homemade stand and base.[602] The laser line and an enclosing border, marked with a one-foot grid, were fixed in the same plane. They were orientated so as to be perpendicular to the longitudinal axis of the model and able to move in the fore and aft direction. Once digitally photographed, the laser line from the images could be plotted by hand to produce a plan, although in practice it was much simpler and more accurate to remove perspective distortion using modern computer software. The end views of the model created with the laser images were used as recorded, with slight adjustments for the differences in dimensions between the model and the survey and to allow for the missing planking below the main wales. The waterlines of the hull were also developed, at intervals of 2 feet 6 inches in height.

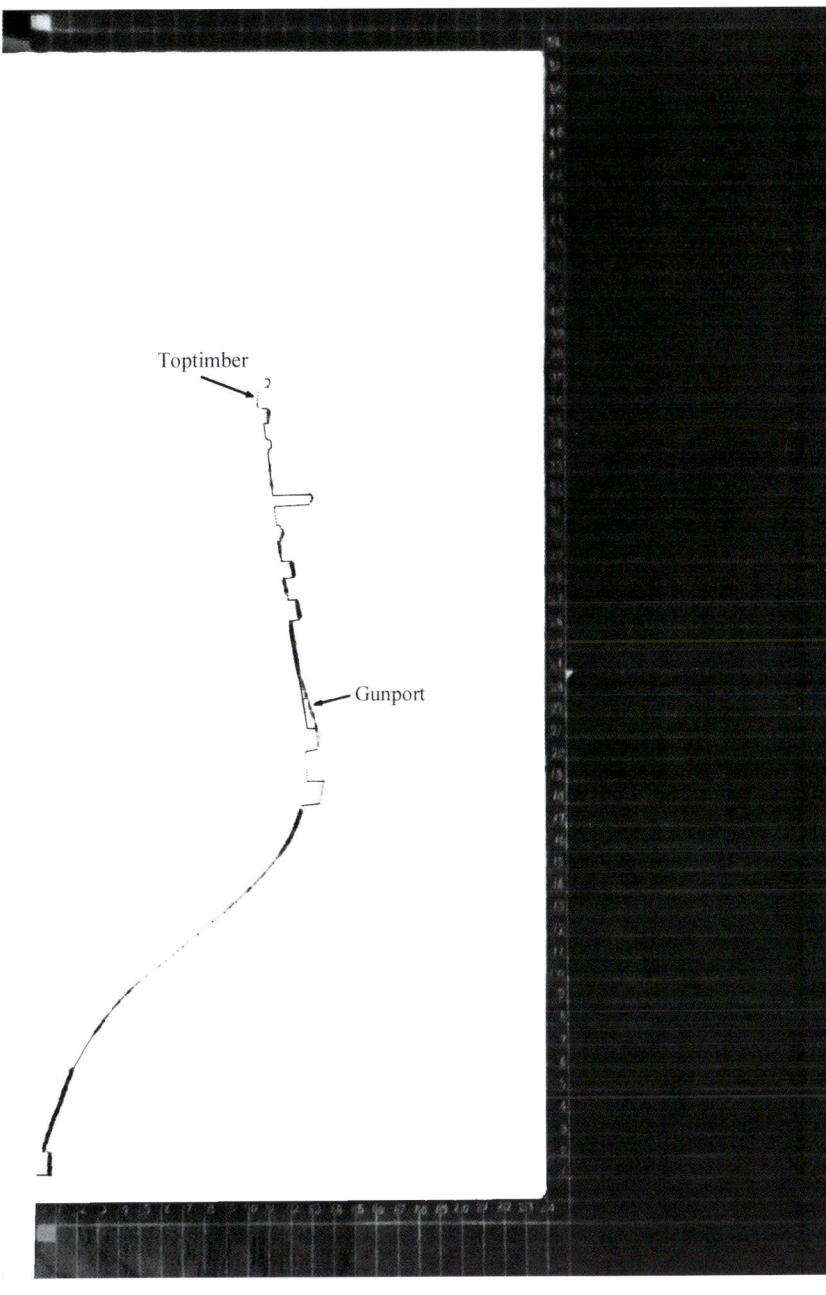

Generally, the model agrees fairly well with the dimensions taken of her by the experts who surveyed the ship in 1682. Their work must be regarded as more accurate than the model and it was used as the primary source above all others. There are a few discrepancies. The gundeck of the model appears slightly too long by 8 inches and the rake of the sternpost is

Above. *The same image after the perspective distortion is removed to produce a true body plan at 92 feet from the keel scarph. Author, by kind permission of the National Maritime Museum.*

Above Left. *A digital image showing the laser device shining a line at a known position along the base board, in this case at 92 feet from the keel scarph. Author, by kind permission of the National Maritime Museum.*

5 feet 2 inches on the model, as against the survey's 6 feet 2 inches. On the model, the steerage bulkhead is two feet further forward than it is in the survey, which restricts the length of the model's main capstan's bars. And to be nit-picking, the model shows no funnel for the cookroom although the survey lists one.

In order to strengthen and simplify the cutwater, the model maker made an enormous keel scarph that incorporates the gripe. The reconstructed draught of the ship uses the profile of the model's cutwater and rabbet, but with a normal ship scarph and gripe. The *Mordaunt* had a gundeck that was almost 20 feet longer than the keel,[603] the difference indicating that she had a slightly less than circular stem, in agreement with the model. Although the survey gives only one height between decks, the height is raised aft to give clearance to the tiller, as was normal practice. The gunports in the reconstructed draught are also more precisely spaced than they are on the model.

The reconstructed draught and the model use the same space and room of 2 feet 1½ inches and the same number of frames in the same position. The midship flat, the largest frame in the ship, is a third of the length of the keel from the bow, as normal, and five flats aft and four forward are shown, which was the common practice, as suggested by William Keltridge in *His Book* of 1675.

The greatest conflict of evidence between van de Velde and the model appears to be in the position of the channel wales forward on. Normally, the distance between main wales and channel wales lessens towards the bow; however, the model has them parallel, and as a result, the upper channel wale is almost level with the upper edge of the foremost upper-deck gunport. Van de Velde, on the other hand, in three clear drawings of the ship, shows the normal practice, with the upper channel wale just touching the lower edge of the gunport. The reconstructed drawing follows van de Velde on the basis that he would not have made the same mistake three times and the model looks very unconventional. Not that van de Velde's drawings can always be regarded as perfect, for one of them[604] has the forecastle deck only about six inches above the same gunport, which would leave a height of little more than four feet in the forecastle. The rails above the roundhouse bulkhead and a funnel for the cookroom are missing in the model and have been added.

The carvings in the reconstructed draught are taken from the model and the van de Velde drawings, which, thankfully, are very similar, the only differences being a pair of additional badges on the stern of the model at the balcony level and a carved frieze on the outside at the level of the half deck. The quarter pieces, carved a little bigger than life size, are almost certainly representations of Prometheus in chains. The model appears to have been refreshed with gilt and paint sometime in its long history and the original black in some panels replaced with bright blue. Two of the van de Velde drawings show the ship with some minor alterations, presumably after being accepted into the Navy. Top hinged port lids have been replaced with side-hung port lids in the wake of the shrouds; lights have been added to the captain's cabins and the steerage; gallows have been fitted in the waist to carry spare masts and spars; and an anchor lining has been added to protect the sides when handling the anchors.

The most guns the *Mordaunt* would ever have been assigned is 46, even though the model, the van de Velde drawings and the survey all show her with 56 gunports. The model has the correct number of ports on the half deck (ten), although the furthest forward port each side is too close to the steps of the steerage bulkhead. The port also has a rail going through the centre, making it useless. The two gunports in the chase ports of the beakhead bulkhead that face directly forward are too small. This is caused by the bulkhead being so close to the stem that its width is severely restricted. In 1684, the planned establishment of the *Mordaunt* was given as follows:[605]

The 1684 establishment of the *Mordaunt*

	Men	Guns	Demi-culverin of 27 or 28 cwt	Saker of 17 or 18 cwt	Light-saker of 6 or 7 cwt
In time of war at home	230	46	20	18	8
In time of war abroad	200	40	18	16	6
Peace at home & abroad	150				

In 1688 the actual situation was somewhat different.[606]

	Guns	12 pounder	8 pounder	Saker fortified
Guns belonging to the ship	40	18	16	6
Ordnance further wanting	6	2	2	2

Using the survey as the principal source but also taking into consideration the model and the van de Velde drawings, a reconstruction was made. It was not difficult to resolve the few conflicts between the sources, giving some confidence that the *Mordaunt* must have been very similar to the reconstruction.

THE DRAUGHT OF THE NEW TYGER

Mordaunt
Stern

The stern. The model and van de Velde drawings closely agree. Scale 1:72. Author.

127

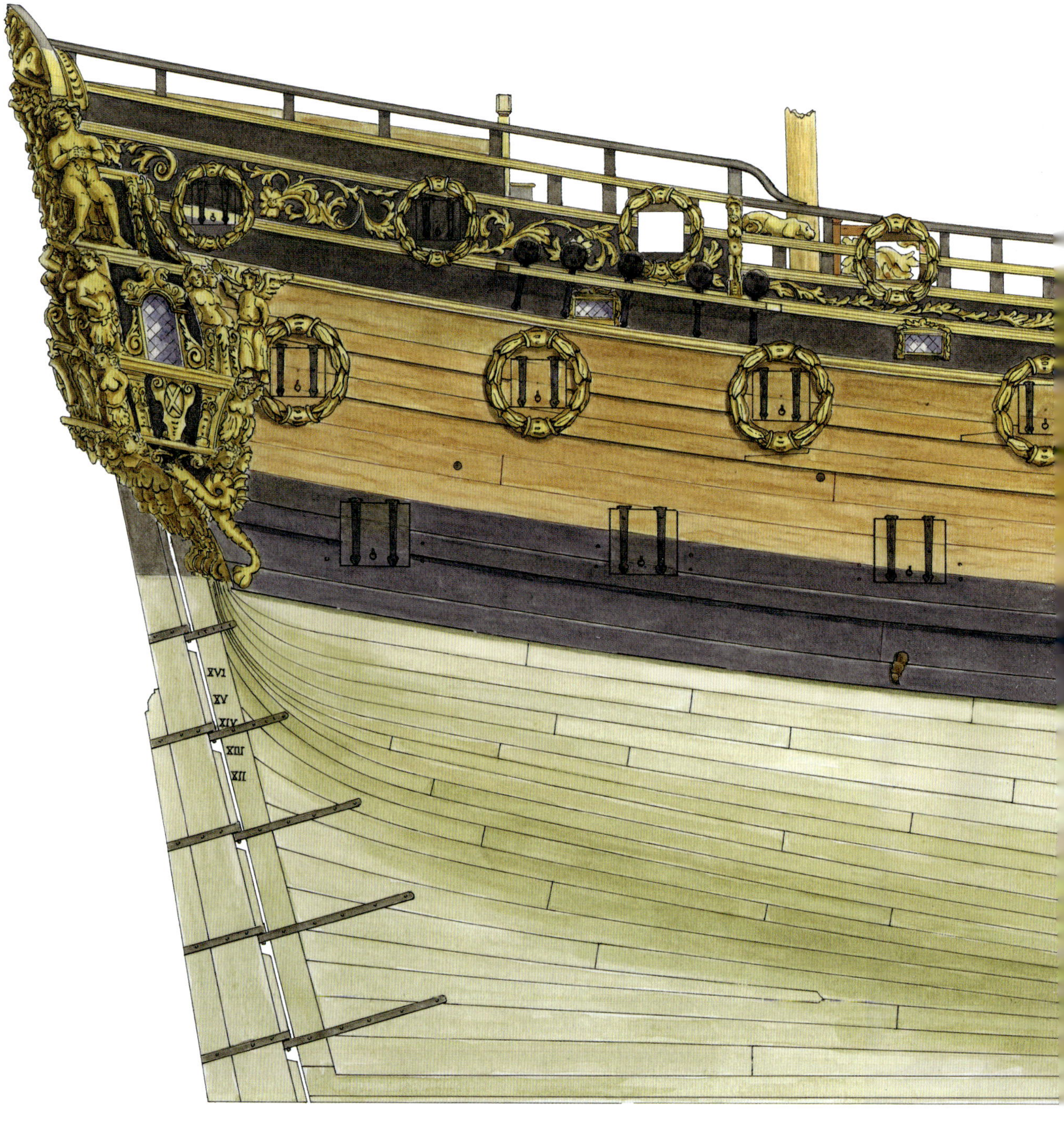
XVI
XV
XIV
XIII
XII

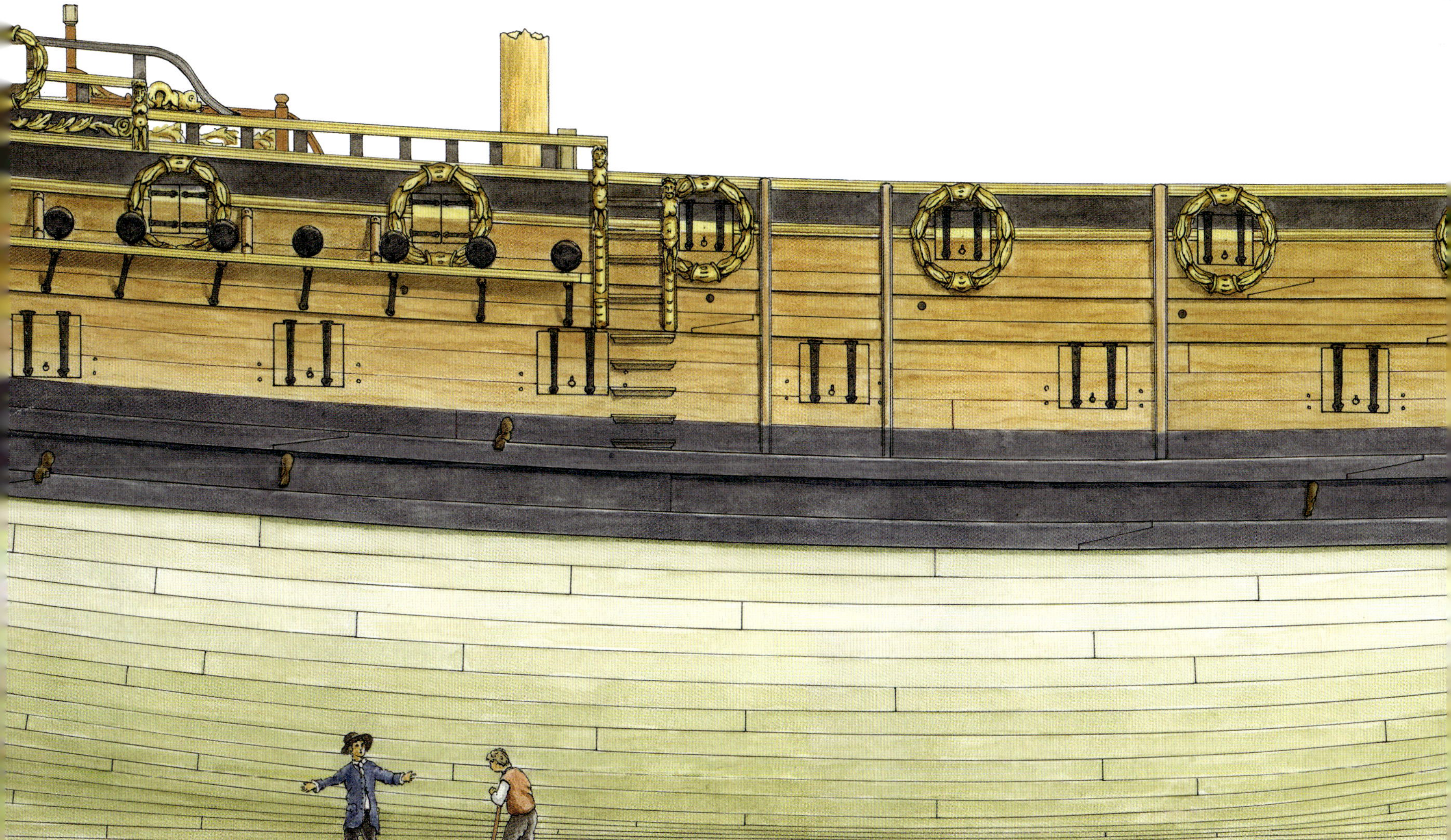

25 22 19 16 13 10 7

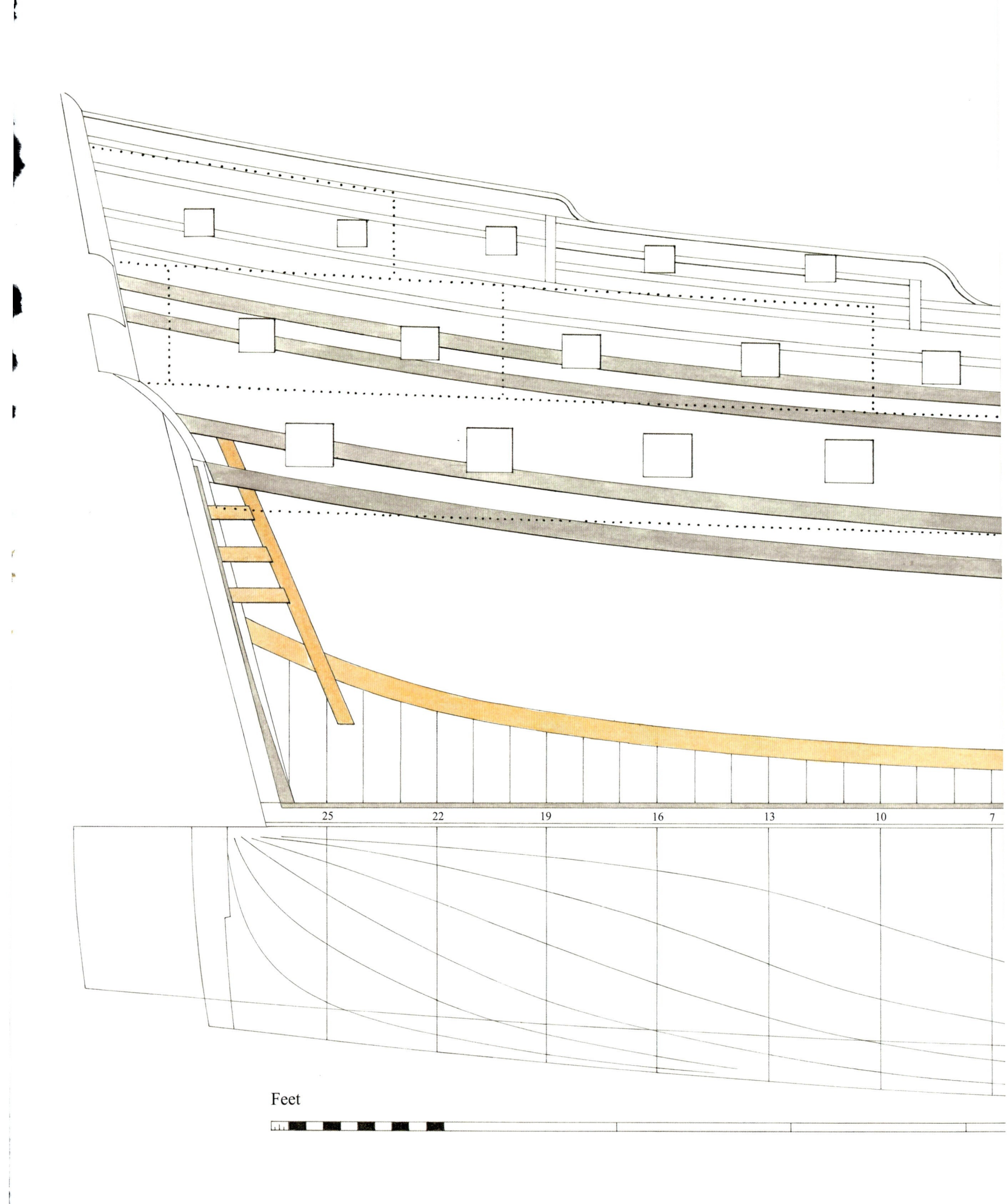

Feet

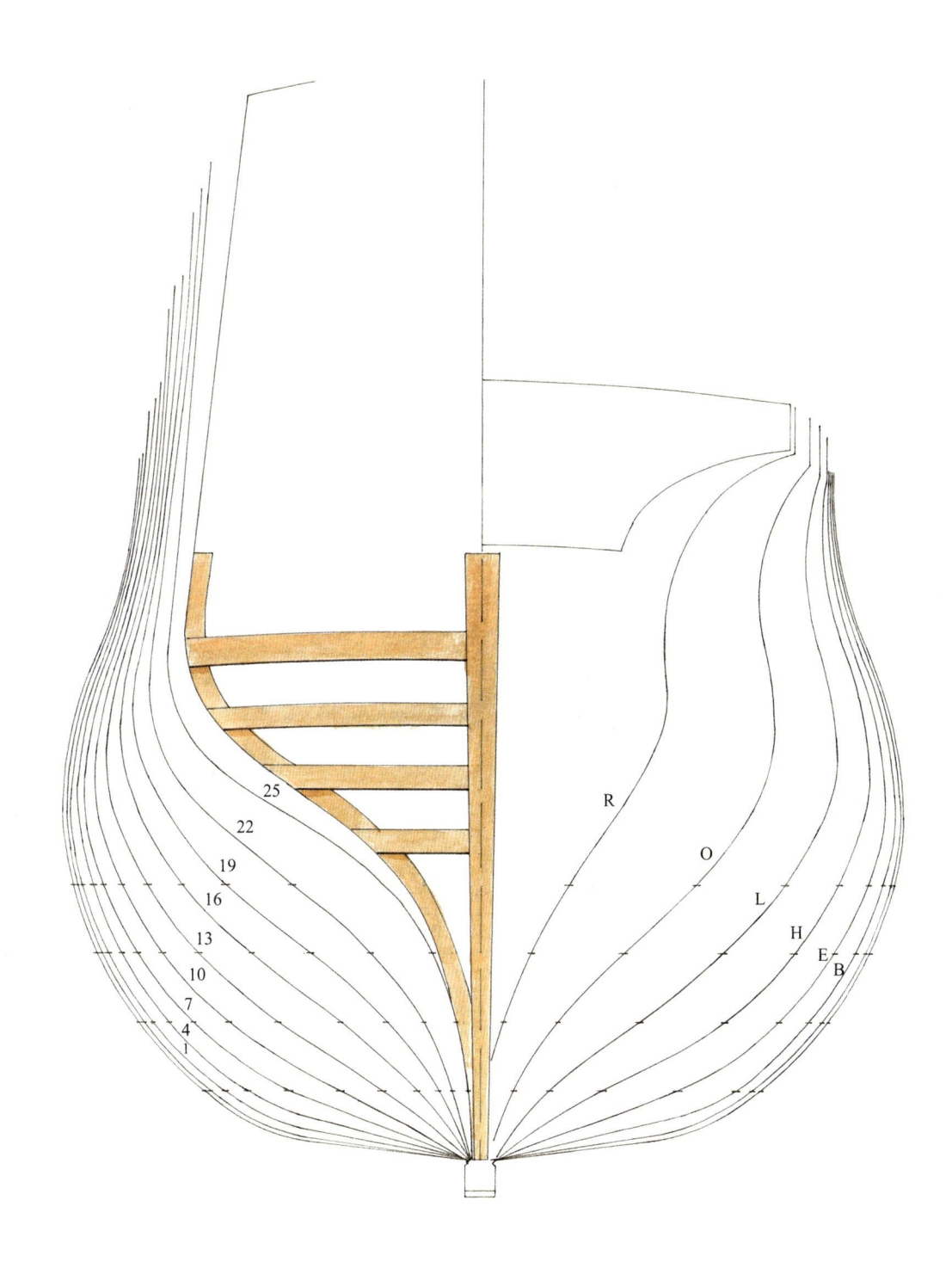

Mordaunt
Draught

The reconstructed draught of the Mordaunt, based upon a detailed survey of 1682, the contemporary model and the van de Velde drawings. The dotted lines that signify the decks are at the top of beams where they meet the frames at the side. Scale 1:72. Author.

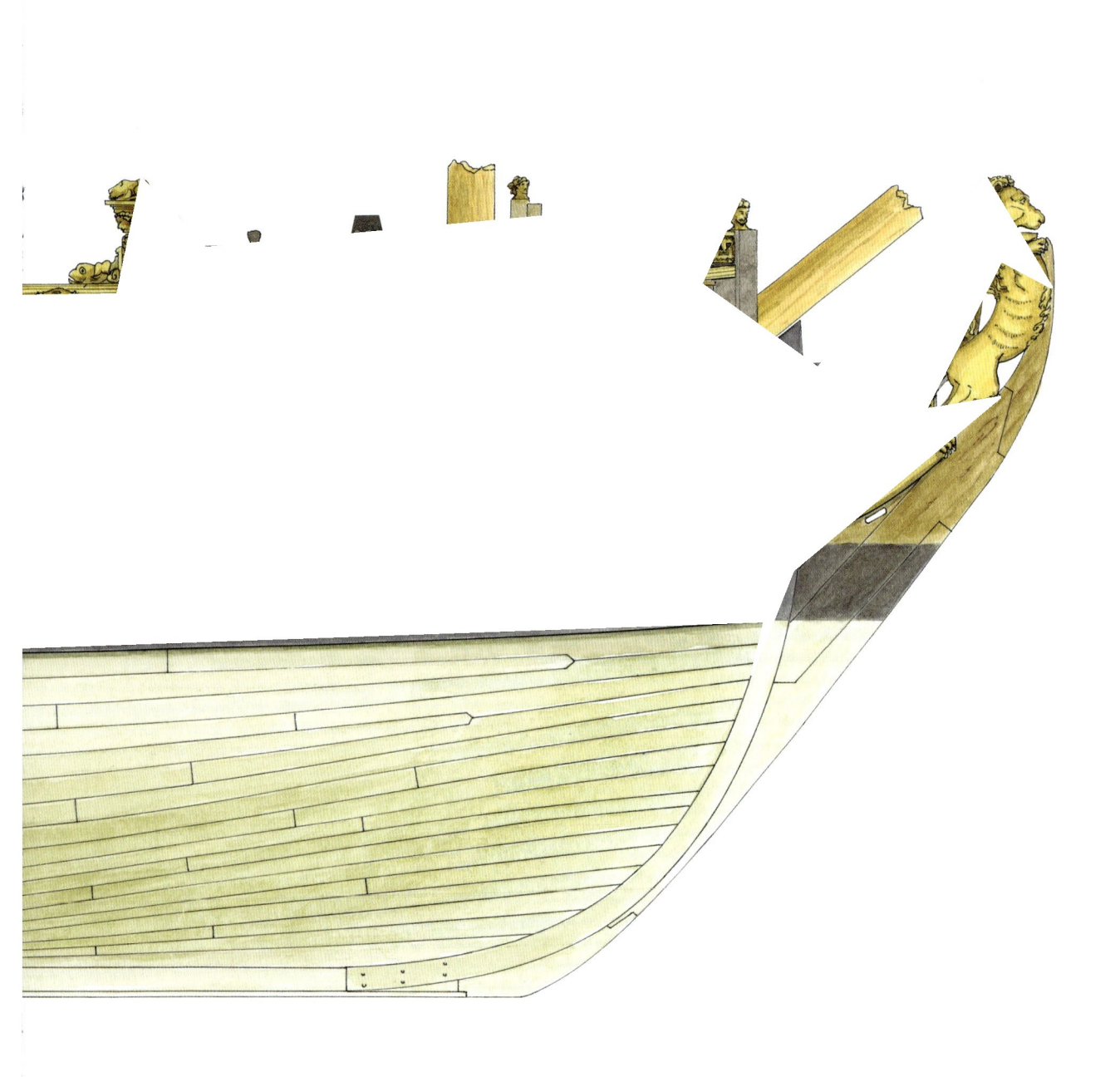

Mordaunt
Broadside

Broadside view. Scale 1:72. Author.

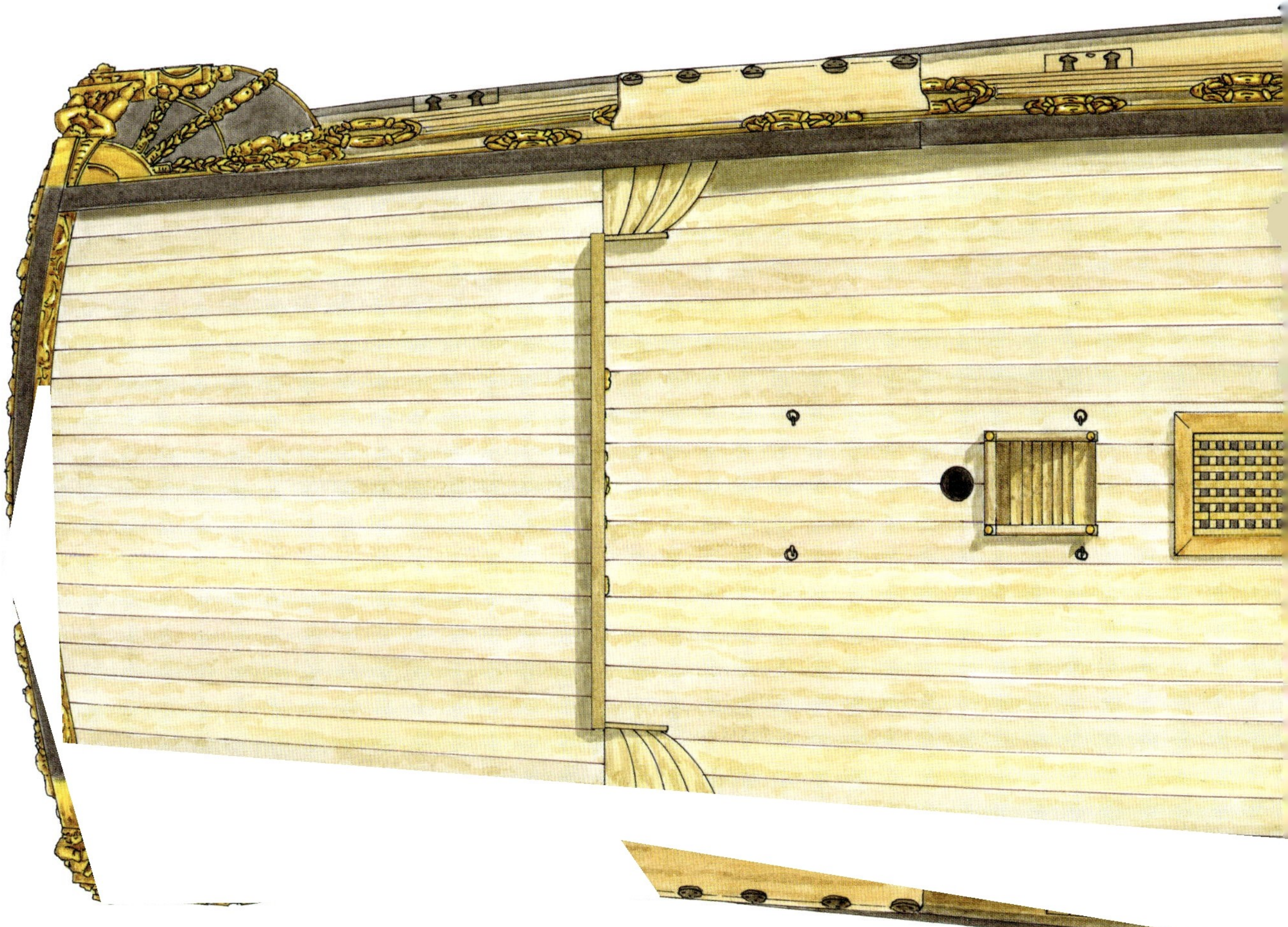

R

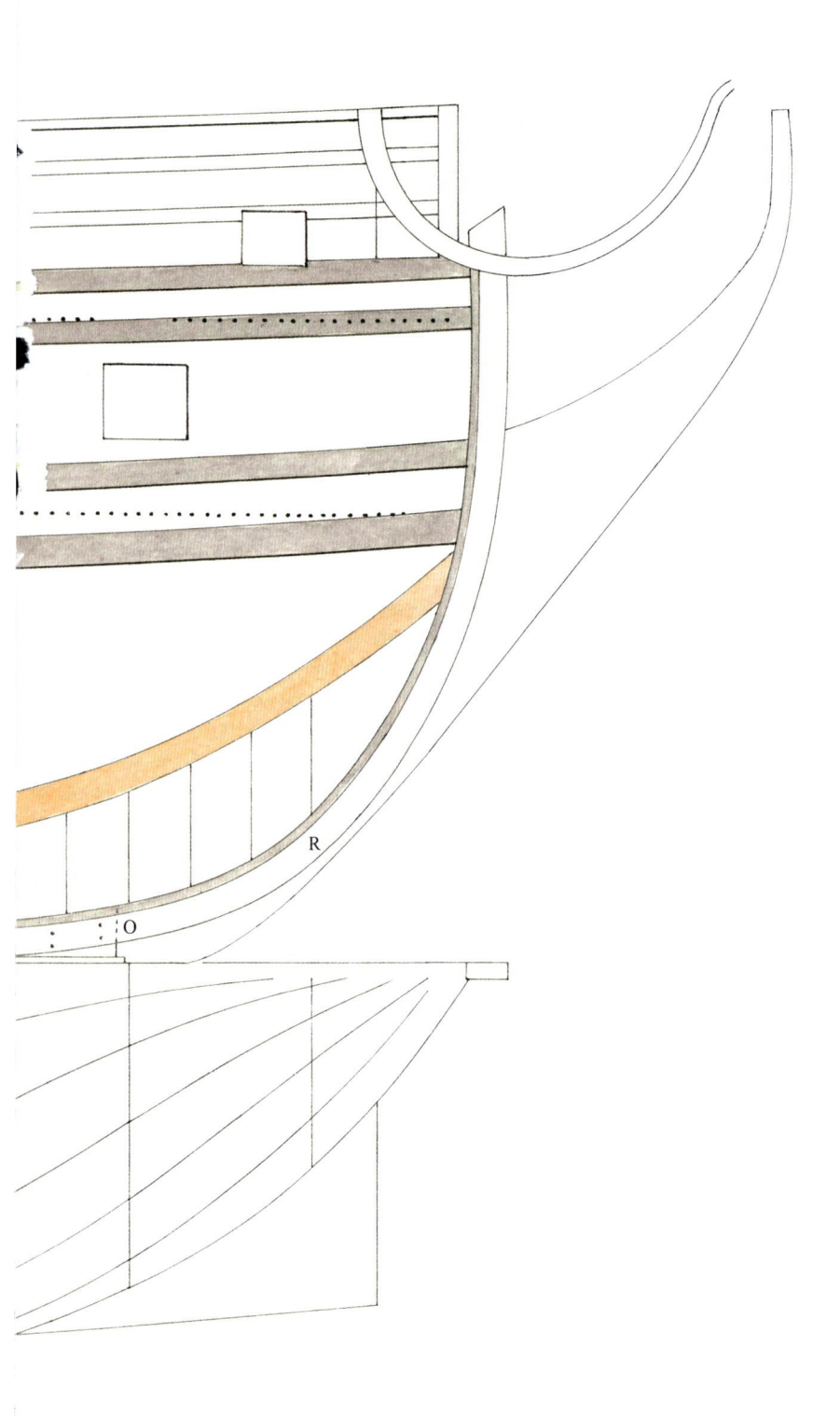

O

4　　　1　　　③　　　Ⓧ　　　Ⓒ　　　B　　　E　　　H　　　L

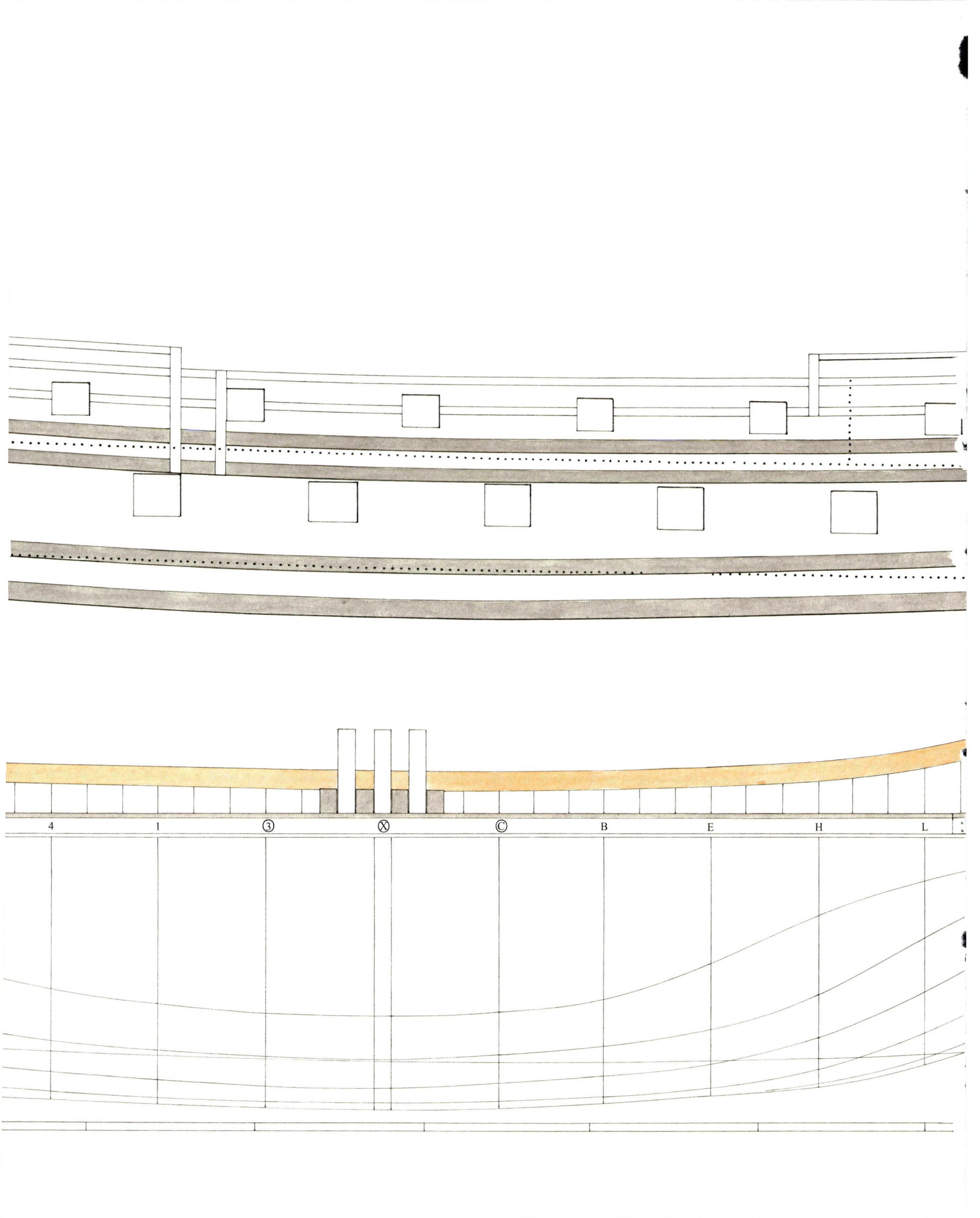

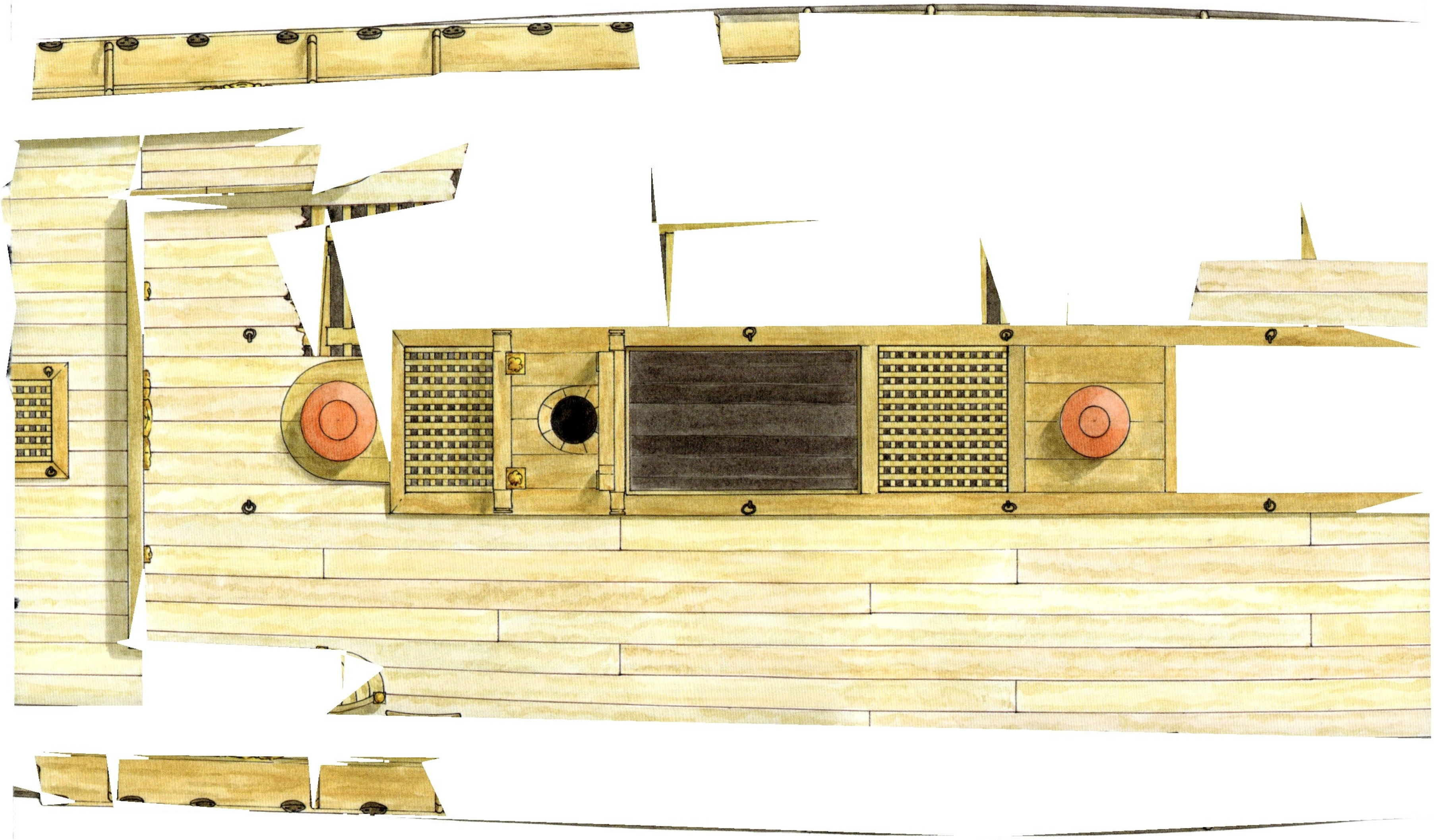

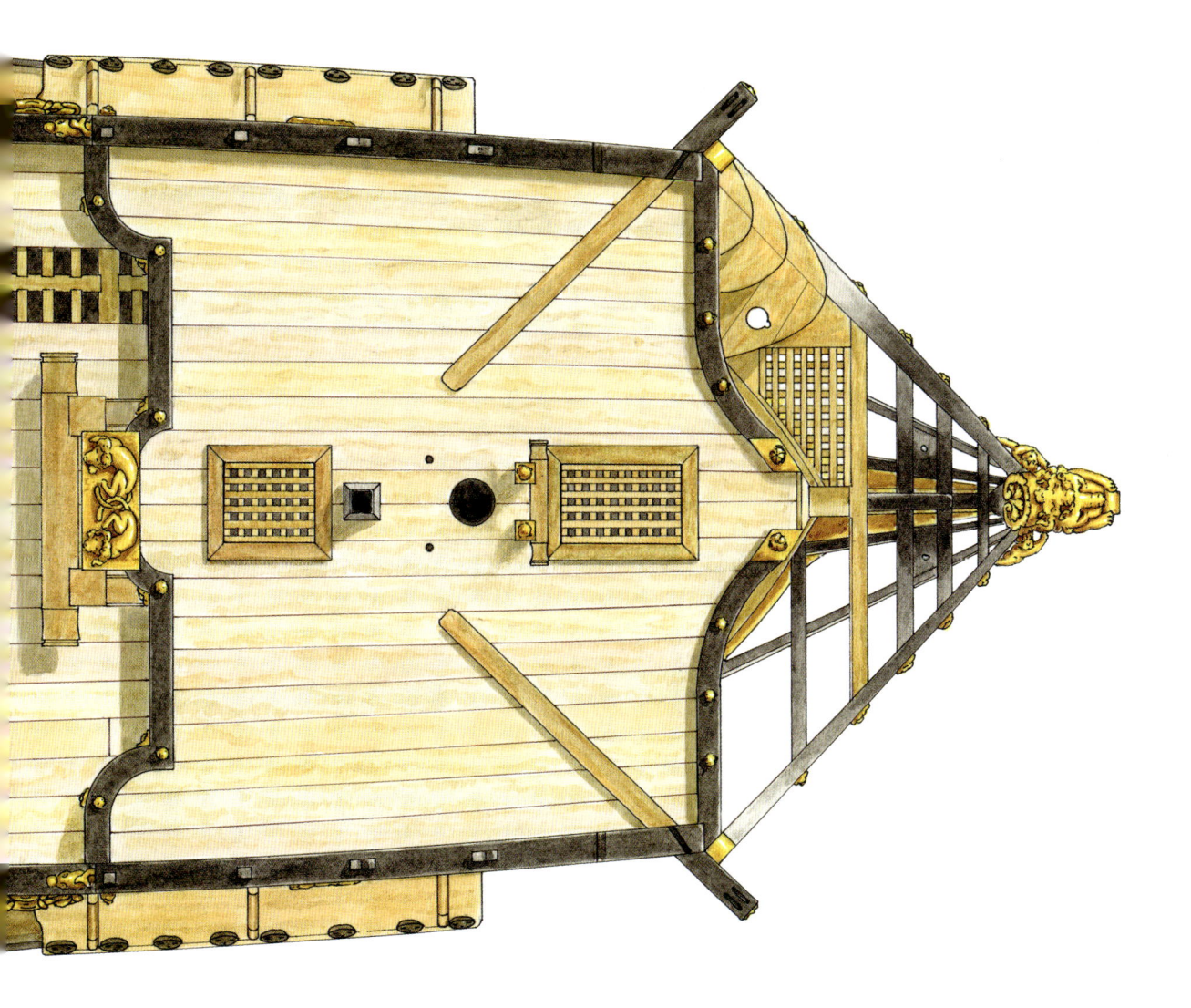

Mordaunt
Decks

The deck plan with some of the upper-deck planking removed to reveal the layout of the supporting structure of beams, carlings and ledges beneath. Scale 1:72. Author.

THE MASTER SHIPWRIGHT'S SECRETS

Mordaunt

The roundhouse (above) and steerage bulkheads. Scale 1:72. Author.

THE DRAUGHT OF THE NEW TYGER

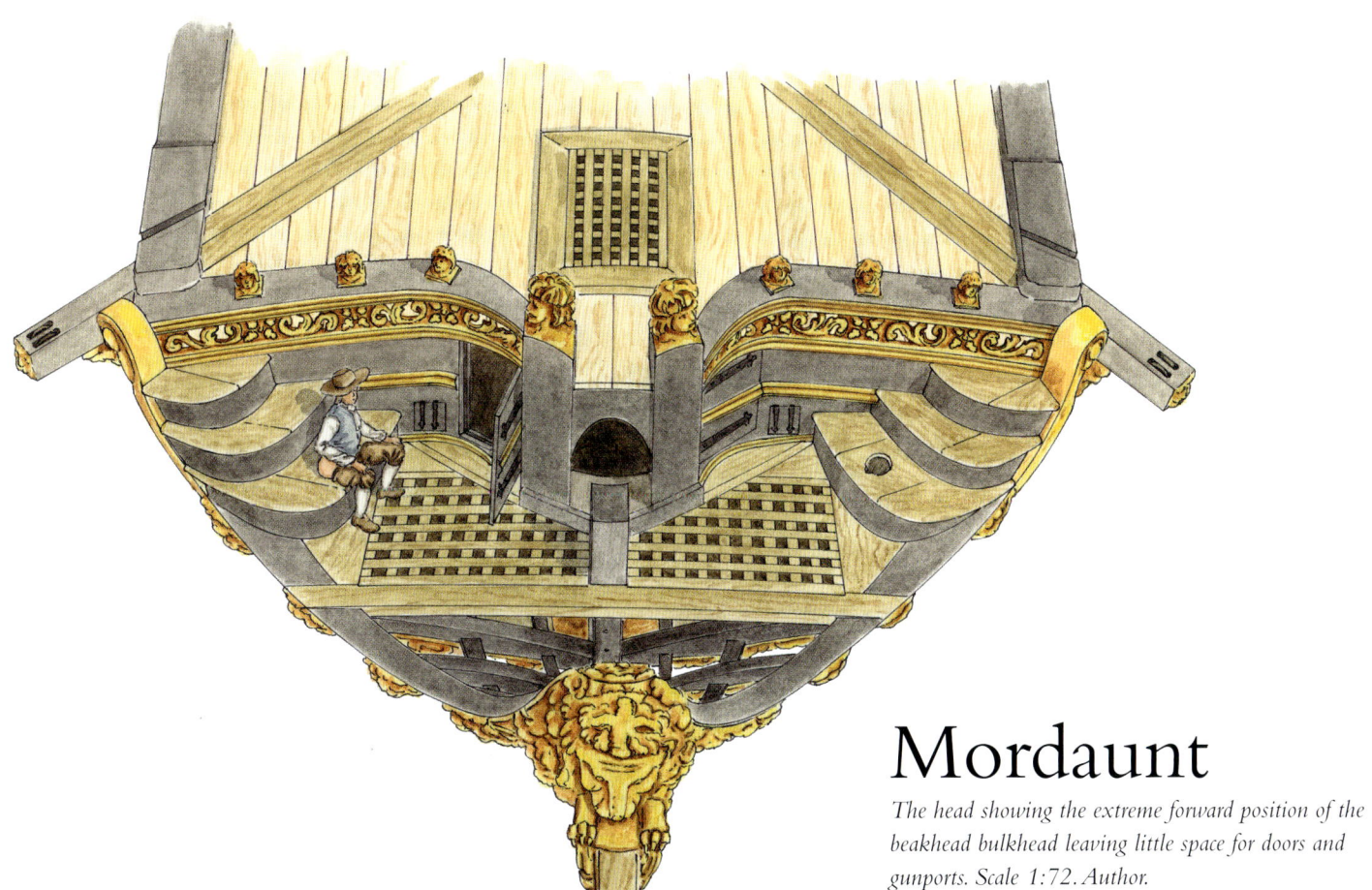

Mordaunt
The head showing the extreme forward position of the beakhead bulkhead leaving little space for doors and gunports. Scale 1:72. Author.

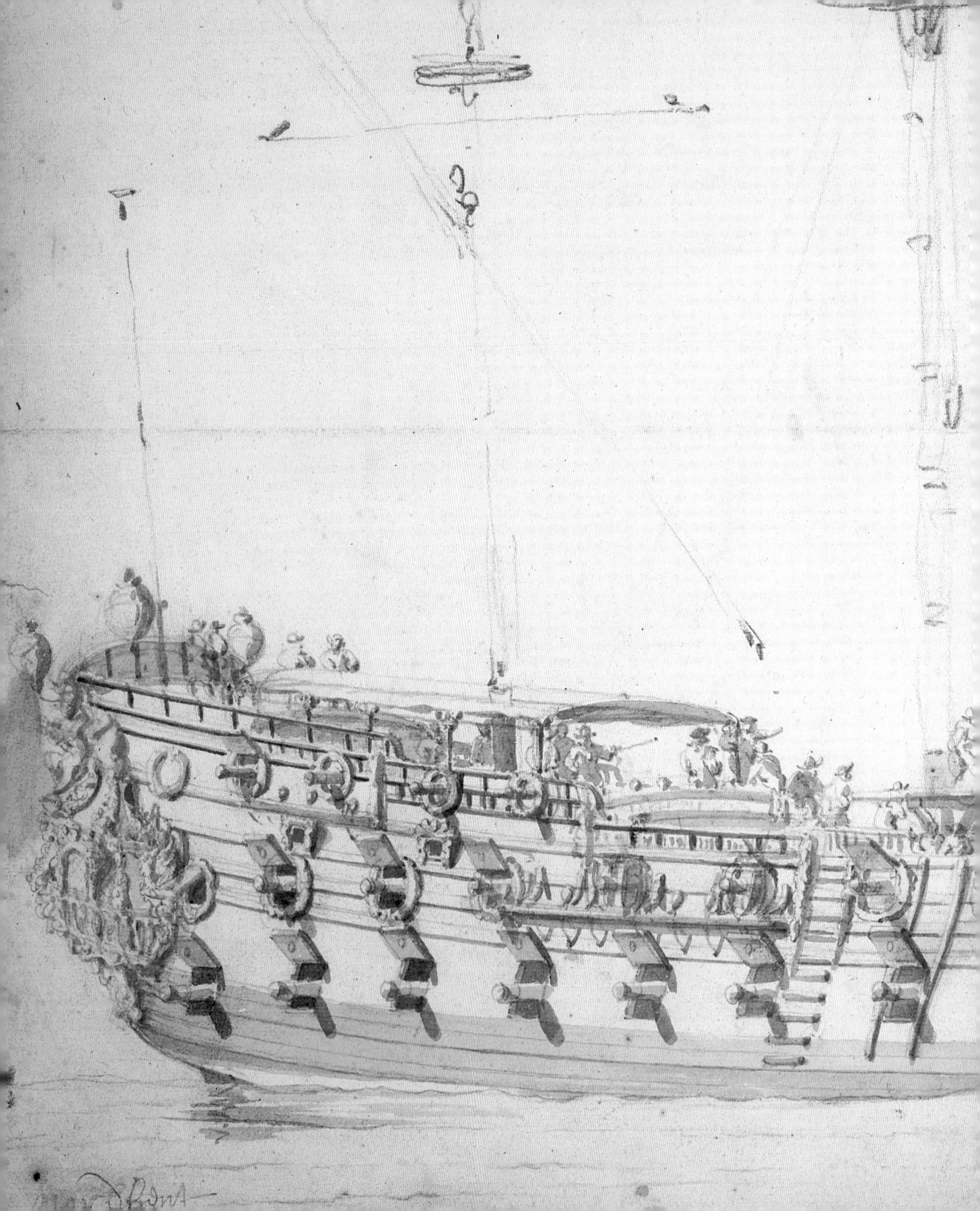

The Mordaunt *viewed from slightly before the starboard beam. Compared with the model, a number of additions have been made to equip her for Navy service. These include an anchor lining, lights for the cabins, gallows and side-hung gunport lids in the wake of the main channels. This view also shows the wales near the bow in a different position to those on the model (Robinson 598). Portrait of the 'Mordaunt'. Willem van de Velde, the Elder, 1681? PAH9367; Photo: © National Maritime Museum, Greenwich, London.*

The *Mordaunt* in the Navy

On 18 April 1684, a little over six months after the *Mordaunt*'s purchase, a survey was carried out with a view to putting all the ships lying in harbour into full repair. An initial estimate of £396 was made for the hull of the *Mordaunt* lying at Deptford,[607] followed two weeks later by a comprehensive estimate for fitting the hull, masts, yards, rigging, ground tackle and sails and for furnishing six months' worth of sea stores:[608]

Above *The bow of the model of the* Mordaunt. *Note the difference between the van de Velde drawing and the model in terms of the position of the channel wale in relation to the forward upper-deck gunport. Photograph by Author, by kind permission of the National Maritime Museum.*

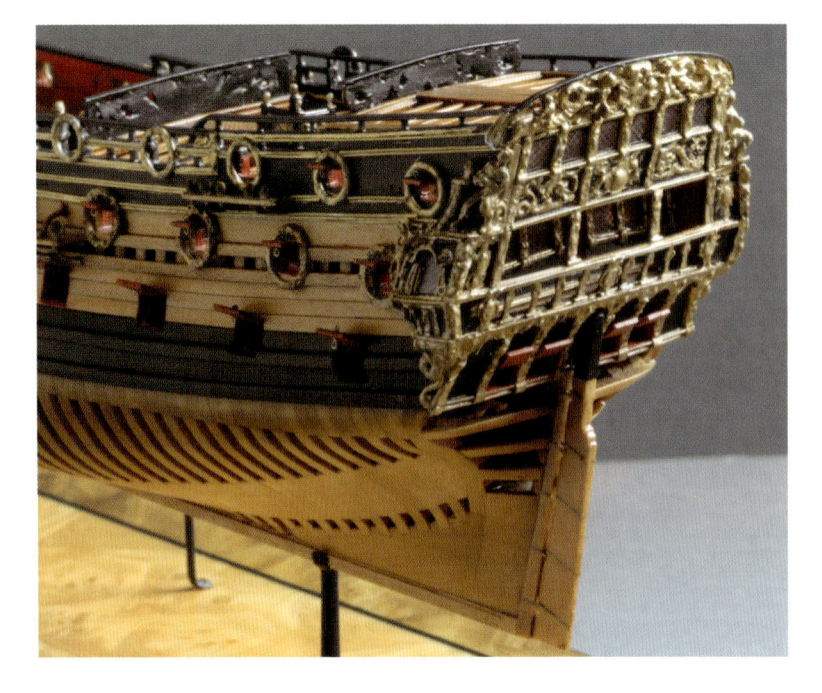

Images on this page are of a model of the Mordaunt *made recently by Phillip Reed and based on the drawings in this book. Courtesy Phillip Reed.*

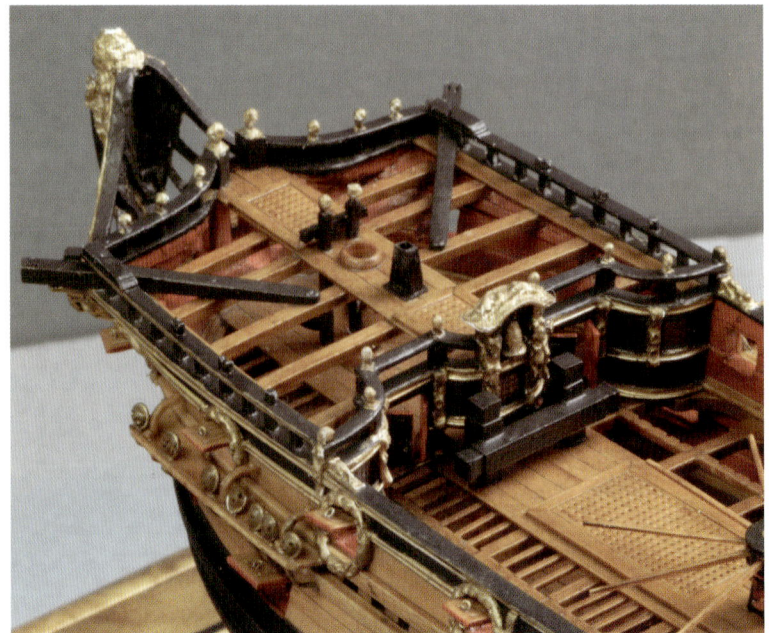

30 April 1864 estimate for the fitting of the Mordaunt with six months' stores

Item	Cost (£)
For timber, plank, boards, masts, yards & carpentry	405
Pitch, tarr, rozin, oyle, & brimstone	50
Reed, broome, ochum, thrumes & other petty provisions	15
Perfecting the carvers, painters, plumbers, glaziers & bricklayers works	20
Completing the Mordaunts sails to three suites of courses & topsails	170
Cables and other cordage & blocks	606
Anchors & other iron work & lead	200
Cotton, colours & hamaccoes	35
Completing the boatswain & carpenter's stores	336
Total	1,837

The figure of £1,837 is almost a third of the original cost of the ship, which was also supposed to have included sea stores. The *Mordaunt* was bought with a complete set of six anchors valued at £150, yet the new estimate requires a further £200 for new ones, other iron work and lead. Other items also seem to have been purchased twice, but no one at the time seems to have queried where the original items went.

On 3 May, the Admiralty wrote to the Major General of the Ordnance to request that the ship be furnished with her abroad complement of 40 guns as it was decided that she would be part of a squadron going to the West Indies.[609] The *Mordaunt* was taken into the dry dock for the repairs, and there it was found that the planking of the buttocks on both sides was rotten and needed to be stripped off. To replace it, John Shish requested four loads of four-inch and six loads of three-inch good Sussex plank.[610] The same day the Navy Board ordered members of Shipwrights' Hall and John Shish to come up with an estimate for the additional work.

The estimate dated 6 May 1684 found the following:

Item	(£) Cost
To shift about six hundred thirty six foot of plank that is decayed and iron work to the same.	100
To bring on about two thousand one hundred eighty six foot of sheathing which is now taken off.	80
To bring on three floor riders and six futtock riders.	60
To fit platforms and store rooms with a shot locker in hold to build one new boat, to perform some joiners, glazier's, plumbers, blockmakers and painters work, to caulk and grave her with other small finishing works.	240
Total	480

It was only 15 months since many of the same experts had found the ship caulked, tarred and well kept. They reasoned that as the ship was not in use, the sudden and surprising defects were caused by the lack of air circulation.[611] Not mentioned was the

Azimuth and brass box compasses of the type used on the Mordaunt. *From* A Naval Expositor, *Thomas Blanckley, 1750.*

Azimuth — Is an Inftrument made in a large Brafs Box, with Imbers and a broad Limb, having Ninety Degrees diagonally divided, with an Index and Thread to take the Sun's Amplitude or Azimuth, in order to find the Difference between the Magnetical Meridian and the Sun's Meridian, which fhews the Variation of the Compafs.

Brafs Box — They ftand in the Bittacle, that the Men at the Steering Wheel may fee to keep the Ship in her right Courfe.

Below. *The Mordaunt was built before the introduction of the steering wheel, but the bittacle, shown here, was of the same type used on ships steered with the whipstaff. From* A Naval Expositor, *Thomas Blanckley, 1750.*

Bittacle — Is a Sort of Locker framed with Deal to hold the Compafs, a Glafs and Candle, and ftands on the Quarter Deck juft before the Steering Wheel, by which, he that Steers the Ship is enabled to keep her in her right Courfe.

extraordinarily harsh winter and the frozen Thames, which caused decay in all ships and not just the *Mordaunt*. It was anticipated that she would remain in dock until 3 June,[612] but when the work started, it was found that the crotch in the breadroom was decayed, and so a new one with six-foot-long arms that spread nine feet at the ends would have to be found. A week later, Thomas Lewsley, the timber purveyor, thought he had found a suitable piece, but after examination it was found that it could not be made to fit.[613] In spite of this extra work, the *Mordaunt* was duly launched according to plan on Tuesday 3 June, and the *Phoenix* was docked in her place the next day.[614]

The *Mordaunt*'s commissioned officers were appointed by King Charles. Henry Killigrew became the captain, William Rigby the first lieutenant and Joseph Lawrence the second.[615] Killigrew was an experienced officer who was first appointed as a lieutenant in 1666 and would later rise to become joint Admiral of the Fleet.[616]

Although the ship was safely launched, John Shish still had 35 shipwrights, 16 caulkers, 12 joiners and 9 ocum boys on board, which required £90 worth of materials and £70 in wages to complete the work in about a week.[617] Meanwhile, Captain Killigrew had not been idle – he was casting a critical eye around his new ship and compared her with the *Tyger*, moored nearby. Unhappy with his ship's masts, he wrote to the Navy Board: '*Her masts I found too small they being two inches less in ye diameter than the English Tyger, to strengthen them a paunch* (probably meaning a strengthening piece fitted to the fore side of the mast) *is brought on upon each. Considering our voyage if ye main mast were a fore mast and we were ordered a new main mast it would be much securer.*'[618] The Navy Board asked John Shish to survey the masts in consultation with his brother Thomas, the master shipwright at Woolwich, and his brother-in-law, Fisher Harding, taking into account Killigrew's criticism. They replied: '*we have been on board of His Majesties Ship ye Mordaunt & surveyed her masts, we find them to be good Gottenburgh masts, clear of knots, well wrought and ye cheeks of good length, there is a substantial fish brought on upon each mast which in our opinion renders them fit masts for any voyage for ye said ship. As for the complaint of the smallness of them they being so good woods with the fishes brought on will answer any New England masts of 2 inches bigger. We judge convenient that a good substantial fish may be brought upon ye bowsprit and have given order for ye doing of it, the dimensions of the masts and yards are as followeth vizt:*'[619]

Masts and yards of the *Mordaunt*

		Masts		Yards	
		Length in yards	**Diameter in inches**	**Length in yards**	**Diameter in inches**
Main Mast		27⅓	22	25⅓	17
	Topmast	16⅔	13½	14⅔	11
	Topgallant Mast	7	6	8	6
Fore Mast		23⅔	19½	21	15½
	Topmast	15	12	12⅔	9½
	Topgallant Mast	6	5	7⅔	5½
Bowsprit		18	21	14	10
Spritsail Topmast		4	5	8⅔	6
Mizzen Mast		23⅓	14	21	10½
	Topmast	8⅔	6½	8	5¾
Crojack Yard		–	–	12⅔	6½

It seems at least some of the masts and yards had a habit of getting lost. Costs for replacements were included in the estimate of April and no one seems to have remembered that they were all present when measured less than two years before, when they were surveyed in September 1682. In spite of his experience, Killigrew seems to have been unaware of the differences between the types of fir used for masts. For example, it was reckoned that a 9-inch Gottenburgh mast was equivalent to a 12-inch mast from New England.[620]

In mid-June, as the ship neared completion, attention turned to some of the smaller but important items of equipment. Ten meridian compasses with pewter boxes were authorised for the ship, though this quantity was queried as being too many, as seven ordinary meridian compasses and an extra one with a brass box had already been supplied.[621] Later, Captain Killigrew complained that no orders had been issued to provide him with an Azimuth compass.[622] Another ship being fitted out at the same time was the fourth-rate *Diamond*. Her captain had two compasses with wooden boxes but complained '*there is but two compasses aboard His Majesties ship Diamond that have brass boxes & of those none of the best; I hope*

Barrecoes — Small Casks of Twenty one Inches long, bound with four Iron Hoops, were formerly allowed to Ships bound on Foreign Voyages for fetching Water.

Above. *Barrecoes. From* A Naval Expositor, *Thomas Blanckley, 1750.*

Anchor Stock — A Piece of Wood fastened together with Iron Hoops and Treenails upon the Square near the Ring, serving to guide the Flook, so as it may fall right and fix in the Ground.

Above. *Anchor stock. From* A Naval Expositor, *Thomas Blanckley, 1750.*

Right. *Types of buoy. From* A Naval Expositor, *Thomas Blanckley, 1750.*

Buoys {

Cann — Are hooped with Iron, and made very strong, in Shape of a Cann; their Use is to lie on Shoals or Sands for Marks.

Nunn — Are made tapering at each End, and filled with Rhine Hoops and some Iron, which being strapped with Ropes, are fastened to the Buoy-rope, so as to float directly over the Anchor.

Wood — Are made out of old Masts, &c. and hath a large Hole made at one End, through which the Buoy-rope is reeved, and serves for the aforesaid Uses: From hence the Word Buoyant, signifies any Thing that is floatable.

}

your Honourables will command ye Storekeeper at Deptford to send two pewter boxes in lieu'.[623] In 1688 it was specified that meridian compasses should have 8-inch-diameter brass boxes and that the wooden boxes should be made from very substantial dovetailed timber, 11 inches square and 7 inches deep with cards of isinglass. Azimuth compasses were supplied with two spare cards and a spare glass.[624]

Rather than send the *Mordaunt* to the West Indies, the Admiralty now decided to send her to Guinea on the west coast of Africa.[625] Among the supplies necessary for the anticipated long voyage were two dozen barrecoes[626] and 42 tuns of beer.[627] It had been planned to provide stores to last eight months, but Killigrew thought the ship could carry enough for ten.[628] Shish agreed that extra space could be found, including enough for 2,000 or more rations of bread stowed in the lazaretto between decks.[629] On 1 July there were four caulkers, two joiners and an ocum boy still working on the ship, which required '*a new drumhead to fix ye jeer capstan, anchors to stock, a table to set up in the steerage with forms round ye same, tables for ye officers, to make and finish some cabins on ye deck of ye cockpit, capstan bars to fit & some caulkers work to be done*'.[630]

A few days later, as the last work was being completed, the *Mordaunt* set sail for the first time and was taken down to Longreach, where she was to take in her guns, some stores and provisions.[631] While there, Killigrew requested a spare gang of short oars for the long boat, four barge oars, two cann buoys, four pendants, more rozin and tallow and a new bell as the one he had was defective.[632] On 9 July the lower tier of guns and some of the carpenter's and boatswain's stores were aboard.[633] By the 22nd the ship was fully laden with all the required stores, except for some belongings of several officers.[634]

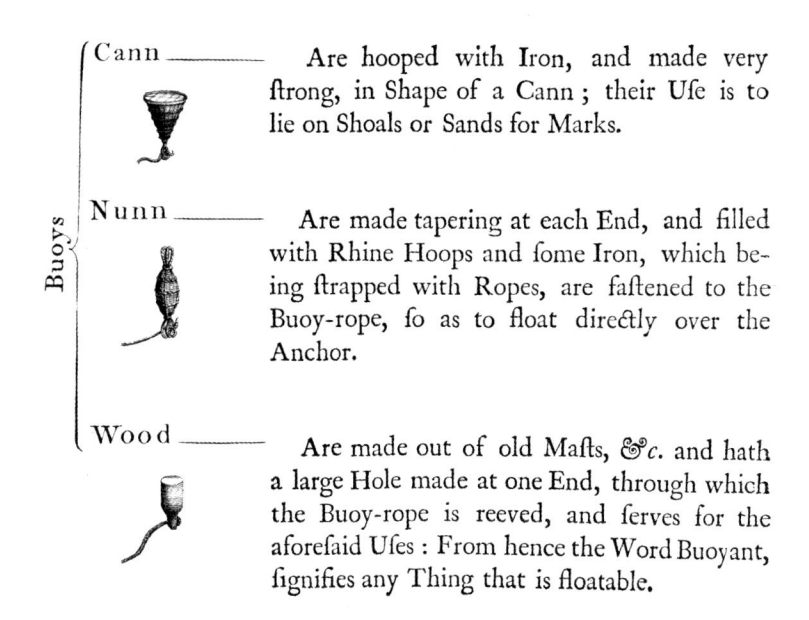

The officers were not the only ones taking care of themselves before they set sail on their long voyage. While moored in Longreach, near Purfleet, the seamen were permitted to go ashore to Aveley and West Thurrock, where they '*pilfered & stole from the said inhabitants several goods, viz sheep, ducks & other poultries to the great grievance & damage of several poor people in the said parishes*'. The inhabitants of the parishes wrote to the Navy to complain and to ask that the offenders be prevented from the like again and for some small restitution.[635] Following a lively career, the *Mordaunt* was wrecked on the Coloradoes shoals, 20 miles off the northern coast of Cuba, on 22 November 1693.[636]

Building the *St Albans*

In 1677, a huge programme was undertaken to build a fleet of 30 ships consisting of one first-rate, nine second-rate and 20 third-rate ships. Parliament voted to provide the necessary money by raising a tax for their construction and ordered that strict accounting be kept to make sure it was spent on the ships and nothing else. When the building programme was almost complete, the Admiralty ordered that a record be taken of the amount of the timber, plank and deals left over which could not be used on the new ships, how much it originally cost, and the present value. The idea was to purchase it for the repair of old ships out of the annual allowance of money provided for that purpose.[637] Six months later, in September 1681, King Charles let the Admiralty know that it was his pleasure that a new fourth-rate ship be built out of the left-over timber at Deptford, but first the Treasurer of the Navy needed to allocate money towards it.[638] Less than a week later, John Shish gave an account of the ship he thought '*fit and proper*' to build.

Amount of timber required to build a new fourth rate at Deptford

Length by ye keel	105 feet
Breadth from outside to outside	33 feet
Straight oak timber	250 loads
Compass oak timber	230 loads
Elm timber	100 loads
Fir timber	40 loads
Knees raking and square	40 loads
Oak plank of all sorts	250 loads

Shish did a calculation for the amount of timber he would need in addition to that lying around the yard left over from the 30 new ships. The shortage included 40 loads of large straight thick stuff for footwaling and wale pieces and about 15 loads of large compass (curved) for harpins, rising pieces, etc., all amounting to only about 5% of the total.[639] The progress made by John Shish in securing work for his yard prompted the master shipwrights at Woolwich and Chatham to submit their proposals to the Admiralty for building similar ships with their left-over timber.[640] As Robert Lee at Chatham put it, *'the greatest part of which timber and plank having lain four years on the ground will soon become sap rotten ... if not in some short time applied to use'*.[641]

Seven months later, on 29 April 1682, King Charles directed the Admiralty to instruct Mr John Shish to build a fourth-rate ship in the King's yard at Deptford to the following dimensions:

Dimensions of a new fourth-rate ship to be built at Deptford

	Feet	Inches
Length on ye gundeck from rabitt of the stem to the rabitt of the post	127	0
Breadth from outside of the plank to outside	32	8
Depth in hold plank to plank	13	4

A month later Shish completed an estimate of the charge of building and fitting the ship,[642] which included the charge of furnishing the ship with two suits of sails and supplying six months' worth of sea stores for the boatswain and carpenter.[643]

Cost of a new fourth-rate ship to be built at Deptford

	£
For timber, plank, treenails, masts, spars, Prutia & ordinary deals	2988
For pitch, tar, rozin, oil, brimstone, ocum, hemp, reed, broom, thrums, etc	80
For all sorts of great and small ironwork	430
For carving, painting, glazing, for plumbers, plasterers & bricklayers works	300
For shipwrights, caulkers & joiners works	1320
Total	5118
For anchors, cordage, sails, blocks, tops, pumps and parrells to complete her rigging, guntackle and sea stores for boatswain & carpenters ye sum of	2086
Total	7204
Materials wanting besides what may be had upon standing contracts to ye value of And what may be had out of the remains of the new ships timbers etc	3060

On 16 September, a second fourth rate was ordered to be built at Woolwich,[644] and a few months later, another at Chatham. Although the three ships originated from similar circumstances, they were not a class built to the same design.

In keeping with the continual desire for innovation and improvement, John Shish made a proposal for fitting one pump cistern abaft and the other before the well. Normally both cisterns for the pumps were fitted abaft the main mast over the well.

What the benefits were or how the arrangement worked remains a mystery, but when Shish presented the proposal to the King and Admiralty, they were impressed enough to order that the proposal be adopted in the new ship.[645] Another innovation Shish introduced was the use of lead ballast, cast specially to replace the chock that was usually fitted between the floor timbers and wide enough to reach the lower futtocks on either side. It would be part of the frame fitting over the keel and under the keelson. In all, 24 lead weights were cast, weighing in total 181cwt 3qtr 6lb. It was an ingenious solution to the problem of how to keep the ballast as low as possible. During March 1685, shortly after the death of King Charles, the Admiralty queried the use of lead, to which Shish replied: '*His late Majesty of ever blessed memory directed that lead ballast should be placed in ye new 4th rate ship... and ye keelson being bolted down those weights cannot be taken out.*'[646] Shish was wrong that the weights could not be removed, for in the absence of the close and creative working relationship that had existed between him and the King, the Navy Board ordered that the lead be taken out.[647]

Due to lack of money work progressed slowly, very slowly; by 1 January 1683, £4,484 worth of work and materials had still to be done, and this had only fallen to £2,400 15 months later.[648] On 1 June 1684 the figure stood at £2,212, with the following work still needing to be done: '*For completing her foot waling in hold and the plank without board under her lower wales. To plank ye ship up to ye gunwales for laying ye decks, quarter deck and forecastle, to perform all her finishing works with carvers, painters, bricklayers, blockmakers, glaziers, braziers and to caulk and grave her.*'[649] A few days later unwelcome news arrived at the dockyard: of the £24,000 needed by the Navy for ships, the Treasury could only supply £12,000. To save money, the three new fourth rates were to '*be no longer proceeded with*' and their frames well tarred for their preservation.[650] The *St Albans* was not finished until mid-1687, having taken nearly five years to build. It's worth mentioning that Shish was able to build the *Lenox*, the first of the 30 new ships and twice the size, in ten months with 194 men at a cost in wages of £2,871,[651] while the new fourth rate cost £1,320 in wages. At that rate, it works out that an average of only 14 or 15 men were working on her at any one time.

There were serious consequences from having a ship so long in building. The Navy Board had an opinion, and wrote during February 1686 concerning the fourth rate at Chatham: '*For better seasoning the upper part of the frame of His Majesty's fourth rate frigate building... these are to direct & require you forthwith to fill up such upper futtocks & toptimbers as are wanting to complete the same which being accomplished you are to desist from any further proceeding on her works until you shall receive directions therein from this board.*'[652] At the same time, members of Shipwrights' Hall wrote: '*On request of the Master Shipwright also taken a view of His Majesty's fourth rate frigate building there and do hereby certify that the long continuance of her frame on the stocks unfinished is very prejudicial thereunto and if not completed & launched sometime within the approaching summer will in our opinions occasion the shifting of her keel with some part of the six strakes of outboard plank now thereon.*'[653] Proof of who was correct followed that summer, for Shish wrote to the Navy Board on 3 August 1686: '*In dubbing down (adzing the frame timbers smooth before planking) ye 4th rate underwater we find ye midship keel piece defective so that I judge it must be shifted.*'[654]

John Shish died in October 1686 before the ship was completed and his position was taken by his brother-in-law, Fisher Harding. As the ship neared completion in March 1687, the Navy Board asked if orders had been given for making the masts, to which Harding replied that John Shish had given the mastmaker the dimensions in 1683.[655]

Dimensions of the *St Albans*'s masts

St Albans		Masts		Yards	
		Length (yards)	Diameter (inches)	Length (yards)	Diameter (inches)
Mainmast		28	24	25½	16½
	Topmast	17	14⅓	14 ⅓	10½
	Topgallant mast	7	5½	7 ⅔	4½
Foremast		25½	22	22	14
	Topmast	15⅓	12½	12 ⅔	9
	Topgallant mast	6	5	6½	4
Bowsprit		18	23½	15	10½
Spritsail topmast		5½	6	8	5½
Mizzen mast		24	13	19	9
	Topmast	8½	6½	8	4½
Crossjack mast		–	–	14	7

The new fourth rate built at Woolwich was constructed with one rowing oar scuttle between the gunports,[656] and it may well be that the Deptford and Chatham ships had them as well, in common with many other fourth-rate ships of the period. King James II followed his late brother's interest in the Navy and approved the launching of the Chatham- and Deptford-built ships for Tuesday 3 May and Wednesday 4 May 1687.[657] He personally visited Chatham on the 3rd for the launch and named the ship the *Sedgemoor*. The event was attended by Edward Gregory, later Commissioner at Chatham, who wrote, '*A crowd of friends both dined & supt with us.*'[658] The next day the festivities moved to Deptford for the launch of Shish's ship, which James named the *St Albans*.[659] On 15 March 1688, the guns belonging to the *St Albans* were listed as 20 fortified culverins (18-pounders) and 30 six-pounders, although she had no carriages to put them on.[660]

William Sutherland wrote about the ships he had sailed in: '*The next ship was the Saint Albans, a ship of 50 guns built by John Shish at Deptford in 1688, her length of keel was 107 foot, on the deck 125 foot, beam 33 foot 6 inches, depth 13 foot 6 inches, she had an excellent character at that time and yet I am well assured her midship timber was described by one radius from the keel to the breadth and yet carried a stout sail and did every other office very well being coveted by most commanders for her sailing and especially when she went directly before the wind. Indeed when she went to windward the* Dover *would out do her.*'[661]

Model Ships Associated with the Shish Family

The three fourth rates were the only ships of the type launched during the reign of James II and they may all have had contemporary models made of them. In 1912, R. C. Anderson visited the Naval Museum of St Petersburg and saw a collection of English model ships. Two of them were fourth rates bearing the initials I.R., for James Regina, and as a third model, identified as the *St Albans*, was still in England, he reasoned the two in St Petersburg must be the *Deptford*, built at Woolwich,

The true arrangement of frames and their component parts as represented in the Tyger of 1681. Almost all other ships, including fourth rates, had the same arrangement. The only exceptions were the largest three-deckers and some small ships. Only one contemporary model with the true late-17th-century fashion of framing is known to exist.[664] *Author.*

Nearly all models were made with this simplified frame arrangement. The three futtocks on each side used in real ships have been replaced by a single futtock. The middle futtock that normally sits between the floor and the toptimber is eliminated and the lower and upper futtocks that overlap them are merged into one. The heel of the toptimber is just visible below the main wale in its correct place. The frames are of constant width as they reach upward in order to simplify the modelmaking. The open frame style was not just easier to make but also very attractive. Author.

and the *Sedgemoor*, built at Chatham. On measuring the models, he wrote that one of them was 'certainly the *Sedgemoor*'.[662]

During the 1920s the *St Albans* model was in the possession of Robert Spence, a collector and talented model maker. At the time it was concluded that the same hand that had made the unusually fine carvings on the model of Fisher Harding's 1692 third-rate *Boyne* had '*obviously*' built the *St Albans* and that both models were Harding's own work. The *Boyne* model is firmly identified as it is inscribed '*Boyne Bt by Mr Harding Deptford*' on the steerage bulkhead. The *St Albans* model has rather indistinct arms in monochrome gold on its steerage bulkhead, which were also taken to be Harding's.[663]

The contemporary models of Jonas Shish's *Charles* of 1668 (for many years wrongly identified as the *Loyal London* of 1666[665] and unfortunately destroyed during the Second World War), John Shish's *St Albans* of 1687, Fisher Harding's *Boyne* of 1692 and the *Royal Sovereign* of 1701 all have an unusual feature: their futtocks are arranged vertically in line, one above the other, with frames between left out completely.

This method of construction is unique to models of ships made by members of the Shish family of Deptford – father, sons and son-in-law. In contrast, all other models' futtocks are overlapping. A model in the National Maritime Museum with *Sheerness* painted on the stern also has this most unusual Shish style, with the frames modelled in almost exactly the same way as they are on the *Charles*. The *Sheerness* model is unlike any known ship of the name, but it has an unarmed waist similar to the ship described in the treatise, although it is somewhat smaller. These similarities make it fairly certain that it has an association with John Shish and his time at Sheerness between 1673 and 1675.

The model taken as the *St Albans* has the characteristic arrangement of simplified futtocks set in line, in common with Shish family models of the *Charles*, the *Boyne*, the *Royal Sovereign* and the *Sheerness*. The theory held in the past that Harding probably made the models of the *Boyne* and the *St Albans* himself is unlikely. A master shipwright's skills were in controlling a considerable work force, finding timber and continually repairing and refitting ships. He spent very little

The style of framing on the model of the Charles *of 1668, built by Jonas Shish, and the model of the* Royal Sovereign *of 1701, built by Jonas's son-in-law, Fisher Harding. The only other model with the same arrangement has* Sheerness *painted on it and dates from c.1670. It is probably associated with John Shish during his time there as master shipwright. The* Royal Sovereign *model in the Hermitage Museum, St Petersburg is slightly different in correctly showing the midship bend with a missing tier of futtocks. Author.*

The framing on the model of John Shish's fourth-rate St Albans *of 1687 and on the third-rate* Boyne *of 1692, built by Fisher Harding. They are similar in style to previous Shish models with every other frame eliminated, but even more simplified. They clearly show the reversal of the floor timbers around the midship bend. Author.*

time making draughts, and concluding that he happened to have an almost uniquely brilliant talent as a model maker, and countless hours available to build them, seems very unlikely to be correct.

In 1926, Robert Spence engaged a talented draughtsman named Herbert Read to measure and draw the model of the *St Albans* on two large sheets of paper.[666] They are reproduced in this work, but as the colour of the paper has darkened to be almost indistinguishable from the brown wash used by Read, they have been digitally enhanced to render the background white again. For interest, the stern in the broadside view is left unaltered. It appears that the purpose of the very fine drawings was to enable Robert Spence to make a copy of the original *St Albans* model. The new model today resides in the National Maritime Museum, while the original is at Trinity House. During his possession of the original model, Spence made a splendid job of its repair and restoration. Worm holes were filled and some of the decayed or missing rising timbers in the floors replaced. The interior retains the look of great age, while the exterior is clean and fresh.

The Read drawing of the original model apparently shows it before restoration, with the outboard work above the great rail painted green. The green paint was probably added by an earlier restorer during the model's long history, which Spence noticed and re-painted with a more likely black. There are other small details: the guns, which are only on the quarter deck, are now black but shown in the drawing painted green as if they are corroded brass. The cathead timber is drawn as being in one piece, but the original model shows that in fact it consisted of two pieces joined in the middle. The drawing is also missing the pawl for the jeer capstan. Apart from these tiny differences, the drawing is remarkably accurate. Spence was unhappy with the appearance of the original stern, which is unusual for a ship of its period, and made his version with unlikely large royal arms in the style of the 1660s.

Although the *St Albans* is a design with many similarities to the *Tyger* and the *Mordaunt*, it has finer lines than either. The largest midship frame, the midship flat, is further aft than usual, in about the same position as that in the Shish treatise. The beakhead bulkhead is in the same extreme forward position but does not curve out to meet the knighthead timbers each side of the stem in the manner of the *Mordaunt* and the *Tyger*. By having a single door to port in the bulkhead, room is available for a single forward-facing chase gunport to starboard. The ship has the usual Navy arrangement for handling the ground tackle, with two pairs of bitts on the gundeck for the anchor cables. An unusual feature is the square transom at the stern rather than a round tuck. It seems old fashioned for the 1680s, and harks back to the earlier years of the 17th century. Unlike the earlier ships, the transom on the *St Albans* ended at

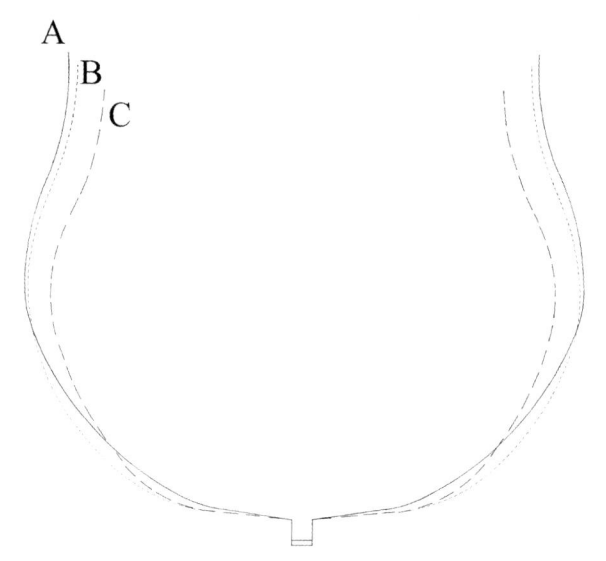

The midship section of the reconstructed new Tyger *('A') compared with the* Mordaunt *('B') and the* Hampshire *of 1653 ('C'), a ship slightly larger than the old* Tyger *of 1647. Note that the narrow floor of the new* Tyger *is similar to the true galley frigates. Although not at all large for their time, the* Tyger *and the* Mordaunt *are considerably larger than the* Hampshire*, built 28 years before. Author.*

about the waterline and would not cause eddies in its wake. It solved a difficult problem, for when planking round-tuck ships, planks had to make a severe curve, especially in fine-lined ships. It was so severe that the planking could not be bent under heat but had to be sawn to shape from English compass timber.[667] One of the most notable features of the *St Albans* is an extremely upright stem, an innovation introduced by King Charles II in 1677 for some of the 30 new ships, and very different to the circular stems used earlier.[668]

Contemporary lists for the dimensions of the ship offer an interesting selection. For the breadth, there is a choice of at least five, ranging from the planned 32 feet 8 inches to 33 feet 6 inches in a contemporary Admiralty list of ships.[669] The most reliable seems to come from the Admiralty Disposition of Ships list, in which there is a greater consensus of opinion and which is in agreement with the model:

Keel length (calculated): 107 feet 0 inches [TNA ADM8/1, f230v]
Keel length touch: 119 feet 3 inches [Taken from model]
Length gundeck: 128 feet 4 inches [TNA ADM180/20]
Breadth from outside the plank: 32 feet 10½ inches [TNA ADM8/1, f230v]
Depth in hold: 13 feet 3 inches [TNA ADM8/1, f230v]
Draught: 15 feet 9 inches [TNA ADM8/1, f230v]
Burden Tons: 615 [TNA ADM8/1, f230v]

All the dimensions, except the keel length, fit the Read drawing of the model as closely as can be measured. The length of the keel is often a problem as there were three ways of defining it. In this case, the measured 'touch' of the model, at 119 feet 3 inches, is 12 feet longer than the keel length of 107 feet given in contemporary lists. Using the 107-feet figure in the standard burden formula, the ship works out at 615 tons, in agreement with most contemporary lists. This is probably because the 107-foot keel length is not the physical 'touch' – the distance historically used, and measured from the back of the keel to the first rising of a circular stem. Instead, it must be an artificial calculated figure that compensates for a long keel in relation to the gundeck, brought about by the introduction of a more upright stem in 1677. If the actual 'touch' keel length of the model were used to calculate the tonnage, it would be a whopping 70 tons larger.

Draught of the Tyger

The primary source used in the reconstruction of the *Tyger* consists of the digital hull lines given in Shish's 1674 *Dimensions of the Modell of a 4th Rate Ship*. Also used were her known dimensions, contemporary building contracts, scantling lists, the numerous van de Velde drawings of the ship and the studies of the *Mordaunt*, the *St Albans* and some contemporary models.

Size

John Shish must have made his draught, or model, of the *Tyger* at about the same time as William Castle made his of the *Mordaunt* and a few years before that of the *St Albans*. The dimensions of the *Tyger* are recorded in a number of sources, the best coming from two letters written by John Shish himself. The first was written in May 1681, a month before she was launched,[670] and the other was written in June 1684.[671] They give the keel length, apparently the 'touch', the breadth to outside the plank, the depth in hold and the draught. The length of the gundeck is given in the Admiralty list of ships generally known as *Dimension Book B*,[672] which is written as if it could belong to the *Tyger* either before or after her rebuilding. It almost certainly refers to the rebuilt ship, as there are few recordings of gundeck lengths ever being made for fourth rates dating from the 1640s.

Dimensions of the new *Tyger*

Keel length: 104 feet 0 inches [TNA ADM106/355 f253]
Length gundeck: 123 feet 8 inches [TNA ADM180/20]
Breadth from outside the plank: 32 feet 8 inches [TNA ADM106/355 f253]
Depth in hold: 13 feet 8 inches [TNA ADM106/370, f342]
Draught: 15 feet 6 inches [TNA ADM106/370, f342]
Burden Tons: 590 24/94 [TNA ADM106/355 f253]

Except for the length of the gundeck, these figures are confirmed in another letter written by Shish in 1684.[673] Some contemporary lists disagree with Shish's figures; one has the ship two inches wider and another two inches less for the depth in hold. These discrepancies are, happily, quite small and very common in contemporary ship lists. They are also understandable, as different people measured the ship at different times, with a choice of ways for measuring the keel. The breadth could be troublesome – it usually meant the dimension outside the plank but not including the thickness of the wales, but in the earlier years of the period it could also mean outside the frames and inside the plank.

The new *Tyger* was to be considerably larger than the original: the keel was five feet longer, the breadth was increased by over three feet and the depth in hold by 1 foot 8 inches, increases that raised her burden from 453 to 590 tons. She was in fact much closer to the ship described in Shish's 1674 treatise, but two feet shorter in the keel, four inches narrower, and eight inches deeper in the hold, making the ships within 24 tons burden of each other. In spite of the increase in size over the old *Tyger*, the new ship was a small fourth-rate ship by the standards of her day and, not counting the galley frigates, smaller than the last eight ships built, the earliest of these dating back to 1659.[674] Her small size was of course necessary because she was sometimes expected to be powered and manoeuvred by oars. The new *Tyger* was clearly designed for speed, for, although increasing in size from 453 to 590 tons, she was built to carry a smaller and lighter armament than the old one.

A lofty height of seven feet between decks, less eight inches for the beams, was preferable, if not essential, for rowing. This height is found in Shish's treatise and in the contract for the 1687 *Mary Galley*, a ship supposed to be '*answerable*' to the *James Galley* of 1676. Such a height would also be appropriate for the rebuilt *Tyger*, and it is apparent in the van de Velde drawings when she is compared with the two other ships she is closely associated with. The recorded deck height of the *Mordaunt* was 6 feet 2 inches,[675] while that of the *St Albans* model is even less. The models of the two ships and van de Velde's drawing of the *Mordaunt*[676] show the upper-deck ports visibly closer to the channel wales than they are in the van de Velde drawings of the *Tyger*, confirming a raised deck height in the latter, suitable for an oared ship.

Whipstaff spindle

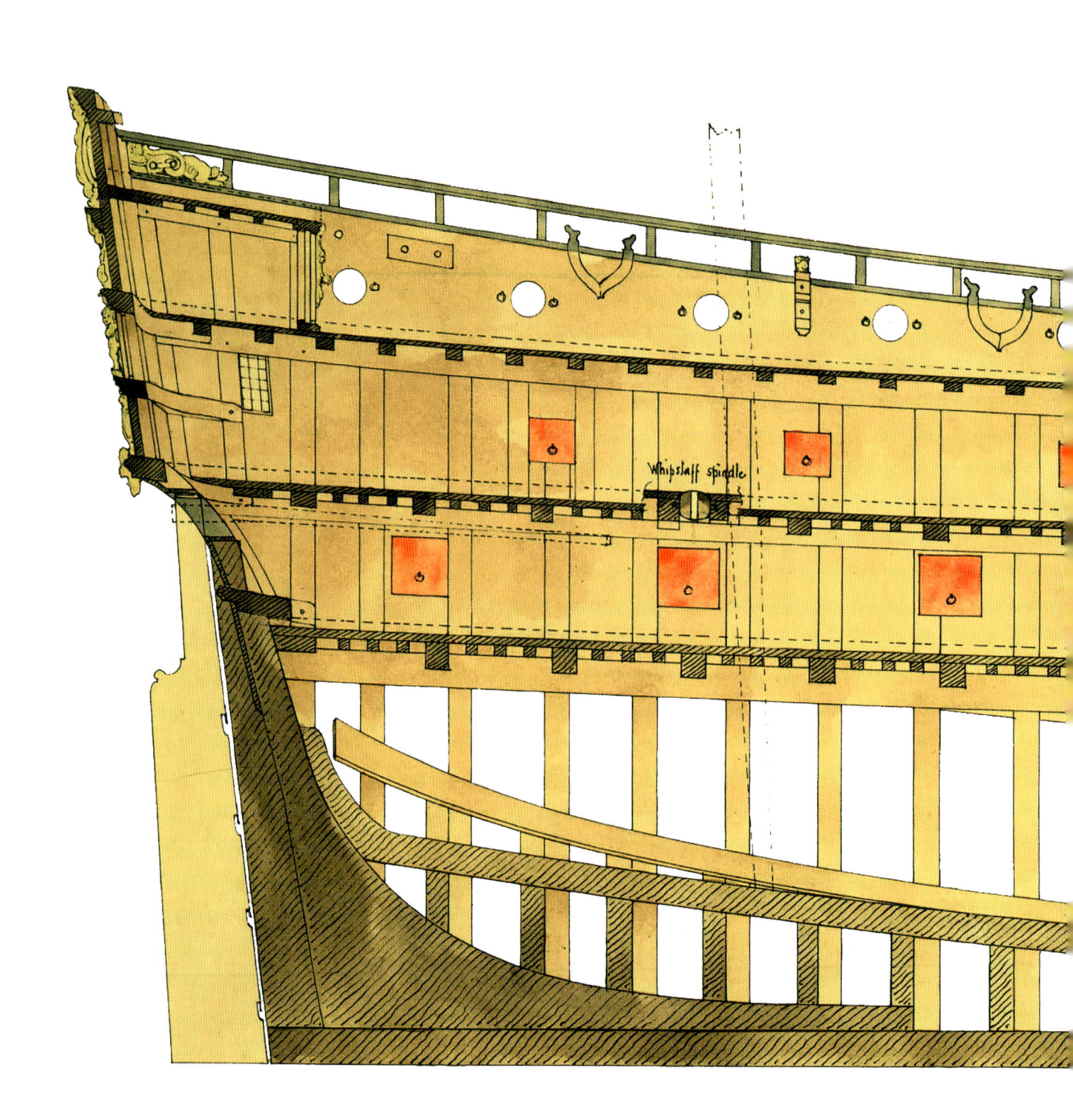

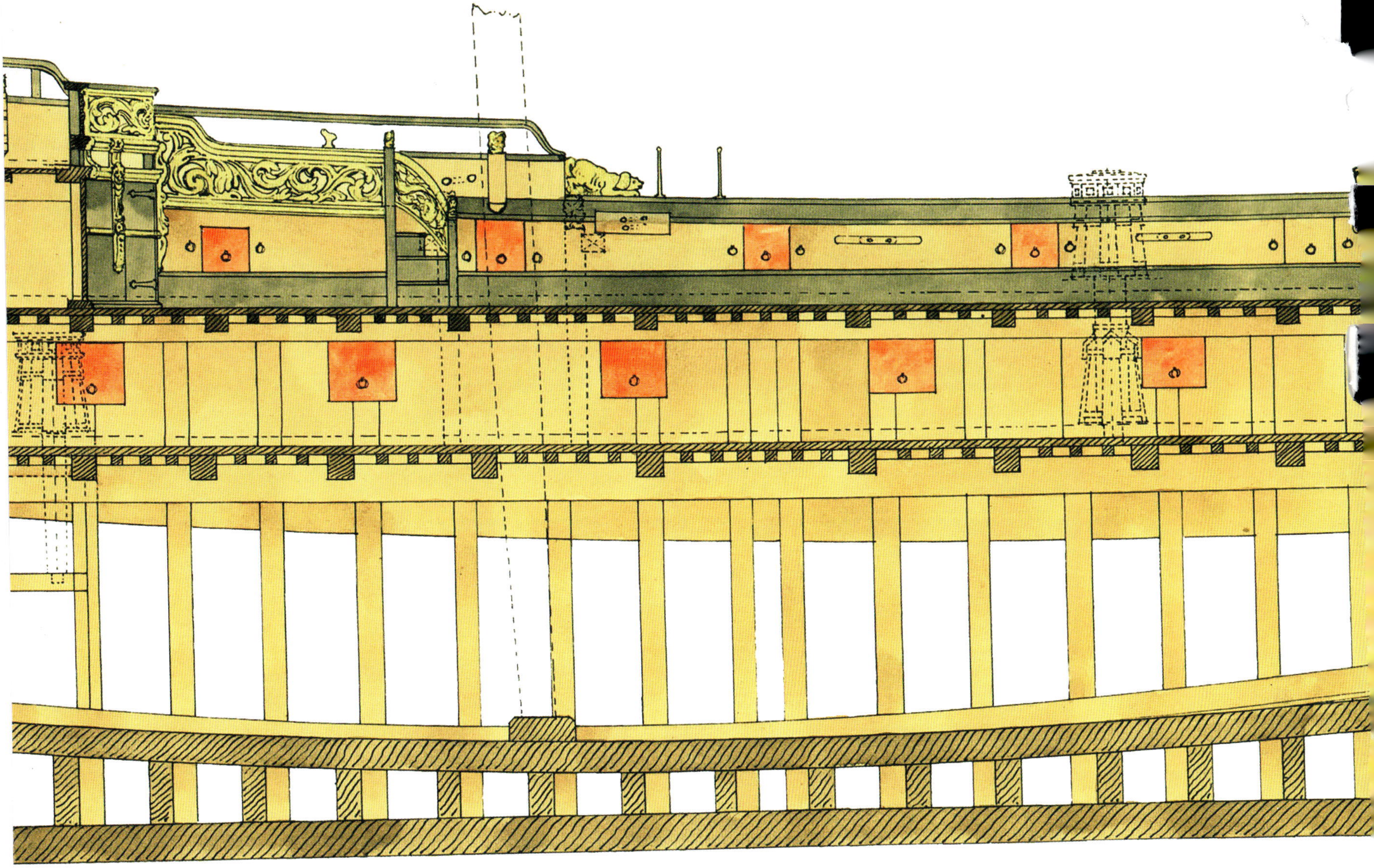

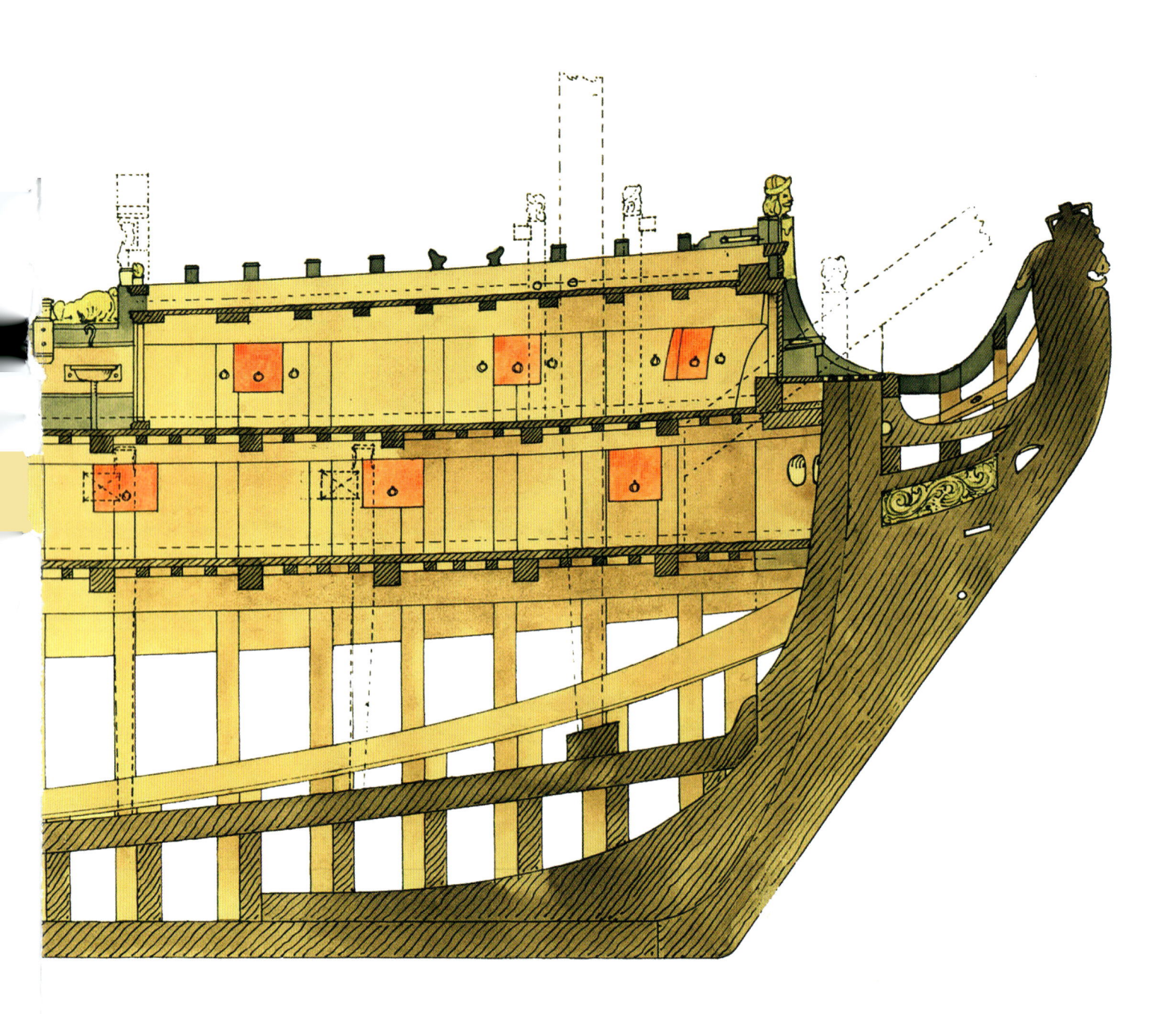

St Albans

The section through centreline showing the conventional layout of the bitts and capstans of the Trinity House model of St Albans drawn by Herbert Read in 1926. Scale 1:72. Courtesy Science Museum / Science and Society Picture Library.

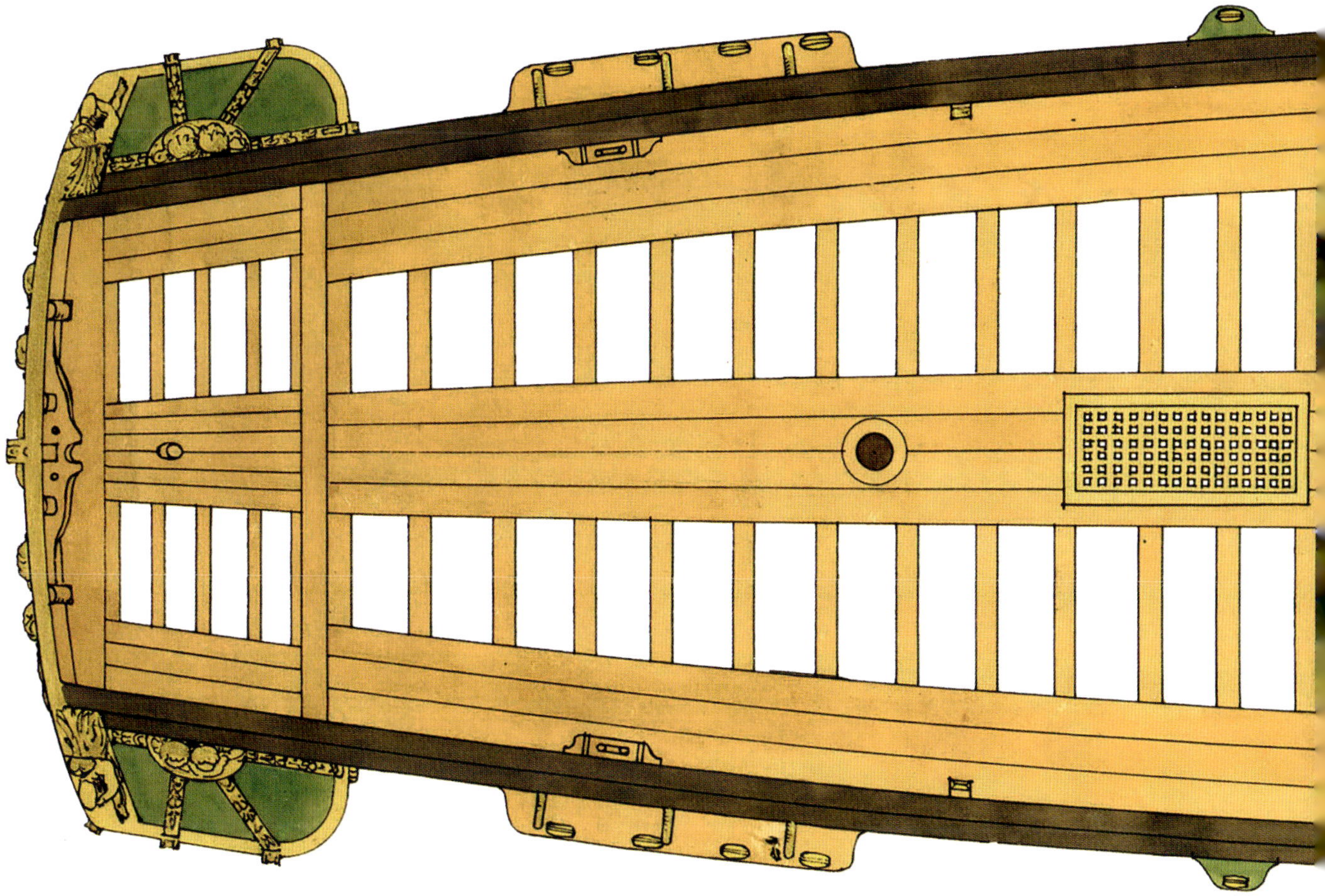

Skid Chesstree.

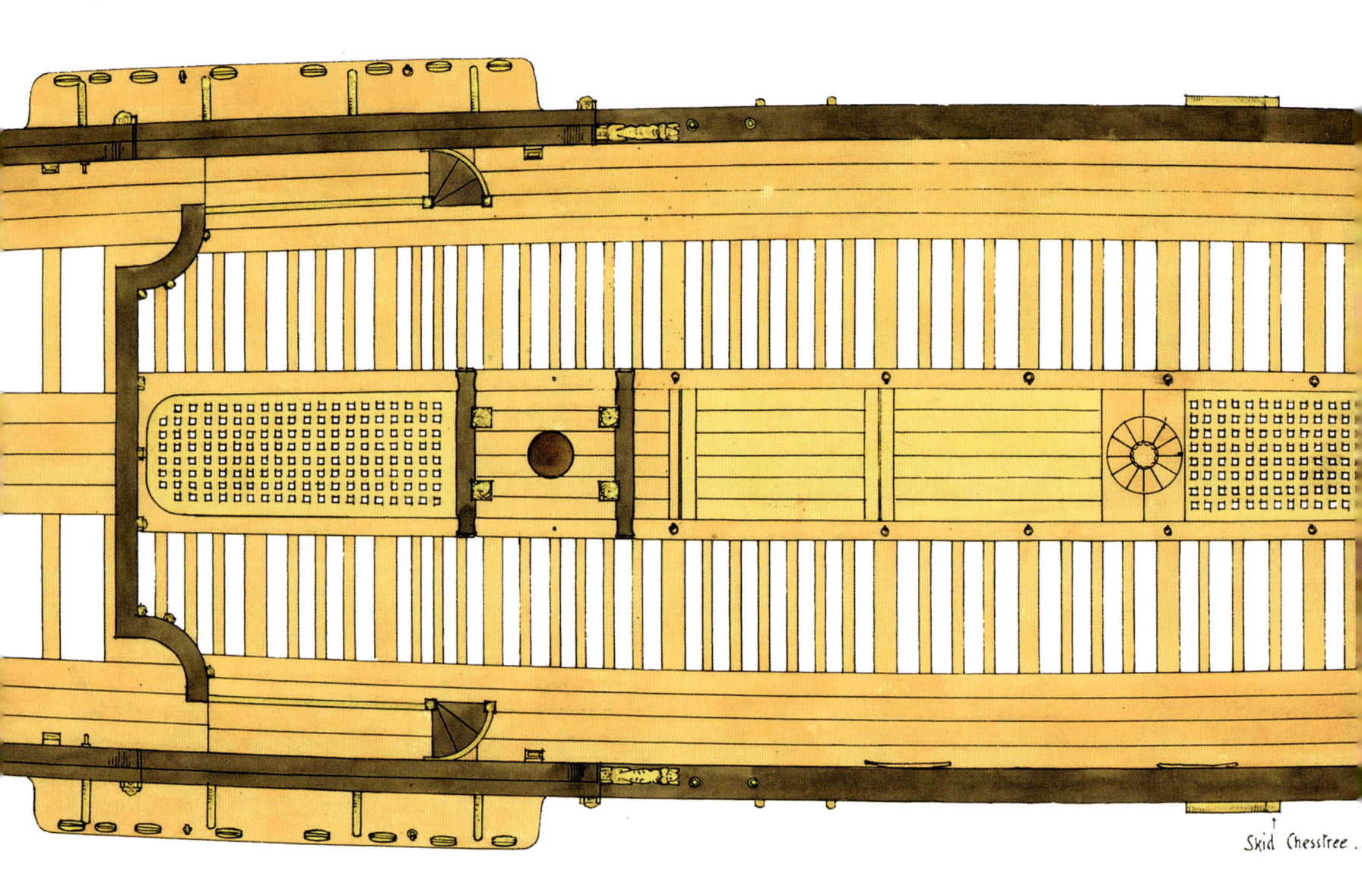

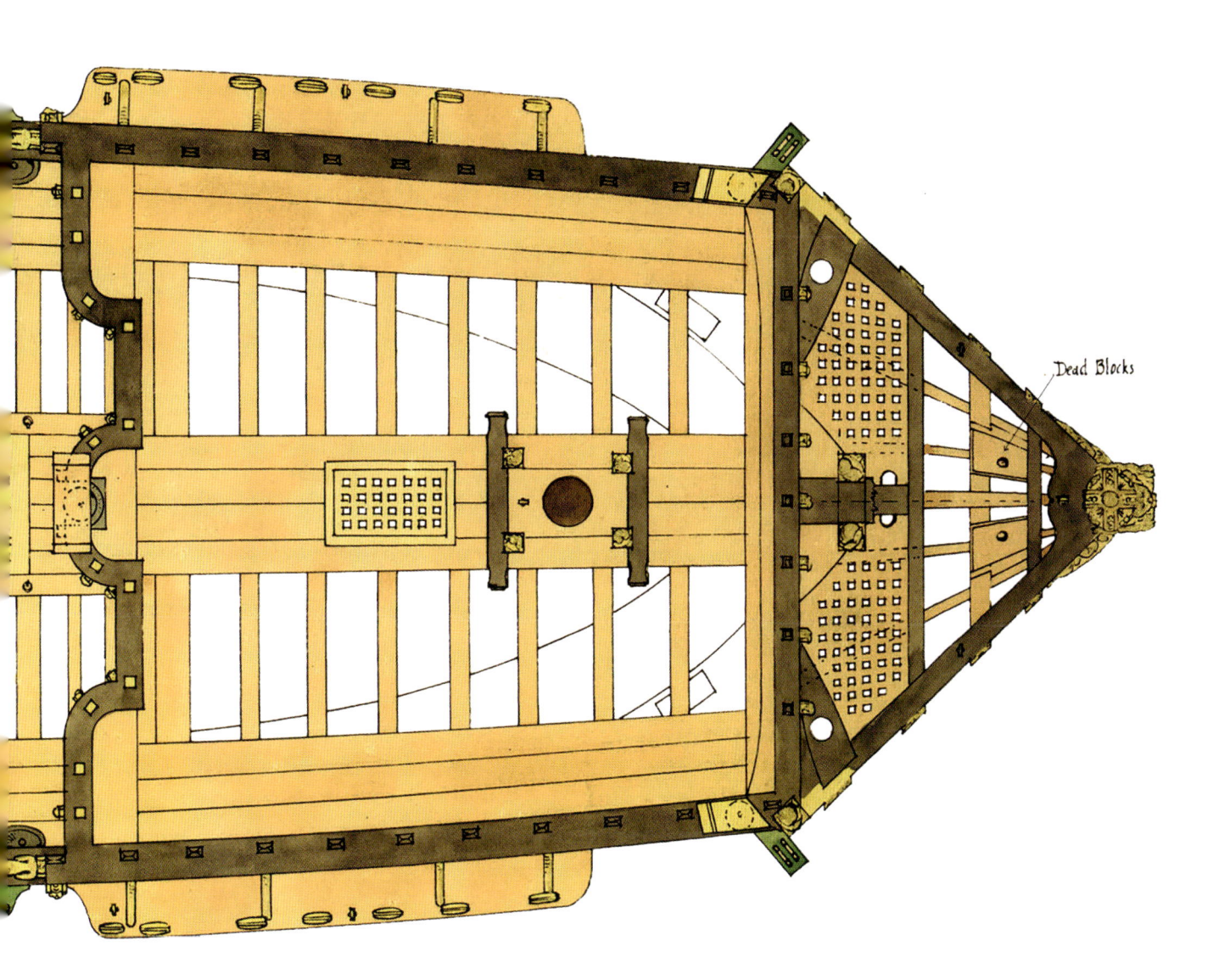

St Albans

The deck plan of the model has some of the planking omitted to reveal the deck beams and carlings beneath. The structure is simplified and is missing the short carlings between the beams. Herbert Read. Scale 1:72. Courtesy Science Museum / Science and Society Picture Library.

THE MASTER SHIPWRIGHT'S SECRETS

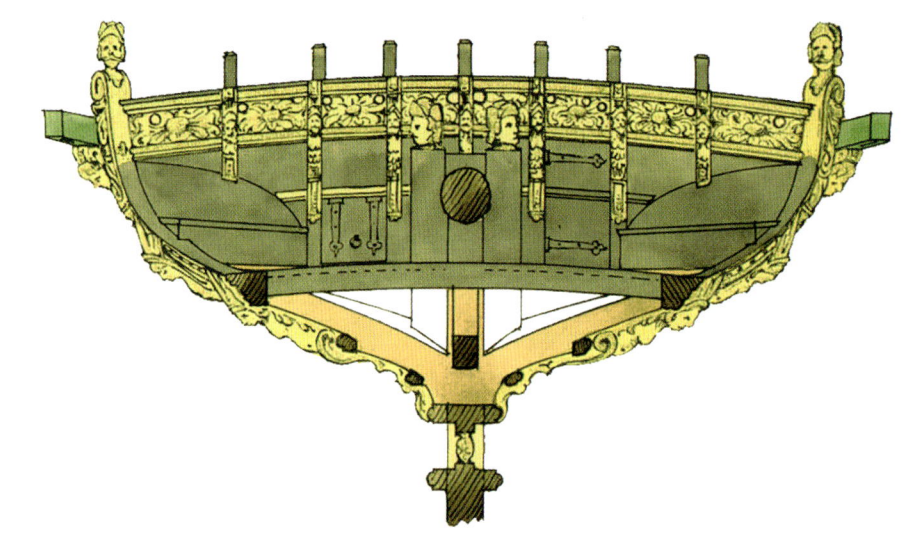

BEAKHEAD BULKHEAD

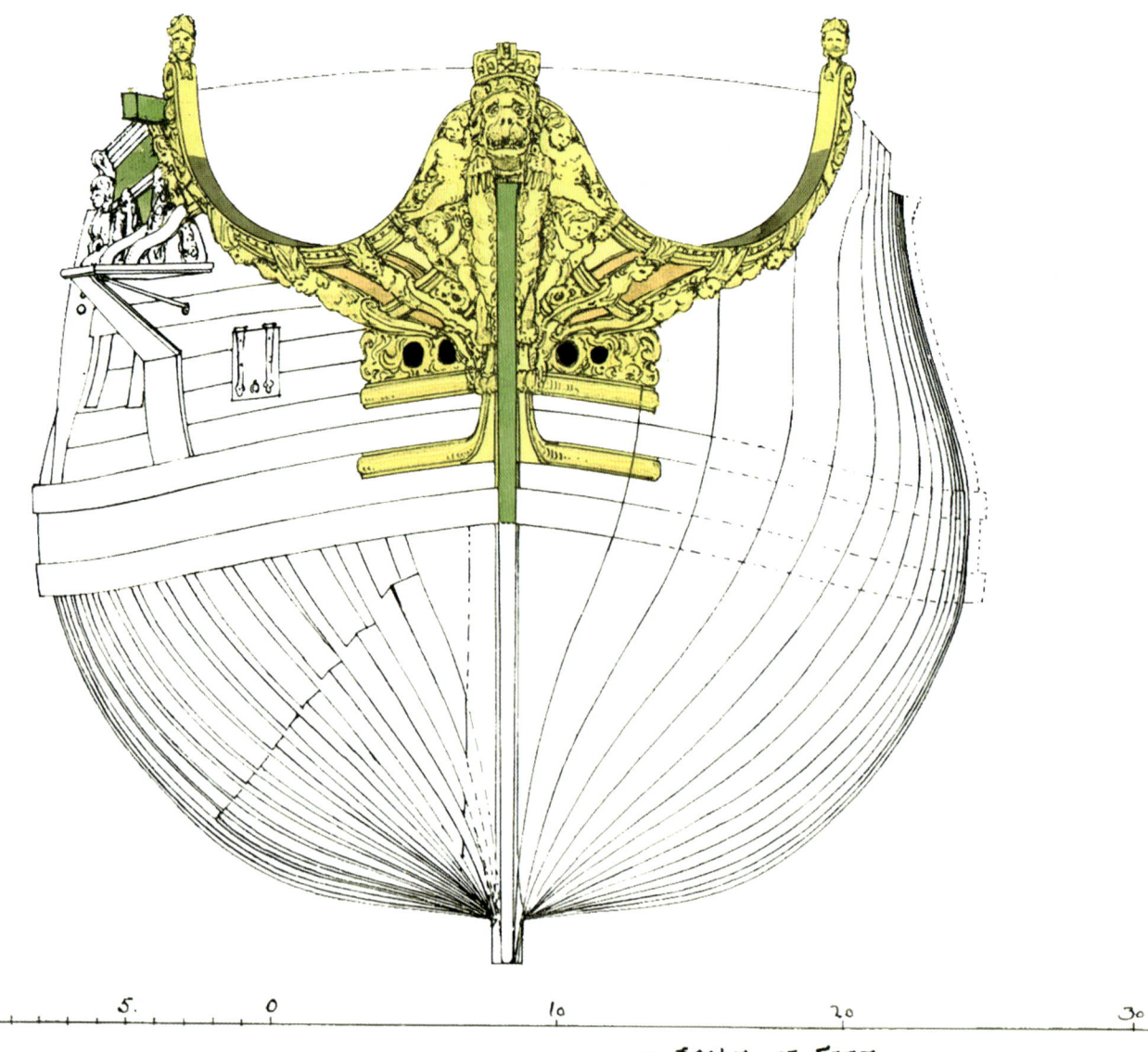

· SCALE OF FEET ·

THE DRAUGHT OF THE NEW TYGER

- GUNS ON QUARTER DECK -

- SECTION AT BREAK OF POOP -

- MIDSHIP SECTION LOOKING AFT -

- SECTION LOOKING FORWARD -

St Albans

The fore and aft sections. Herbert Read. Science Museum / Science and Society Picture Library.
The roundhouse (above) and steerage bulkheads. Scale 1:72.

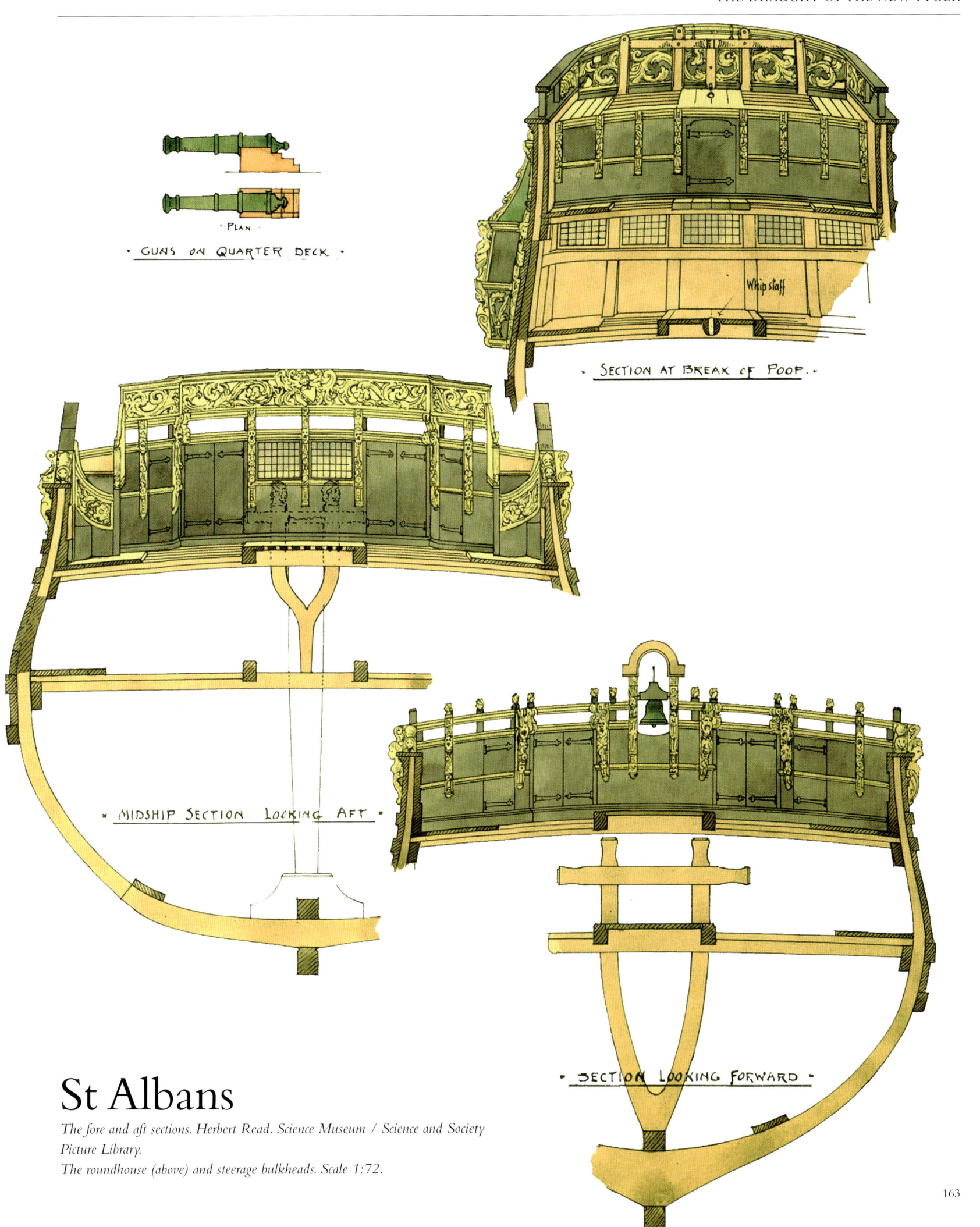

163

Hull lines

Most contemporary authorities, as well as the draught of the midship section of the fourth-rate *Hampshire* of 1653,[677] have a floor-to-breadth ratio of between 0.3 and 0.333 of the beam. Shish's treatise has a much narrower floor of seven feet, and with a beam of 32 feet 6 inches, the ratio is only 0.215. The narrow floor resembles the galley frigates and increases the probability that the ship described in the treatise was intended to have oars. It would be reasonable to suppose that the same narrow floor would have been used by Shish in his design of the new *Tyger*.

There are other indications that the *Tyger* had a narrow floor. A typical warship's hold had enough space for stone ballast, while iron and lead were used for performance-related vessels such as the King's yachts. As he was getting ready to launch the *Tyger*, Shish wrote to the Navy Board for permission to use three old demi-culverin cannon *'for part of her ballast which will be a great advantage for making of room in hold for ye stowage of her provisions'*.[678] Once in the water, the ship was given 90 tons of ballast, of which 33 tons were iron and 57 tons shingle.[679]

It was normal practice for the largest frame, the midship flat, to be placed two-thirds forward from the aft end of the keel and marked with an 'X'. Although Shish marked the 'X' frame at this place in his treatise, the largest in section is four frames further aft. The model of Shish's *St Albans* also has the largest section frame three or four frames further aft than usual. It is easily seen on the model as the floor and futtock arrangement is reversed at this point, in common with actual shipbuilding practice, to maintain the bevelled-under condition.

At the time the *Tyger* was being built, a number of innovations were introduced to the shape of hulls above water; some would become a permanent improvement while others were fashionable and transient. One of the most noticeable changes was a reduction in the narrowing at the top of the side, now usually called the tumblehome. It was around 3'6" for fourth-rate ships of the period, but the models of the *Mordaunt*[680] and the *St Albans*, both closely associated with the *Tyger*, have noticeably less, measuring only 2'6" for the *St Albans* and 2'9" for the *Mordaunt*. Their example in respect of the narrowing was almost certainly followed by the *Tyger*. The other short-lived and fashionable change seen on models and drawings from the period, including the *Tyger*, was the moving of the beakhead bulkhead forward to within a couple of feet of the stem.

A fourth rate draught near ye largest dimensions. (Pages 166–167)

A large fourth rate from An History of Marine Architecture *by John Charnock. Although the print dates from 1802, it is a fairly accurate copy from one of a series of drawings dated 1684 by William Keltridge. It probably, and loosely, shows one of three ships then under construction that were to become the* St Albans, *the* Deptford *and the* Sedgemoor. *Scale 1:96. Keltridge lists the dimensions, which broadly agree with his drawing, below.*

		Ft. Ins.
Length on ye gundeck from ye rabbet of ye stem to ye rabbet of ye post.		127 0
Main breadth of to the outside of ye outboard plank.		35 0
Depth in hold from ye ceiling to ye upper side of the beam.		14 8
Breadth at ye after side of ye main Transom.		21 0
Height on ye gundeck from plank to plank.	Afore.	5 9
	Midship.	6 0
	Abaft.	6 6
The center of the masts from ye rabbet of ye stem.	Fore.	13 6
	Main.	69 0
	Mizzen.	102 0
Draught of water afore 14½ feet. And abaft.		15 10
Number of tuns 664. Tuns & tunnage 885.		
Number of men in war 260. And number of guns 50.		
Burthen in tuns, what she will really carry. 433tuns 16cwt 2qr 8lb		

A fourth rate draught near ye dimensions of ye Adventure. (Pages 168–169)

A smaller fourth rate, from An History of Marine Architecture *by John Charnock. It is a copy taken from a William Keltridge drawing who described it as 'A fourth rate draught near ye dimensions of ye Adventure'. The* Adventure *was a contemporary of the old* Tyger *and much smaller than the ships built in the 1680s. Scale 1:96. Keltridge listed the dimensions below.*

Although the gundeck length of 116 feet agrees with other sources, the main breadth is about three feet too wide. This may be explained if the dimension was taken to the outside of the girdling known to have been fitted to the ship.[681] Another problem is the length of the keel, which was measured at Chatham in 1689 as being 95 feet,[682] as against the drawing, which gives about 104 feet. The late-17th-century style of decoration is shown with two rows of stern windows, which replaced the earlier single row and large coat of arms. The alterations were performed in 1678 at Jamaica Dock by contract with Jonas Shish junior, the youngest son of old Jonas Shish.[683]

Several Dimensions of this Draught.		Ft. Ins.
Length on ye gundeck from ye rabbet of ye stem to ye rabbet of ye post.		116 0
Main breadth of ye to ye outside of ye outboard plank.		32 0
Depth in hold from ye ceiling to ye upper side of the beam.		13 2
Breadth at the after side of the main	transom.	18 4
	Afore.	6 0
Height between decks from plank to plank.	Midship.	6 0
	Abaft.	6 3
	Fore.	12 9
The center of ye masts from ye rabbet of ye stem.	Main.	62 0
	Mizzen.	96 9
Draught abaft of water 15 feet. And afore.		13 6
Number of tuns 435. Tuns & tunnage 580.		
Number of men in war 180. And number of guns 43.		
Burthen in tuns 324.		

A change in ship design occurred at about this time with the rising line of the breadth at the bow moving upward from between the wales to finish above them, a change probably associated with the movement of the beakhead bulkhead. Shortly afterwards, evidence appears of a second rising line of breadth. The new upper rising line was directly above the lower rising line and connected by a vertical straight line. A draught of an English fourth rate[684] with dimensions similar to those in the treatise and to those of the *Tyger*, and with a burden of about 610 tons, clearly shows a second rising line of breadth. The draught is in the Venetian State Archives and probably dates from the time of Edmund Dummer's visit there in 1684. This second rising line became standard practice for wooden warships.

Shish used 2¼ inches of backward lean for the sternpost for every foot of height,[685] and this would have been appropriate for the new *Tyger*. The old ship of 1647 would have had a circular stem in common with most ships of the period. This stem, or at least part of it, was reportedly used again in the new ship, which fits in well with the her known dimensions.[686]

The methods and formulae used by Shish for obtaining the rising and narrowing lines in his treatise easily and appropriately fit the new *Tyger*, with a few small considerations to be taken into account. Small adjustments were necessary as the keel length of the *Tyger* was two feet shorter than that in the treatise, and the gundeck nearly four feet shorter. It was also necessary to allow for the *Tyger* being four inches narrower but eight inches deeper in the hold.

Appearance

The appearance of the *Tyger* and the arrangement of the gunports, carvings and the like are exceptionally well recorded by Willem van de Velde the Elder, the finest marine draughtsman

A British Fourth Rate in 1684.

Fore Body

Sheer Draught

Horizontal Plan

Stern and After Body

A Fourth Rate of the Second Class 1684.

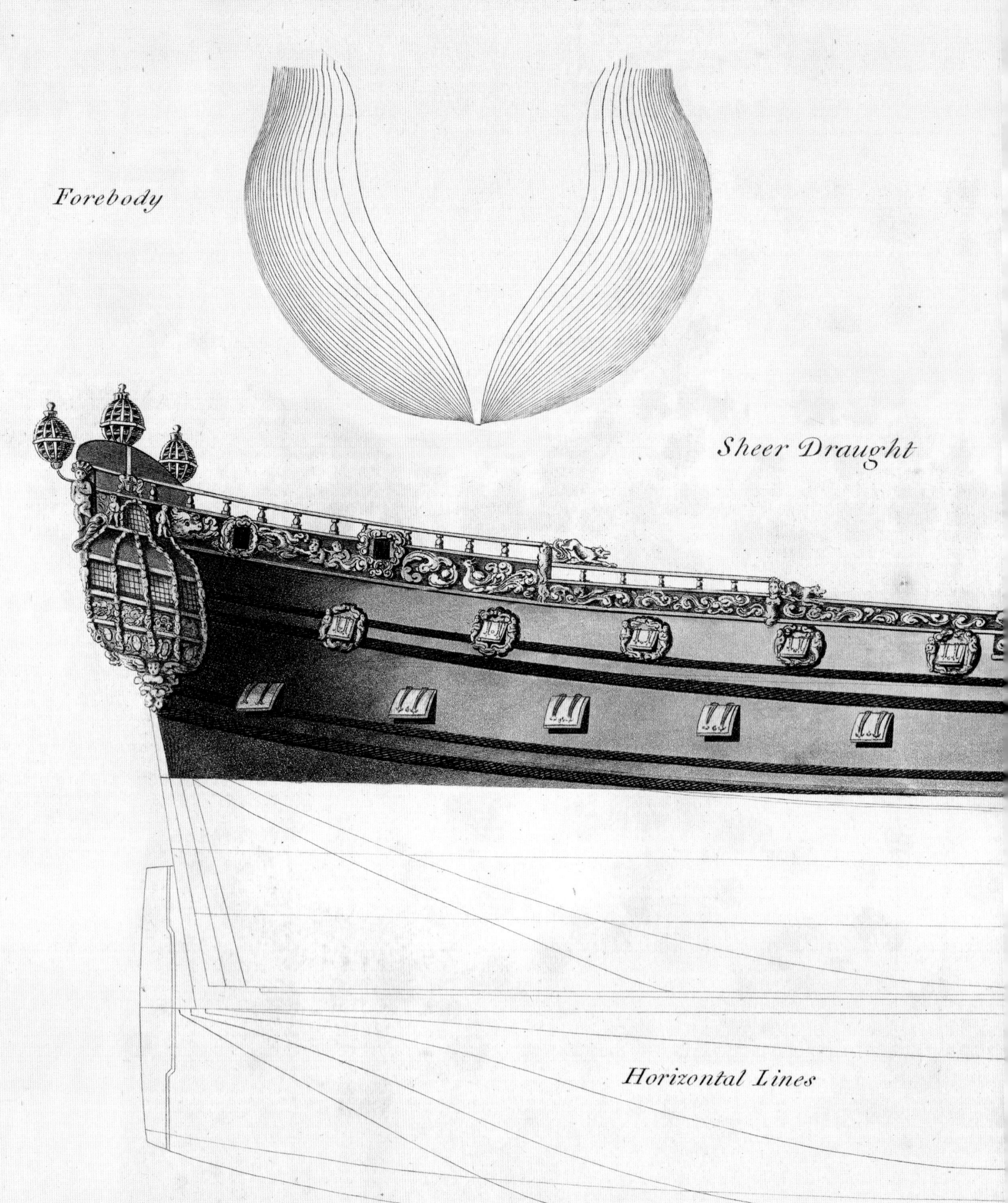

Forebody

Sheer Draught

Horizontal Lines

Stern and Afterbody

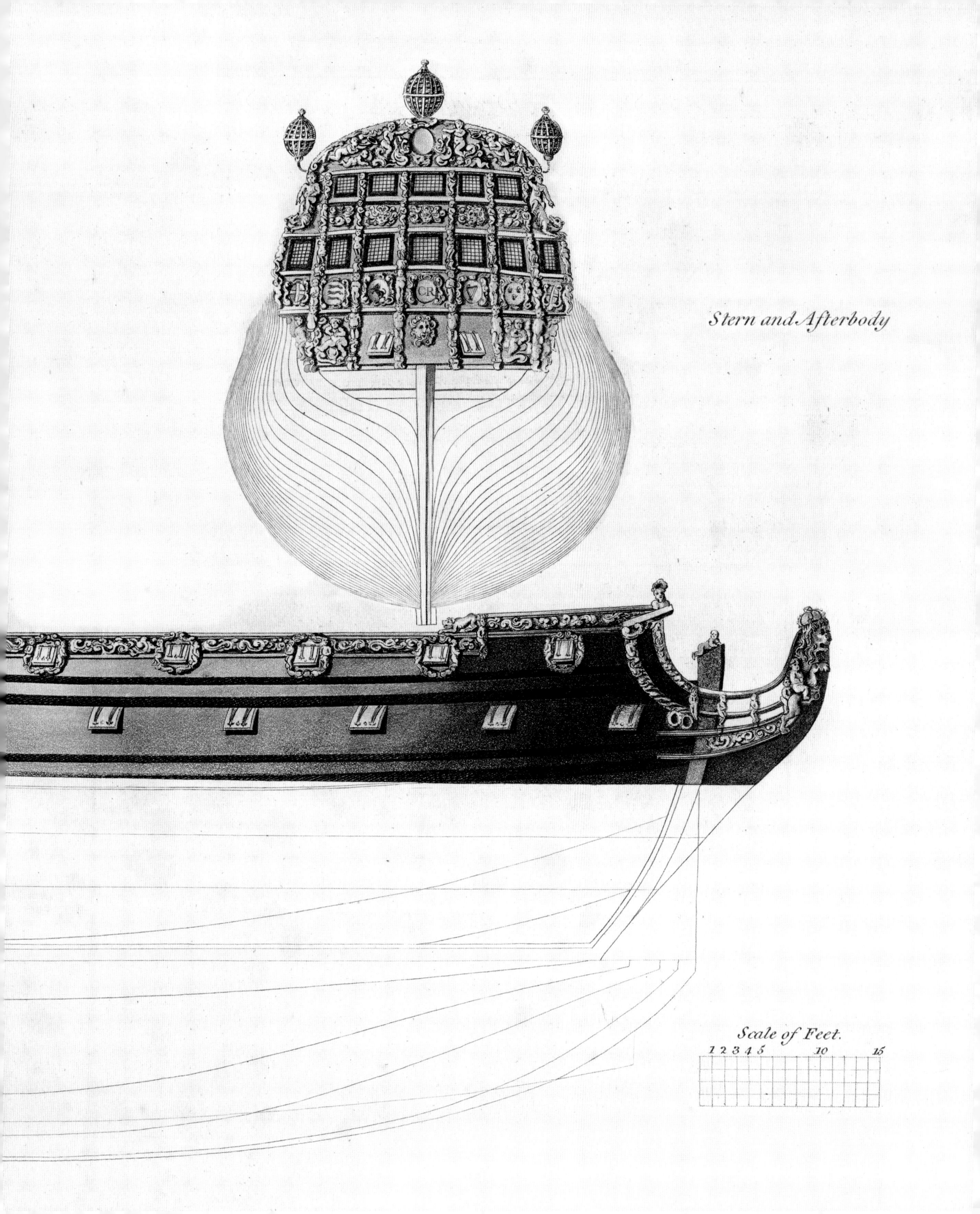

Scale of Feet.

THE MASTER SHIPWRIGHT'S SECRETS

Below. *The broadside of an English fourth-rate ship with a keel length of about 107 feet, from the Venice State Archives, probably drawn by Edmund Dummer. It has wings added to the English lion to change it into the St Mark's lion. Scale 1:96. Archivio di Stato di Venezia. P.P.A. one roll, dr.13, ex S.M. LXXXV-17, neg.14972, print 196.*

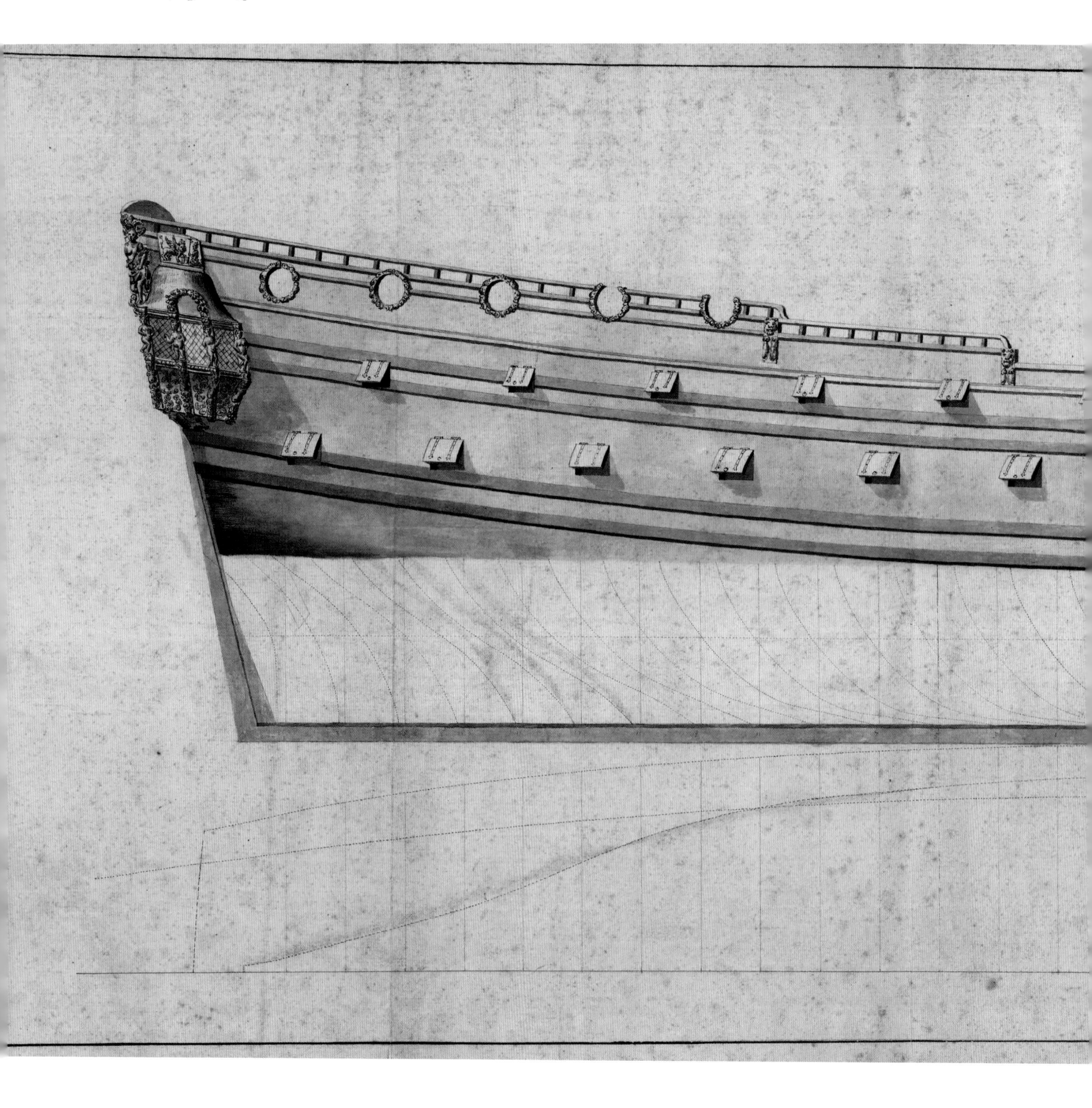

THE DRAUGHT OF THE NEW TYGER

Next Pages. *Section through an English fourth-rate ship, from the Venice State Archives. Note that the midship section shows a vertical line at the extreme breadth between the upper and lower arcs. This was a very recent development that would soon become universally adopted in ship design. Scale 1:96. Archivio di Stato di Venezia. Senato communicate del C.X. 1619-28-1, ex S.M. LXXXV-22.*

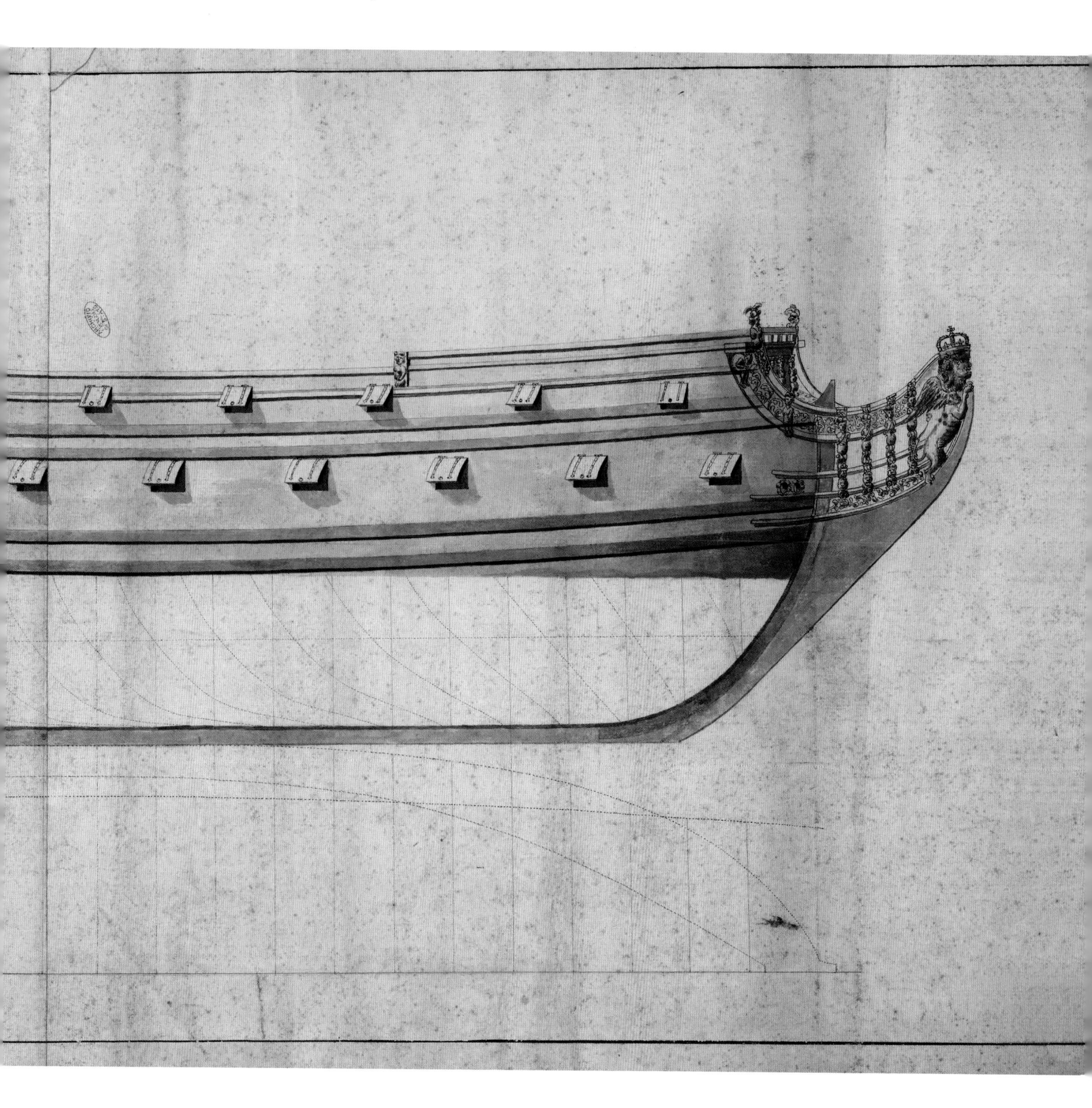

pies d'angleterre

pies de Roihye

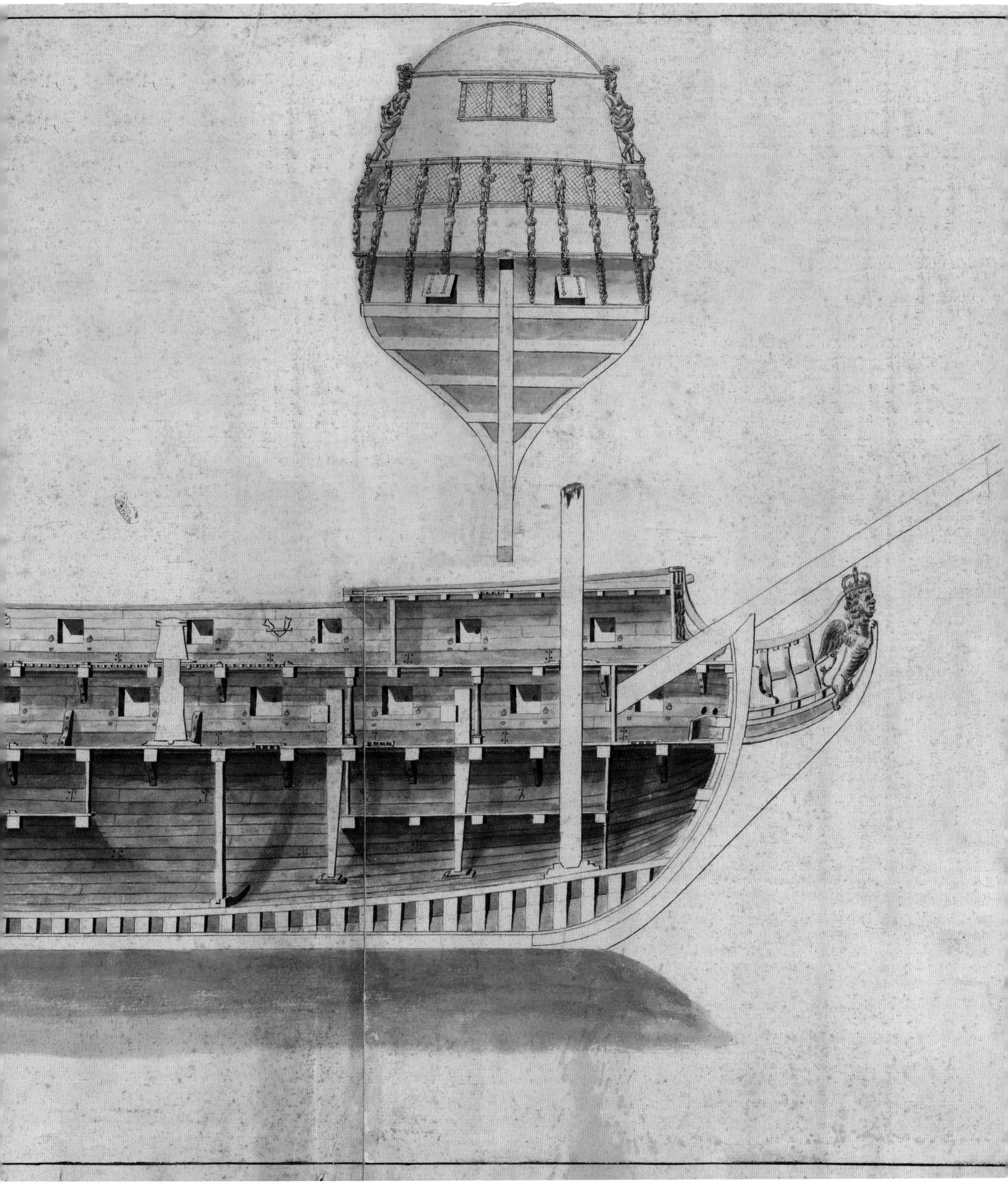

who ever lived. His drawings and Shish's treatise giving digitally recorded lines are as informative as any draught, had one survived. For a couple of reasons, van de Velde quite possibly made as many drawings of the *Tyger* as he did of any other ship. The first reason is because he worked in the Queens Building at Greenwich, only a few hundred yards away from Deptford. The second is because the first commander of the rebuilt ship was Charles, Lord Berkeley, who was joined by King Charles II as he took her down the Thames on her first voyage. The Dutchman was stimulated in his labours by receiving a salary of 25 pounds paid by the Navy each quarter year.[687] The drawings broadly fall into two categories: first, there are the architectural ship portraits, and second, there are the event scenes, many of which show the ship sailing down the Thames. The portraits must have been drawn shortly after the ship's launching as some show the *Tyger* with a flagstaff, which was used at such events in place of a foremast. This idea is supported by the artist's alterations to the waterline which suggest that the drawings were made at the launching and then altered as the ship was loaded with ballast and stores.

Van de Velde's drawings of the foremost bulkhead of the *Tyger* close to the bowsprit, the beakhead bulkhead, are difficult to interpret. It appears impossibly too far forward as it is attached to the knights' heads on each side of the stem. The knights' heads also appear to extend far higher than normal. The configuration suddenly becomes clear when comparing the drawings with the model of the *Mordaunt*. It is an extreme design, with the bulkhead pushed forward within two feet of the stem, leaving space for only a tiny deck. The bulkhead is curved forward to join the knights' heads in an arrangement that must have been fashionable at the time among the shipwrights of Deptford. It made the space inside the forecastle greater, but the curved toptimbers forming the ship's sides must have been difficult and therefore expensive to make as they had a very severe bend where they spread out to the width of the forecastle deck above. It also resulted in the width of the bulkhead at deck level being severely restricted, with hardly enough space for gunports and doors. Just as difficult was the lack of space for the hawse timbers, and awkward arrangements must have been made to accommodate them.

Van de Velde shows the *Tyger* with a strong-looking head with short rails. This is partly due to the forward position of the beakhead bulkhead. The 'beast', as the figurehead was commonly known, is typical of the period, being a crowned emaciated lion. Other drawings give clues regarding the gangways in the waist. They appear to stretch forward from the steerage bulkhead to cover the fourth and fifth upper-deck guns from the stern. Although the gangway itself is not visible, the shadow it cast can be seen through the two gunports.

Opposite Top. *The* Tyger, *just abaft the starboard beam (Robinson 595). Details of interest are the spare topmast in the waist between the gallows and the forecastle. A flagstaff forward confirms that the drawing was made shortly after the launching at Deptford. The guns would not have been aboard, but have been added by the artist for effect. 11½ x 23 inches, probably a worked-up offset. Portrait of the 'Tiger'. Willem van de Velde, the Elder, 1681? PAG6250; Collection: National Maritime Museum, Greenwich, London.*

Opposite Bottom. *The* Tyger, *viewed slightly before the port beam with an awning rigged (Robinson 1219). She is directly before the wind with the main topsail set and the other sails in various stages of being loosed or furled. The large alterations to the waterline may well indicate that the drawing was made at her launching, when she was very lightly ballasted and high out of the water, and then altered to show her at a later date with masts, sails, rigging, ballast, stores and guns aboard. 11 x 18⅞ inches. Portrait of the 'Tiger'. Willem van de Velde, the Elder, 1681. PAH9366; Collection: National Maritime Museum, Greenwich, London.*

Frame

Perhaps the most important architectural feature the van de Velde drawings illustrate is the sheer, or curve, of the wales. They can be seen in relation to the gunports as they curve upward at the stern to cross them. Some of the drawings show corrections to the wales as the artist makes them as accurate as possible, but even so, there are some contradictions. For instance, in the Boymans T356 drawing the upper-channel wale crosses the aft upper-deck port about six inches higher than the average for the drawings, while NMM595 has it about the same amount lower. All the other drawings have it nicely in between, making it statistically safe to assume the average condition is the correct one. Plotting a curve as near as possible to that drawn by van de Velde at the top edge of the upper main wale gives a radius of about 465 feet. Similar inferences can be made for the curve of the upper rails when examining the quarterdeck gunports. Most of the drawings have the aft ring port carving just overlapping the gunwale; however, some drawings have it a rail lower and one has it a rail higher. In common with other ships of the period, it is noticeable that the sheer of the main wales and those of the channel wales and great rail are not parallel but closer toward the bow. The sheer is noticeably less than in ships built just a few years earlier, resulting in a more efficient-looking lower stern.

The toptimbers, the uppermost timbers of the frames, are another interesting architectural feature seen in the drawings of the *Tyger* . These are drawn most clearly by van de Velde along the forecastle, between the gunwale and the planksheer. By dividing the length of the forecastle by the number of visible toptimbers, the distance between the ship's frames, known as the space and room, can be obtained. This is hardly precise science as the length of the forecastle can only be

THE DRAUGHT OF THE NEW TYGER

25 22 19 16

10   7   4   1   ③   Ⓧ   A   D

K   N   Q   T

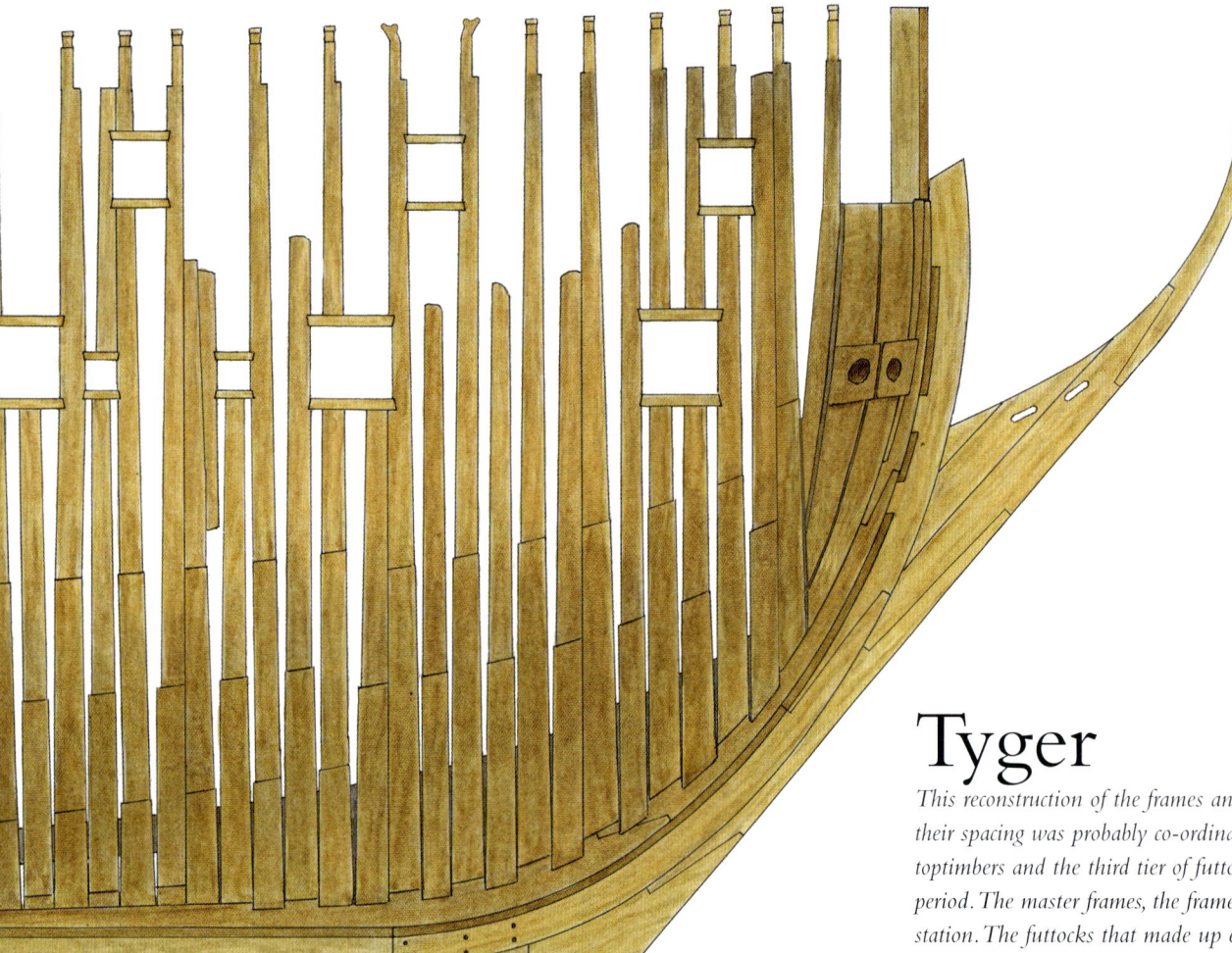

Tyger

This reconstruction of the frames and numerous ports of the Tyger shows how their spacing was probably co-ordinated. The necessary local adjustments of the toptimbers and the third tier of futtocks to form the port sides are typical of the period. The master frames, the frame bends, are placed at every third frame station. The futtocks that made up each frame bend were joined together before erection, while the timbers between are not joined to any other frames but only to the inboard and outboard planking. Between frame bends 'X' and 'A' there is an extra half frame so that all the floors, middle futtocks and toptimbers in the ship are bevelled under. The frame timbers from N forward and 20 aft were much straighter along their profile than the midship frames, allowing the number of futtocks to be reduced from three to two. Shish called the forward timbers 'harping timbers' and those aft, 'long timbers'. The extreme forward position of the beakhead bulkhead meant that the foremost toptimber must have been a very awkward shape. Scale 1:72. Author.

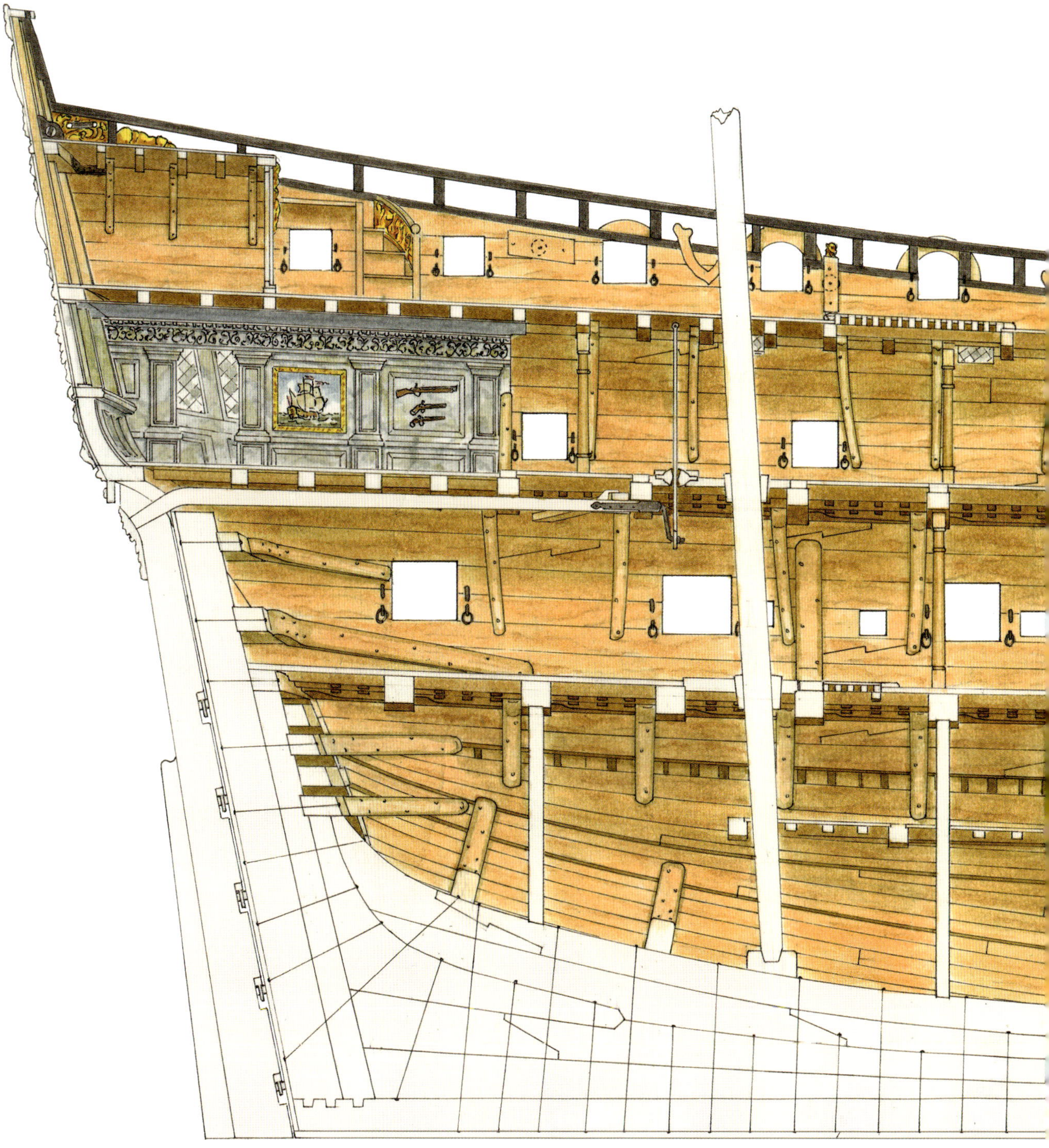

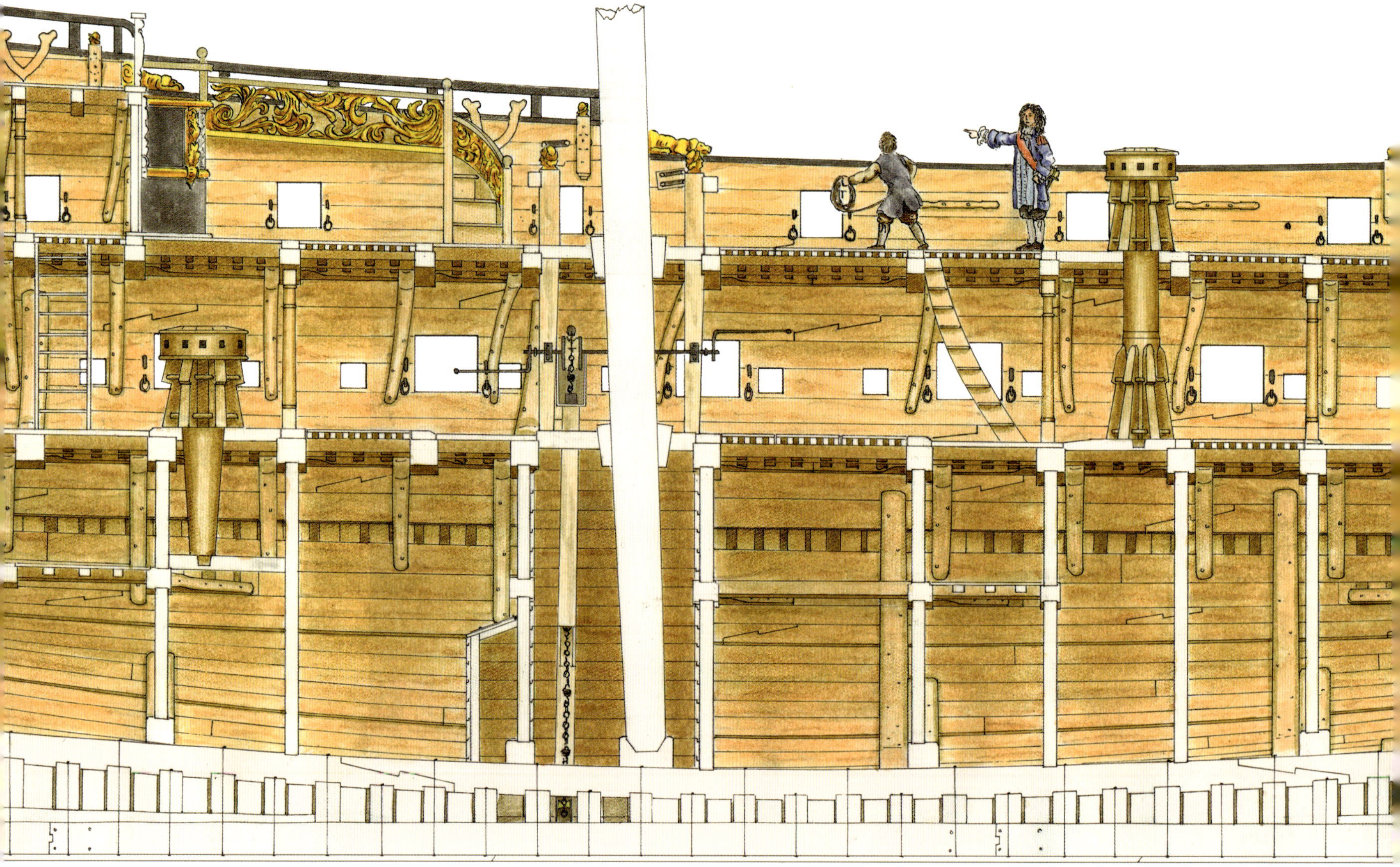

Tyger

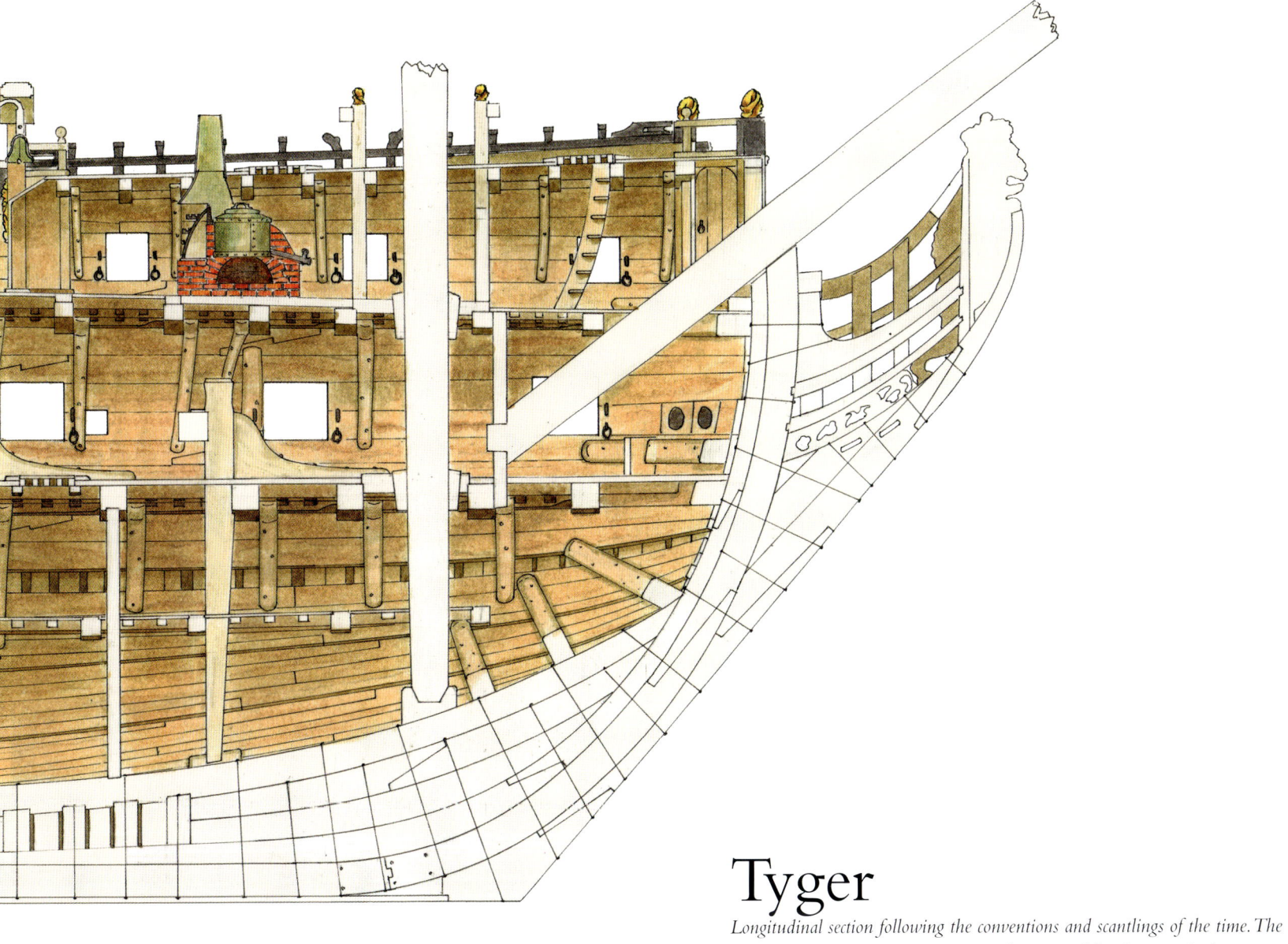

Longitudinal section following the conventions and scantlings of the time. The internal cabins and partitions are not shown, as if the ship were laid up in 'ordinary' to allow the free circulation of air. Cross pillars were not generally fitted to fourth-rate ship at this time. The orlop is shown in the usual manner as a series of removable platforms to allow better access for rummaging in the hold. The forward jeer double capstan was for everyday use. When it was necessary to use the main capstan, the nearby pillars and ladders had to be removed to allow space for the capstan bars. Scale 1:72. Author.

THE MASTER SHIPWRIGHT'S SECRETS

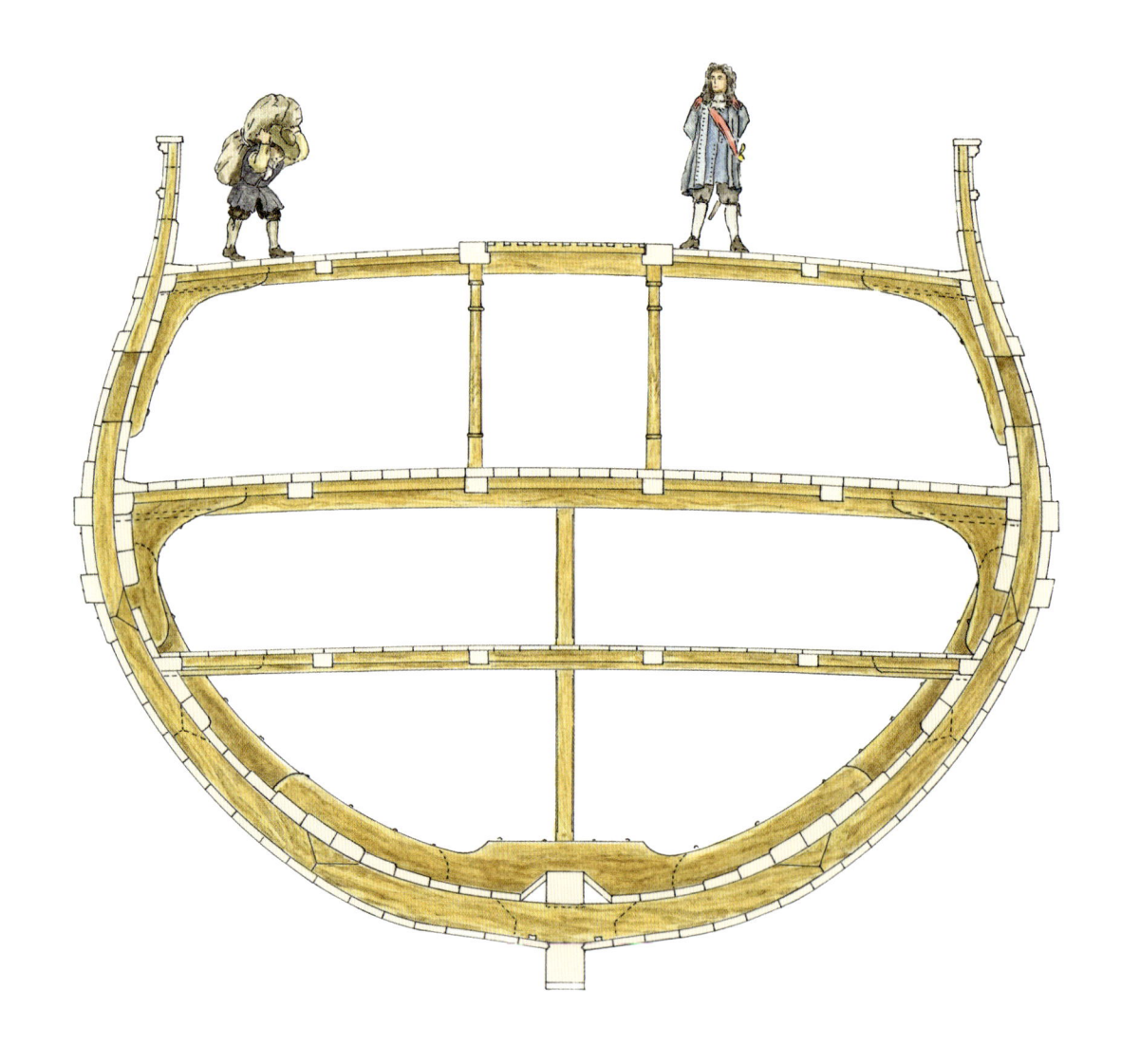

Tyger
Midship cross section viewed from frame "X" looking aft. Scale 1:72. Author.

Above. *The* Tyger, *before the port beam. In common with many of the other drawings, the davit and awning are rigged. 14½ x 28½ inches, Boyman's MB1866/T355. Museum Boijmans Van Beuningen, Rotterdam.*

Below. *The* Tyger, *slightly before the starboard beam; the davit is shipped on the forecastle for handling an anchor. 14 x 28¾ inches, possibly a worked-up offset (Robinson 1220). Portrait of the 'Tiger'. Willem van de Velde, the Elder, 1681? PAI7304; Collection: National Maritime Museum, Greenwich, London.*

The Tyger, viewed from the port quarter, and armed with more guns than she ever carried (Robinson 1221). There would have been no room in the actual ship for guns in the aft-most broadside and the stern port at the same time. The drawing is marked with a stylus, indicating that it was used during the production of a painting of the ship now hanging at Berkeley Castle. 11 x 13¼ inches. Portrait of the 'Tiger'. Willem van de Velde, the Elder, 1681? PAG6251; Collection: National Maritime Museum, Greenwich, London.

estimated and the number of toptimbers recorded along its length by van de Velde varies between 13 and 19. However, three of the six drawings that illustrate them show 14 toptimbers, and this number gives a best estimate for the space and room of 26 inches.

It was essential to have a solid upright timber frame forming the gunport sides as the breeching rope ring bolts that took the recoil of the guns were bolted through them. Contracts for merchant-built 17th-century fourth-rate ships, from the *Foresight* of 1650[688] to the *Norwich* of 1693,[689] give the space and room as 26 inches, qualified in the case of the *Foresight* as being '*at the most*'. The spacing of the gundeck gunports was not usually co-ordinated with the spacing of the frames to form the sides of the gunports as the frames were easily moved or specially made to fit during building. This applied to all ships, including those with one oar scuttle fitted between the ports, these scuttles often being added after the ships were built. An example of this is the case of the fifth-rate *Garland* lying at Portsmouth in August 1686. The Admiralty wanted the ship fitted with oars and the Commissioner at the yard thought it no great matter to cut out the necessary scuttles.[690]

The ad hoc arrangement of altering timbers to suit each gunport normally worked well enough, but in building the *Tyger* difficulties arise. Two oar scuttles between the gunports would require a huge rearrangement and cutting through of the frame timbers to make space for them. The structural problems of the *Charles Galley*, built a few years previously, had emphasised the need to consider frame spacing, '*the row ports having cut & mangled them so that ye ships is so weak*'.[691] The best structural solution was achieved by co-ordinating the gunports with every fifth or sixth frame so that they conveniently formed the gunport and oar-scuttle sides.

This concept is demonstrated by a model which has *Bristol 1666* painted in the panel below the centre stern window. The model is eight feet longer on the gundeck than the measurement given in the recorded dimensions of the *Bristol* of 1653, and it has only ten gunports on the gundeck instead of the 12 shown in van de Velde's drawings. Whichever ship the model represents, it is the only contemporary model with two oar ports between the gunports. Although the model is about the same length on the gundeck as the *Tyger*, it has one gunport a side less, and these are spaced rather imprecisely every sixth frame at 12-feet centres co-ordinating with a space and room of 24 inches.

Most of van de Velde's drawings showing the *Tyger*'s oar scuttles have them level with the gunports and equally spaced between them. This is at odds with the *Bristol* model, which has the scuttles slightly above the lower edge of the gunports. Fortunately, in one carefully made drawing, van de Velde does show the scuttles in the same position as they are on the model.[692] The model also has the scuttles unequally spaced around the gunports. Aft of the gunports they are 1 foot 10 inches away, but they are only nine or ten inches from the forward side, nine or ten inches being the width of the frame timbers at this level. This unequal spacing allowed enough room for the oarsmen between the guns, as they were, of course, standing at the forward side of their oars. The *Tyger*'s oars were double handed, with a probable position for a third oarsman at the end handle.[693] Other van de Velde drawings appear to show the scuttles unequally spaced, one way or the other. In his defence, he would have had no idea that one day his drawings would be scrutinised so closely for such small details. It has to be concluded that the unequal arrangement shown on the *Bristol* model has to be the correct one.

The end view of the ship draughts shows the profile of the frame bends. They were the master frames erected first to establish the hull shape. In large ships, every third frame was a frame bend, although sometimes it was every fourth frame in fourth-rate-sized ships.[694] The frames near the midship bend changed little in shape and were called the 'flats', with their identification marked inside a circle. Draughts made about this time began to have 'diagonals' showing the heads and heels of the individual pieces of frame timbers. Towards the ends of the

The Tyger, *viewed from the starboard bow with davit and spare topmasts in the waist but without the awning. 16⅞ x 31½ inches, Boyman's Catalogue MB1866/T357. Museum Boijmans Van Beuningen, Rotterdam.*

ship, the futtocks and half timbers were much straighter than the other timbers, allowing the number of futtocks in each frame to be reduced from three to two as long straight timber was easier to find than long curved timber.

There are many sources that seem to give the number of futtocks required to make up each frame. Unfortunately for us, however, they are open to interpretation, although at the time custom and practice would have made the correct number obvious to all shipwrights. This uncertainty has led to a great deal of debate among shipbuilding historians. Most contemporary ship models show a simplified form with a floor, one pair of futtocks and a pair of toptimbers.[695] Contracts for Navy warships built privately outside the King's yards give considerable detail about the structures. The earliest specification yet known for a fourth rate relates to the *Assurance* of 1646.[696] For the frame, it mentions the floor timbers and that the gaps between them are to be filled with timbers overlapping – or scarphed, to use the shipwright's language – by six feet. This could be interpreted as meaning that only one frame timber each side reached from the floor to the top of the side, or it could mean that the frame was made up of more

The Tyger, *viewed from the port quarter with a flagstaff forward, indicating that the ship had just been launched. The stern windows appear a little too small when compared with a similar view of the stern in NMM 1221. 14½ x 32⅜ inches, Boyman's MB1866/T356. Museum Boijmans Van Beuningen, Rotterdam.*

The arrangement of the frame timbers around the gundeck ports and oar scuttles of a contemporary model ship which has Bristol 1666 painted on the stern. The frames in the model are of equal breadth from the keel upward, in model-making tradition rather than following shipbuilding practice as shown here. Author.

than one tier of scarphed frame timbers not mentioned in the specification. The vague theme continued with the 1649 contract for the *Foresight*, that of the *St Patrick* of 1665, and a 1673 contract for building some fourth rates in Ireland, all reproduced in chapter 10. They are more detailed, and all mention an additional tier of timbers each side, the gundeck timbers, apparently implying a floor and only two tiers of frame timbers. However, hidden in the payment clause of the 1665 contract is mention of another tier, the toptimbers, while the 1673 contract mentions them appropriately in the specification. All of these contracts imply a floor, two or three pairs of futtocks and a pair of toptimbers.

Clearly the authors of the contracts did not feel the need to indicate precisely how many tiers of timbers were used in the ship's frame. Later, in 1692, the frame timbers were described in the contract for the *Norwich*,[697] a ship that was slightly smaller than the ships of 1665 and 1673. It mentions only the floor, lower futtocks, upper futtocks and toptimbers, but further in the contract, when mentioning the ceiling, there is a concealed reference to a middle tier of futtocks, meaning that there must have been three tiers of futtocks as well as the toptimbers. Further confirmation that there were indeed three tiers of futtocks is contained in building reports for the *Winchester*, another fourth-rate ship built by a contract signed in 1692. On 12 August that year, the following progress report was given: '*Floor cross & keelson bolted, the frame bends up & some lower & middle futtocks got in*'; a week later, the report stated: '*Second tyre of futtocks all in and most of the third tyre & fast.*'[698]

There are other sources, too. In 1670, Anthony Deane, the famous shipwright, listed the timbers for all rates of ships in his *Doctrine*,[699] and mentioned only the floor, lower futtocks, upper futtocks and toptimbers. This could be interpreted as meaning that huge first-rate ships had only two tiers of futtocks. This was certainly not the case though, because building reports for the ships of 1677 confirm that even second-rate ships had four tiers of futtocks.[700] The seeming lack of clarity over the number of futtocks is understandable because the dimensions of the middle futtock in three-futtock ships did not need to be mentioned at all as they were defined by the head of the floor timber and the heel of the toptimber. Other evidence that fourth-rate ships of the period had three tiers of futtocks comes from William Keltridge in *His Book*, dated 1675.[701] He gives, in tabular form, first- and second-rate ships with four tiers of futtocks, third-, fourth- and fifth-rate ships with three tiers and only sixth-rate ships with two tiers. Later, in 1684, he made a detailed end-view drawing of a small fourth rate, showing her with three tiers of futtocks[702] in agreement with the Thomas Fagge prints of a third rate. Also in 1684, Edward Battine listed the size of ship's timbers and gave the same number of futtocks as Keltridge, except that now, even sixth-rate ships had three tiers of futtocks.[703]

Finally, a repair report was written in 1675 concerning the fourth-rate *Princess* of 1660, a ship the same size as the new *Tyger* and built by Daniel Furzer.[704] It specifically mentions that she had three tiers of futtocks, and this may indicate that around 1660 was the time of change from two to three pairs. At the same time, the *St David* of 1667, another fourth rate of the same size, required 16 new lower futtocks, 40 middle futtocks and 30 upper futtocks, together with 20 new toptimbers.[705] It is therefore safe to assume that each frame of the *Tyger* had a floor, three pairs of futtocks and a pair of toptimbers.

THE DRAUGHT OF THE NEW TYGER

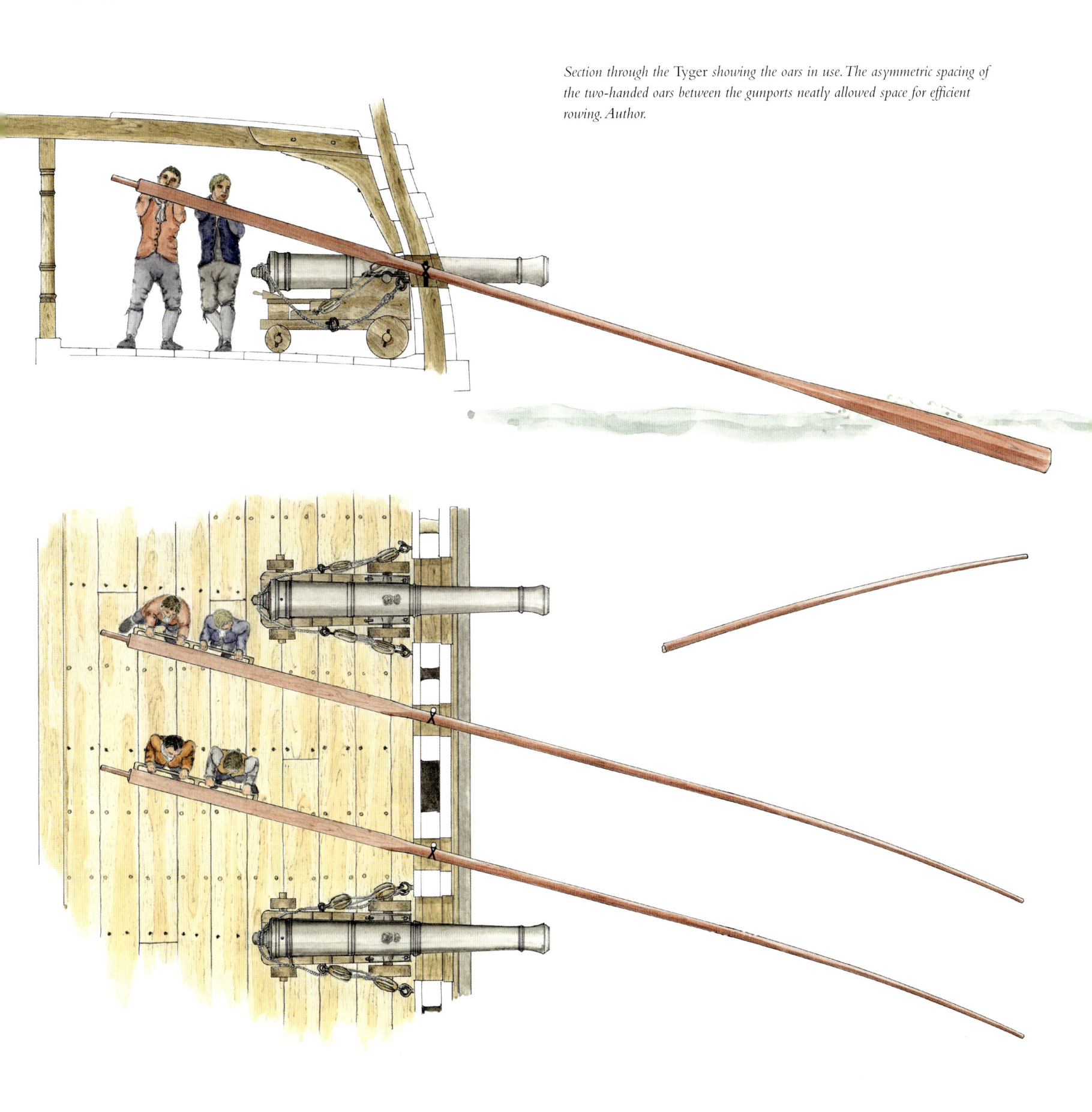

Section through the Tyger showing the oars in use. The asymmetric spacing of the two-handed oars between the gunports neatly allowed space for efficient rowing. Author.

THE MASTER SHIPWRIGHT'S SECRETS

The end views of a third-rate ship, showing every third frame – the frame bends. Also shown are the joints between the individual timbers that form the frames, called 'diagonals', and the sire marks necessary for lining up the timbers to each other during assembly. Note the reduction in futtocks from three to two towards the ends of the ship. From Thomas Fagge, The Bends of a Ship c.1680. Private collection, via author.

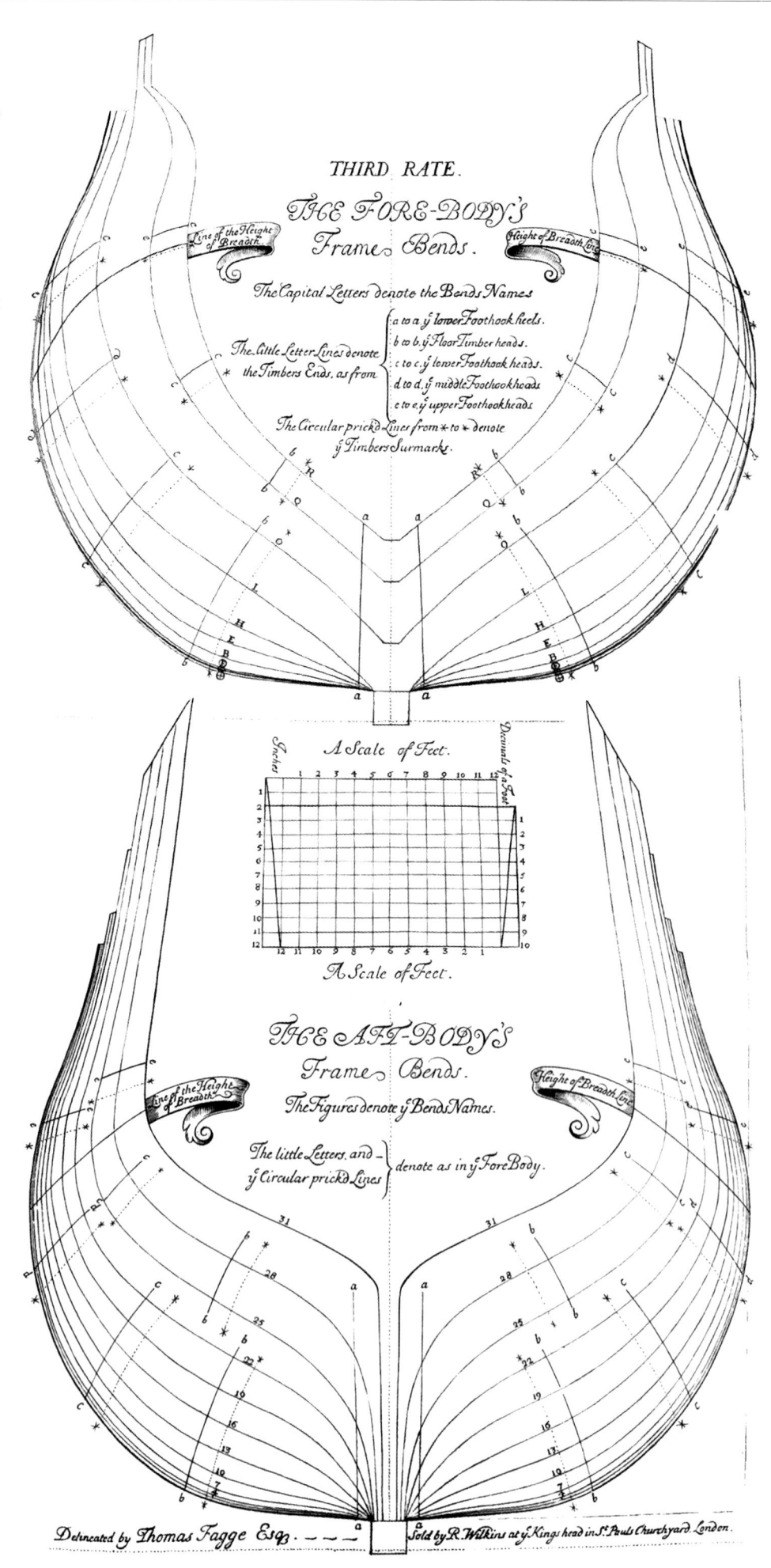

THE DRAUGHT OF THE NEW TYGER

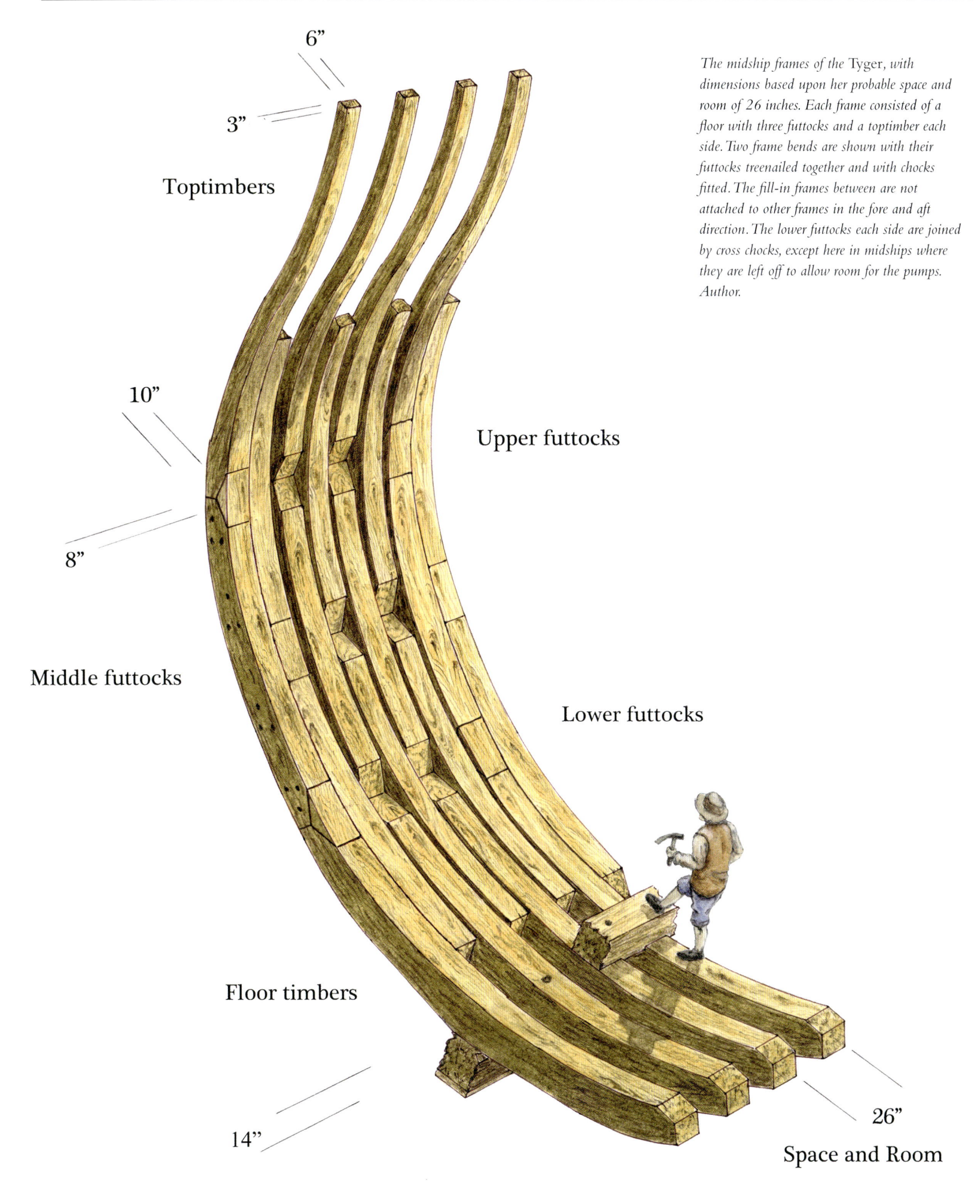

6"

3"

Toptimbers

10"

Upper futtocks

8"

Middle futtocks

Lower futtocks

Floor timbers

14"

26"

Space and Room

The midship frames of the Tyger, *with dimensions based upon her probable space and room of 26 inches. Each frame consisted of a floor with three futtocks and a toptimber each side. Two frame bends are shown with their futtocks treenailed together and with chocks fitted. The fill-in frames between are not attached to other frames in the fore and aft direction. The lower futtocks each side are joined by cross chocks, except here in midships where they are left off to allow room for the pumps. Author.*

CHAPTER 7

BUILDING THE NEW TYGER

The old *Tyger* arrived at Deptford for repair in 1674, but following a series of mishaps, she was '*broake up at Deptford in 1676*'.[706] That was about the time the process of taking her apart began. During this period, regular reports were made detailing the amount of work necessary to repair every ship in the fleet. The last report for the *Tyger* is dated October 1677, but after that there were no more estimates as she was beyond repair or rebuild. By April 1678, nothing remained except for a few pieces of disarticulated timbers. They were necessary only to keep her name on the books as part of the King's

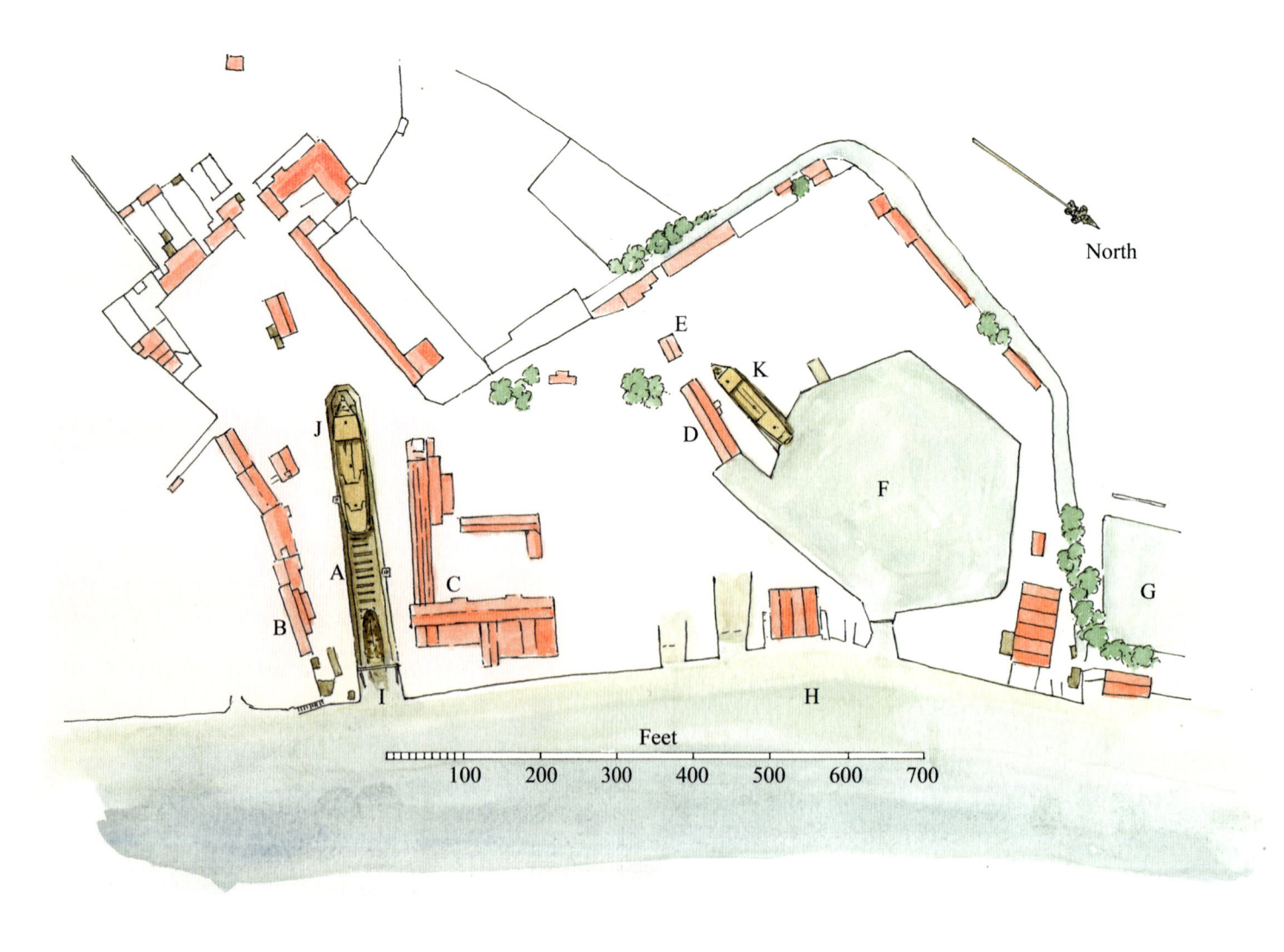

BUILDING THE NEW TYGER

Above. *The foundation stone of the Great Storehouse, with the inscription 'A.X.H.R. 1513' (Anno Christo, Henricus Rex, 1513). The storehouse served the whole of the Navy during the 17th century. It was still in place in 1952, but the building was demolished shortly afterwards. The stone was preserved, and it is hoped to bring it back to the site as part of the Lenox Project, a heritage project to construct a replica of the* Lenox, *which was built in the head of the dock at the time the old* Tyger *was broken up. The drawing was made by C. R. B. Barrett and is dated 1894.*

Above Right. *One of the original single light windows of the Great Storehouse, drawn by C. R. B. Barrett, and dated 1894.*

Opposite. *Deptford dockyard during the period the new* Tyger *was built. Author. A. The double dry dock; B. The Master Shipwright's House, occupied by John Shish; C. The Great Storehouse; D. The mould loft; E. A saw house; F. The wet dock; G. Mast pond; H. River Thames; I. The old* Tyger *on 6 June 1677. She is partly sunken at the stern of the dock, where she was taken completely apart; J. The new third-rate ship the* Lenox, *built at the head of the dock. The remains of the old* Tyger *were cleared away for her launching by 12 April 1678; K. The slipway where the new* Tyger *was built between 12 May 1679 and 24 June 1681. Author.*

benevolent scheme for paying deserving warrant officers and because of his distaste for abandoning old ships. He must have planned the design for the new ship with John Shish at the launching of the third-rate ships *Lenox* and *Hampton Court* at Deptford during the same year. The King's most recent visit was at the launching of the *Duchess* on 12 May 1679, and it was probably on this occasion that he gave Shish instructions to start the 'repair' of the *Tyger*. The earliest mention that work had started comes less than two weeks after the launch of the *Duchess* in a letter written by Henry Loader, the supplier of heavy ironwork to Deptford. On 26 May, he wrote to the Navy Board complaining that he was owed £3,000 by the Navy, '*and now the Foresight, Tiger and the fireships are under repair and other works daily coming… I have run into debt*'.[707] The secretive nature of the work continued, for while extensive and detailed reports were made for other ships being repaired at Deptford, including those mentioned by Loader, nothing is mentioned of the *Tyger* for a long time.[708] The King and John Shish maintained the secrecy about repairing the *Tyger* because, of course, it was no such thing. It was a new ship built under the disguise of a repair, and the supposed use of a few fragments from the old ship were intended to conceal the fraud.

197

The Mould Loft

If any remaining timbers of the old *Tyger* were used in the new ship they would have been taken less than 200 yards across the yard to the launch, usually called a slipway, that led into the wet dock. According to a plan of the dockyard made by Edmund Dummer before 1688, it was a little to the north, or upriver, of the mould loft.

The process of building the new *Tyger* began in the conveniently nearby mould loft where the moulds, or patterns, were marked out. It had a very smooth and even platform with a spacious floor lit by large windows. It was washed in black size so that the plan of the ship could be marked out full size in chalk.[709] If the mould loft was big enough, the procedure followed that for drawing the 1/48 scale paper draught. Often compromises had to be made, such as drawing the end views on top of and overlapping the side view. This may not have been necessary for the *Tyger* as it's safe to assume the mould loft at Deptford would have had much better facilities than most.

The first lines chalked in to produce the three views were the keel and vertical centreline to establish a starting point for the ship's dimensions. These, and other straight lines such as the frame stations in the side view, were probably made using chalked twine stretched between two fixed points. By plucking the twine, it would vibrate against the floor to leave a perfectly straight line. After that, the co-ordinates of the point data from the tables for the rising and narrowing lines were added.[710] Their dimensions would have been previously marked out on staffs which may also have been used later when erecting the finished frame timbers in the hull. Taking the data created by geometric mathematical formulae ensured that the hull lines did not rely on a method of crude scaling up from a paper draught. With no errors, they would be exactly what John Shish expected them to be. The rising and narrowing lines were then completed by drawing curves through the rows of points. It is most probable this sophisticated practice was used on all ships built for the Navy in the King's yards.

Attention then turned to the end view below the waterline, where the arcs were drawn using instruments called sweeps.[711] Sweeps were large compasses used to make the two arcs controlled by the point data of the rising and narrowing lines. A larger arc, the reconciling sweep, was then drawn to connect the two smaller sweeps. For the *Tyger*, this arc had a constant radius, estimated at 17 feet 10 inches. Using a compass that large would have been difficult and would have also involved the very awkward problem of finding its centre. A simple wooden pattern made with a fixed arc was used instead.[712] Other details were added to the side view, such as the curve at the top of the floor timbers directly over the centreline of the keel, known as the cutting-down line.

Once the full-size draught was chalked out, the waterlines on the half-breadth plan were plotted using data from the other two views. They were checked to make sure they followed even curves without sudden bumps. This was only a problem at the bow and stern, where the system of tangential joined-up arcs broke down as the waterlines joined the stem and sternpost. Any necessary changes were made and the offending chalk lines rubbed out. Above the waterline a similar process was carried out to the hull shape. The beakhead bulkhead of the *Tyger* was so far forward that locally it created a particularly complex shape that needed resolving. Further aft, the toptimber sweep gradually increased to become a straight line at the stern.

Making the Moulds

Once the chalk draught was perfected and approved by John Shish, the process of making the moulds was started. This would have been carried out under the supervision of Owen Bagwell, Shish's foreman for repairs. John Shish had such confidence in him that during the building of the new *Tyger* he spent only a tenth of the time supervising her that Bagwell spent.[714] The moulds were made from dry and seasoned deal board planed smooth.[715] Apart from one or two references saying that a set of moulds was, or had been, made, contemporary naval records hardly ever mention them. The early works on shipbuilding are of limited help, but when they are studied together with later works, such as William Sutherland's *Shipbuilders Assistant* of 1711 and the 1754 *Treatise on Ship-building and Navigation* by Mungo Murray, a reasonably clear picture emerges. Unfortunately, Sutherland shows unlikely features of his own invention, such as a curved sternpost and a circular hull – not that any particular standard method was used, as the sources are full of references such as '*generally made*' and '*usually*'.

The Floor Mould

The most important moulds were for the floor timbers. It was important to make them as accurately as possible as the rest of the frame was built up upon them. They were made using an ingenious old method that was still in use towards the end of the 18th century.[716] The beauty of it was that only one simple and practical mould was needed to make many of the floors, the exceptions being the awkward ones near the bow and stern.[717]

BUILDING THE NEW TYGER

The previously calculated co-ordinates that controlled the rising and narrowing of the floor and breadth were marked on staffs and used for making the chalk draught in the mould loft. The rising staff of breadth for the Tyger would have been 18 feet long. From William Sutherland's book The Ship-builders Assistant, *1711.*

Drawing the full-size chalk end view of the aft frames of the Tyger. Note the points of the rising and narrowing lines of the breadth, marked in red as 'A', and the floor, 'B'. From them the sweeps, or arcs, at the breadth and the floor are drawn. The reconciling sweep that connected them is shown being drawn using a pattern. The broken line is the sirmark and the parallel line near it marks the frame timber heads. The sirmarks were marked on the mould and later transferred onto the frame timbers themselves to ensure they were accurately assembled in their correct positions. For clarity, only every third frame (these are known as the frame bends) is shown here, but in practice all the frames were drawn.[713] *Author.*

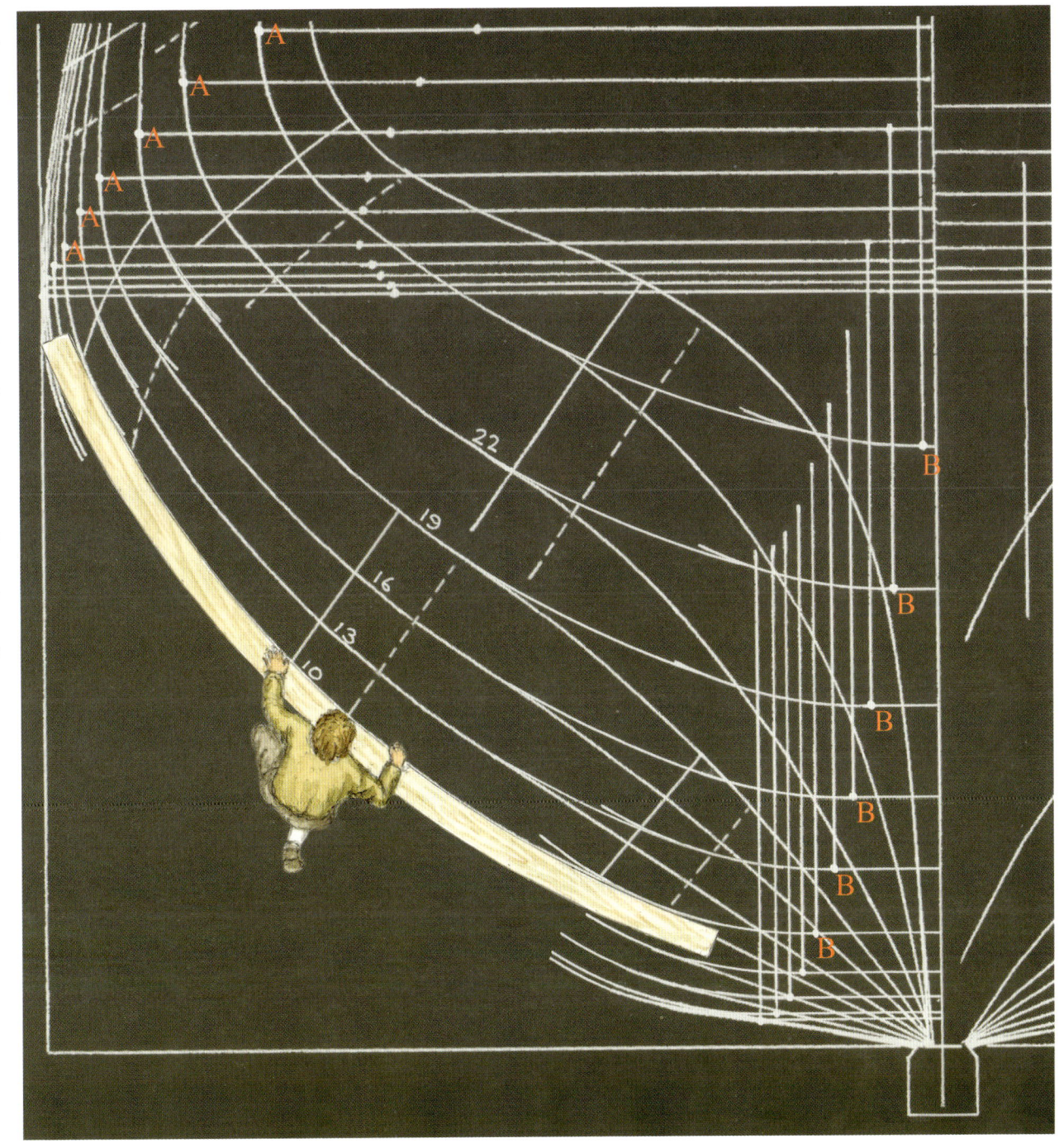

THE MASTER SHIPWRIGHT'S SECRETS

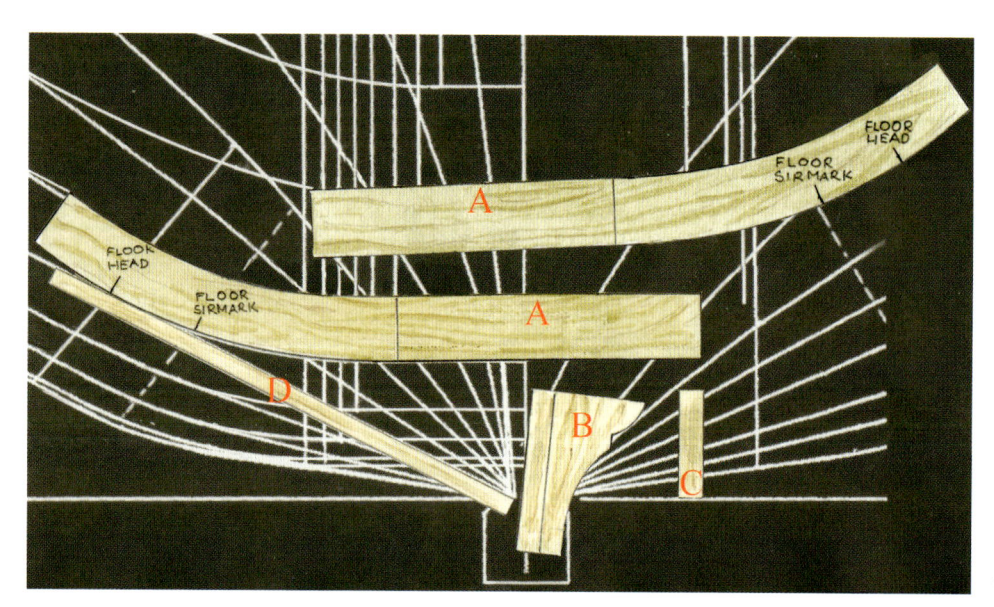

Left. *The individual components of the universal floor mould before the frame stations were marked on them. For the* Tyger, *they could be used to mark out all the floors between frame 10 aft and frame G forward. Author. A: The two 'bend moulds', one for each side of the ship. They were made to overlap and accommodate all the widths of floors. They were also made to the in and out dimension of the frames, the moulded dimension. B: The 'rising square'. While the bend moulds were used to set the correct width of the floor, the rising square was set to control its height. The straight edge represents the side of the keel, while the marked line is the centreline of the ship. C: The 'cutting-down batten'. It was used to mark the depth of the floor timbers above the keel. D: A batten to mark the floor timbers from the floor sweep to the keel both inside and out. On some ships this was a curved 'hollow mould', marked to give its precise position.*

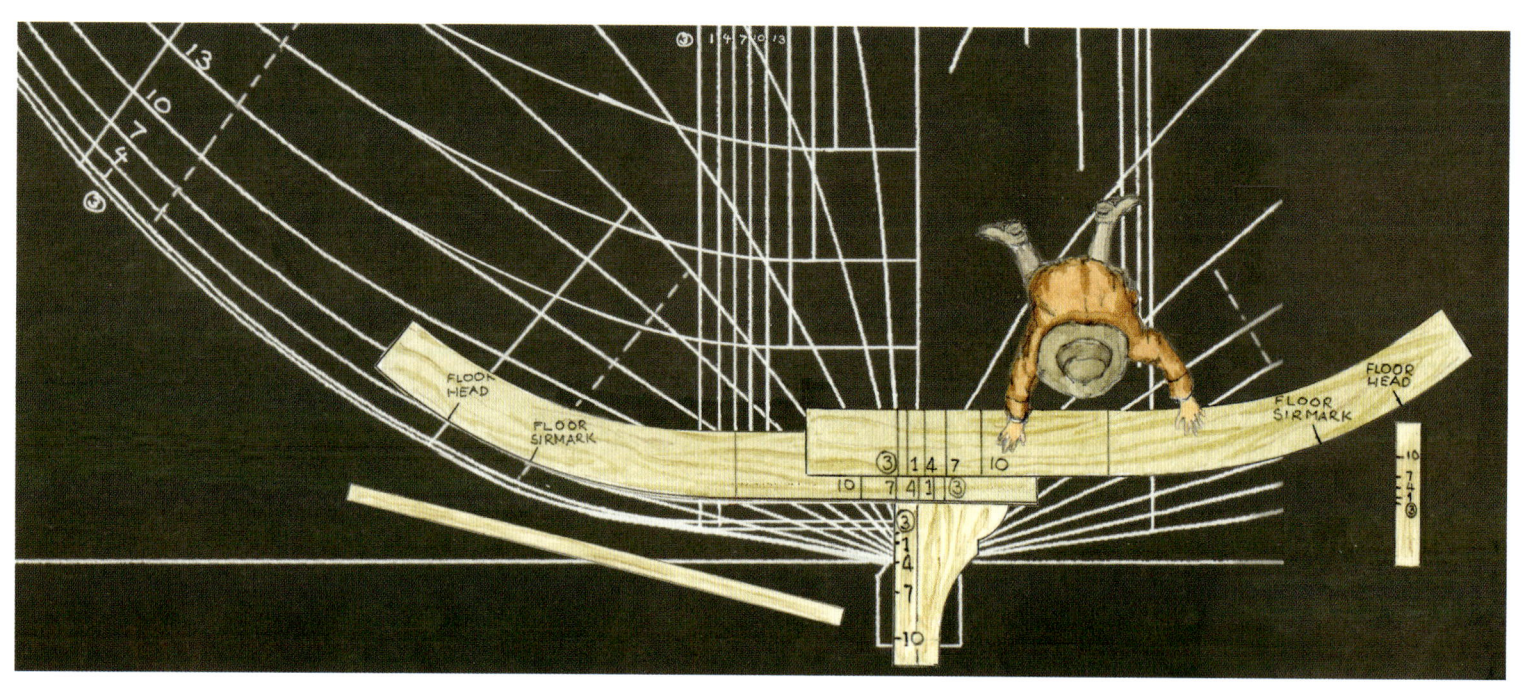

Above. *The floor moulds being assembled at frame station 4. The left-hand bend mould is set at the rising and narrowing of the floor at station 4. Once in place, the bend mould and the rising square were marked with the frame station numbers as shown. The moulds for the forward end of the ship would have used the same principle. Author.*

Opposite. *A shipwright adding the final nails during the construction of the mould that would be used for marking out the complex floor and half timbers from frame 10 aft, where every frame was different. The height at which the half timbers joined the rising wood is also added to the vertical section above the keel. To mould the opposite side, it could be turned over and re-marked. Only the outside profile is shown here, although in practice the inside may also have been marked on the mould. Alternatively, the inside profile may have been measured and marked directly onto the timber after the outside profile was drawn. In practice, this mould may have been made in two or three separate pieces so that it would not be so large and clumsy. A similar mould would be made for the bow timbers from frame G forward. Author.*

200

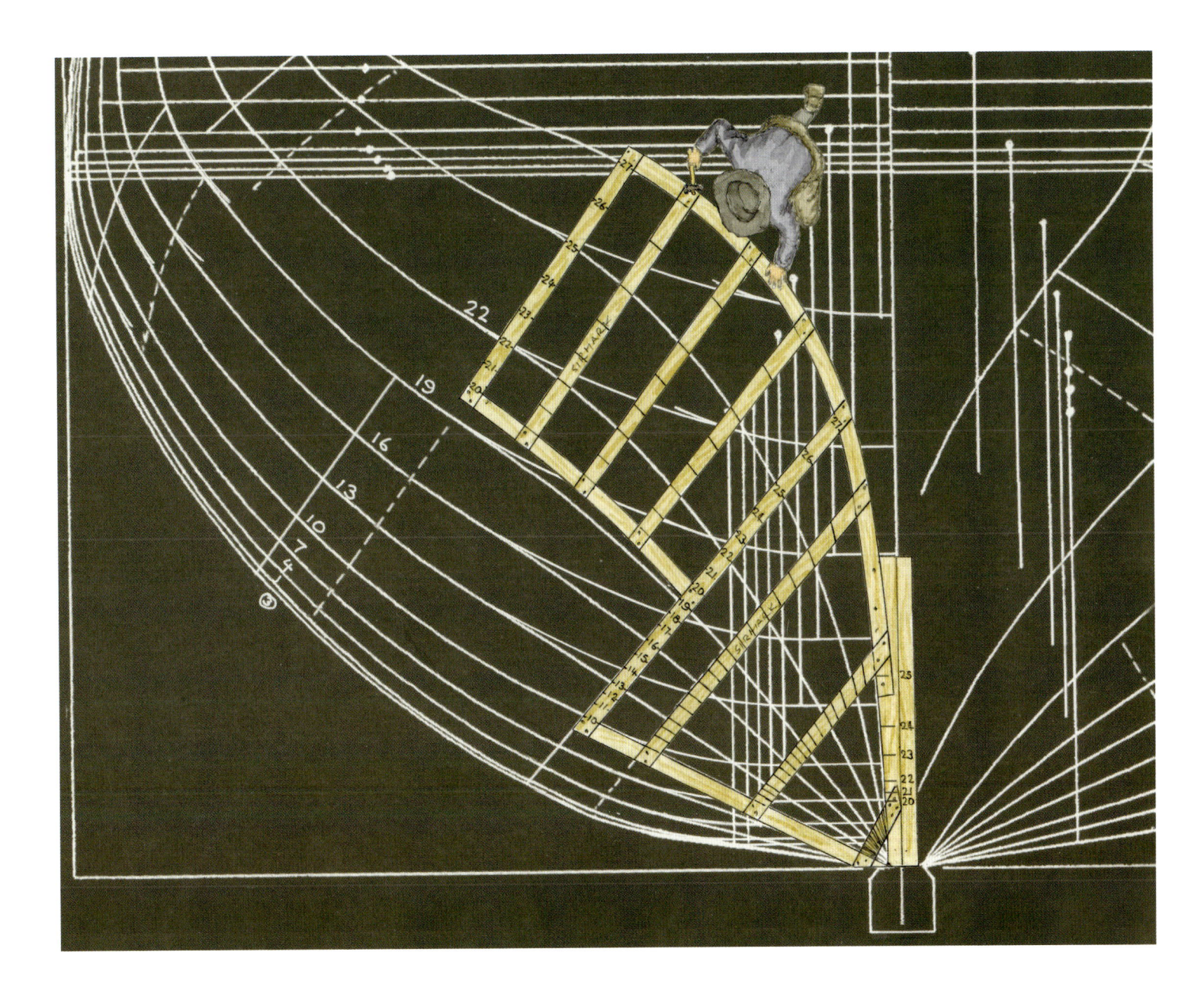

The Bevelling

As the floor timbers of the *Tyger* were 13 inches wide in the fore and aft direction (known as sided), the curvature of the hull had to be taken into account when cutting the inside and outside moulded profile. Allowance for the tapering, or bevelling, of the floor, middle futtock and toptimber was made so that even if the outside profile was cut square to that marked out by the moulds there would be a safe excess of material to remove. This was termed 'bevelled under'. On their inner moulded profile and the outside profile of the lower and upper futtocks an offset allowance had to be added to account for the curvature of the hull, termed bevelled standing. To maintain the outside profile bevelled-under condition, fore and aft of the biggest frame station 'X' for the floor, middle futtock and toptimber a frame timber was added or subtracted to reverse their orientation. This reversal is seen clearly on the model of the *St Albans*.

There were a number of ways to determine the bevel. The most sophisticated involved drawing other views on the half-breadth plan in the mould loft. They were similar to the waterlines, but instead of being viewed perpendicularly to the water level, they were perpendicular to the 'diagonals'. The angle generated by the curve could then be taken over the 13-inch-wide timbers. These angles were then marked onto a 'bevel board' for use in marking out the timber.[718]

This sophisticated and time-consuming method of taking the bevelling by projecting diagonals may not have been used, and probably was not, in the 17th century. The method relied on the diagonals lying in a flat plane, but 17th-century diagonals, such as those shown on the draughts dating from *c.*1680 by Thomas Fagge, were curved. To allow for the curves, the bevelling was probably taken directly from the end views.

THE MASTER SHIPWRIGHT'S SECRETS

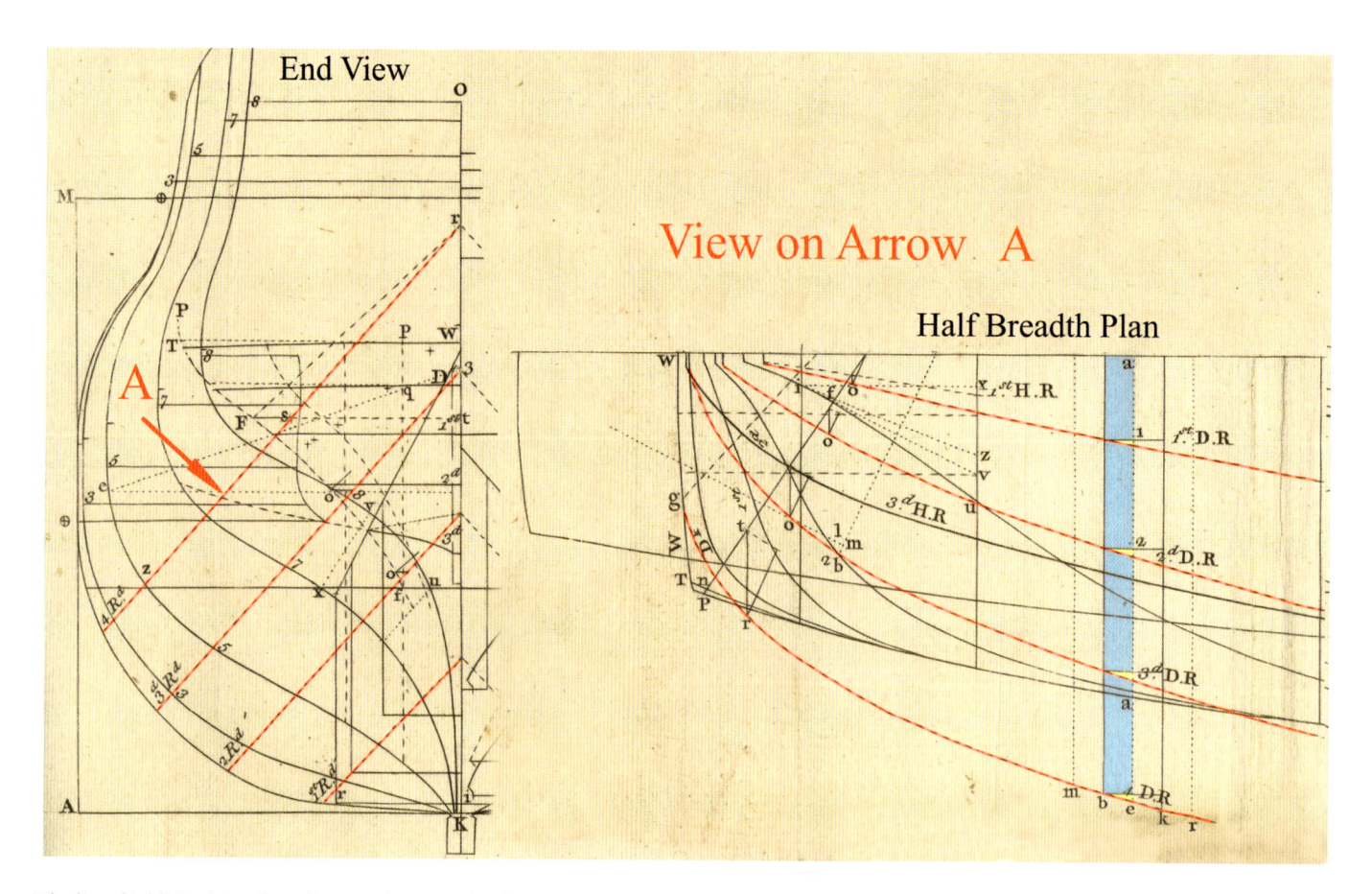

Above. *The lines highlighted in colour illustrate the principle of plotting the bevelling. The diagonals viewed in the end view from arrow 'A' and shown in red are plotted onto the half-breadth plan, the distance r z in the end view being the same as a e in the half-breadth plan. The frame, including the floors, is shown in blue and the bevel in yellow. The bevels were then marked on a 13-inch-wide bevelling board for use on the ship. This system allowed not just the angle to be recorded on the bevelling board but even the slight curve over the width of the frame piece. The image is taken from* A Treatise on Ship-building and Navigation *by Mungo Murray, 1754, plate VII.*

Right. *The method probably used in the 17th century for taking the bevelling directly from the end view in the mould loft – a method which did not rely on the diagonals lying in a flat plane. The outside surface of frame number 19 is shown in blue between the white lines at its fore and aft edges. The distance between the white lines is marked on the bevel board. In this case, the bevel has been taken between the floor head and the floor sirmark where the ribband would be placed. The ribbands were temporary fore and aft timbers that secured the frames before the outside planking was fitted. For the outside surface it was only necessary to cut the frame timbers accurately in those positions. Otherwise, it was prudent to allow a safe amount of excess timber to be left on the bevel, which would be trimmed to its correct profile after all the timbers were in place, a process known as dubbing. The floor timbers, the middle futtock and the toptimber were always arranged so that they were 'bevelled under' on the outside surface and 'bevelled standing' on the inside. On the inside of the ship the bevel allowance had to be added to the moulded profile, where accuracy was not quite so important. Bevels were probably taken at three places for each piece of frame timber. Because of the large number of bevels taken, a separate bevel board would be made for every tier of futtocks. The slight disadvantage with the simple method shown here is that the bevel is a straight line. Author.*

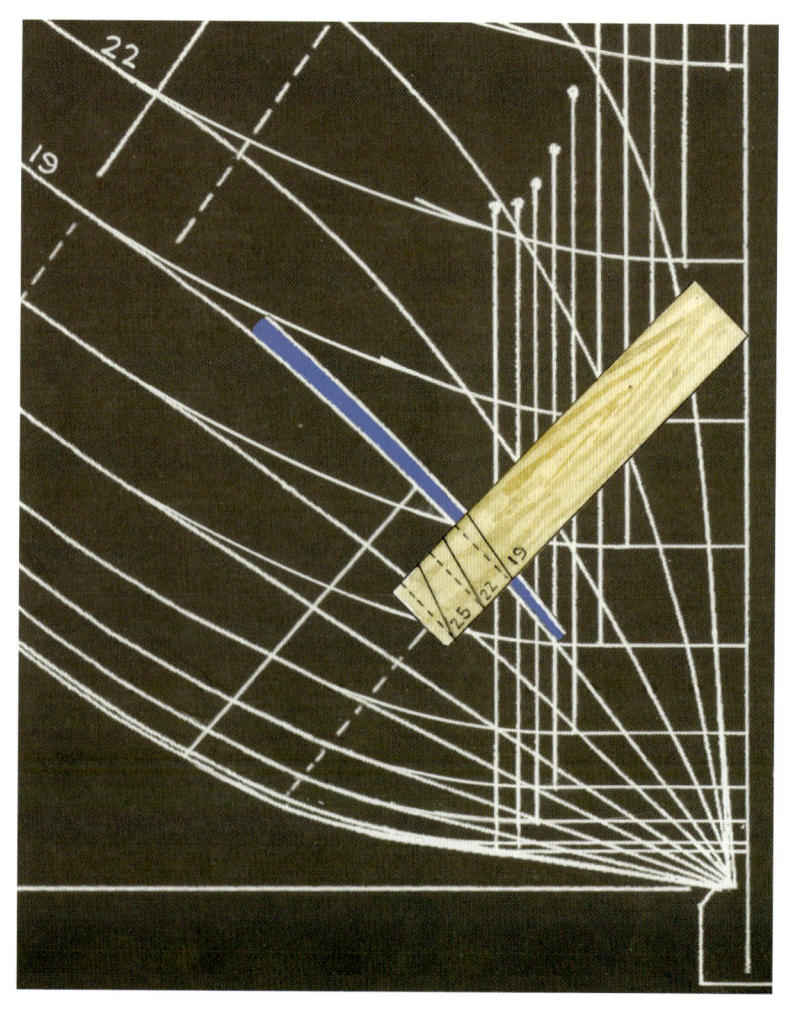

The Lower Futtock Mould

After the floor moulds were completed, attention turned to the next timbers higher up in the frame, the lower futtocks. Their lower end stopped short of the keel, leaving a gap between it and the opposite lower futtock. This was filled by a chock, except amidships, where it was left open to allow access for the pumps which had to reach down to near the outside planking.[719]

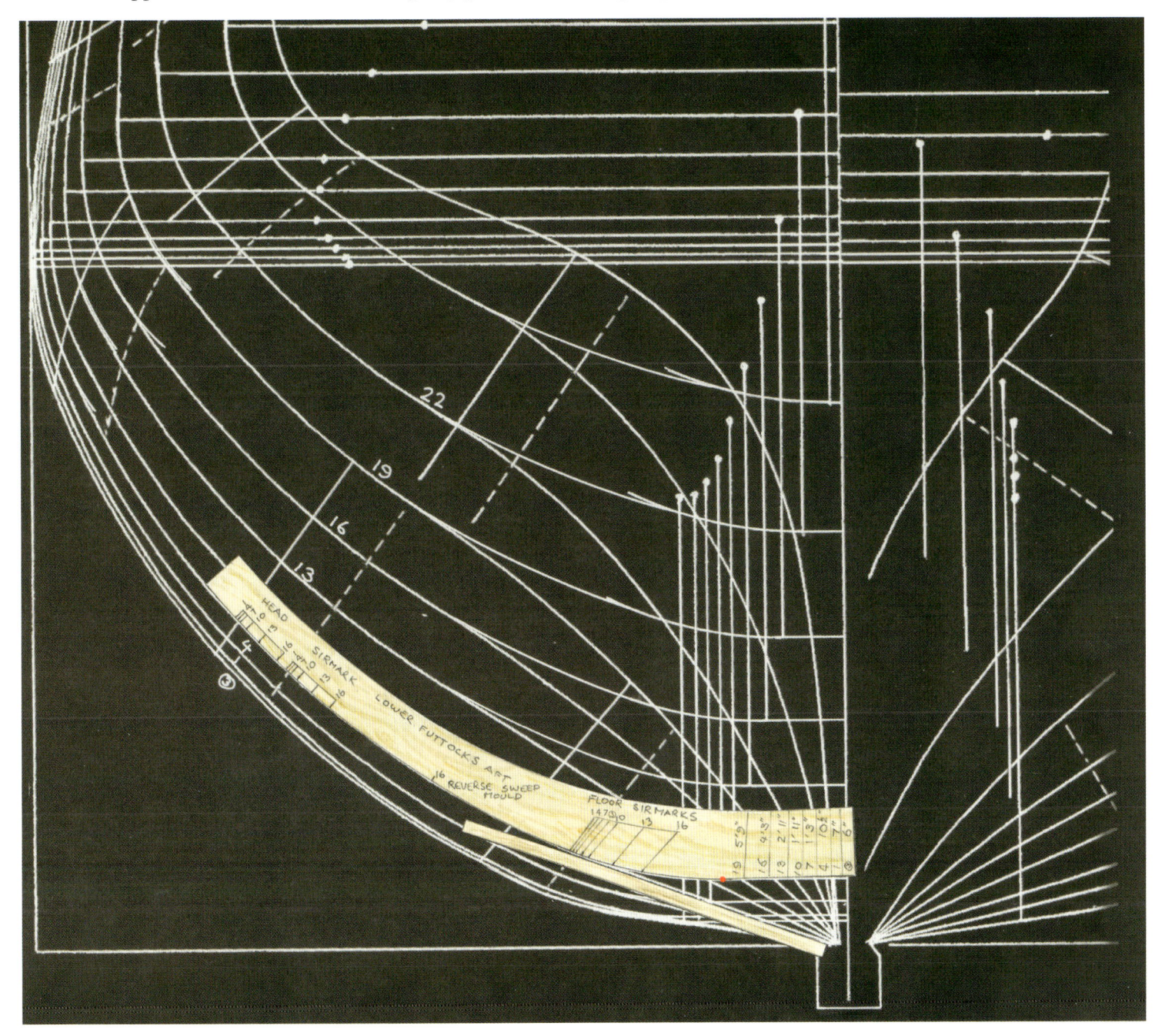

The lower futtock timbers, one each side, reached inboard to within about a foot short of the keel, although the mould itself was made to reach the side of the keel. Most of the lower futtocks could be made with a single mould as the floor and reconciling sweeps both had a constant radius. The first stage was to make a mould to fit the biggest frame amidships from the futtock head to the height of the rising floor in much the same manner as the floor mould was made. Its height of 6 inches above the keel was marked on it, and the edge where the frame reached the keel was then drawn on the mould parallel to the vertical middle line of the chalk drawing. While in this position, the floor sirmark, the lower futtock sirmark and the head of the lower futtock were all marked. The mould was then moved upward and inward to the next frame station, frame 1, so that the two sweeps were matched with the chalk drawing, and the new height of 7 inches was added. At the same time, the new positions of the sirmarks and the head were also marked. This view shows the mould at frame 7. Note the small amount of rotation lifting the inner side of the mould. When marking out the futtock timber, the outside profile below the mould to the keel was marked with a batten. From frames 16 to 19 a hollow, curved mould would be used which would have its tangency point marked on the mould. How far the inboard end of the futtock stopped short of the keel may have been dependent on the size of timber available to make it. For the opposite side of the ship, the markings were transferred to the back of the mould. The bevel allowance was added to the lower and upper futtocks' outside profile. As an alternative method, an open frame mould similar to that used for the fore and aft floor timbers could have been used.[720] Author.

The Middle Futtock Mould

The middle futtocks moulds were the least troublesome to make as most, if not all, are a simple arc. They could take on a number of forms depending on the preference of the master shipwright.

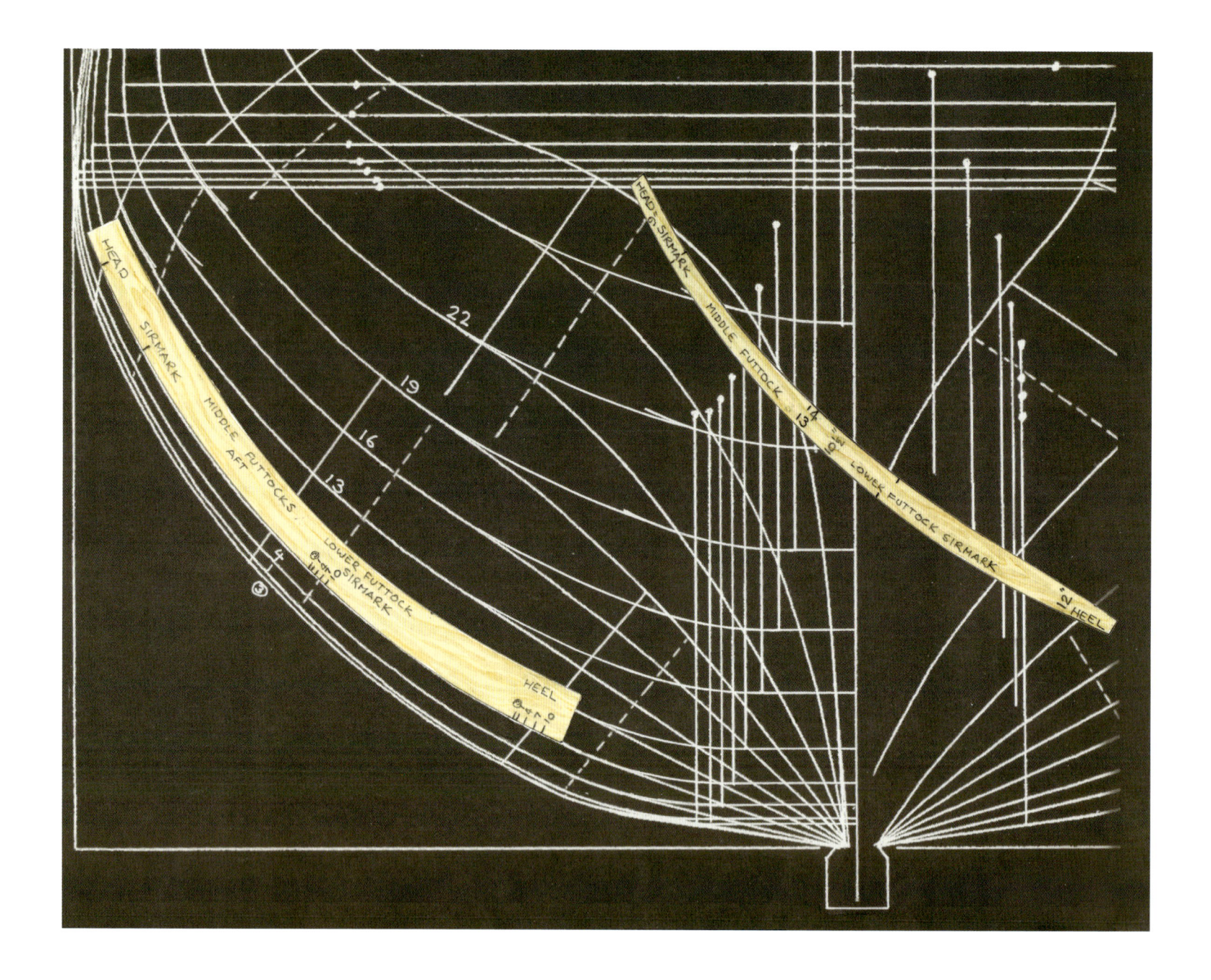

A simple mould could have been used to mark out most of the middle futtocks of the Tyger as it needed to consist only of the reconciling sweep. This was made to the width of the futtock scantling, with the head, heal and sirmarks marked on it. It is shown here at frame station 7. The middle futtocks at the fore and aft ends which had a component of a reverse curve could have been made using an open frame mould similar to that used for the fore and aft floor timbers. Alternatively, a slight mould could have been made that defined two futtocks. The mould shown here to the right is for marking out futtocks 13 and 14. To help the shipwrights when marking out the timber, the scantling width is added as shown, 9 inches at the head and 12 inches at the heel.[721] There were many types of moulds that shipwrights may have used, depending on their personal preference. One shipwright is said to have made a separate deal mould for every floor and futtock up to the breadth.[722] Author.

The Upper Futtock Mould

Most of the upper futtocks at the widest part of the ship had a different sweep radius below the breadth. They were *'of a very different shape from any of the rest, they are in consequence more troublesome to mould. Some artists employ more stuff and time about these moulds than all the others in the ship.'*[723] There were at least three ways they could have been moulded. A separate mould the full width of the scantling may have been made for every futtock. A slight mould that made two futtocks, similar to that described

This simple open-frame method of moulding the upper futtocks saved a considerable amount of time, which is being appreciated by this shipwright. In use, the mould required a certain amount of interpretation to mark out the lines and to draw the inner side to the correct scantling if it had not been marked on the mould. Also shown is one of the long double futtock moulds at frame station 22 that may have been made the whole size of the timber. This critical frame timber was placed aft of its frame station in the bevelled-under condition. The shipwrights had to take care they cut the bevel the correct way. It would seem prudent to mark on the moulds what the bevel condition was. Author.

Opposite *An example of how the toptimber moulds may have been made. The lower part of the mould contained the variable sweep below breadth and is made in the same manner as the upper futtock mould. Above breadth it blends into the sweep above breadth, which, in the case of the Tyger, was a fixed radius of 15ft 2in. The mould for the largest frame amidships was made to extend to the full height of the highest toptimber. When in place in the mould loft, the height of the toptimbers was marked together with the distance of their offset using a square. The second part of the mould was the toptimber hollow, consisting of the reverse sweep radius, again 15ft 2in, which led into the straightening upper part of the mould. It was made full width to the correct scantling. It was placed in position on the loft floor with the two sweeps blending and the frame station positions marked. It was temporarily tacked together for use. The further aft the frame station, the less of the reverse sweep was used. The toptimbers were moulded 8 inches at the lower end and only 3 inches at the top. Author.*

for moulding the middle futtocks 13 and 14, may have been used, saving a great amount of timber. Saving even more timber was the open-frame type. It may not have been quite so precise, but this mould could be used for all the aft upper futtocks.

It was wise to make the long double futtock moulds at the fore and aft ends of the ship the whole size and to the correct in and out dimension. They were probably the only moulds made like this as these futtocks were the longest and the most expensive in the ship. They were also more complex geometrically as they finished above the keel and against the rising wood.[724]

The Toptimber Mould

The toptimber mould had to take account not only of the variable radii of the sweep below breadth but also of the constant reverse sweep above. Towards the stern the tumblehome gradually reduced to become a straight line. At the bow, the toptimbers of the *Tyger* were even more complex. But as the toptimbers formed the hull shape above water they were not as critical as those below. There is evidence that the moulds were made in an economical manner using a variation of the open-frame method.[725]

Other Moulds

Moulds were probably made for parts such as the forefoot at the front of the keel and the fore and aft rising wood that fitted on top of it.[726] This may have been unusual as earlier works advised that this was seldom done except for round parts, such as the stem.[727] For the straighter parts, it is possible to mark them out using dimensions taken directly from the mould loft. Moulds would have been necessary to make the transoms that fitted onto the sternpost. These must have been made quite simply by plotting out waterlines at the height of each transom. Making the mould for the fashion piece required a little more thought as it did not lie in any of the three views laid out on the draught or in the mould loft. However, it could be plotted out using a combination of two views, the side view and the half-breadth with waterlines.

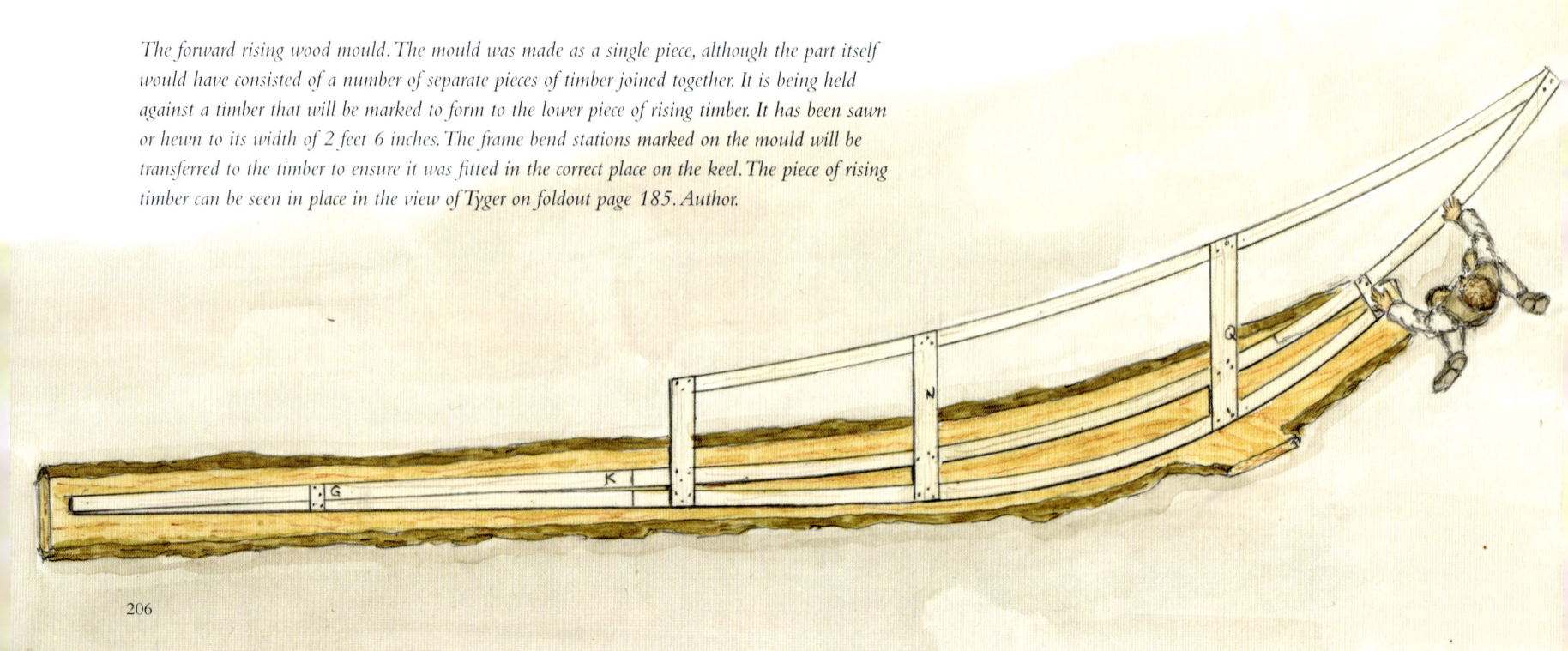

The forward rising wood mould. The mould was made as a single piece, although the part itself would have consisted of a number of separate pieces of timber joined together. It is being held against a timber that will be marked to form to the lower piece of rising timber. It has been sawn or hewn to its width of 2 feet 6 inches. The frame bend stations marked on the mould will be transferred to the timber to ensure it was fitted in the correct place on the keel. The piece of rising timber can be seen in place in the view of Tyger on foldout page 185. Author.

BUILDING THE NEW TYGER

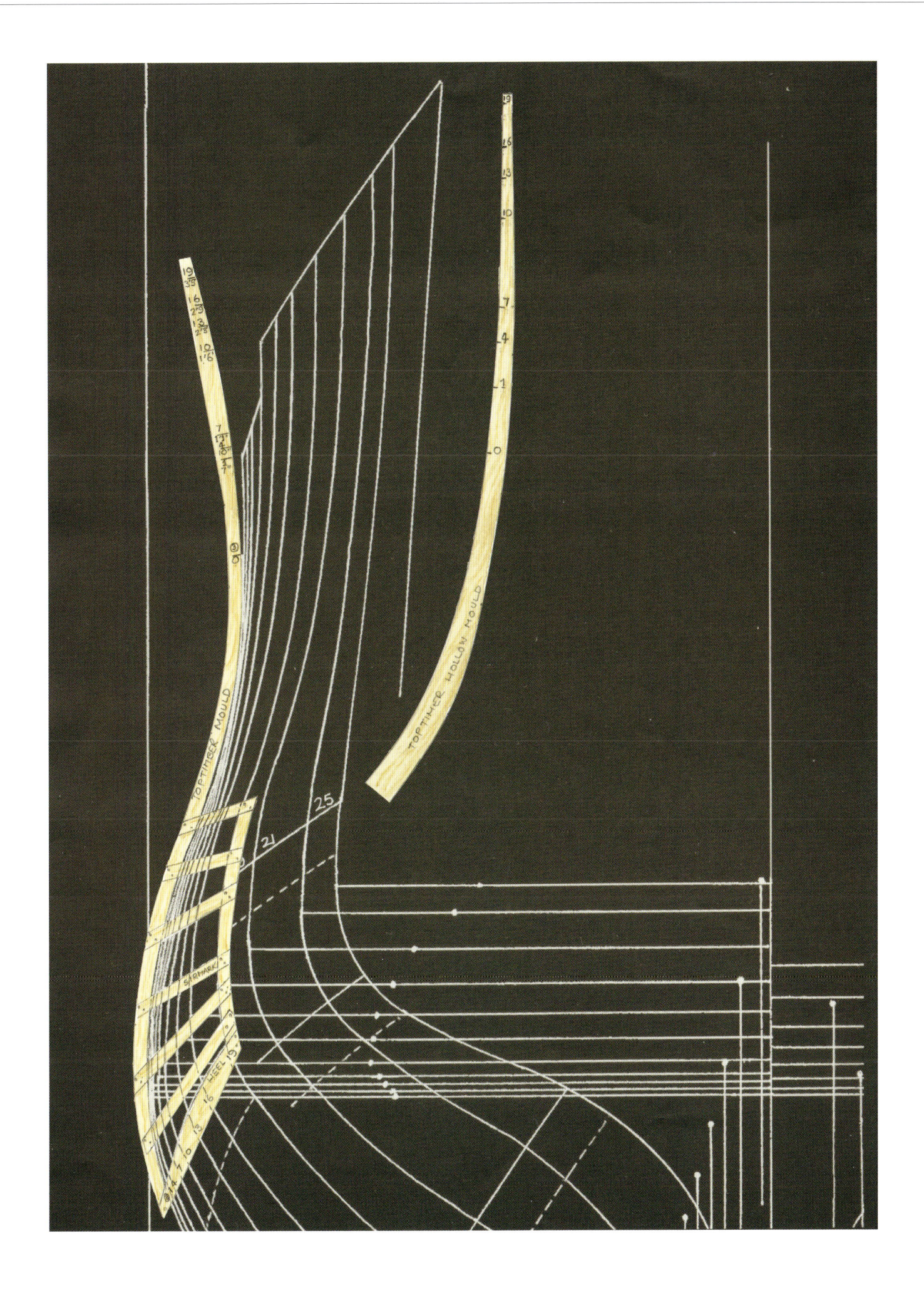

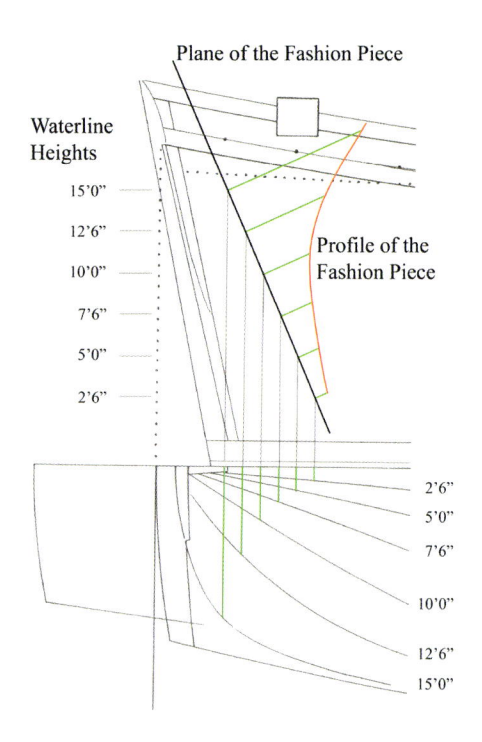

The principle of plotting out the fashion piece. In practice, both the fore and the aft side of the piece would probably have been plotted as it was a complex surface and made from awkwardly shaped and difficult-to-find timber. The lengths of the green lines on the waterline view, below, are transferred to plot out the curve of the fashion piece on the side view above. Author.

Timber for the Frames

While the draught was being plotted out full size in the mould loft, consideration was given to finding timber to build the ship. The amount needed would have been much the same as Shish estimated two years later in September 1681 for building the similar-sized *St Albans*.[728] Each load consisted of 50 cubic feet of hewn timber, about the same amount found in a grown tree.[729]

Estimate of Timber for Building the *Tyger*

Type of Timber for building the *Tyger*	Amount in Loads
Straight oak timber	250
Compass oak timber	230
Elm timber	100
Fir timber	40
Knees raking and square	40
Oak plank of all sorts	250

It was a considerable amount of timber to find, but as the new *Tyger* was built at a modest rate over a two-year period, it no doubt came from a number of sources as the opportunity arose. As work started during May 1679, 30 large ships were being built as part of an extremely well-funded and unprecedented building programme. The third rates *Lenox* and *Hampton Court*, both twice the size of the *Tyger*, had been launched at Deptford the year before. The *Duchess*, a great second-rate three-decker, had just been finished and another third rate and a second rate were under construction. Parliament had voted for a tax to pay for the ships and the money raised was not supposed to be used on anything else. But, of course, the Admiralty found good reason to question this, and just as work on the *Tyger* was getting under way, it wrote to the Navy Board about building some fourth, fifth or sixth rates '*as also what is ye sum at which the timber that is already or that may happen hereafter to be bought for ye 30 new ships & that will not be wrought up in or upon them by reason of its smallness or otherwise & that you will prepare & send hither your answer hereunto with as much expedition as conveniently you may*'.[730] As no records can be found of timber being supplied exclusively for the new *Tyger* during the first year of her construction, it is reasonable to assume that some of it came under devious circumstances from timber intended for the 30 new ships.

Framing

John Shish described in detail his procedure for building ships in a series of 41 weekly building progress reports for the *Lenox* and the *Hampton Court*, made during 1677 and 1678.[732] They show that the most difficult timbers to find were large elms to make the keel. Only those with no visible defects were selected to be sawn to within an inch of their final scantling. At this stage, hidden defects were often exposed, resulting in a large percentage of keel pieces being rejected. Those of good enough quality were then carefully hewn to their final dimensions and finished by the most skilled shipwrights using the axe and the adze,[733] taking care not to tear the rabbet, a groove into which the bottom outside plank fitted. The scarphs of the keel pieces were bolted together and the three-inch-thick elm false keel treenailed to the underside for protection. The completed keel was then lifted onto blocks to keep it clear of the ground so that men could work beneath the ship. The sternpost was then assembled; it consisted of the post itself, the false post within, transoms and usually the fashion pieces attached to the edge of the transoms. At the bow, the two-part stem was also constructed and the two assemblies raised at each end of the keel. Middle lines were marked on them to check they were perpendicular to the keel and without twist. They were then securely shored to make sure they remained perfectly aligned and did not move.[734]

The midship floor timbers were then put in across the keel. Working towards the ends of the ship, they became more curved in shape and the timbers to make them more difficult to find. To resolve the problem, vertical pieces forming the rising wood were placed over the keel and half timbers bolted to each side. The knee of the main post was then bolted in the corner between the false sternpost and the rising wood. Following standard practice, Shish then bolted alternate floor timbers through the keel and floor.

With the floors, the stem and the sternpost in position, the floor ribband, a temporary supporting band, was nailed at the head of the floor timbers where the bevelling had been accurately cut to the hull profile. The ribbands were generally made from old masts and had to be securely shored as they not only made sure the floors were all in line, but they took the weight of all the other frame timbers that fitted in between them. The shores were placed on timber foundations called sholes and well secured with treenails at the heels.[736]

After the floors were in place the master frames, the frame bends, which each consisted of three futtocks, were assembled on the ground, making sure the sirmarks were carefully lined up. They were treenailed together and erected in the ship as a pair. The frame-bend pairs were set at the correct distance between them as previously calculated for the rising and narrowing of the breadth. Once in their correct place, they were fixed in position by shores and temporary cross beams known as crosspawls. In all probability, the centreline and the breadth were marked on the crosspawl beforehand.

Once enough frame bends were in place to establish the shape of the hull, other bands of ribbands were nailed to them. Starting with the lower futtocks, the other tiers of fill-in futtocks were laid in place against the ribbands. The lower futtocks overlapped, or scarphed, the floor timbers by about six feet. Work erecting the toptimber frame bends then proceeded in a similar fashion. The position of the port holes was marked on the ribbands so that the upper futtocks and toptimbers could be placed or worked to form their sides. This was especially important with the *Tyger* as she had oar ports as well as gun ports.

At about this stage the keelson was fitted onto the floor timbers and scored down by an inch. Then the alternate floors not already bolted were themselves bolted through the keel, floor and keelson.[738]

Below. *Marking out floor timber number 4 using the two bend moulds and the rising square. The bend moulds have been overlapped at frame 4 and tacked together in position, taking care they are in line. The rising square is then set on the centreline so that the top and the side of the keel can be seen marked at 4. When the shipwright is satisfied that the moulded profile fits inside the timber, their profile is marked out. Author.*

Bottom. *The depth of the floor timber above the keel has been marked out using the cutting-down batten, shown lying in place. The scribed profile markings were probably made with an awl or knife point, while the identification mark '4' was made with some form of race knife with a hooked cutting blade.[735] The right-hand side has been marked out from the keel to the floor head. To complete the marking out, the bevelling allowance from the bevelling board at three positions has been added to the inside profile and connected by a dotted line. The floor head has been drawn with provision for a wedge so that straighter timber can be used, avoiding the sapwood. The sirmark is marked, and, on assembly in the ship, this will be lined up with the first futtock sirmark to make sure they are in their correct positions. Author.*

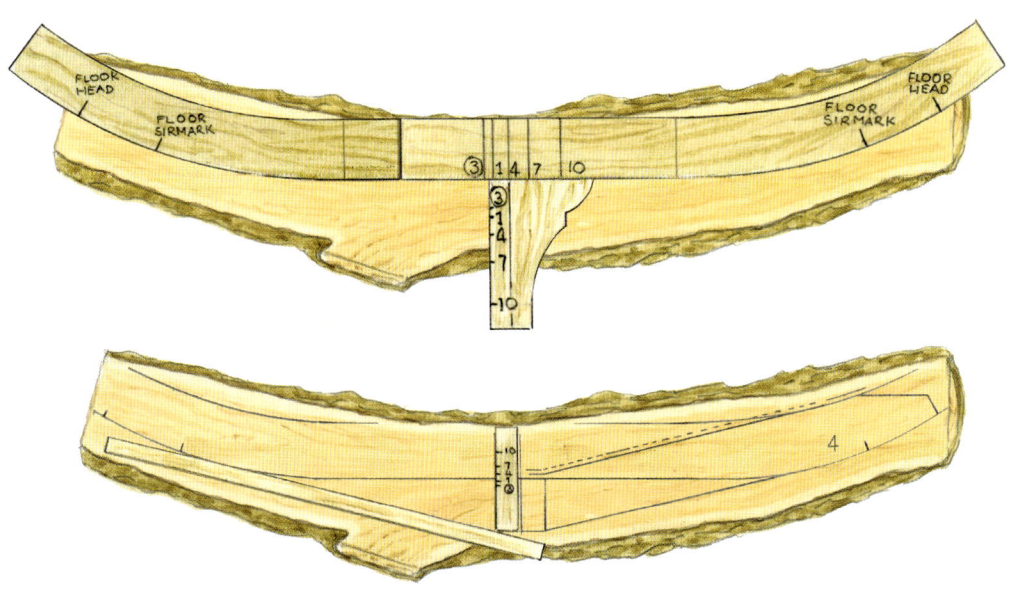

THE MASTER SHIPWRIGHT'S SECRETS

A shipwright cutting the bevel to suit the ribband at the head of the floor timber. The timber has already been sawn to its sided fore and aft dimension and marked out using the floor mould. Notice the second line on the inside profile which has been added to take account of the tapering, or bevel, of the ship. The outside profile where the shipwright is working has been sawn square and is now being cut 'bevelled under' according to the angle from the bevel board. This bevelled-under condition allowed the outside profile to be cut square to the marked-out moulded line in the safe knowledge that excess material could be dubbed off later with an adze when the timbers were assembled in the ship. The floor, the middle futtock and the toptimber were all in line, one above the other, and all bevelled under. In practice, the whole profile of this floor would have been sawn before bevelling. Author.

BUILDING THE NEW TYGER

Building Sequence of the Tyger

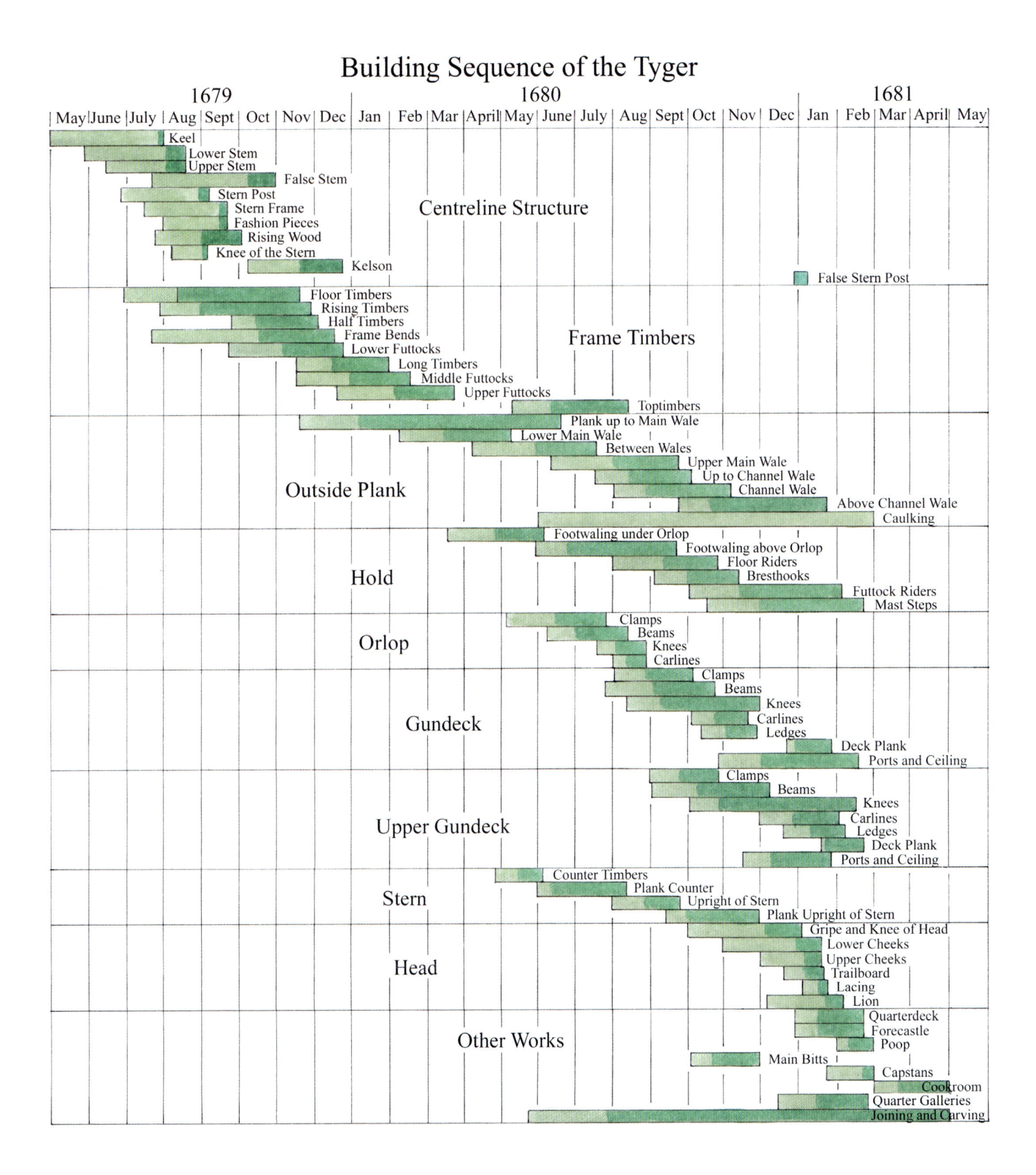

1679	1680	1681
May, June, July, Aug, Sept, Oct, Nov, Dec	Jan, Feb, Mar, April, May, June, July, Aug, Sept, Oct, Nov, Dec	Jan, Feb, Mar, April, May

Centreline Structure
- Keel
- Lower Stem
- Upper Stem
- False Stem
- Stern Post
- Stern Frame
- Fashion Pieces
- Rising Wood
- Knee of the Stern
- Kelson
- False Stern Post

Frame Timbers
- Floor Timbers
- Rising Timbers
- Half Timbers
- Frame Bends
- Lower Futtocks
- Long Timbers
- Middle Futtocks
- Upper Futtocks
- Toptimbers

Outside Plank
- Plank up to Main Wale
- Lower Main Wale
- Between Wales
- Upper Main Wale
- Up to Channel Wale
- Channel Wale
- Above Channel Wale
- Caulking

Hold
- Footwaling under Orlop
- Footwaling above Orlop
- Floor Riders
- Bresthooks
- Futtock Riders
- Mast Steps

Orlop
- Clamps
- Beams
- Knees
- Carlines

Gundeck
- Clamps
- Beams
- Knees
- Carlines
- Ledges
- Deck Plank
- Ports and Ceiling

Upper Gundeck
- Clamps
- Beams
- Knees
- Carlines
- Ledges
- Deck Plank
- Ports and Ceiling

Stern
- Counter Timbers
- Plank Counter
- Upright of Stern
- Plank Upright of Stern

Head
- Gripe and Knee of Head
- Lower Cheeks
- Upper Cheeks
- Trailboard
- Lacing
- Lion

Other Works
- Quarterdeck
- Forecastle
- Poop
- Main Bitts
- Capstans
- Cookroom
- Quarter Galleries
- Joining and Carving

The sequence of activities for building the new Tyger, compiled using the weekly building reports made by John Shish for the Lenox *and the* Hampton Court *during 1677 and 1678. The two ships were built using the same methods, with little difference between them. The chart is configured to fit the time taken to build the* Tyger *and the known dates of activities, such as Shish obtaining the gundeck beams on 12 May 1680. The Shish building reports mention the most important activities and how a multitude of tasks were carried out at the same time. The light green shows when work started and the dark green when fitting into the ship took place. They reveal some surprises; the main wale was started before the planking beneath had reached its height, and the toptimbers were put in after. The masting and rigging are not included as the finishing of the ship was delayed as she neared completion. Author*

THE MASTER SHIPWRIGHT'S SECRETS

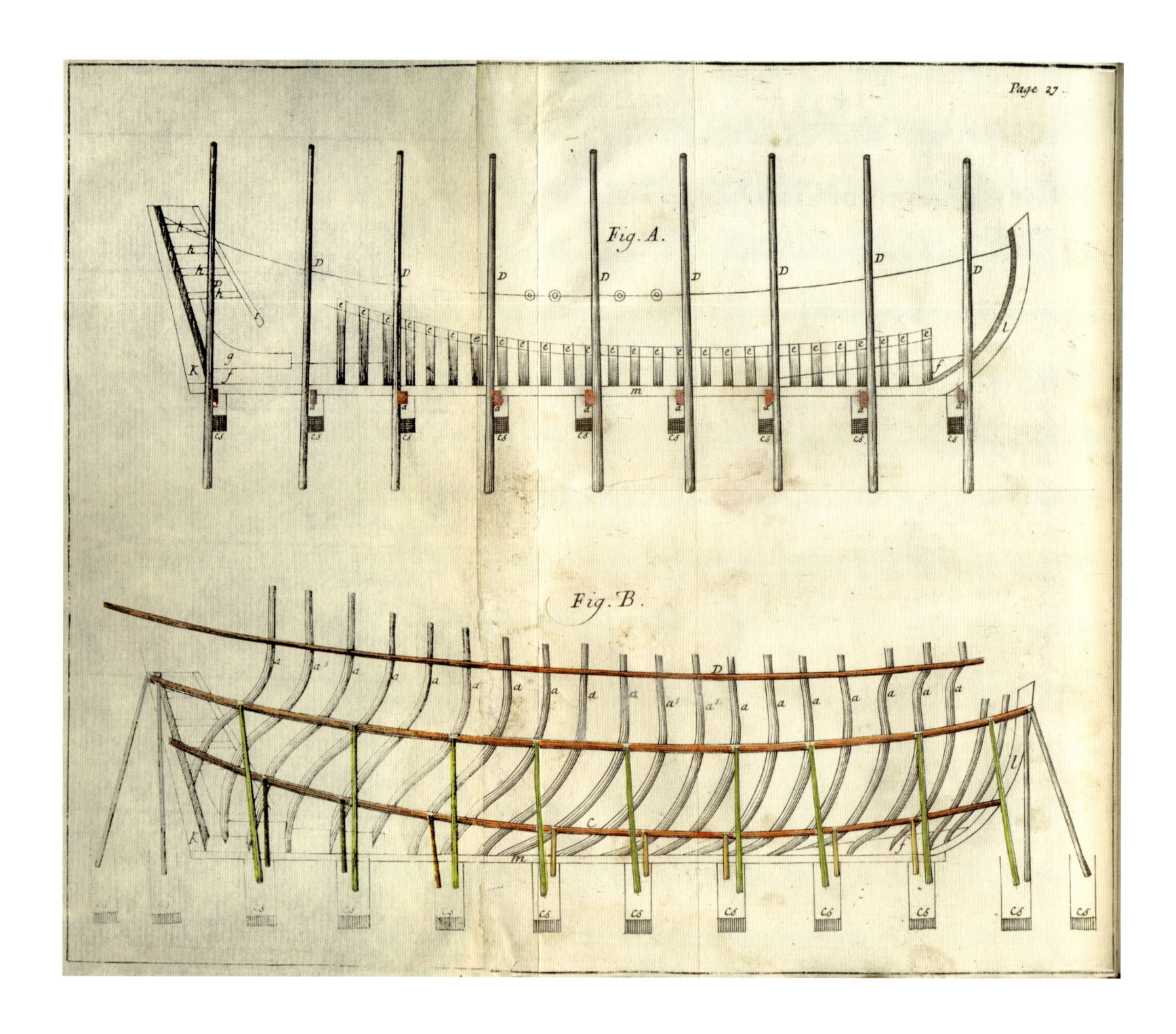

A ship during the early stages of construction, from William Sutherland, The Ship-builders Assistant, 1711, p.27. It shows the ribband in place to support the frame during construction. The ribbands were removed when the outside planking reached up to their position.

Figure A: e: the floor timbers placed on the keel

O: a ram line made fast on the stem and sternpost, and weighed by some device or other to steady it

D: standers to raise the stages

f: the dead wood afore and abaft (Shish called the dead wood 'rising wood')

g: the knee on the dead wood; h: transoms; i: the fashion pieces

K: the post; l: the stem; m: the keel; a: in the red, these are splitting blocks

Figure B: Resembles the whole frame bends, which are indicated by the letter a; the ribbands are marked in red; c: the floor

D: the double-depth ribband, or the ribband at the sirmark of the toptimber. The shores at the breadth are marked in yellow. The shores at the floor sirmark are marked in brown.

BUILDING THE NEW TYGER

This is how the Tyger *would have looked towards the end of October 1679. The floors are across the keel and secured by ribbands and shores. The first frame bend, or master frame consisting of the lower, middle and upper futtocks treenailed together, has been erected and set at the correct width of the ship by means of its temporary crosspawl (or cross beam). Another frame bend is in the process of being raised using lift gear made up of sheer legs, blocks and tackle, with power provided by shipwrights turning a crab capstan. Print by John Cleveley the Elder.*

1. Keel
2. Sternpost
3. Rising wood and knee of the stern
4. Floor timbers
5. Frame bends
6. Crosspawl
7. Ribband
8. Shores
9. Crab capstan

The Tyger *at about the end of January 1680. The frame bends, floor timbers and lower fill in futtocks are all in place. In reality the progress would not have been quite so orderly, as timbers were put in place as they became available. Print by John Cleveley the Elder.*

1. The frame, consisting of the floor and lower futtocks
2. Frame bends
3. Stem
4. Transoms
5. Fashion pieces
6. Shores
7. Lower futtock ribband
8. Crosspawls

Shipwrights Discharged

Towards the end of 1679, as the frame of the *Tyger* was taking shape, John Shish must have been disturbed by events happening outside the yard. Money for building the 30 new ships was running out, while the scurrilous Popish Plot had resulted in the naval administration that had overseen their construction being replaced by inexperienced opposition politicians. Of the five new ships built at Deptford, four had been launched, while the last one, a three-deck second rate, had just been started in the double dock and would not be completed until 1683.

Many of the numerous shipwrights who had been recruited at the start of the programme in 1677 were no longer needed and many were now discharged. Some who had been conscripted into the yard would have been relieved to be able to return to work at merchant yards where the pay was slightly higher. Even though the daily wage of 2s 1d in the King's yard was many times more than a farm worker received, earning the king's money and getting paid were two very different things. In November 1679, the shipwrights working on the *Tyger* and other repairs were a year's pay in arrears, while those working on the new ship were six months in arrears.

Shipwrights were supposed to be paid every quarter year. That was a long time to wait, and the added delays made it customary for them to obtain credit around the town. But with such huge arrears in pay, this credit was now being denied, and with winter approaching, the shipwrights petitioned the Admiralty to release at least some of their pay to relieve them, their wives and their children. William Staines was a shipwright who had been born at Deptford and who had served his apprenticeship there. He was married with two children and had another on the way. He was among those discharged along with all the '*foreigners*' who did not come from Deptford. He spoke to Jonas Shish, who suggested that a mistake must have been made and advised him to write a petition pleading to be reinstated. In December, Sir John Tippetts, the Surveyor of the Navy, was in the yard and saw a labourer, Richard Lawrence, splitting a log to make chips that he could take home as fuel to warm his house. Tippetts immediately discharged him – an act that would mean the loss of the 20s in wages owed him. It would be the utter ruin of him, his wife and child and, unable to pay his debts, he would be thrown into debtor's prison. Fortunately for him, however, the Navy sympathetically reconsidered his case and decided his name should be put on the list of those discharged in the normal manner so that he would receive his wages.[739]

Planking

It was important to add planking to the individual unattached frame timbers as soon as possible, for the planking held the frame together as much as the frame held the planking. It was, in modern terminology, a stressed skin construction. The first task was to smooth off any excess allowance of timber from the frames by dubbing with the adze. Work must have begun at the start of 1680 on the first strake of three-inch-thick plank that ran the whole length of the ship. This, the garboard strake, fitted against the keel. The next strake above it was left off for now, allowing a gap through which rainwater and wood chips could be easily disposed of.[740]

The three-inch-thick underwater planking between the keel and the main wale required careful consideration. The girth between them wale on the *Tyger* ranged from 17 feet at the bow to 22 feet at the stern. In order to keep the same number of strakes throughout, a 17-inch-broad plank at the bow, would have to be 22 inches broad at the stern. This was usually not possible as much of the planking was not cut specially for the ship but was brought in from the Baltic in lengths ready cut to thickness from the tall, straight oaks grown there. Even after making sure the widest end of the plank was toward the stern, it was often necessary to add in extra plank at the stern and to drop some planks at the bow to ensure the planks remained reasonably even in width. Less expert or perhaps more careful shipwrights measured the girth at every frame bend to plot the run of strakes, while others used the run of the ribbands as a guide. However, an expert workman thought it "*no difficult matter to birth up a ship's bottom exact and genuine from experience*". Another important consideration was

Opposite. *A theoretical view of the* Tyger *showing the frame bends and how they were made to suit the narrowing hull towards the stern. Frames 4 and 7 are standard midship frames consisting of a floor timber, three pairs of futtocks and a pair of toptimbers. The cross chock that connects the heels of the first futtocks was necessary to complete a nearly solid bed of framing into which the plank treenails could be driven. Further aft, frames 10 and 13 are similarly made, but as the floors are more 'V'-shaped here, they are called rising timbers. They sit on top of the most forward piece of the rising wood. It was difficult to find such large curved timber to make the rising timbers, and they were often fabricated from a number of smaller pieces bolted together. Frames 16 and 19 have the single rising timber replaced by two half timbers, recessed and bolted into the rising wood. Frames 22 and 25 are not as curved as the other frame timbers, allowing the number of futtocks to be reduced from three pairs to two. They were referred to by Shish as 'long timbers'. Author.*

BUILDING THE NEW TYGER

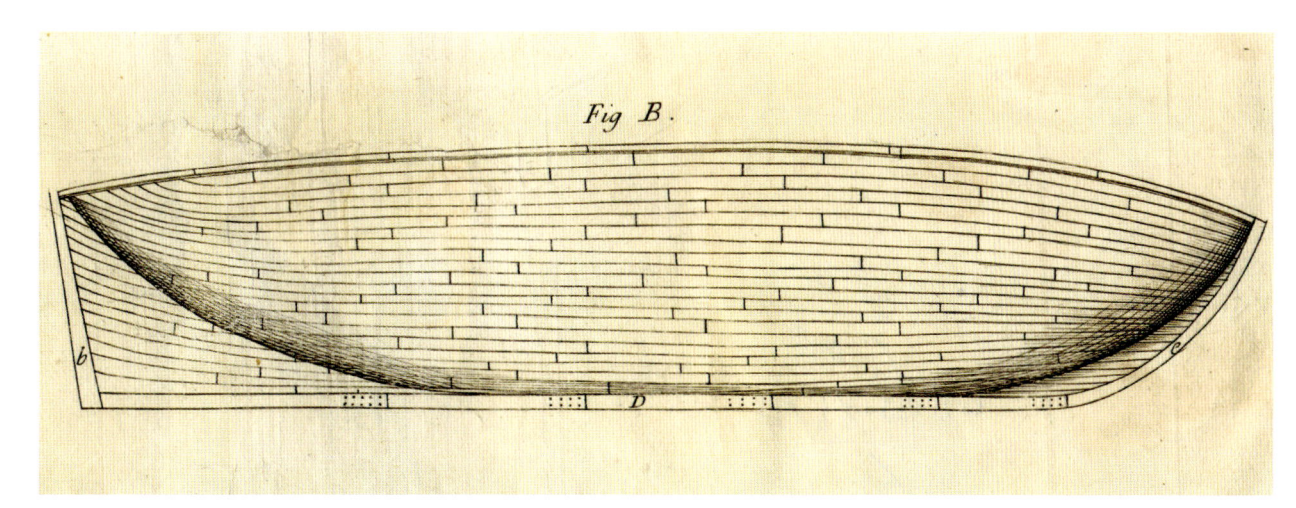

Left. *The planking between the keel and the lower main wale, from William Sutherland's book* The Ship-builders Assistant *of 1711. b is the sternpost, D the keel and e the stem. The illustration shows the positioning of the butts, which were to be six feet clear of each other.*[741]

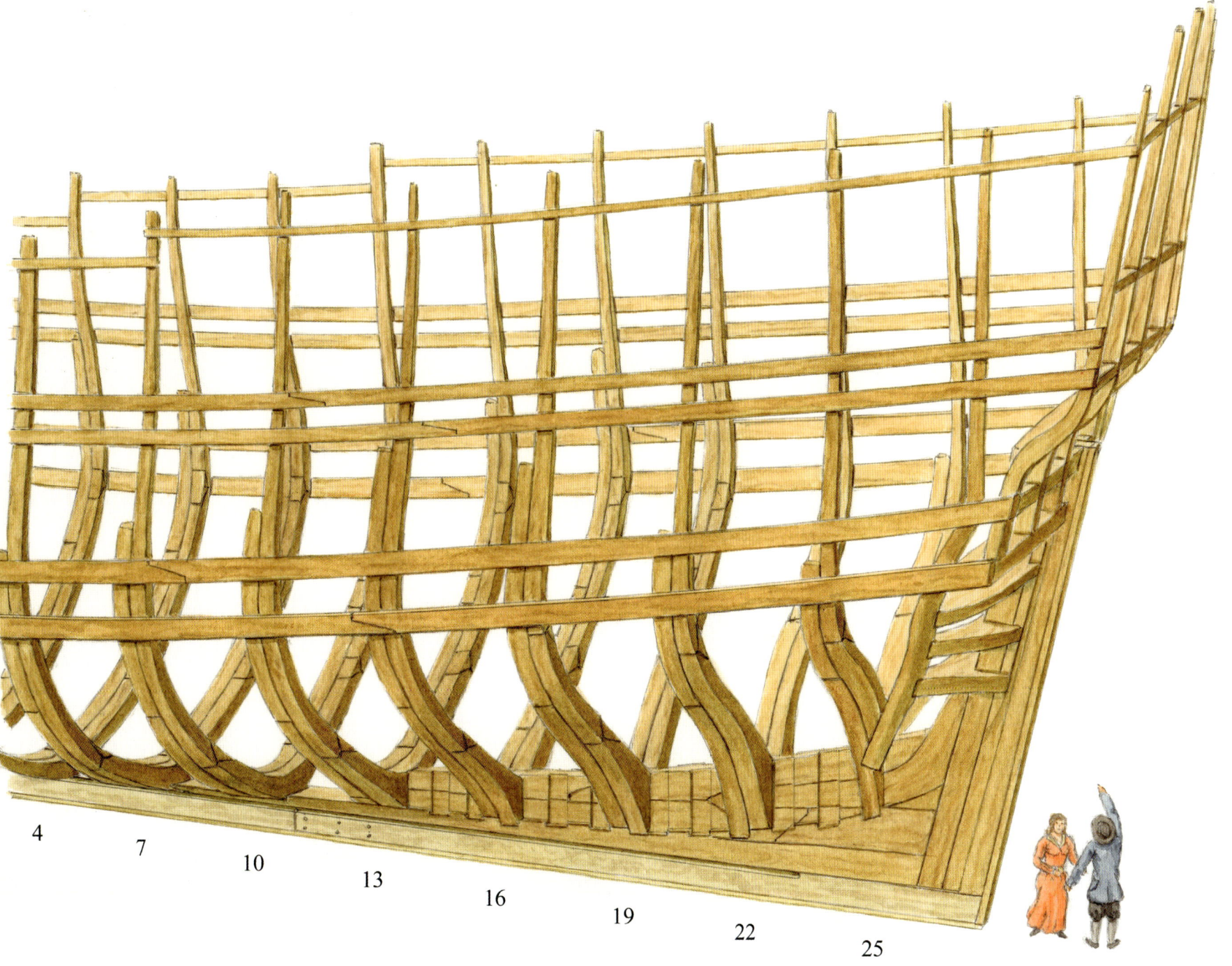

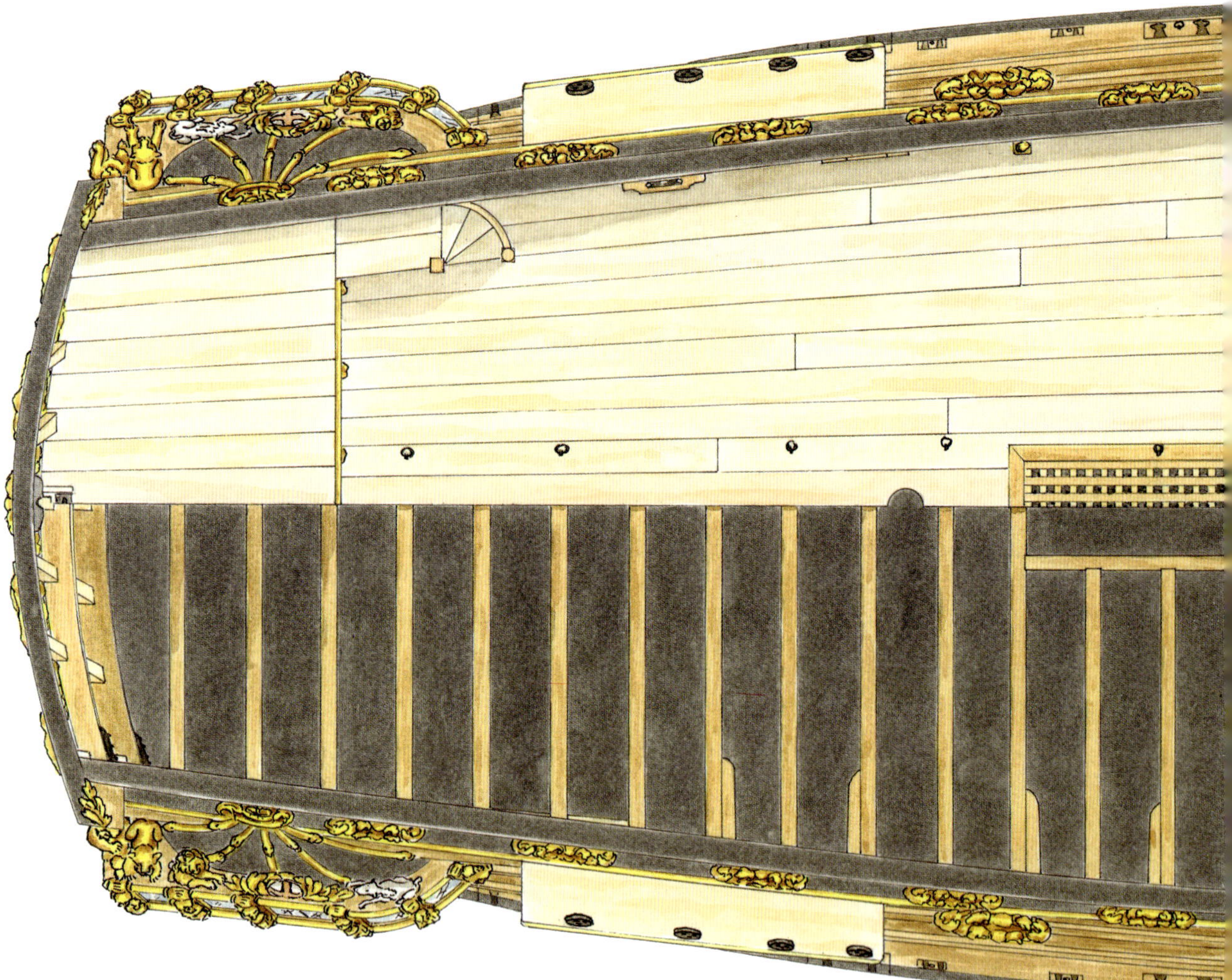

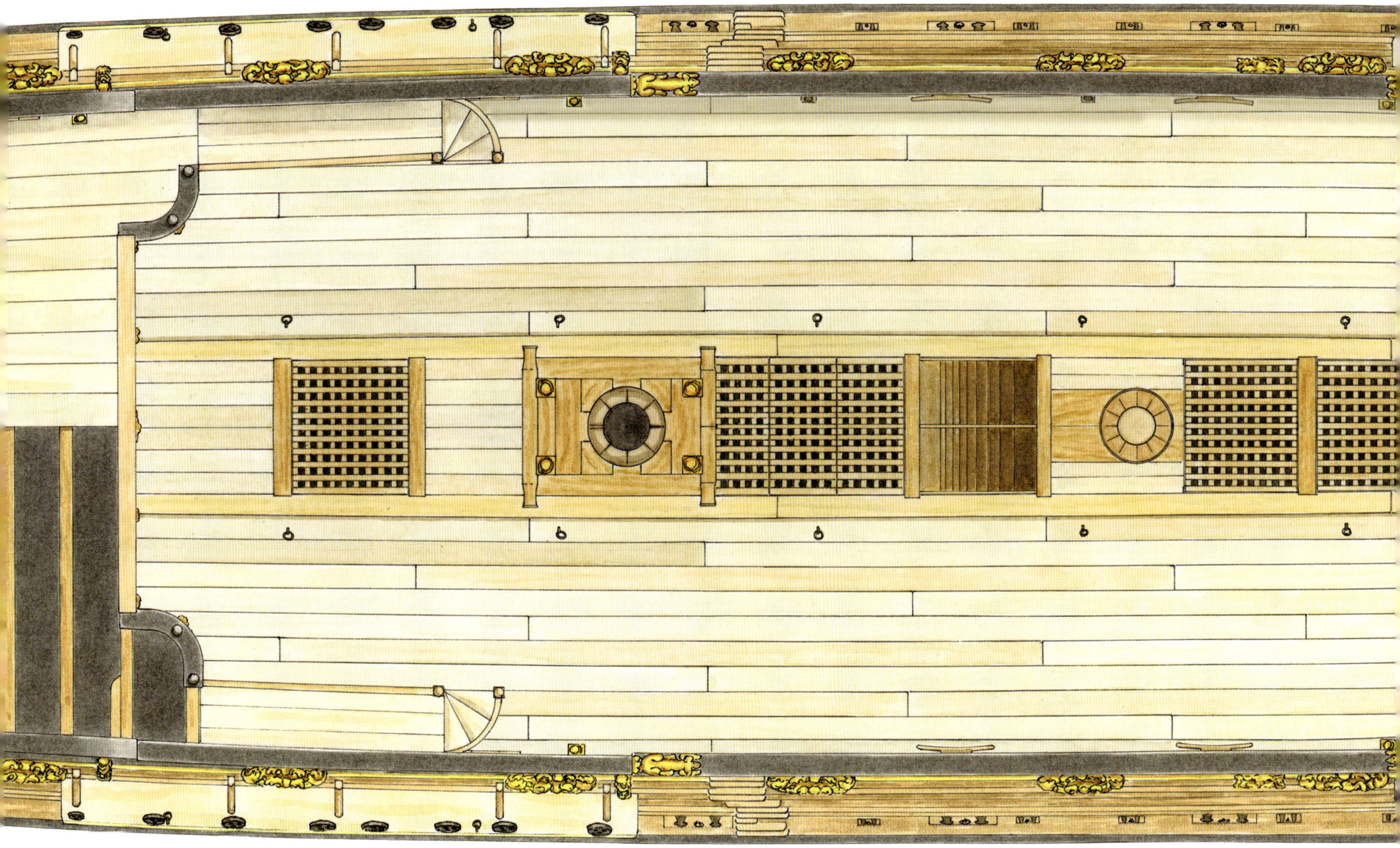

Tyger

Topside plan. Scale 1:72. Author.

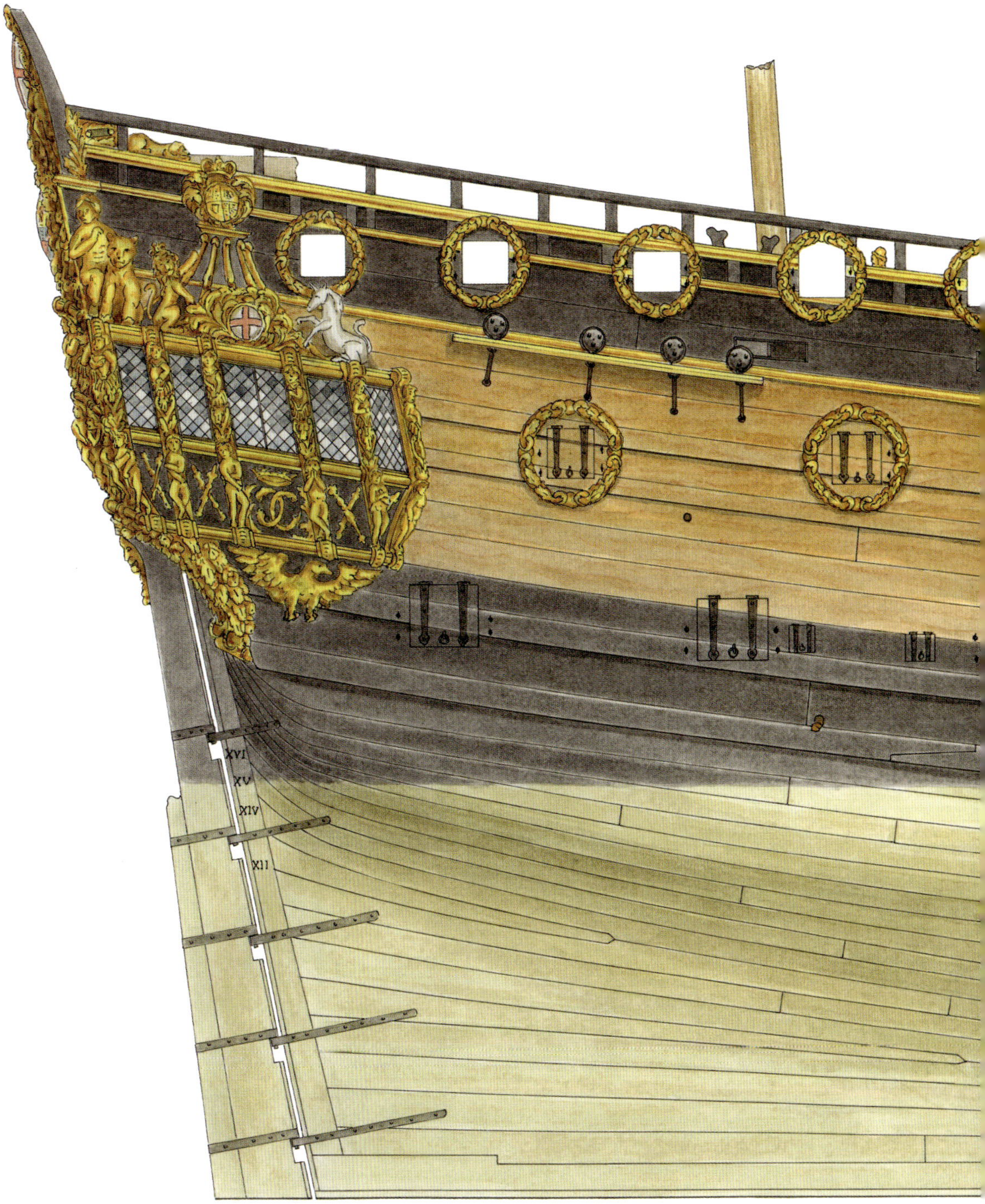
XVI
XV
XIV
XII

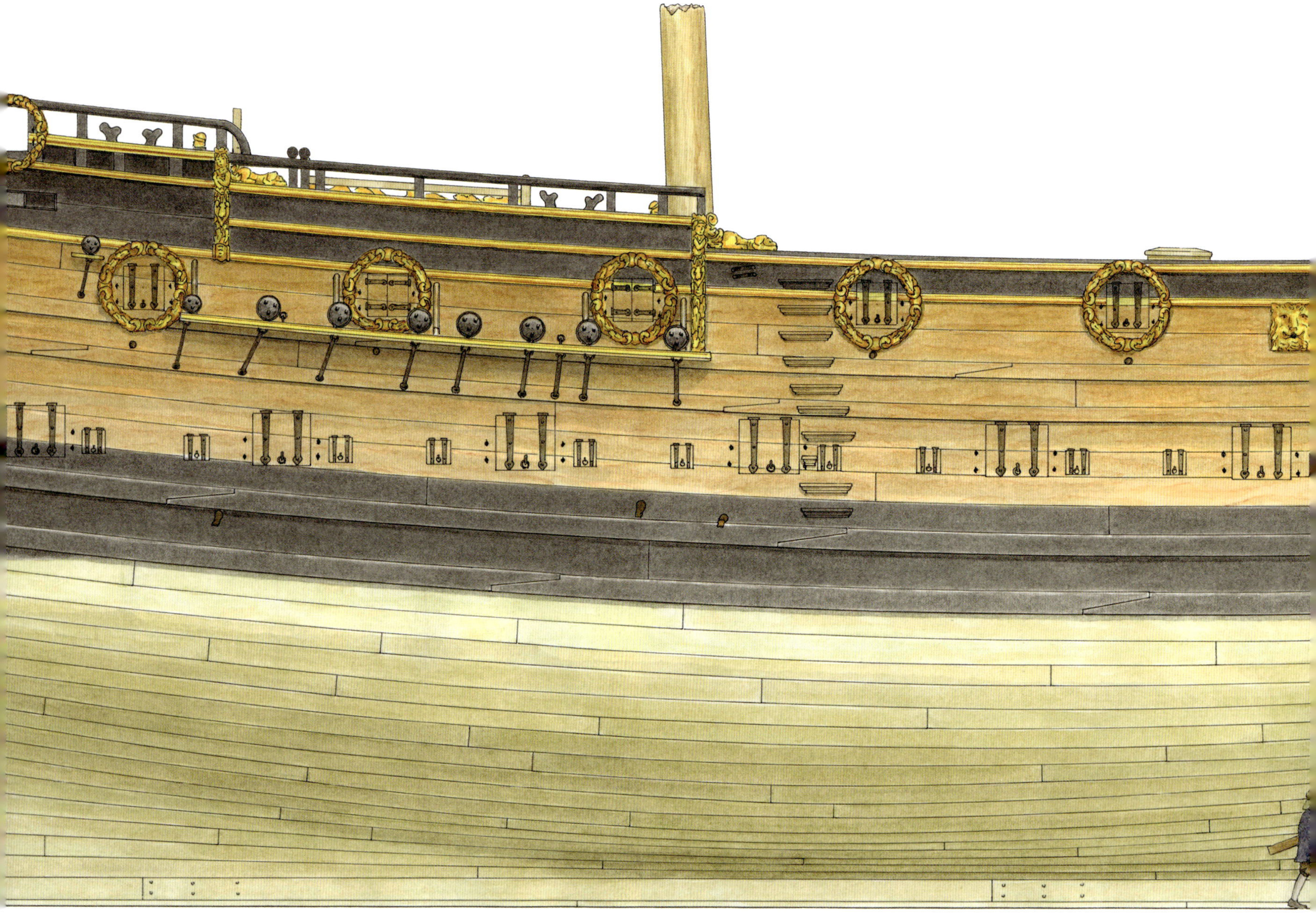

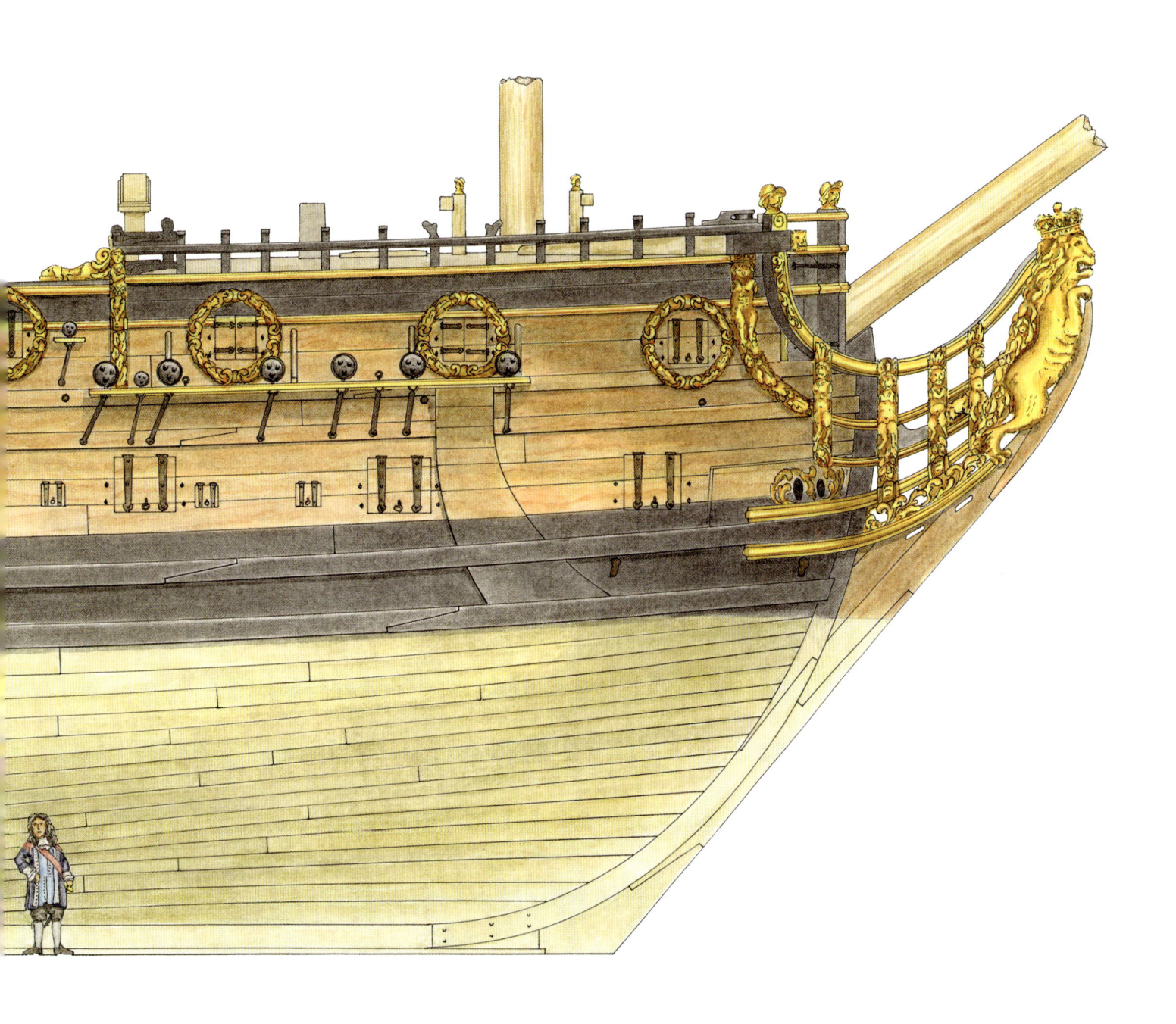

Tyger
Broadside. Scale 1:72. Author.

THE MASTER SHIPWRIGHT'S SECRETS

Tyger

Right. *The stern of the Tyger. The drawings and paintings of the ship by van de Velde allow an accurate reconstruction of the stern to be made. Although many ships, such as the* Mordaunt, *had two rows of windows, the* Tyger *retained the earlier style, with one. The Royal Arms have been reduced in size as in the later style allowing space for a panel each side with prominent carvings of a tiger and a phoenix. The phoenix is of course the mythical bird that dies but rises from the ashes and is cyclically reborn. This carved statement on the stern of the ship is an amusing comment graphically telling the story of the ship herself. Scale 1:72. Author.*

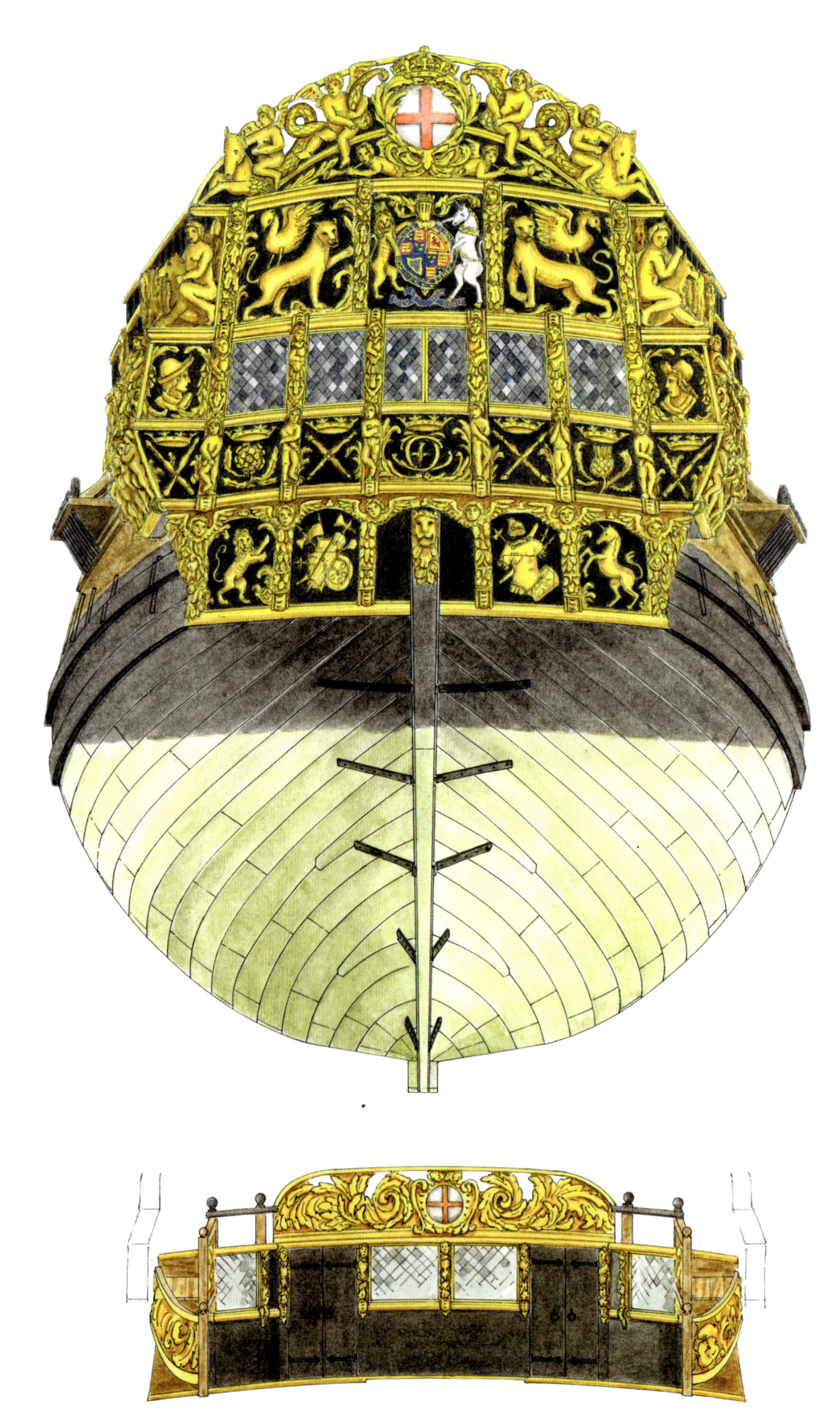

BUILDING THE NEW TYGER

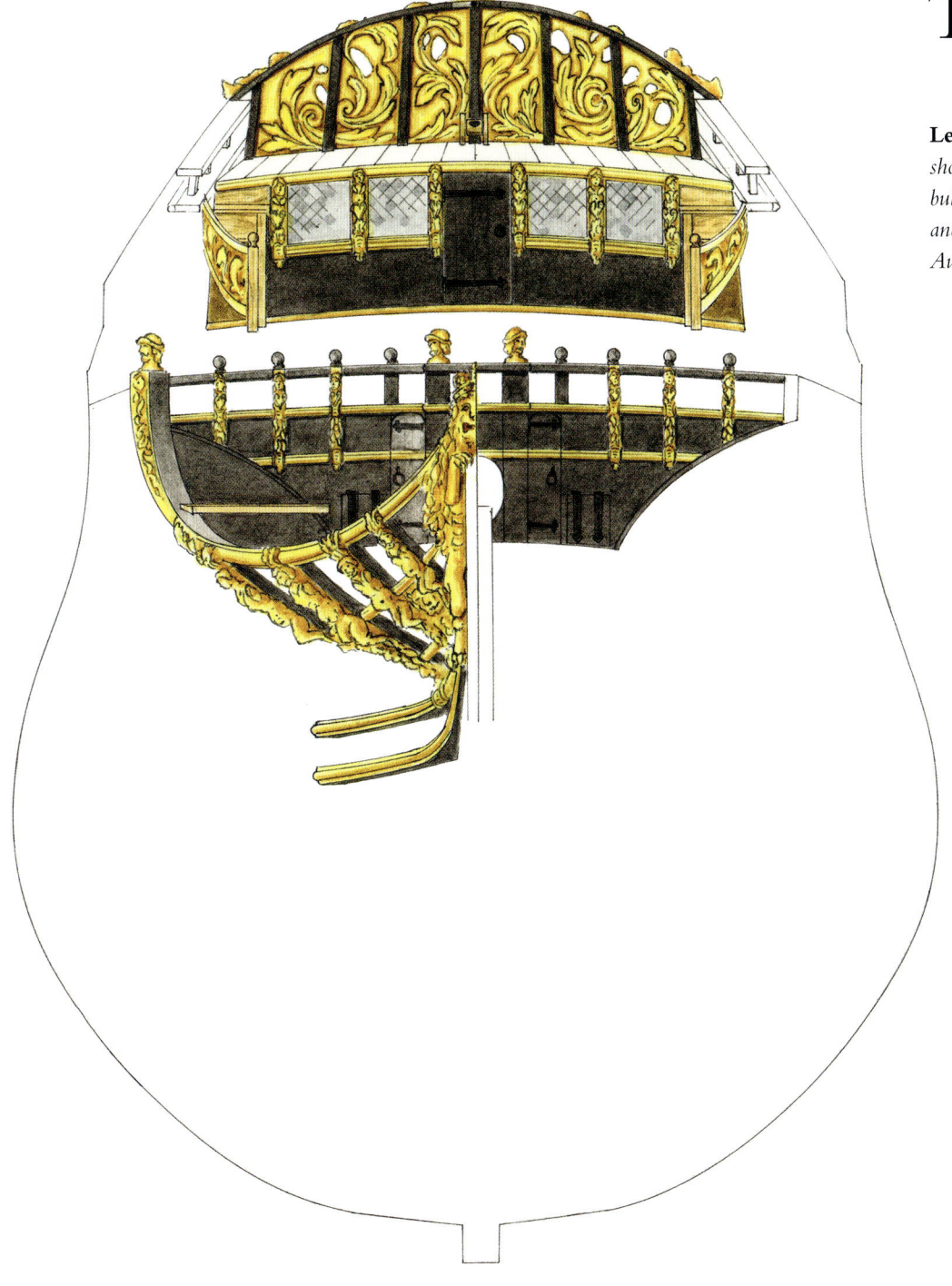

Tyger

Left. *A view from ahead showing the head, beakhead bulkhead, coach bulkhead and stern. Scale 1:72. Author.*

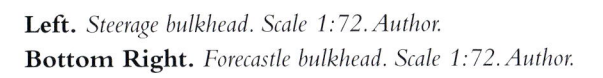

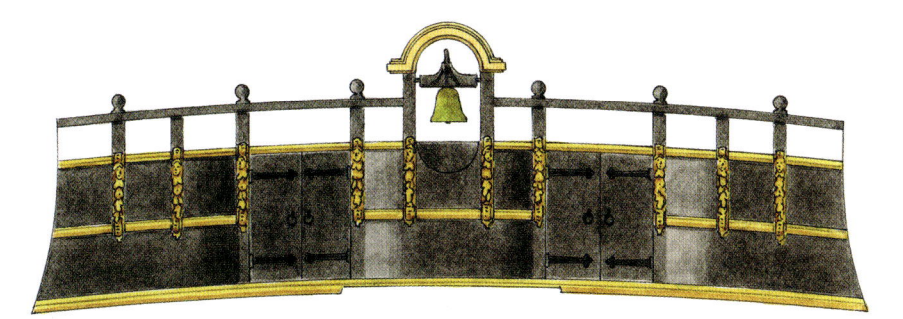

Left. *Steerage bulkhead. Scale 1:72. Author.*
Bottom Right. *Forecastle bulkhead. Scale 1:72. Author.*

227

THE MASTER SHIPWRIGHT'S SECRETS

Below. *The framing of the stern of the Tyger. The starboard side, furthest away, shows only the frame bends but includes both the full-length transoms and the half transoms attached to the sternpost, the inner post and the fashion piece. The half transoms complete an almost solid frame of timber into which to drive the planking treenails. The port side shows all the frame timbers. Note how the main transom and the port helm transom above it are in line with the main wales. The wales and transoms were joined firmly together by large fore and aft knees on the inside of the ship. The two transoms also form the upper and lower sills of the stern gunports. Author.*

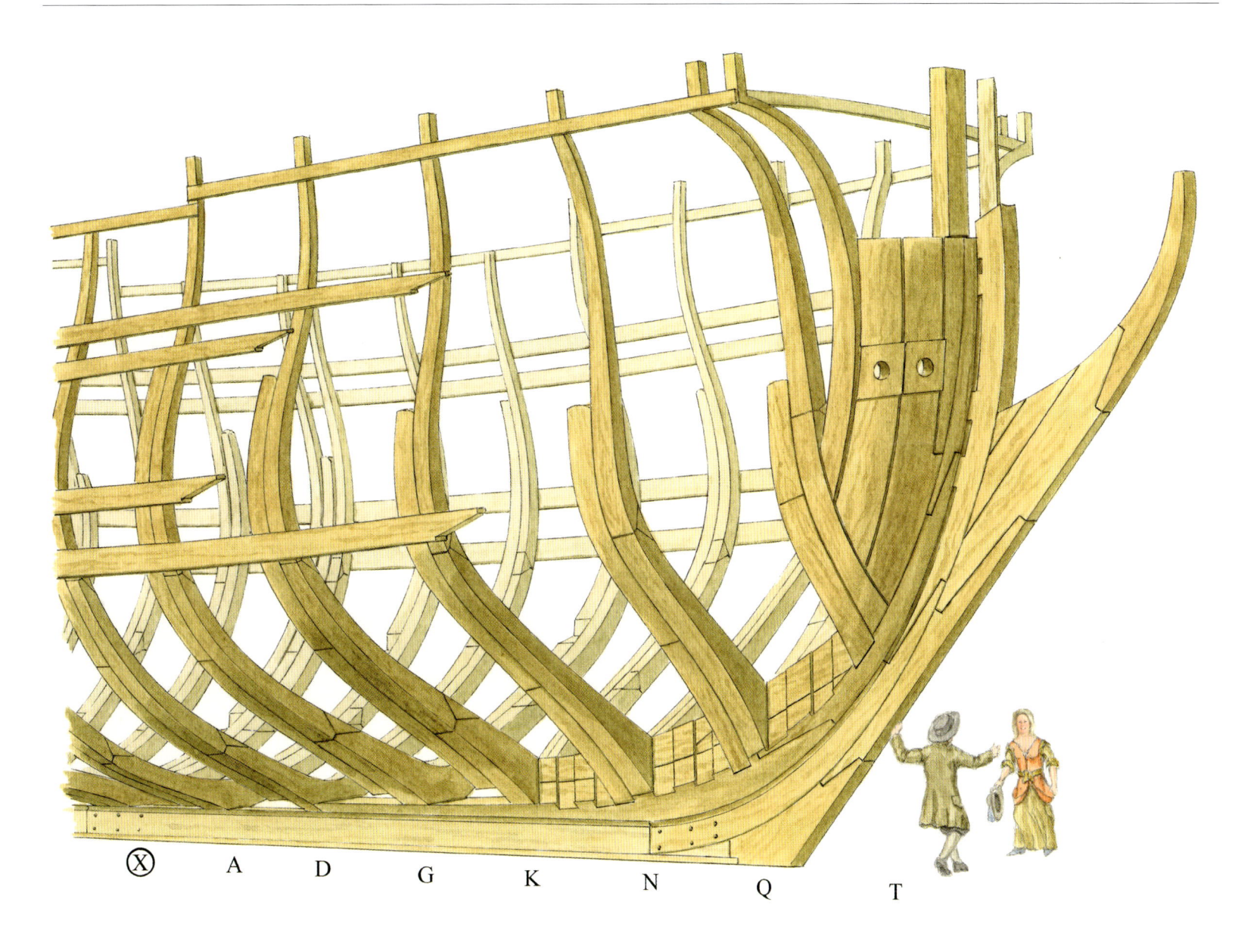

The lettered forward frame bends of the Tyger *and the probable arrangement of the hawse, shown here without planking. To reduce damage where the cables passed through the hawse holes, the area known as the boxing was made to the thickness of the frame and the inner and outer planking. Frame X is the midship bend, where the arrangement of the futtocks is reversed. Frames X and D consist of a floor timber, three pairs of futtocks and a pair of toptimbers. Between frames D and K the rising timbers sit on top of the furthest aft piece of rising wood. Between frames K and N the rising timbers have been replaced by half timbers bolted to the rising wood. The futtocks from N forward are much straighter than the other frame timbers, making it possible to reduce their number from three pairs to two. These are called the harping timbers. Note how the rabbet of the stem is interrupted where the wales fit.*[743] *Author.*

keeping the ends of the planks, the butts, clear of the keel scarphs and away from where the pumps were situated. For strength's sake, the butts also had to overlap each other by about six feet or else there needed to be three whole planks between two butts.[742]

The straight plank had to be made pliant before it could be treenailed to the frame timbers. This was done by heating over an open fire of bavins, which were brush bundles tied together in three-foot lengths and about eight inches in diameter.[745] The cost could be considerable. A medium buttock plank could consume 700 bavins at a cost of £4 40s when the plank itself cost £2 at the most. It was not only expensive, but the edges of the plank were burnt even though soaked with water. Softening a wale piece for a moderate bend could take eight hours, consuming 50 bavins an hour tended by five men.[746] It was not until 1722 that the method of heating planks over an open fire was improved, when Captain Cumberland successful experimented with laying the planks in a stove covered in wet sand.[747]

It was relatively easy to bend the planks in and out, but very difficult to bend them in an up and down direction. Many of the planks had to have their edges trimmed to shape before bending. This was done by spiling, where a straight but thin board was placed in the position of the plank, making sure it would only bend in and out. The up and down offsets to the

THE MASTER SHIPWRIGHT'S SECRETS

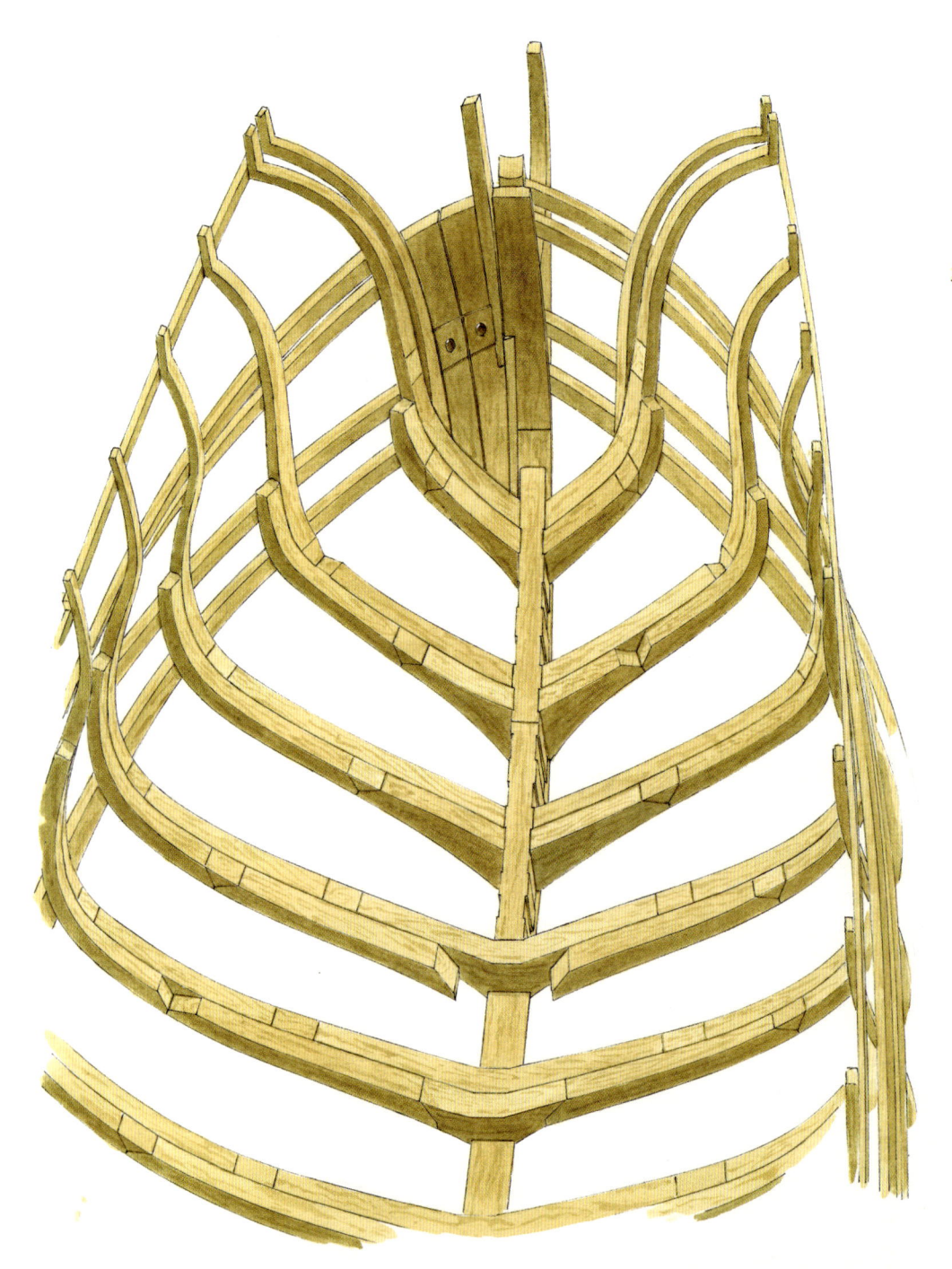

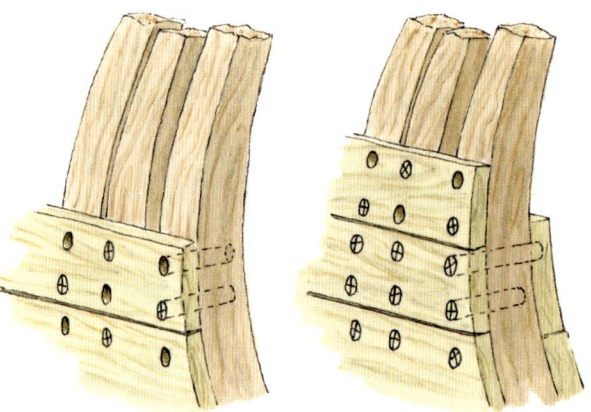

Above. *All the treenail holes are bored through the outside plank and frame timbers but only half the treenails have been fitted.* **Right.** *When the ceiling on the inside is in place the remaining open treenail holes are bored through and treenails fitted to them. Author.*

Below. *The hot and pliable planks were bent to the frame timbers by using wrain staves and bolts. The wrain bolts were passed through empty treenail holes in the frame and bolted on the inside. The wrain stave was then passed through the rings and a tapered wedge driven between it and the plank. 1: Wrain bolt; 2: Wrain stave; 3: Wedge; 4: Hot plank set against the frame timber; 5: The bolt forelock; 6: Chock to suit length of bolt; 7: Frame timber. Author.*

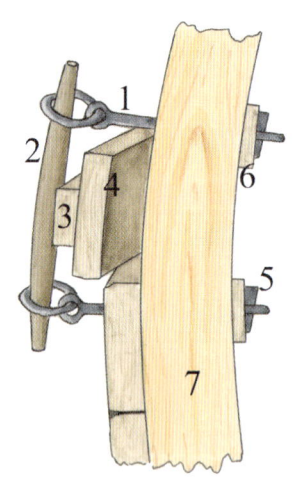

The arrangement of the frame bends and hawse timbers from the inside. The keelson that fits over the floor timbers and runs parallel to the keel is not shown. Note a small triangular fillet that smooths the step between the hawse and the inner stem,[744] providing a seat for the internal ceiling plank. Note also the cross chock fitted between the first futtocks on the nearest complete frame. It is not certain that all ships were fitted with them, but if not, then the gaps between the floors would be the only area of the ship supporting the outside plank that was not almost solid timber. A few of these chocks were always left off near the mainmast to provide access for the pumps. Shipwrights were careful to make sure there were no plank butts in the vicinity as, if there were, the oakum caulking could be sucked out of them. Author.

plank in its true position were then recorded and the plank trimmed to that amount. The angle of the plank edge, the bevel, was also measured and cut to suit. Once the heated and pliable plank was hauled up against the ship's frame it was clamped in position with wrain staves to hold it in place until it was permanently fastened with treenails.

Everywhere a plank crossed a frame timber in the lower hull of the *Tyger* there were probably two treenails of 1½ inches diameter driven through to attach them together.

The treenails were made from the best straight-grained, seasoned oak from the top of the tree. After being roughly cut to size, they were finished by being passed through a round die, a process known as mooting. The outside plank was the first to be fitted. When laying a plank or wale, all the treenail holes were first bored through but only half the treenails were driven home. The hole was bored $7/32$ inch undersize and tallowed to ensure a good tight fit. Once the treenail was driven home, the outside end was cut off flush, then split in a cross and caulked. These treenails were then cut off flush on the inside so that when the inside planking, the ceiling or footwaling was later fitted, the empty hole could be augured through and a treenail driven to hold the outside plank, the frame and the ceiling together. Towards the keel, the number of through treenails was increased three or fourfold.[748] The footwaling would have been started on the *Tyger* in about the middle of March 1680. Nails or bolts were not used to hold the planking in place as they were more expensive and damaged the oak as they corroded.

Towards the ends of the ship, especially near the fashion piece, some of the curves were so severe that the planking could not be bent by heat. Instead, the plank was sawn with the necessary curve from compass timber. In 1684 Thomas Shish could not plank the buttocks of the fourth-rate *Greenwich* until he was supplied with suitable compass timber lying at Guildford.[750]

Decks

On 12 May 1680, John Shish and his brother Thomas, the master shipwright at Woolwich, went to Rainham in Essex to view timber lying there that belonged to the timber merchants Sir William Warren and Sir John Shorter. They measured 189 large tree trunks, '*ends*' averaging 94 cubic feet each, and knowing there were 50 cubic feet in each load, they reckoned there was a total of 357 loads 48 cubic feet, all useful for the second-rate ships the brothers were both building. While there, they also viewed '*timbers for repairs for the rebuilding of the Tygar, & repairing any other 4 rates*'. For this use, there were 153 ends averaging 72 cubic feet each, making a total of 221 loads 22 cubic feet. The brothers carefully listed what parts the timber would be used for. It included '*thick stuff*' – that is, planking above four inches thick used for the internal ceiling in the hold. They also listed curved compass timber for the forward clamps, futtock riders and harping pieces. There was also straighter timber for clamps and wale pieces. Thirty-one ends were suitable as gun deck beams and four as upper-deck beams.[751] The list does not include the rising wood that fitted on top of the keel or any frame timbers as by that date they had already been found, even if they were not all in place.

The 31 ends suitable for gundeck beams seen by Shish at Rainham averaged 67 cubic feet in content and were just the right length and size for the *Tyger*. Gundeck beams were placed one under and one between each port, and as the ship had 11 gun ports each side, she would have needed only 21 beams. The remaining ten beams must have been used for repairing other fourth-rate ships. From this it seems that all of the *Tyger*'s gundeck beams were made in one piece and not from a number of shorter pieces scarphed together, a practice often carried out when timber of sufficient size could not be found. Scarphed beams had a number of advantages. Trees large and old enough to form a single beam may well be past their best, while smaller, younger trees aged between 80 and 120 years old would be in their prime. There were significant disadvantages too, however. If the beams were made from two pieces, then each piece needed to be two-thirds the length of the beam to allow for an overlap of one third. The scarph itself involved complex coaking[752] or tabling with hook and butt joints[753] before being bolted together. The workmanship and ironwork cost significantly more for a scarphed beam: in 1683, a large scarphed beam cost £20, as against £15 or £16 for a whole beam.[754]

In terms of the time taken to construct the new *Tyger*, John Shish was just halfway through building her. Judging by the productivity rate he achieved for building other ships, an average of 39 men were working on the *Tyger*. at that time.[755]

One of the last tasks in completing the decks was '*shutting in*', which involved attaching two-inch-thick deal planks to the upper-deck beams. These were bought by Deptford's purveyor of timber, Thomas Lewsley, from a Mr Body on 23 December 1680.[756]

At about the same time work started on the decks, work would also have started on the stern and the head. These structures were not completed until the ship was nearly finished, indicating that they were probably each made by a small band of specialist shipwrights. Work also started in May 1680 on the carvings, by John Leadman and Joseph Helby, the appointed carvers to Deptford, at a cost of £139 17s 10d. The work took until 1 May 1681, taking almost a year to complete.[757] The wood for the carvings was supplied by the yard and often consisted of defective or crooked fir masts[758] or elm.[759]

By the start of 1681 most of the large construction work was completed and an estimate of £880 to finish her was made on 21 January. This would indicate that the ship was about 85% complete in terms of cost.[760] By this time holes had been bored through the sides of the ship and John Shish was awaiting the supply of 26 lead scuppers of three inches in diameter to fit through them. Eighteen of the scuppers were to be two feet long and the other eight, 20 inches long.[761]

THE MASTER SHIPWRIGHT'S SECRETS

John Shish and Susanna Beckford on the gundeck of the Tyger, *looking towards the stern. Susanna Beckford ran her own business as the approved supplier of ironwork to Deptford dockyard. The nearest double frame is frame bend 'G' and the next aft is frame bend 'D'. The fore and aft timbers against the frames and underneath the beams are the clamps into which the beams are dovetailed by two inches. The vertical knees are called the hanging knees, while those in the fore and aft direction are the lodging knees. Between the gundeck beams are the fore and aft short carlings. The fir long carlings on the upper deck form the sides of the hatches. The small beams fitted into the carlings and parallel to the main beams are the ledges. On the outside, the lower pair of fore and aft timbers are the main wales and the upper pair, the channel wales. Author.*

Cabins and Store Rooms

During February 1681, John Shish received a letter from the Navy Board requesting him to discharge as many workmen he could spare. Apart from three bricklayers, he was reluctant to let any go as the days were lengthening and the men could be in the yard from six in the morning till six in the evening. He also had enough timber to keep the men busy working on the new second-rate ship, finishing the *Tyger* frigate and repairing old boats. '*And the joiners cannot be conveniently spared till the works of the Tygar be finished.*'[762] The joiners were needed to complete all the cabins, bulkheads and store rooms as it was intended that the ship should go to sea as soon as she was finished.[763]

Fourth-rate ships came in a great variety, making the assignment of cabins, as well as their size, position and layout, always a matter of expediency. The cabin arrangement listed in the survey of the *Mordaunt* at the time she was bought by the Navy demonstrates how different they could be. In earlier years, among other things, their proliferation led officers '*to neglect their duties and misspend their time in drinking and debauchery*'.[764] To try and bring some order, an establishment of cabins was drawn up and adopted in 1673,[765] and this was confirmed with small alterations in 1686.[766] Nevertheless, most fourth-rate ships were so different to each other that special arrangements must have been made to suit each one.

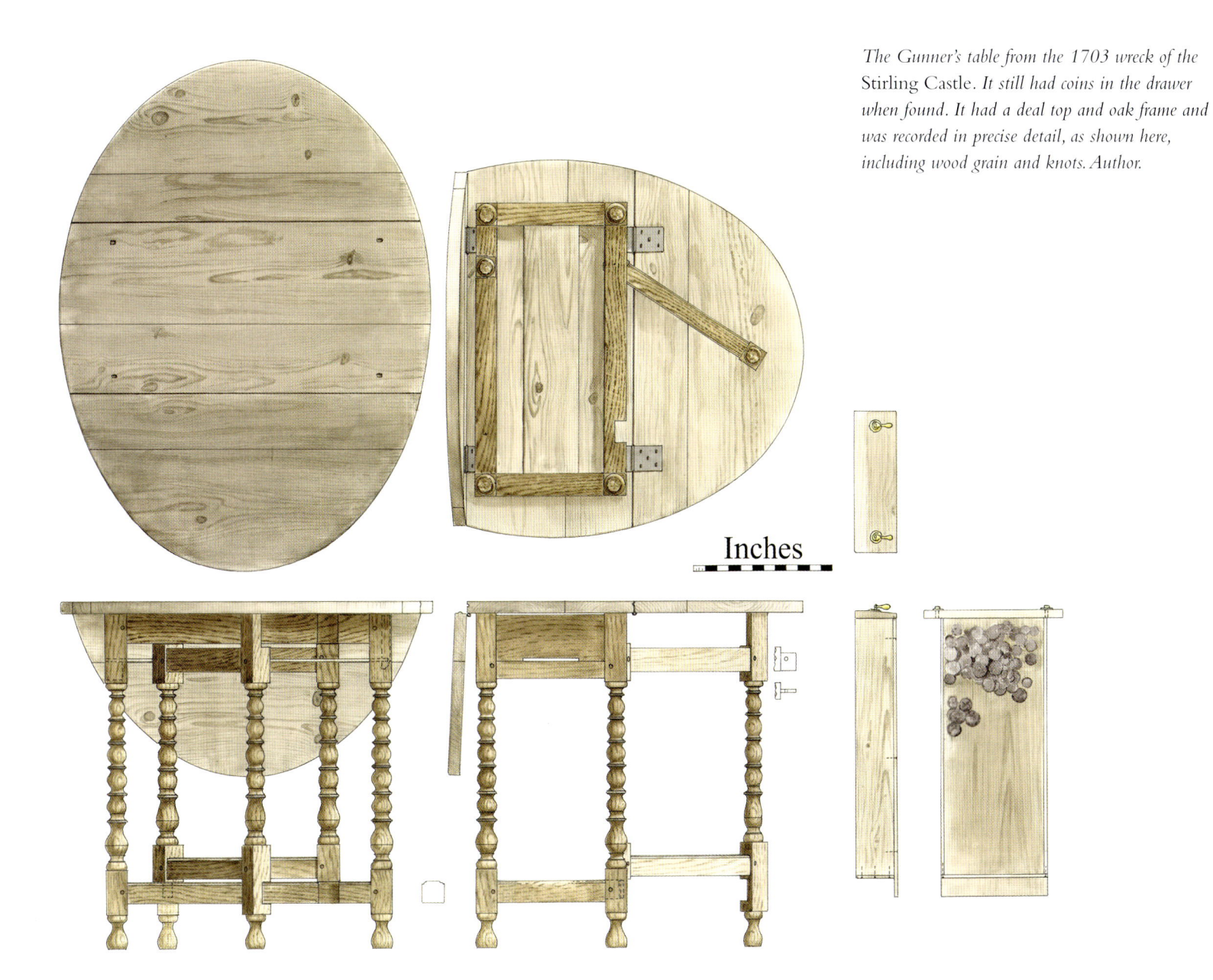

The Gunner's table from the 1703 wreck of the Stirling Castle. *It still had coins in the drawer when found. It had a deal top and oak frame and was recorded in precise detail, as shown here, including wood grain and knots. Author.*

At the time the new *Tyger* went to sea in 1681, fourth-rate ships had only one lieutenant, and a sketch dated 1674 survives showing how the roundhouse was divided down the middle to suit the master and the lieutenant, with a small entrance hall called a cuddy.[767] Towards the end of 1684, second lieutenants were added to the ship's complement, and naturally enough, they expected to have their own cabins. This inevitably caused some accommodation problems. When two lieutenants turned up to serve aboard the *Charles Galley* when she was being fitted out at Deptford, the first lieutenant thought that he should have the whole of her rather small roundhouse and that the second lieutenant should be allocated a cabin in the steerage below, displacing the master. Hearing this, the unhappy master had the captain write a letter to the Navy Board for advice in how to resolve the debate.[768] This was not the only dispute; during the same period there was a similar occurrence aboard the *Phoenix* at Spithead.[769] Accommodation was not such a problem aboard the *Jersey*, which was described as '*not only very roomy but hath good accommodation and is a ship of large storage*'.[770] Captains often made requests to have changes made to their own cabins. In 1678, the captain of the *Oxford* wrote that '*she hath the longest cabin of any ship in England… (and it) is very large cold and naked*'. He wanted a bulkhead fitted to divide it in two.[771] The Captain of the 45-year-old *Foresight* wanted a similar bulkhead erected in his cabin and alterations made to the staircase from the steerage to the quarterdeck. The Navy Board thought this unnecessary, and wanted to know '*why you should not be contented with the same accommodation your predecessors had*'.[772] The formal allocation of cabins built by joiners was complemented by the erection of canvas cabins as the need arose. In 1681, for example, the *Sweepstakes* was found to have four on the gundeck.[773]

funnels for the cookroom which Shish had requested two weeks previously.[792]

Normally, the Admiralty and the Navy Board had complete control over the building of a warship. They would authorise and have copies of the warrant which included details of the ship's specification, and any slight alteration or addition would require their written approval. It is truly astonishing that the naval administration had not known that a new ship had been built for the Navy and thought the old *Tyger* had only been subject to a repair. Now that the ill-informed Admiralty was becoming aware that the new ship would be very different to the original, it issued the following order on 10 May: '*The Navy Board be directed to consider what number of men & guns are judged fitting to be established for the Tiger with respect to her oars and additions made to her dimensions over & above what she was in her former build.*'[793] The officers of the Navy Board, having themselves little idea about the nature of the new ship, wrote to John Shish for some details. He made a full reply on 16 May which included the number, type and weight of guns that in his opinion the *Tyger* should carry.[794] His proposed armament would make her the most lightly armed English-built fourth rate in the fleet, except, of course, for the two galley frigates and the oldest and smallest fourth-rate-sized vessel, the *Constant Warwick* of 1645,[795] which, at only 315 tons, was almost half the burden of the new *Tyger*. Not wishing to obediently follow Shish's proposal, the Navy Board had Sir Richard Haddock prepare an establishment of guns for the ship on its behalf.[796] This establishment was very similar to Shish's, and the Admiralty resolved that Lord Daniel Finch show it to Charles for his approval before any orders for the ship be given.[797] Charles approved the establishment on 25 June 1681.[798]

The celebrations attending the launching of a ship at Deptford. Print by John Cleveley the Elder.

CHAPTER 8

THE NEW TYGER COMMISSIONED

Launch

At the beginning of June 1681, the building works on the *Tyger* were finished and she was ready for launching. It was important to get the new ship in the water before the summer weather dried out the planks and caused the seams between them to open. In the meantime, the best way to prevent this happening was the continual watering of the ship's sides.[799] King Charles II had the ship built exactly as he wanted and was now ready to hand her over to his naval administration. In mid-June orders were given through the Admiralty and the Navy Board for the ship to be launched and fitted out to escort a victualling ship bound for Admiral Herbert's Mediterranean fleet at Gibraltar.[800] John Shish made hurried preparations to construct the bilge ways for her to slide down into the water and found it necessary to purchase 16 large planks to build them with from the timber merchant and ship builder Sir William Warren.[801] Having sold them, Warren later pleaded for the loan of the same bilge planks, as he wanted to launch a ship of his own at Wapping. He promised to return them safely and make good any damage that may occur.[802]

Following Admiralty instructions, the new *Tyger* was successfully launched into the wet dock on Friday 24 June.[803] It was customary for master shipwrights to be rewarded with a fine piece of silver plate on the successful building of a new ship. Seeing the striking new *Tyger* in the river, John Shish reasonably supposed he was deserving of such a reward and wrote a suitably pleading letter to the Navy Board. He remembered to add that a precedent had been set when his father received a chased bowel for rebuilding the *London* some years earlier.[804] The Navy Board agreed that the *Tyger* really was a new ship and granted Shish a piece of plate, probably a tankard to the value of £15.[805] Shish's reward established a precedent that other shipwrights did not fail to notice. In 1684, Robert Lee at Chatham applied for a piece of plate for repairing the *Sovereign*.[806] An enquiry was held which found that he did not qualify as '*the works done to her were not equal to those done to the London & Tyger*' and '*The Tyger was taken all in pieces & wholly made a new ship*'.[807]

After her launching on the Friday, the *Tyger* stayed in the wet dock until the following day, when she was taken out into the River Thames. By the Monday morning she was alongside the mast hulk with her main and mizzen masts set in place. During the afternoon, on the next flood tide, they expected to set the foremast and bowsprit. The masts would have been made under the supervision of Humphrey Cadbury, the master mastmaker at Deptford.[808] A small problem occurred when a lighter full of ballast was brought alongside but they had no shovels or brooms to heave the ballast in, despite earlier demands for some to be supplied.[809] In spite of that irritating delay, work proceeded swiftly. The *Tyger* remained alongside the hulk, and only two days later the tops, together with all the standing rigging, were set up, securing the masts. During the next spring tide they planned to heave her on shore to grave her, but were delayed as they waited for a supply of broom.[810]

THE MASTER SHIPWRIGHT'S SECRETS

The two boats needed for the Tyger at the time of her launch. The moulds to make them used the same principle as those used to make the floor timbers of large ships, a principle later known as whole moulding. Top: The longboat, the all-purpose heavy-duty boat used to fetch supplies and carry out anchors. In service it was usually towed behind the ship for ready use. Bottom: The pinnace, used by the officers and for transporting men. It was stowed in the waist on top of spare spars which were laid between the forecastle and the quarterdeck. Scale 1:72. Author.

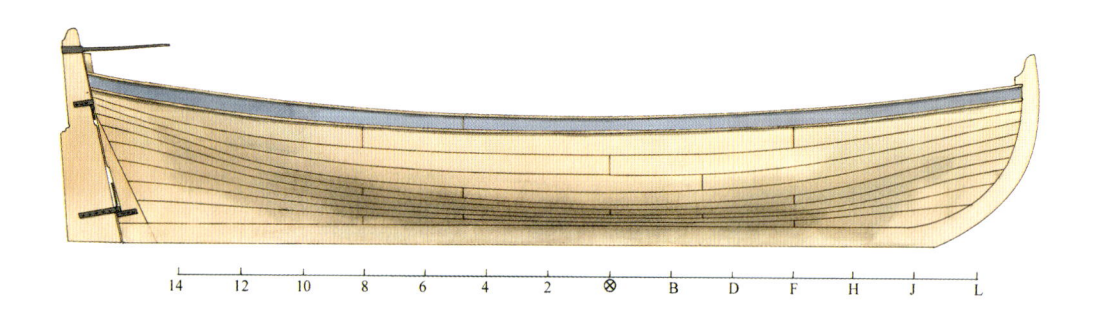

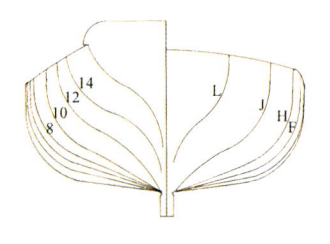

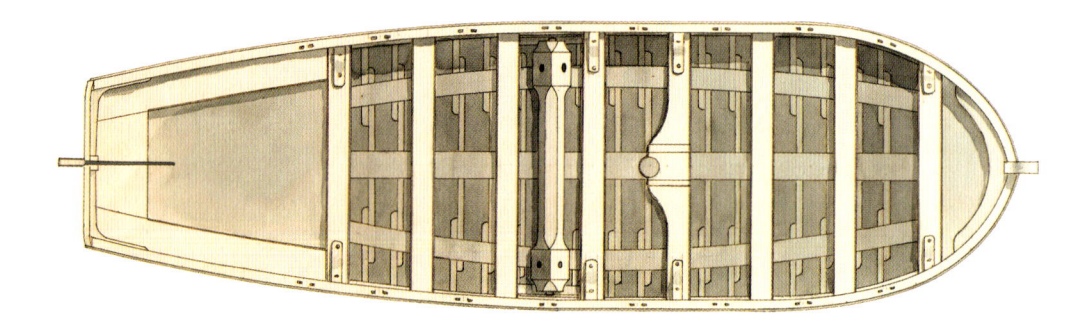

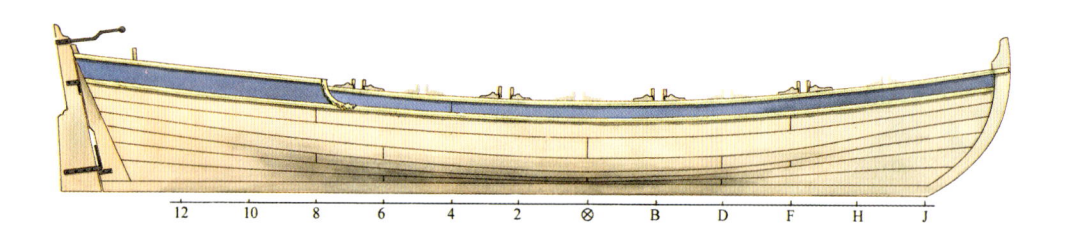

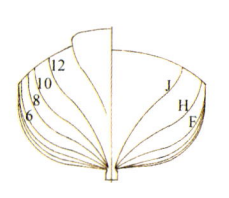

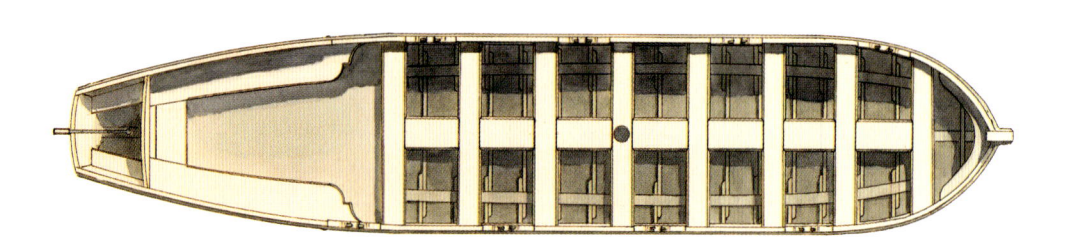

Feet

Sails and Rigging

As the ship was being made ready, the master attendant at Deptford wrote to Sir Richard Haddock to inform him that they needed 3,297 yards of Holland duck canvas, 558 yards of Ipswich canvas and 450 yards of Broad Lincoln to complete the *Tyger* with three suits of great sails and one suit of small sails.[811] At the same time an estimate was made for making the ship ready for the voyage to the Mediterranean:[812]

Estimate for fitting the *Tyger* for sea

Her standing and running rigging with tops, blocks etc and fitting it	£395 0s
Her sails repaired and completed to three suites of courses & topsails and one suite of small sails	£376 5s
Six cables of 16 and 15½ and a stream cable	£477 0s
Four great anchors, a stream and a kedge anchor	£213 4s
Cablets, hawsers and other cordage for stores	£138 15s
Two boats, furnaces, canvas, colours and other Boatswain's stores	£330 0s
Carpenter's stores	£76 0s
Graving her with white stuff, stocking her anchors, wedging and cleating her masts and yards, fitting her half ports and other finishing works	£59 10s
	£2065 14s

On 1 July it was reported that the *Tyger* frigate would be fully rigged by noon.[813] The work continued to progress rapidly, for just a week later the cost to complete her with nine months' stores had reduced from £2,065 to £950.[814]

Fitting Out

The *Tyger* would have been launched with a just a few tons of ballast[819] but would need about 100 tons more before she could sail. Stone shingle was normally used, but although cheap, it consumed a considerable amount of space. Instead, John Shish used three old redundant iron demi-culverin cannon at Deptford[820] and 33 tons of other iron supplied by the Ordnance Board.[821] This brought her total to 90 tons, but as soon as she went to sea it was found to be inadequate, with her lieutenant stating: '*we finding the ship to be much out of the water we got on board by judgement 17 tons of Deal beach*'.[822]

On 12 July, the master attendant at Deptford reported that of the five anchors required for the *Tyger*, she was short of the three largest, a 26-hundredweight sheet anchor and the bower

The sizes of the *Tyger's* five anchors

Weight in Hundredweight	Length of Shank m–h	Length of Arm c–x
26	13ft 6in	4ft 10in
24	13ft 1in	4ft 8in
24	13ft 1in	4ft 8in
8	10ft 0in	2ft 8in
4	8ft 6in	2ft 2in

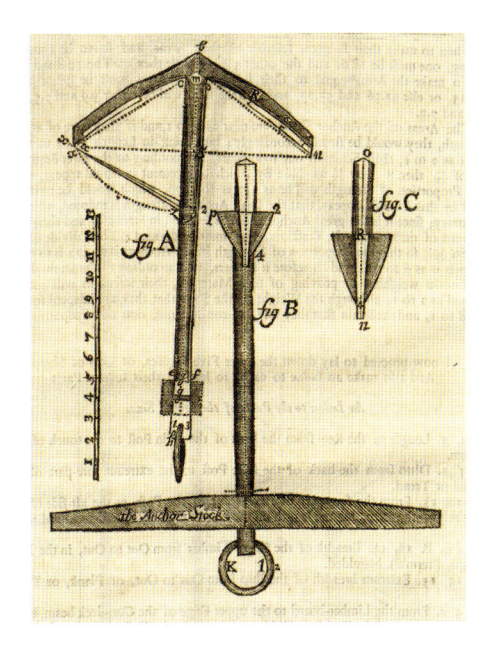

Above. *The five anchors of the Tyger were all the same type but different in size. From the known weights of three of her anchors, their dimensions were obtained from a near-contemporary list.[823] The weights of the two anchors not mentioned for the Tyger are taken from another contemporary list.[824] From* William Sutherland, Shipbuilding Unveiled, *1717, p23.*

THE MASTER SHIPWRIGHT'S SECRETS

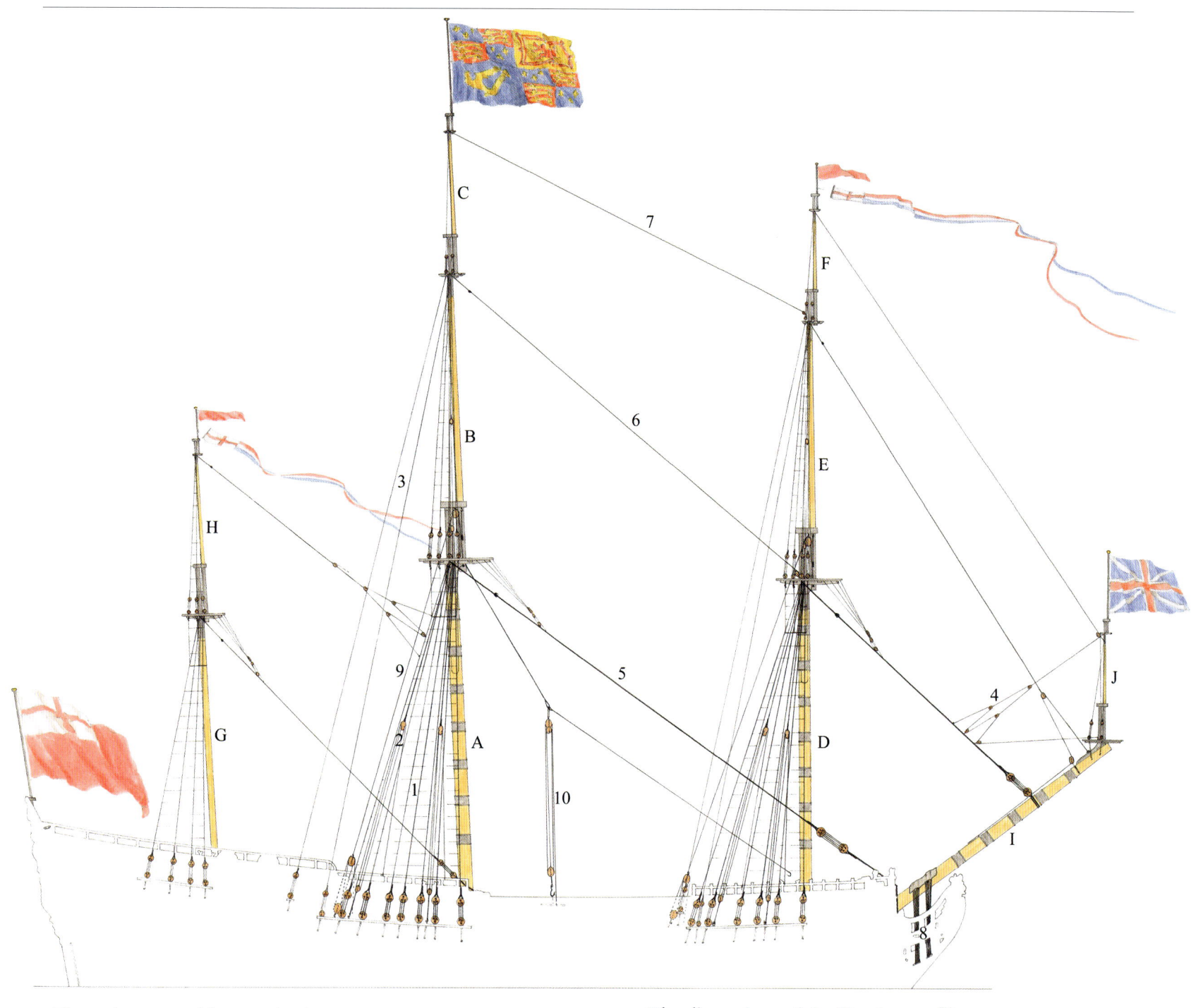

The standing rigging of the Tyger, based on contemporary lists, van de Velde paintings, drawings and the model of the St Albans. The sources for some of the rigging, such as the complex spritsail topmast backstay, are satisfyingly all in agreement, while others, such as those for the masthead tackles, are contradictory. Author.[815]

1: Shrouds
2: Masthead tackles
3–4: Backstays
5–7: Stays
8: Gammoning
9. Top-rope for heel of mast
10. Garnet.

The dimensions of the *Tyger's* masts[816]

	Mast	Length in yards	Diameter in inches
A	Mainmast	27½	24
B	Main topmast	16½	15½
C	Main topgallant	8	7¼
D	Foremast	24	21½
E	Fore topmast	14	13½
F	Fore topgallant	6½	5
G	Mizzen mast	23½	13½
H	Mizzen topmast	9	7
I	Bowsprit	18	24
J	Spritsail topmast	6	6

THE NEW TYGER COMMISSIONED

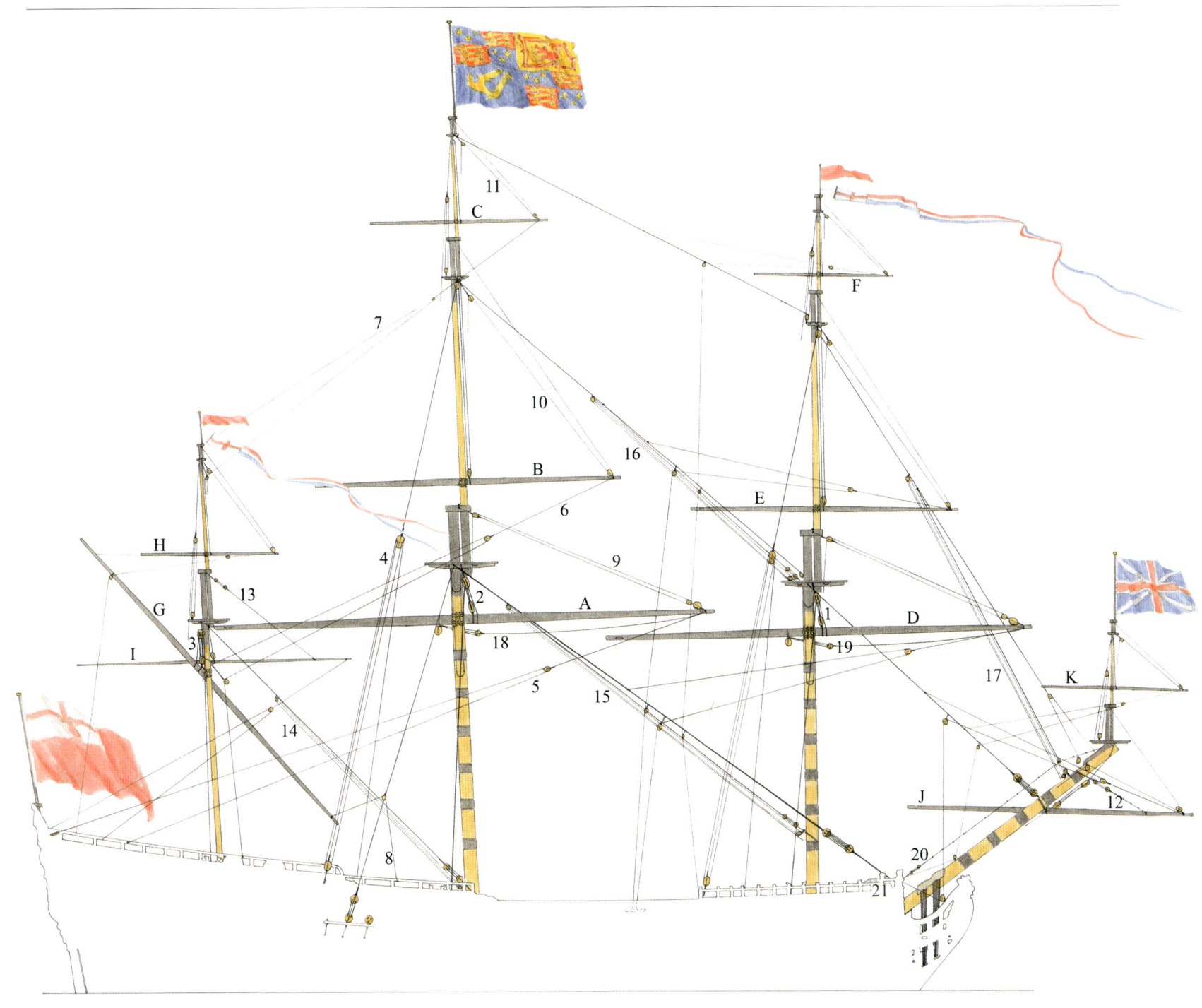

The dimensions of the *Tyger's* yards[817]

	Yard	Length in yards	Diameter in inches
A	Main yard	25½	17
B	Main top yard	15½	11
C	Main topgallant yard	9	5½
D	Fore yard	21½	14½
E	Fore top yard	13½	10
F	Fore topgallant yard	7	4½
G	Mizzen yard	19½	10½
H	Mizzen top yard	7	4½
I	Crojack yard	14	7
J	Spritsail yard	14½	10
K	Spritsail top yard	7½	5

The yard rigging of the Tyger. Although ships' rigging was very much standardised, differences did occur. A slightly unusual feature of the Tyger's rigging is the crojack brace, which van de Velde shows leading aft when it was usual for them to lead forward to the main shrouds. Author.

Rigging for raising and lowering yards:
  1–3: Jeers
  4. Halliards (all yards except the crojack).

Rigging for swinging and trimming yards:
  5–7: Braces (all square-rigged yards)
  8: Bowline (similar function to the square yard braces)
  9–11: Lifts
  12–13: Standing lifts.

Other:
  14: Staysail halliard
  15–17: Staysail stay and halliard
  18–20. Horses (for health and safety)
  21: Snatch block for cathead ropes.

A Table of the Names of the Ropes.

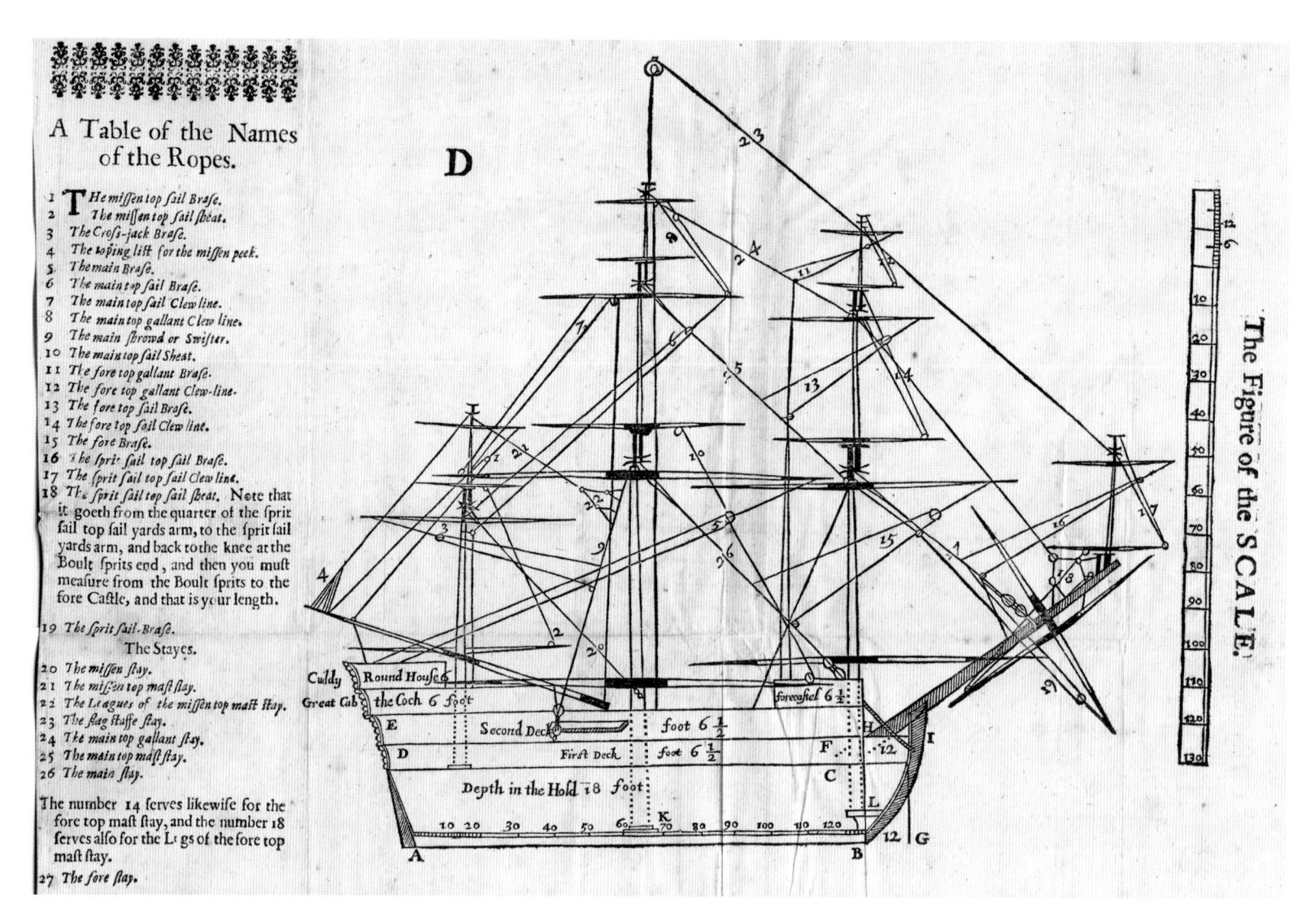

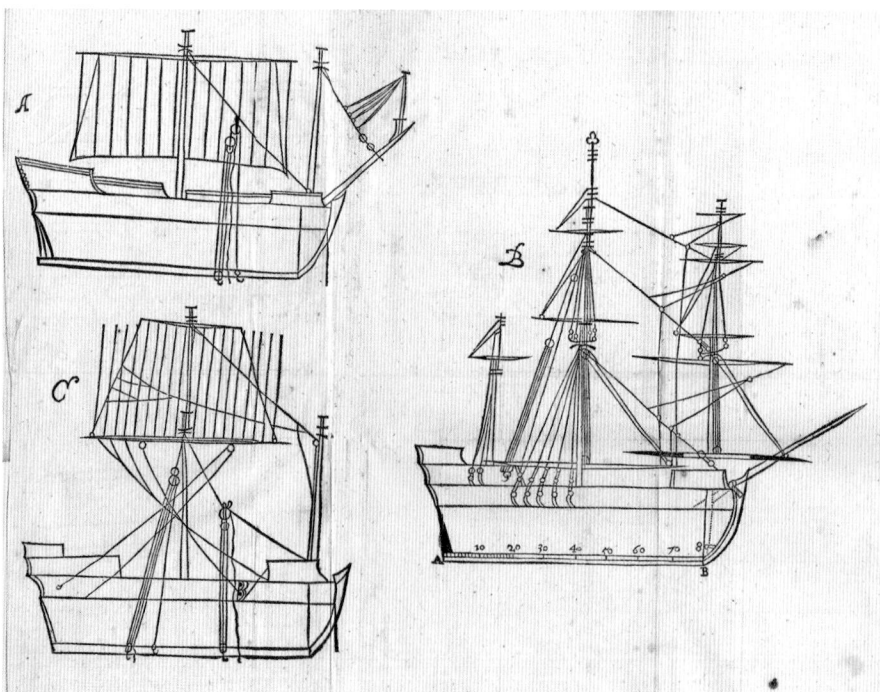

1. The missen top sail Brase.
2. The missen top sail sheat.
3. The Cross-jack Brase.
4. The toping lift for the missen peek.
5. The main Brase.
6. The main top sail Brase.
7. The main top sail Clew line.
8. The main top gallant Clew line.
9. The main shroud or Swifter.
10. The main top sail Sheat.
11. The fore top gallant Brase.
12. The fore top gallant Clew-line.
13. The fore top sail Brase.
14. The fore top sail Clew line.
15. The fore Brase.
16. The sprit sail top sail Brase.
17. The sprit sail top sail Clew line.
18. The sprit sail top sail sheat. Note that it goeth from the quarter of the sprit sail top sail yards arm, to the sprit sail yards arm, and back to the knee at the Boult sprits end, and then you must measure from the Boult sprits to the fore Castle, and that is your length.
19. The sprit sail Brase.

The Stayes.

20. The missen stay.
21. The missen top mast stay.
22. The Leagues of the missen top mast stay.
23. The flag staffe stay.
24. The main top gallant stay.
25. The main top mast stay.
26. The main stay.

The number 14 serves likewise for the fore top mast stay, and the number 18 serves also for the Legs of the fore top mast stay.

27. The fore stay.

Above. *A rigging plan showing the yards in their raised and lowered positions, from Thomas Miller's* The Compleat Modellist *of 1664. It reveals that a number of changes had taken place between its publication and the building of the new Tyger in 1681.*

Opposite. *The main course and topsail of the Tyger, based on a contemporary drawing of a sail in the British Library.[818] 1: Earing for securing the sail to the yardarm cleats; 2: Reef cringles for hauling the sail up to the yardarm by the reef tackles; 3: Leech cringles for gathering up the sail; 4: Bowline cringles for holding the weather leech of the sail open to the wind; 5: Bowline pieces put on double to strengthen the sail; 6: Buntline cringles for gathering up the sail; 7: Clews attached to the sheets, tacks and clew garnets on the course and to the sheets and clewlines on the upper sail. Author.*

Left. *Figures A and C show the garnet and winding tackles. The heavier-duty winding tackle was temporarily rigged to handle heavy objects such as the guns. Figure B shows jeers and halliards. From Thomas Miller's* The Compleat Modellist *of 1664.*

THE NEW TYGER COMMISSIONED

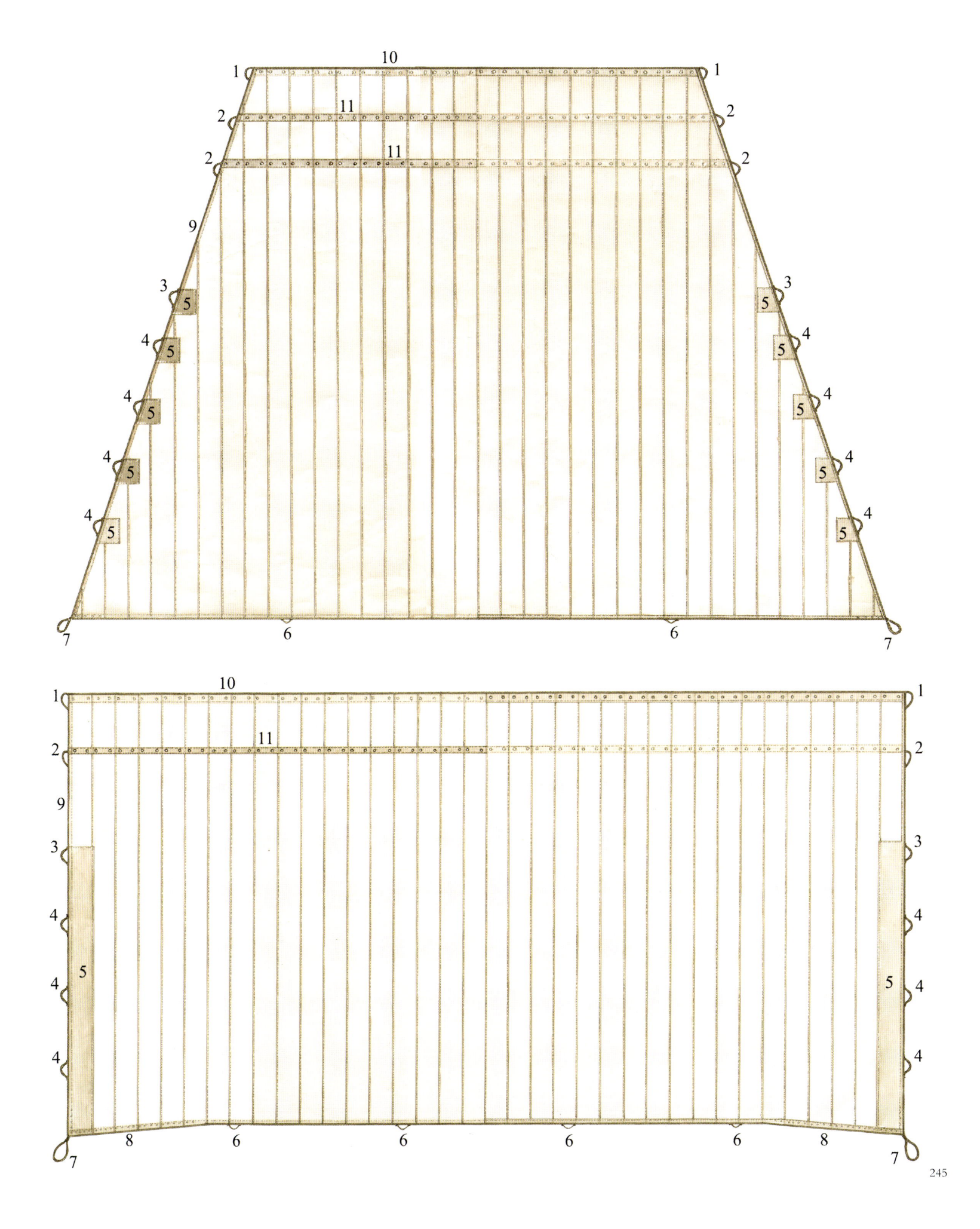

and best bower anchors, 24 hundredweight each. Also wanting were three cables of 15 inches circumference.[825] These were all supplied and safely brought aboard over the next few days.[826] As this was progressing, the ship was brought ashore upon ways for her delayed graving. Organic matter was burnt off on her bottom, which was then scraped clean and payed with new 'stuff', after which she was taken out to her berth and moored once more against the hulk.[827] The burning, or breaming, blackened some of the painted carvings and outboard work. It had also been some time since the ship was painted, and since the recent hot and dry weather had worsened it, it was all in need of refreshing. On the inside, the joiner's work about the great cabin had shrunk and also needed repair and repainting.

Captain Charles, Lord Berkeley

The request to refresh the joinery and paintwork was made to John Shish by the ship's newly appointed captain, 19-year-old Charles Berkeley, the 2nd Baron Berkeley of Stratton.[828] His father was a prominent royalist soldier who went into exile with the Stuarts and became closely associated with the Duke of York, serving with him in campaigns against Condé and against the Spaniards in Flanders. For his services with the Duke he was raised to the peerage as Baron Berkeley of Stratton, in Cornwall, by a patent dated 19 May 1658 at Brussels. Among his achievements was being a co-founder of New Jersey. He died in 1678 and was succeeded by his eldest son, Charles, the 2nd Lord Berkeley of Stratton, who was now captain of the *Tyger*. In spite of his youth, Charles had years of military service behind him, having entered the French army as a volunteer at the age of 14.[829] Returning home two years later, he joined the Guards under the Duke of Monmouth, becoming a major in October 1677. On 2 December 1679, he became cornet and major of the King's own troop of horse guards, and finally lieutenant colonel on 27 October 1681, even though he was serving in the Navy at the time.[830] Perhaps seeking more adventure, he joined the fourth-rate *St David* as a volunteer in August 1679, and it was reportedly there that he learnt navigation and became skilful in maritime affairs. To make life more comfortable afloat he took a gentleman, Mr Henry Hudson, and a retinue of three others with him.[831] In spite of his apparent abilities, he had not advanced or qualified through the recently introduced lieutenant's examination of 1677. Even so, the enthusiastic young lord with hardly a year's experience of the Navy turned up at an Admiralty Board meeting on the morning of 1 July 1680 to express his desire to command of one of His Majesty's ships.[832]

His request reached the ears of the King. Although the Navy had adequate and reliable bureaucracy to maintain its customs and rules, there were occasions when royal hegemony stamped its authority over them all. King Charles understood the benefits of having aristocrats fight alongside others of the lower classes. Having apparently met the young lord, he wrote to Daniel Finch, First Lord of the Admiralty, *'tho' I am not for employing of men merely for quality, yet when men of quality are fit for the trade they desire to enter into, I think 'tis reasonable they should be encouraged at least equally with others, and I assure you, this young man has been so industrious to improve himself and so successful in it, as he deserves some partiality in his case, to encourage others to do the like...'*[833] With such a splendid testimony, Lord Charles Berkeley was duly appointed captain on 28 June 1681. At the same time, Albion Thompson was appointed his lieutenant.[834]

On 2 July, the Navy Board informed the Admiralty that the *Tyger* was ready to be manned. Straight away orders were given for the commissions of her captain and lieutenant to be signed and for her to receive four months' victuals for 200 men.[835] Three days later, Lord Berkeley and Lieutenant Thompson,

The hulk at Deptford dockyard. From the Thomas Milton print of a geometrical plan of Deptford, 1753.

Two views of the new Tyger *under the command of Charles Berkeley, the 2nd Baron Berkeley of Stratton, who died aboard her. The commemorative painting is recorded by Willem van de Velde the Elder as being paid for by Lord Berkeley, almost certainly Charles Berkeley's younger brother, John Berkeley, the 3rd Baron, for £20. With Grateful Thanks to Berkeley Castle Charitable Trust.*

along with all the other senior officers, were entered on the ship's pay book.[836] The most important of the warrant officers was the master, John Rounsevall. Berkeley had recommended him for the position as he truly thought him "*so good a man*", adding that captains Churchill and Aylmer of the *Dartmouth* gave him a very good character.[837] Berkeley also brought along five retainers, one of whom had served with him earlier in the *St David*. Another, Henry Hudson, was now made a midshipman.[838] The purser, the gunner and the carpenter, who had been appointed warrant officers of the *Tyger* up to three years before with no duties to perform, now found themselves very busy indeed.[839] Another warrant officer, Cuthbert Sparks, the one-legged boatswain appointed to the *Tyger* in April 1678, reminded the Admiralty of the great hindrance his condition would be if he ever went to sea. They agreed he could exchange places with Henry Morgan, the boatswain of the *Grafton*, aboard which he could resume his peaceful duties.[840] The only warrant officer position left open was that of the lowly cook. One of those seeking such a position was William Munday, the late cooper of the *Charles Galley* who had petitioned for the post five months earlier. He was not appointed, however, and instead the post went to John Brisco, who perhaps wanted to go to sea as he was at the time being legally prosecuted by Valentine Castillion for a debt.[841] As King Charles would have wished, a number of gentlemen of some status volunteered to join a ship captained by a peer of the realm. These included Sir Palmes Fairborne's son. Fairborne had been governor of Tangier and had been killed there the previous October.

THE MASTER SHIPWRIGHT'S SECRETS

CHARLES L.d BERKELEY *eldest*
Son of IOHN L.d BERKELEY *of* STRAT
-TON, *who died unmarried in* 1682 *on*
board His Majesty's Ship the TYGER,
which he commanded.
by Sir PETER LELY.

Charles Berkeley, the 2nd Baron Berkeley of Stratton. This memorial painting by the leading portrait painter of the day, Sir Peter Lely, shows him as captain of the Tyger, *wearing armour, something he would not have worn in real life. With Grateful Thanks to Berkeley Castle Charitable Trust.*

# THE NEW TYGER COMMISSIONED

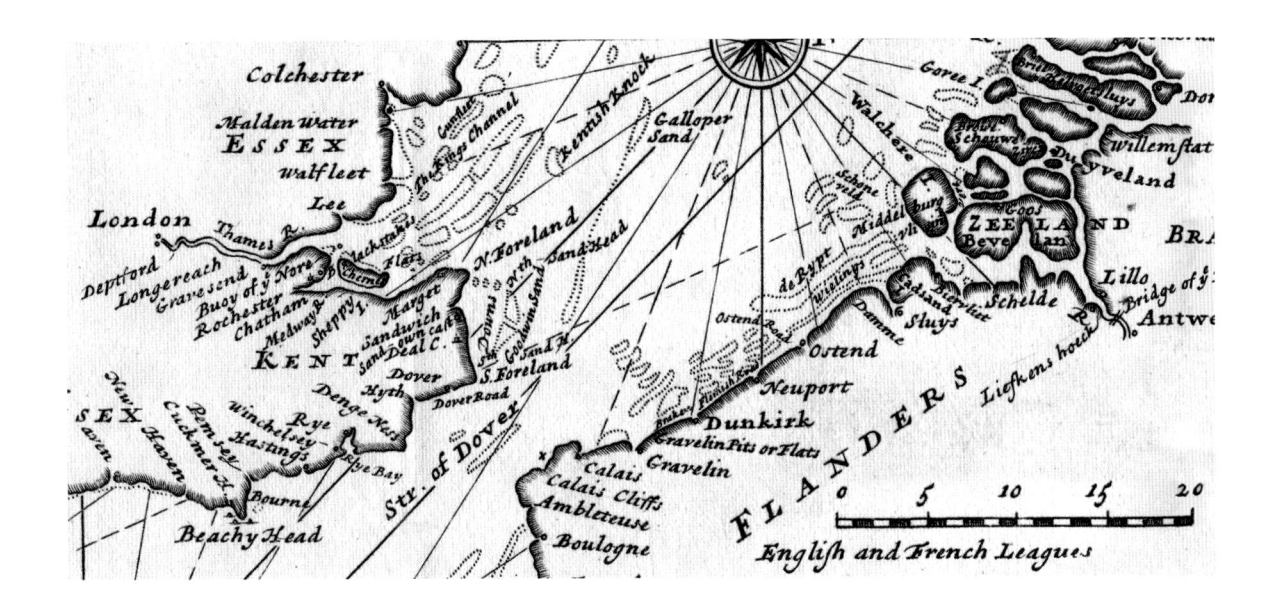

Left. *A contemporary map of the English Channel by H. Moll showing the places visited by the Tyger during the first weeks of her career.*

Below. *The Tyger under way, with some of her oars visible (Robinson 578). Probably drawn by van de Velde near the bend in the river at Greenwich where the artist lived. The 'Tiger' with sweeps out. Willem van de Velde, the Elder, 1681. PAH3921; Photo: © National Maritime Museum, Greenwich, London.*

THE MASTER SHIPWRIGHT'S SECRETS

A new ship being commissioned near London in times of peace was greeted as a welcome opportunity by the maritime community. On 5 July the pay book opened and 55 men's names were entered that day. Three days later, 205 men had enlisted, more than enough to complete the complement. Astonishingly, within a month, and while the ship was still in the Thames, 92 had deserted. The men were encouraged to make up their minds very quickly as to whether to stay or leave as if they left they would not receive any pay for their time spent aboard. At one point they were leaving faster than they could be recruited.[842] It seems the reason for their desertion was down to Lord Berkeley himself. Captains had tremendous authority over the lives of the men under their command, consequentially becoming either very popular or unpopular. Popular captains had seamen who would stay with them whatever ship they commanded. It was a delicate balance. Sir William Booth told Samuel Pepys that it wasn't possible to command an English seaman only with good words and without blows. But he went on to say that blows had to be administered with justice and great care, which some gentlemen captains did not understand, particularly Lord Berkeley and Captain Layton. They would always beat their men in the round, a brutal form of punishment in which the victim was put into a small boat and flogged in front of every ship in the fleet or in the vicinity.[843]

Setting Sail

There were many other concerns for Lord Berkeley besides finding men. He wrote to the Navy Board on 21 July to say the ballast was in and stores were being brought aboard, although he still needed kersey waist cloths, a suit of pennants and French and Dutch flags.[844] Every man had an allowance of eight pints of beer a day, and to satisfy this thirst, 25 tons of beer arrived.[845] The next day, orders arrived for the *Tyger* to sail, wind and weather permitting, the ten miles downriver into Longreach, where she was to take in her guns and victuals. From there she was to go out of the river and into the Downs to receive further orders.[846] The same day a messenger from Deptford went up to Trinity House to request a pilot, but as he didn't have a warrant from the Navy Board, the Trinity House clerk told him no one could be appointed.[847] In spite of the delay a pilot, Captain William Godfrey, was aboard the next day, but as there was only little wind from the south-west he would not set sail.[848] This prompted Lord Berkeley to write to the Navy Board the same day complaining of the difficulties and saying that His Majesty's service was being neglected.[849] The departure of the *Tyger* to the Mediterranean and the protection she could offer merchant ships had traders in London asking for and receiving permission for five of their ships to sail with her.[850]

The wind picked up from the north-west on the 30th, and the *Tyger* left the hulk to begin the difficult passage down the Thames. They briefly anchored in the bend at Greenwich but had to warp their ship round the river to head northward towards Blackwall. After spending a day marooned by light winds, they got under way again on 1 August, reaching Longreach that afternoon. After safely mooring, they got in their guns and carriages, and then, over the next week, the victuals arrived in the form of bread, beef, pork, butter, cheese, peas and oatmeal.

Opposite Top. *Visit of Charles II to the 'Tiger': The King leaving one of the yachts (Robinson 580). Willem van de Velde, the Elder, 1681. PAH1871; Photo:* © *National Maritime Museum, Greenwich, London.*

Opposite Centre. *The 'Tiger' with yachts astern in light airs (Robinson 585). Willem van de Velde, the Elder, 1681. PAH3922; Photo:* © *National Maritime Museum, Greenwich, London.*

Opposite Left. *The 'Tiger' and yachts at anchor off the Isle of Sheppy (Robinson 1210). Willem van de Velde, the Elder, 1681. PAH4118; Photo:* © *National Maritime Museum, Greenwich, London.*

The King's Visit

The *Tyger* remained in the Longreach for two weeks before setting sail on 13 August and making her way to about a mile and half past Gravesend.[851] Although she had all her allocated stores aboard, Lord Berkeley had been complaining that he had not been sufficiently supplied. He received an answer from John Brisbane, the Secretary of the Admiralty, on the 16th saying in no uncertain terms that the ship had all the stores appointed to her and that some of Berkeley's requests such as '*an extraordinary suite of pennants are not usually allowed*'.[852] The day before, Berkeley had returned to Berkeley House in Piccadilly, where twice during the morning Brisbane had unsuccessfully tried to visit him, bringing the momentous news that King Charles intended to make a visit to the *Tyger*, and wanting to explain to Berkeley how he should proceed and govern himself.[853] On learning of this, Berkeley must have quickly returned to his ship, for she was cleared and brought to order as they made their way to the Buoy of the Nore, where they anchored. The pennants Berkeley had recently requested were flown at every yardarm in a uniquely flamboyant greeting.[854]

The *Henrietta* yacht with the King aboard and the five others accompanying her were seen at distance anchoring near Lee Road at five in the afternoon of the 17th. It was blowing a fresh gale, and as the yachts were in close proximity to each other, the *Mary* yacht's middle lanthorn was hit and broke into two battered pieces, beyond repair.[855] At six'o clock the following morning the King came down with the yachts and boarded the *Tyger*. With the Royal Standard flying at the main they unmoored at about eight, hove short on the best bower anchor and weighed it at noon. As they set sail, King Charles must have made a tour to inspect the ship which he had had such a close personal involvement in building. He also took the opportunity to dine on board, no doubt taking the opportunity to get to know the officers and other appointees better.[856] With a gale blowing from the east-north-east, they turned down and by six in the evening were a mile and half below the Skew Beacon. King Charles then commanded them to bear up, and at seven left in his own boat when they were just short of the Ooze Edge Buoy near the Isle of Sheppey. As he left, the *Tyger* fired a 21-gun salute.[857]

King Charles coming alongside the Tyger *to visit the ship whose design he had personally influenced. Author.*

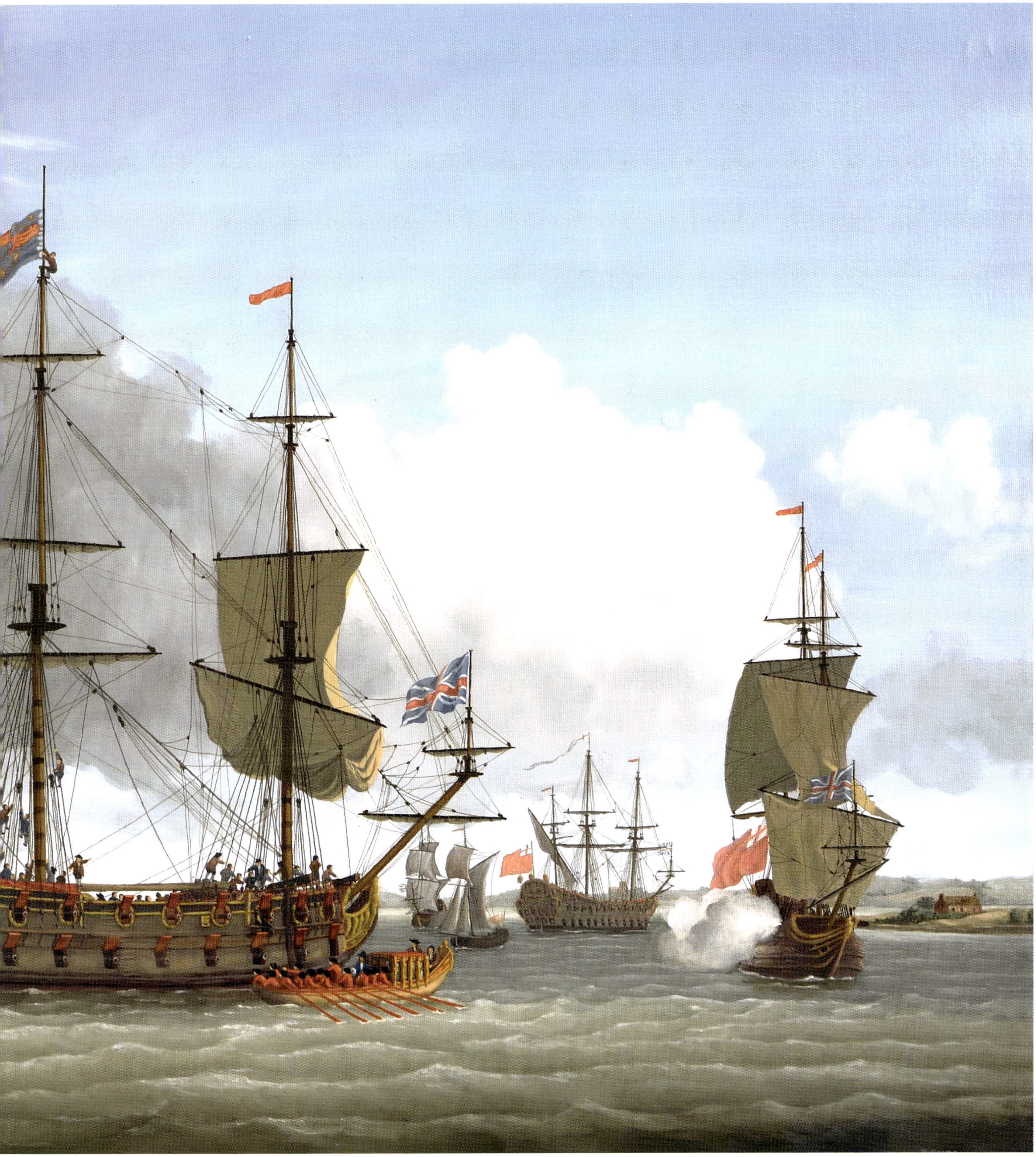

THE MASTER SHIPWRIGHT'S SECRETS

The TYGER commanded by
CHARLES Lord BERKELEY of STRATTON
in the year 1681. and who died on board that Ship in 1682

Near Disaster

Now alone, the officers and the pilot began the awkward task of taking the ship past the numerous sandbanks of the Thames estuary and into the Downs. As the sand banks generally lie in a north–south direction it was safer to sail north past them, then turn east before heading south towards the Downs. Early in the morning, just 32 hours after the King had left, the *Tyger* came about at the north end of the Gunfleet sandbank, struck the sand and lay fast aground. It was an incredibly dangerous situation. Less than a year later a similar-sized ship, the *Gloucester*, ran aground on a sandbank off Great Yarmouth with the Duke of York aboard. She beat on the sand for some time, which staved in some strakes of planks. When the tide came in, the *Gloucester* filled with water and over 130 of her crew were drowned.[858]

The tide went out around the stranded *Tyger*, leaving her in only two feet of water at the stern and the sea bed completely dry some 200 feet off her bow to the east. In these conditions, the ship could easily settle on her side so that the incoming tide would flood her, while any sort of waves would beat in the planks as she rose and fell. The crew acted quickly, and the ship was shored to remain upright using the spare topmast, the mizzen yard and the spare fish, a spar intended to strengthen a broken or damaged mast or spar. Having done this, they reckoned the ship settled three feet into the sand. Then two anchors were attached to cables and carried out ahead in boats. All was ready for the next flood tide, and when it came they hauled in the anchors and managed to pull the *Tyger* off, probably helped by two or three colliers that came to help. They had done well to save the ship, and were very lucky to have done so.[859]

News that the *Tyger* was fast aground reached John Brisbane at the Admiralty early the next morning. Not knowing the ship would be safe and not waiting for instructions from their Lordships, he assumed responsibility himself. He sent a letter to Captain Clements with the instruction from Sir Richard Haddock, the Controller of the Navy, to take the *Kitchen* yacht and find the *Tyger* with all possible speed in order to give her what help he could. Brisbane then had an incredibly anxious wait until the following day, when the son of Captain Wetwang of the Navy burst in to see him with the good news that the *Tyger* had got off the sands with the help of some colliers. Brisbane explained his actions to the Lords of the Admiralty, saying he had taken it upon himself to act as he heard the *Tyger* had *'been on ground Saturday morning upon the spit near the Gunfleet & too much reason to fear for her'*.[860] To find out exactly what happened and apportion blame, Captain Berkeley received orders to leave behind any fitting witnesses as well as William Godfrey, the pilot, for him to be tried at a court martial for running the *Tyger* aground.[861] Surprisingly, Godfrey received a certificate for his services and was paid £11 10s.[862]

The Tyger *under the command of Charles Berkeley, with yachts in attendance at the time of King Charles II's visit on 18 August 1681. The painting commemorating the visit and the death of Berkeley is recorded by Willem van de Velde the Elder as being paid for by the captain's mother, Christiana, for £23. With Grateful Thanks to Berkeley Castle Charitable Trust.*

Trouble for Lord Berkeley

The weather became calm as the *Tyger* safely entered the Downs on 22 August, giving an opportunity to adjust the mainmast by setting it forward 15 inches. To do this they probably took down the upper masts and yards and reset the timbers and chocks of the mast partners. Once everything was secured, the shrouds were set up again. Over the next few days, ships they would convoy to the Mediterranean arrived and exchanged salutes. Eight East Indiamen also came to join the convoy, intending to stay with them as far as they needed. During the first few days of September, 28 of *Tyger*'s 44 guns were replaced by guns of better quality brought to them in a hoy.[863] While the guns were being exchanged, the gunner of the *Happy Return,* anchored nearby, went to the *Tyger* in the long boat to see his opposite number. When the boat came alongside, Berkeley had three of the *Happy Return*'s men forcibly taken out of their boat, claiming that they had deserted the *Tyger*. He then sent Lieutenant Thompson over to the *Happy Return* with a note claiming that another 13 men aboard the *Happy Return* had also deserted the *Tyger* and should be returned. Captain John Wyborne of the *Happy Return* was an old tarpaulin of the Dutch wars[864] and reasoned that he had not manned his ship simply to lose those men to another. He also asserted that all the men who joined his ship had sworn they had not run from the *Tyger*.[865] The incident is symptomatic of the huge division that existed among officers. On the one side the old, experienced tarpaulins, and the other, the young, brash gentlemen. In this case, the ordinary seamen had a very definite preference for whom they preferred. Adding to his problems, Captain Berkeley began to fear that the ill will among his men was spreading. Edward Solby, the purser, was one of the *Tyger*'s five warrant officers, a position he had held at the benevolence of King Charles for two and a half years before the *Tyger* was built. Alarmed at the situation developing aboard the ship, he rather bravely, or stupidly, wrote a letter of complaint about his captain to the Admiralty. Hearing of this, Berkeley wrote to John Brisbane saying the claims were false and couldn't be proved. Captain Berkeley had Solby locked in his cabin *'for fear he should breed a mutiny which I fear he has been designing to do though he could not effect it, my seamen & officers having a greater love to discipline, loyalty & me than ye worst are pleased to think'*. Investigating the purser's accounts, Berkeley also claimed he had embezzled 205 sized fish.[866] Hearing of this through Brisbane, the Lords of the Admiralty directed Berkeley to release Solby from his confinement and keep him aboard to be tried by a court martial held by Admiral Herbert once they had reached the Mediterranean.[867] The young and inexperienced captain, his small band of retainers and perhaps his lieutenant must have felt uneasy in the ship.

A few days later an accident occurred when one of the sakers in the waist was fired during an exchange of salutes. The vent, or touch hole, had not been properly stopped and while the saker was being reloaded the powder exploded and blew off a great part of the hand of able seaman David Dryborough. His awful injury healed, however, and he stayed with the ship for many months. While the ship was in the Downs, her draught was measured at 15 feet 5 inches aft and 13 feet 5 inches afore. Just before they finally left the Downs on 11 September, Lord Berkeley and others went ashore to take the sacrament before their long voyage. The *Happy Return*, an old fourth rate dating from 1654, set sail at the same time as the *Tyger*. The ships' captains, probably influenced by their recent encounter, sailed against each other, with the *Happy Return* being described as extremely '*wronged*' by the sailing performance of the *Tyger*.[868]

Once the *Tyger* reached the British base at Tangier she was heeled both ways to allow some of her bottom to be cleaned. At the end of October she left Gibraltar with the newest fifth rate in the fleet, the *Sapphire*, built by Anthony Deane in 1675. In a fresh gale both ships sailed so exactly alike that no difference could be found between them, even though the *Sapphire* had the advantage of a clean bottom after being careened only 12 days before.

On 8 January 1682, Lieutenant Albion Thompson made a last entry into his journal before dying two days later.[869] He was followed eight weeks later by the captain, Lord Charles Berkeley, who died on 6 March, reportedly from smallpox. His body was embalmed and brought back to England. The funeral took place on 21 September and King Charles provided two troops of horse guards to escort his body from Berkeley House in Piccadilly to St Mary the Virgin's Church at Twickenham for burial. He was only 20 years old.[870] Deaths during long voyages in the Mediterranean were not uncommon and were often caused by the 'bloody flux', or dysentery. When this happened, many men would die in a relatively short space of time. This did not happen on the *Tyger* though, for although she lost seven men, including her captain and lieutenant, out of her crew of 200 between January and March 1682, it was not the type of brutal epidemic that sometimes occurred. In view of the relationship between Charles Berkeley and his men, and even some of his warrant officers, it is tempting to treat the deaths with a little suspicion. The most obvious suspects would be the warrant officers – Solby, the purser, had been locked up and was to be tried by court martial, and John Brisco, the cook, was prosecuted by the law for debt. Who else could make men so ill they would die but the cook? He would never have to answer any questions about this himself, however, as he became the tenth man to die on the *Tyger* at the end of December 1683, after two and a half years in service.

Lord Charles Berkeley had been made captain of a prestigious warship at the age of 19 with only a year's experience at sea and never having sat the lieutenant's examination. But for his advancement by King Charles he would have spent more time gaining experience as a volunteer beforehand. As he was skilled in navigation and maritime affairs he would surely have qualified as a lieutenant before becoming a captain. If he had followed this sensible career path he may well have known better how to handle his men. He was unlucky in being promoted so quickly, and even more unlucky to have died so young in the service of his king and country. He was certainly missed by his family. As well as giving him a very grand and ceremonial burial, they commissioned a painting of him from the leading portrait artist of the day, Sir Peter Lely, and two paintings of his ship by Willem van de Velde. Charles Berkeley was succeeded by his younger brother, John, who became the 3rd Lord Berkeley of Stratton. He was already a volunteer in the Navy and had served in a number of ships before becoming a lieutenant in 1685 when he was 22 years old, and a captain the following year. He was probably as forthright as his brother, as he supposedly murdered a pub landlord. In spite of this, he became a very successful rear admiral before dying in service in 1697.[871]

As for Edward Solby, the purser, he was tried by a court martial and dismissed from the *Tyger* two years later in October 1683, losing his pay to the seaman's charity the Chatham Chest.[872] However, his dismissal only came in the form of an exchange of positions with Stephen Jull, purser of the *Newcastle*, by order of the Admiralty. Unfortunately for Jull, the *Tyger* sailed to join Lord Dartmouth's squadron before he arrived. Fortunately for the *Tyger*, the commander, Captain Francis Wheeler, had a clerk, John Powis, who was '*well certified to be in every respect fitly qualified*' for the position. Powis thus became purser until the ship returned to Deal on 14 April 1684, when Jull at last caught up with her.[873] John Powis then returned to his duties as Wheeler's clerk, but when Stephen Jull left two months later when the ship was in the Hope, Powis became the purser warrant officer on a permanent basis. He was trusted by the seamen as he collected many of their wages. The ship continued in service until the end of the 17th century when, being worn out, she was taken apart and once again rebuilt.

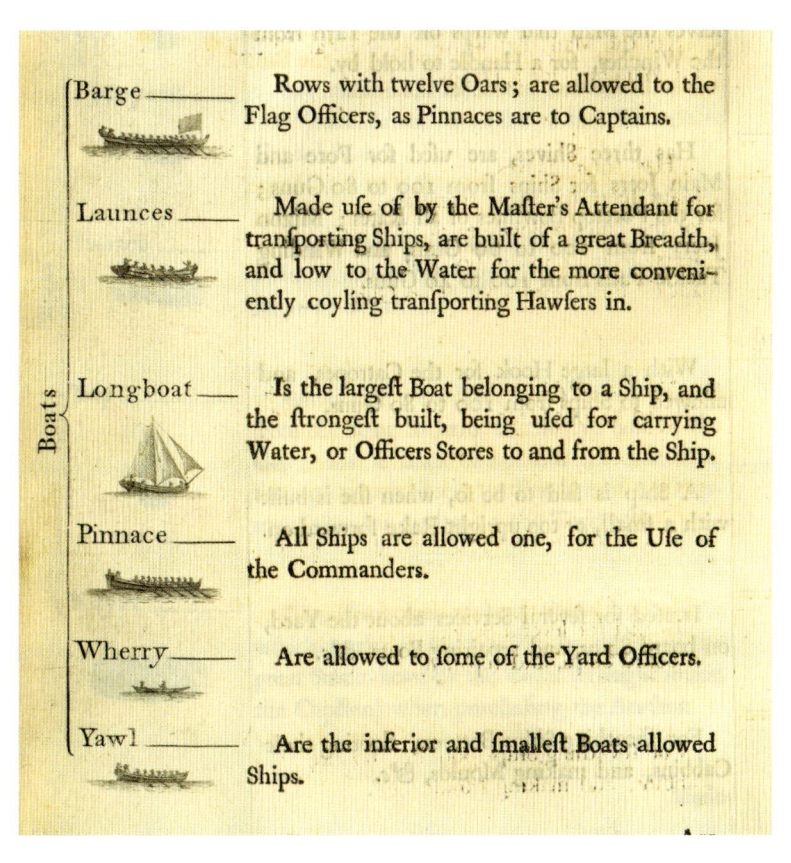

The type of boats used by dockyards and warships to carry out a multitude of tasks. From A Naval Expositor, *Thomas Blanckley, 1750.*

CHAPTER 9

THE TYGER'S GUNS

The Old *Tyger's* Guns

The *Tyger* and three similar fourth rates built to the same requirements in 1647 were true lightly built frigates designed for speed rather than gun power. They had a complete tier of gunports on the gundeck, but fewer on the upper deck, with no guns at all in the open waist. Each ship was given a new set of 30 high-quality 'fine metal' iron guns supplied by the famous gun founder John Brown. They consisted of ten culverins (18 pounders), 14 demi-culverins (9 pounders) and six sakers (5¼ pounders). They were of the newly introduced lightweight 'drake' design which altogether weighed 31 tons and cost £920.[874] Drakes had a tapered chamber to take a reduced charge of two-thirds the amount of powder used in the usual 'fortified' guns. This allowed the guns to be made lighter, but in use they were found to be fragile and to possess a violent recoil.[875]

Between 1647 and the end of the old *Tyger's* career there were many changes to both the type and number of guns she carried. The introduction of the line of battle and fleet actions against the Dutch brought about a change in philosophy as firepower became more important than speed. Periodic 'establishments' were compiled giving the number and type of guns each ship in the fleet was expected to carry. Very often they were idealised visions which in practice it was impossible to fulfil as little or no attempt had been made to reconcile them with the guns actually available. In 1664, many fourth rates were short of their allocated establishment of demi-culverin cutts (short 9 pounders) and a new establishment was proposed which took into account the guns actually available. The *Tyger* was allocated 38 guns consisting of 12 culverins, 16 demi-culverins, eight sakers and two saker cutts.[876] By 1666 her armament was further increased to 40 guns with the addition of two more sakers.[877] The upward trend continued into the Third Dutch War, and in July 1673 the Admiralty recorded her as having 46 guns.[878] They were probably an entirely different set of guns from the originals, for, in common with many other ships, her original unpopular drakes were replaced by regular fortified but smaller-calibre guns. The allocation of 46 guns was the *Tyger's* 'war at home' armament and it would have been reduced during peace or when there was a war abroad. In 1663, for example, her allocation for operating in the narrow seas during peacetime was only 26 guns.[879]

Gun Establishment of 1677

On 16 March 1674, Samuel Pepys, the great administrator of the Navy, set about producing an establishment of men and guns for the whole fleet. Unfortunately, the Ordnance Board and the Admiralty were separate bodies, making it difficult and time consuming to get them to agree on anything. Their uneasy relationship was not made clear until 1679, when an order in council gave the Admiralty precedence. For the future, the Master General of the Ordnance was to comply with all directions issued by Admiralty letters, but to retain a semblance

Opposite. *Although there were continual changes and improvements to ships' ordnance, the gunner and his servant seen here on board the Mary Rose in 1545 would have soon found themselves at home aboard the* Tyger *140 years later. Author.*

of civility, they were to start with a polite '*pray and desire*'.[880] By organising conferences between senior officers of the Navy and members of the Ordnance Board, Pepys intended his establishment to be far ranging and definitive. It was to give the numbers and types of gun each ship should carry in peace and war, as well as the numbers of men and the size of each ship. At the time, the *Tyger* was awaiting repair and was listed as being 457 tons, the size of the old ship. The establishment was finally signed over three years later by the King and the Lords, on 3 November 1677,[881] and was supposed to be strictly observed. On 22 December a version was sent to the Master of Ordnance which included the weight of guns.[882] The Admiralty's official journal showing the disposition of ships for January 1678 gives the same establishment of guns for the old *Tyger*, even though she then existed only in the form of a few pieces of timber.[883]

The *Tyger's* guns according to the establishment of 1677

	Number of guns	Weight (tons)	Demi-culverin	Weight (tons)	Saker	Weight (tons)	Light saker or cutts	Weight (tons)
War at home	44	42	22	32½	18	15½	4	2
War & peace abroad, peace at home	38	36¼	18	27½	16	14½	4	2

The *Tyger's* men

Men, war at home	Men, war abroad	Men, peace abroad & at home
190	160	120

The establishment was not Pepys's greatest achievement. In spite of the number of conferences and years it had taken to produce, there were surprising and fundamental errors. Much of the simple mathematics was wrong: from the list of the *Tyger*'s guns, it can be seen that the total weight of her 'war at home' guns should be 50 tons, not 42. Worse, the guns listed were not matched by those available in store, so that at least one 70-gun third rate went to sea with only 26 guns that matched her establishment.[884] Another problem was its failure to recognise the different lengths of guns of the same type. The common and usual practice was to have specialist chase guns in the most forward or aft ports as they could be moved to fire at those being pursued or at the pursuers. For accurate long-range shooting, these chase guns were longer and therefore heavier than normal broadside pieces. Often, they were so heavy that their length and weight could only be accommodated by downsizing them to guns of smaller calibre. This was not taken into account. Predictably, the establishment was soon in trouble as practical changes had to be made to the fitting out of ships, which had Pepys complaining that it had been '*broken into*'.[885]

The *Tyger*'s 22 demi-culverins are listed as weighing 32½ tons which would make them on average 29½ hundredweight each. Demi-culverins of this weight would have been slightly larger than average for nine-foot-long guns. Her sakers appear typical, weighing on average 15 hundredweight, and they would have been seven feet long. The light sakers weighing half a ton each would have been six feet long.[886]

An interesting example of the different types, lengths, weights, as well as positions of guns can be found in a list for the *Dunkirk* from 1652.[887] It is noticeable that her heaviest guns weighed 38 hundredweight, which was about the biggest that iron foundries could produce at the time.[888] The demi-cannon (32 pounders) may have been drakes and would have been fired using a reduced charge. Within a few years the technology improved so that demi-cannon were cast weighing 50 hundredweight.

Guns of the *Dunkirk*, 1652

Gun deck	Type of gun	Length feet	Weight cwt	Number
Lower deck	Demi-cannon	9	36	20
Fore and aft chase	Culverin	11	38	4
Upper deck				
Forecastle chase	Demi-culverin	10	38	2
Forecastle	Demi-culverin	8	18	4
Waist	Demi-culverin	8	18	6
Under quarterdeck	Demi-culverin	8	18	6
Great cabin	Demi-culverin	8	18	2
Quarterdeck				
Cuddy	Saker cutts		8	2

The Guns of the New *Tyger*

On 12 March 1681, as the new *Tyger* neared completion, the Admiralty ordered her to be fitted out for service in the Mediterranean, with the ordnance and gunner's stores to be according to the establishment.[889] Unfortunately, Pepys's flawed establishment related not to the new *Tyger* but to her predecessor. The mistake was quickly realised, and the following new orders were sent a couple of days later: '*In pursuance of His Majesties pleasure we do hereby pray and desire you to forebear putting any guns or ammunition on board His Majesties ship the Tyger until you hear further from us.*'[890]

It was not only the Admiralty that was uncertain as to what guns the new *Tyger* should carry. A few hundred yards down the Thames from Deptford dockyard at Greenwich was the studio of the van de Veldes, father and son. They did a series of drawings of the new ship at Deptford which the elder used to produce the two paintings which today hang at Berkeley Castle, showing her under the command of Lord Charles Berkeley. The *Tyger*, like all new ships, was launched without guns, so, making an optimistic guess at the number she would carry, van de Velde drew many more than he should and in one drawing had a gun at every port,[891] giving her a total of 56 guns instead of the correct number of 44. Guns are shown in the aft gundeck broadside port and stern chase port, where they would have overlapped each other's space. The paintings are a little more considered for the stern chase guns, and a pair of quarterdeck guns are removed. Even so, the 52 guns shown are eight more than she actually carried. Van de Velde was of course an artist aware that gunports with guns pointing out of them look better than blank gunports.

The only men who knew the details of the new *Tyger* and who were thus able to judge accurately what guns she could carry to the Mediterranean were the King and her builder, John Shish. Shish was duly commanded by the Navy Board on 13 May 1681 to provide details, including how many guns and men she could carry. He promptly replied with his recommendations.[892]

John Shish's suggestion for the *Tyger*'s abroad armament

	Number of guns	Type of gun	Weight of each gun
Guns on ye lower deck	18	Demi-culverin	27 cwt
Guns on ye upper deck	18	6 pounder	17 cwt
Guns on ye quarter deck	8	3 pounder	6 cwt

The officers of the Navy Board considered Shish's suggestion for a month before writing to the Admiralty with their own proposal for her establishment. They probably noticed the similarity in weight between the old *Tyger*'s guns and those in Shish's suggestion. It seems they were so close it was decided to keep the old guns already belonging to the ship but to reduce the number to 40 rather than the 44 guns suggested by Shish for her abroad armament. They also refined their proposed establishment by adding 'peace at home or abroad' columns[893] before sending it to the Admiralty for acceptance.[894] The official establishment of 3 November 1677 was amended on 1 November 1684 to take account of new ships that had come into the Navy. It included the new *Tyger* and followed the Navy Board's proposal of 1681.[895] A proposed establishment originating from the Ordnance Office two months later, which only listed the 'war at home' guns, gave the same armament.[896]

The June 1681 Navy Board proposal for the gun establishment of the *Tyger*, confirmed in 1684

	Number of guns	Demi-culverins of 27 or 28 cwt each	Sakers of 17 or 18 cwt each	Light sakers of 6 or 7 cwt each
War at home	46	20	18	8
War & peace abroad, peace at home	40	18	16	6

As the *Tyger* was being fitted out for Mediterranean service, she received her establishment-sanctioned 'abroad' complement of 40 guns. However, Captain Berkeley thought his armament inadequate and received permission from the King to have an additional two sakers for the upper deck and two saker cutts for the quarterdeck, bringing the total number of guns up to Shish's earlier recommendation of 44.[897] Before they were brought aboard, the gun carriages were repaired. On 18 July, Edward Silvester, the master smith at the Ordnance Board, provided ironwork, including 27 exbolts, 20 pairs of forelock keys, 22 cap squares and some small spikes; he also mended and fitted seven pairs of ringbolts, and repaired and riveted

The 1681 disposition of the Tyger's 'war and peace abroad' guns for service in the Mediterranean, including the additional two fortified and two light sakers obtained by Lord Berkeley. Author. A: Demi-culverin 9 pounders; B: Fortified saker 5¼ pounders; C: Light saker cutt 5¼ pounders; D: Position of the chase ports.

one side bolt.[898] While this was being done, Thomas Moore, the carpenter for repairs, worked on all 44 gun carriages.[899] During the next few days, round shot weighing 5 tons 11 cwt 1 qtr was drawn out of the shot yard, and the *Tyger*'s 44 guns, weighing 43 tons 8 cwt 0 qtr 21 lb, together with their carriages, were drawn out of the Warren store. They were taken aboard the ship as she lay in the Longreach by Isaac Bennett, master of the *George of London* hoy.[900] One of the guns, the biggest demi-culverin, was recorded as weighing over 30 cwt.[901] Six of the guns had been fired so often that their vents had worn to become oversize, and they were refurbished by being drilled out and re-bushed by John Clare, a labourer, who also cut markings into 19 other guns.[902] Interestingly, when her guns were recorded years later in 1698, she still had one demi-culverin weighing over 30 cwt,[903] which had probably remained with her all that time.

The *Tyger* made her way out of the Thames estuary and into the Downs fully armed and ready for service in the Mediterranean. Lord Charles Berkeley suddenly perceived another problem and wrote to Samuel Pepys on 21 August complaining that he had 28 'bad' guns and telling him that he desired orders allowing him to stop at Spithead for them to be exchanged. His request was initially granted,[904] but when discussed again by the Admiralty a few days later,[905] it was decided to send his new guns directly to him in the Downs. Nevertheless, it was considered more important that he set sail at the first opportunity, and he was not to delay his departure by waiting for them to arrive.[906] With commendable speed, Nicholas Whitaker, a clerk at the Ordnance Board, and John Clare, the labourer,[907] hired a boat and went aboard the *Tyger* the very next day to inform her captain that his new guns were on their way. They stayed with the ship for ten days overseeing the exchange[908] while two other labourers, Edward Snapes[909] and Edward Barnes, together with an Ordnance Board clerk, Thomas Townsend, went to Woolwich to make the arrangements for shipping the guns.[910] They were taken from Woolwich to the *Tyger* in the Downs by John Piercy of Brighton, master of the *Edward and John* ketch.[911]

Although there is considerable correspondence concerning the exchange of guns, there is nothing to say why Captain Berkeley considered them bad. Judging by their weight, the existing guns were not the unpopular and dangerous lightweight 'drakes' but were regular 'fortified' guns. Nevertheless, if they were the old *Tyger*'s guns, they were old, and from the evidence of the work recently carried out on them, they were also considerably worn. Bearing this in mind, and hearing tales about the fabulous Rupertino nealed and turned iron alloy guns, Lord Berkeley may have used his influence with the King to exchange most of his guns for them. The ship's lieutenant, Albion Thompson, recorded the 28 new guns as being ten demi-culverins, ten nealed sakers and eight nealed saker cutts.[912] The saker cuts that he described as '*nealed*' have been identified as Rupertino guns, making it very likely the others were as well.[913]

Lord Berkeley was not the only captain who struggled to accurately describe a problem with his guns. In 1689 the captain of the *Plymouth* wrote to the Admiralty to complain that her guns were '*unfit*' for her. Perplexed, the Admiralty passed the complaint to the Navy Board for its opinion, receiving the following sensible reply: '*But having no account of the reason of their unfitness whether they were too large or too small, too heavy or too light, or what was the occasion thereof; and presuming they would be changed of course by the Officers of ye Ordnance upon the Captains desire at ye ships coming into Portsmouth harbour …* The Navy Board sensibly advised *that if the said ship be not gunned according to ye establishment made on that behalf the Officers of ye Ordnance should have orders to do it.*'[914] Demonstrating how a complaint should be made, Captain Wright of the *Mary* gave an account of the size and heavy weight of his guns, saying he had '*several twelve pounders of thirty three, thirty four & thirty five hundredweight in her upper tier*', which the Navy Board considered and found '*by the establishment should be demi culverin (9 pounder) of about twenty four hundredweight. We are of opinion the said guns ought to be changed as he proposes.*'[915]

The next change to the *Tyger*'s guns was rather unfortunate and occurred during a violent storm on 9 October 1684. Having suffered the loss of her main mast, the *Tyger* was sailing under a jury mast when that too gave way and went by the board. As it came down, the sail and rigging took a turn round the muzzle of one of the sakers in the waist, tearing it out of its carriage and flinging it over the side.[916] A few years later, on 15 March 1688, an account of ships' guns and carriages[917] reveals that although the *Tyger* was still allocated 44 guns, the missing saker lost over the side had still not been replaced. She still had the empty saker carriage but was short of seven saker cutt carriages, possibly because the old ones were worn out. On 6 June 1688, one of her quarterdeck saker cutts was found to have a defective vent and the gun was exchanged the same day.[918] During 1689, the *Tyger* operated in home waters, cruising in the Soundings before visiting Hamburg and Holland. For this brief period she carried an armament of 48 guns, but before the year was out she had reverted to her reduced armament.[919]

The Survey of 1698

A list of the fleet dated 1 December 1694 shows that the *Tyger*'s highest gun establishment was 42 guns.[920] This was confirmed after she was launched from Deptford dockyard in October 1695 and took aboard 42 guns, which her then captain recorded was '*my allotment at present*'.[921] These guns remained with the ship and were surveyed in the great ordnance survey of the fleet in 1698.[922] This survey shows that the reduction from 44 to 42 guns was accomplished by the removal of two saker cutts from the quarterdeck. The survey contains a great deal of information about the guns, including their weight, their condition, important dimensions and to which ship they belonged. The survey number and weight were cut into each gun, allowing them to be compared with the manuscript survey description. By checking the weights of guns recorded in the survey against the weight records of new guns delivered to the Ordnance Board, it was found that one of the *Tyger*'s saker cutts, survey number 10133, was delivered on 21 September 1675 and the other five were delivered on 24 July 1676. They are all recorded as being five and a half feet long and were Prince Rupert's fabled Rupertino cannon.[923] These comprised the *Tyger*'s final set of guns.

Plan of the Tyger's decks, showing the probable position of her guns from 1694 onwards. The positions were determined by assuming that the two longest demi-culverins, 10104 and 10105, would have been the furthest aft broadside pair that could be moved to fire through the stern ports while the next longest pair, 10095 and 10096, would have been the most forward pair. Similarly, on the upper deck, 10121 and 10122 would have been the furthest forward, from where they could be moved to fire through the beakhead bulkhead. Author.

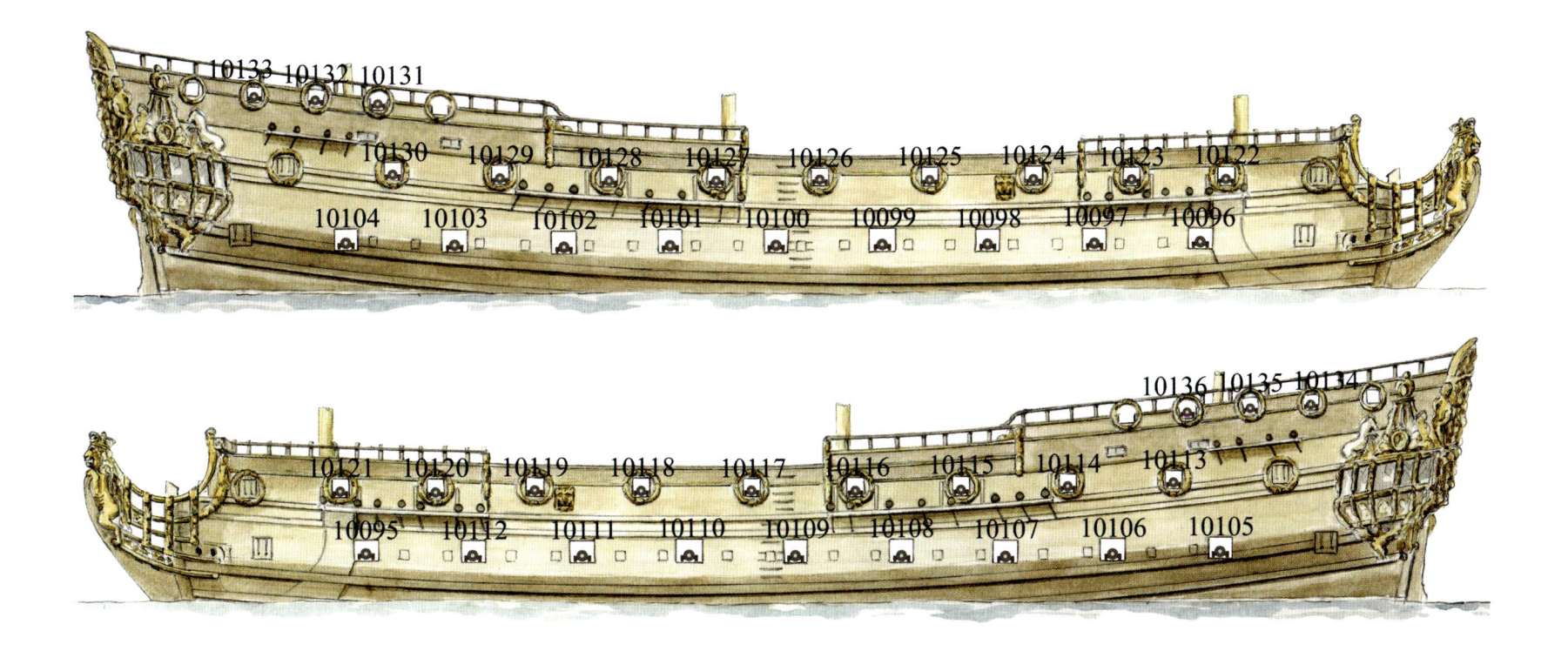

The *Tyger*'s guns recorded in the 1698 survey. After repeated firing guns tended to crack at the muzzle end. When this happened, the affected area was sawn off and these guns are identified in the last column of the survey as "To be cut at muzzle"

	Number cut	**Weight**			**Length**	**Diameter Trunnions**	**Diameter at trunnions**	**Taper bored**	**Wants venting**	**To be muzzle cut**
		Cwt	Qtr	Lb	Feet	Inches	Inches			
	10095	24	2	00	9	3¾	12½	0	0	0
	10096	24	3	01	9	4½	12½	0	0	0
	10097	27	3	06	8½	4½	13	0	0	0
	10098	28	2	15	9	4¼	13¼	0	0	0
	10099	29	2	03	8½	4½	14	0	0	0
	10100	28	3	19	8½	4¼	14	0	0	0
	10101	27	1	04	8½	4½	13¼	0	0	0
	10102	24	0	00	8½	4½	13¼	0	0	0
Demi-culverin	10103	18	3	16	7	4½	12½	0	0	0
	10104	32	0	17	10	4½	13¼	0	1	0
	10105	29	3	21	10	5	12¾	0	0	0
	10106	19	0	12	7	4¼	12¾	0	0	0
	10107	26	2	17	8½	4¼	13	0	1	0
	10108	27	3	00	8½	4½	13¼	0	0	0
	10109	28	1	06	8½	4½	13½	0	0	0
	10110	27	0	17	8½	4½	12½	0	0	0
	10111	27	1	14	9	5	13¼	0	0	0
	10112	26	0	00	8	5	13½	0	1	0
	10113	2060	-	-	7	4	12¼	0	0	0
	10114	19	0	00	8	3¾	11½	0	0	0
	10115	21	2	13	8	4	12¼	0	1	0
	10116	18	0	26	7½	4	11¾	0	0	0
	10117	18	0	20	7½	3½	11¾	0	1	0
	10118	18	0	12	7½	4	11¼	0	0	0
	10119	18	0	12	7½	3¾	11	0	0	0
	10120	17	0	00	7½	3¾	11	0	1	0
Saker	10121	25	1	08	9½	4	12¼	0	0	0
	10122	25	0	04	9½	3¾	12¼	0	1	0
	10123	16	0	05	7	4 ¾	11¾	0	0	0
	10124	18	2	05	7½	4	11½	0	0	0
	10125	18	0	05	7½	4	11½	0	0	0
	10126	16	2	26	7½	3 ¾	10¾	0	0	0
	10127	17	1	24	7½	4	10¾	0	0	0
	10128	17	3	12	7½	4	11½	0	0	0
	10129	17	0	02	8	3 ¾	10¾	0	0	0
	10130	18	2	17	8	4	11¾	0	0	0
	10131	8	2	14	5	3 ¼	9⅓	0	1	0
	10132	8	3	14	5	3 ¼	9½	0	0	0
Saker cutts	10133	8	2	20	5	3 ½	9¼	0	0	0
	10134	8	3	14	5	3 ½	9¼	0	0	0
	10135	8	0	06	5	4	9½	0	0	0
	10136	8	0	02	5	3½	9¼	0	1	0

Manning

Gun establishments were intrinsically linked to the manning establishments. During the period when the *Tyger* carried 42 guns, her middle complement of men was 197, but in practice this number could vary considerably each way by at least 25 men.[924] In 1680, Cloudesley Shovell, commander of the fifth-rate *Sapphire*, reckoned his ship was undermanned and thought his allowance should be increased by 25 men.[925] His well-reasoned request is interesting because the *Sapphire*, in common with the new *Tyger*, had demi-culverins on the gundeck and sakers on the upper deck. The *Sapphire* was built in 1675 by Sir Anthony Deane and, at 333 tons burden, was much smaller than the *Tyger* of 590 tons, a discrepancy that emphasises how the *Tyger*'s sailing performance took precedence over carrying heavy guns.

An exact account how the men are quartered aboard His Majesties Ship Saphire in time of fight.

BETWEEN DECKS (GUNDECK)
For the 18 guns 4 men per gun — 72
For carrying of powder — 04
The Lieutenant & Gunner to look after those guns and men — 02

IN THE HOLD
The surgeon, his mate, boy and barber — 04
The Carpenter, his mate, 2 of his crew & his boy — 05
For the powder room afore — 03
For the powder room abaft — 02
For the Boatswains store room — 01
For the bread room & store room abaft — 01
The cooper for drawing of drink — 01

UPON THE UPPER DECK & STEERAGE
For the 10 guns, 3 men per gun — 30
For carrying powder — 03
To steer the ship — 01
To cund (conn) ye ship — 01

QUARTER DECK
For the 2 guns, 1 man per gun — 02
To carry powder — 01
The Captain & Master upon that deck — 02

The present complement employed as above said is in all 135

And for small shot, trimming of sails & mending of rigging it is humbly prayed there may be allowed more 25 men
Clowdisley Shovell

Gun Carriages

In Windsor Castle, at the top of the Curfew Tower behind an embrasure, there is a 17th-century saker measuring 9 feet 3 inches long and mounted on a naval gun carriage. It points directly at the bridge spanning the Thames and connecting Windsor to Eton 300 yards away. A painting of Windsor Castle from the north painted by Johannes Vorsterman in about 1678 shows the tower with the gun in the same position as it is in today.[926] In 1669, Edward Wise, the storekeeper at the castle, made an inventory which included a nine-foot-long saker of 24cwt, 0 qtr, 2lb mounted on a ship's carriage at the Eton battery. Considering the difficulty involved in getting the gun in and out of the tower, it is reasonable to assume that the same gun on its carriage has remained on duty in the same place for all the intervening hundreds of years.[927] The gun and its carriage give a very good idea of what those aboard the *Tyger* would have looked like. The carriage sides are three and a half inches thick, which is the same size as the saker carriages bought in 1677 for twenty-two shillings and nine pence each, a small difference being that the Windsor carriage has a bed that is four inches thick instead of three and a half inches.[928] The carriage is made from elm, the wedges securing the sides to the bed are oak, and the axletrees are ash.[929]

In June 1668, John Knight, the approved carriage maker to the Office of His Majesty's Ordnance at Portsmouth, repaired or replaced the old *Tyger*'s carriages in preparation for her joining the summer fleet. The work and cost were recorded as follows:[930]

This list of work mentions '*live*' trucks (wheels) that revolved and fixed '*dead*' trucks that helped stop recoil. These are the same design as the carriage at Windsor Castle. This type of carriage is shown on many contemporary ship models and was common for demi-culverins and sakers during the period in which the old and the new *Tyger* was in service.

Tygar	£	s	d
For one Demy Culvering carriage new	1	2	0
For six saker carriages new at 17s a piece	5	2	0
For ten pair of live trucks 18 inches high & upward at 4s a pair	2	0	0
For three pair of live trucks of 15 inches at 2s 4d a pair	0	07	0
For eight pair of dead trucks at 2s a pair	0	16	0
For 9 new extrees at 12d a pair	0	9	0
For 11 foot of 4 inch plank at 7d a foot	0	6	5
For 5 new sides for saker carriages at 5s 8p a pair	0	14	02
For 5 days work for 3 men and his servant	2	5	0
For boat hire to and from work	0	5	0

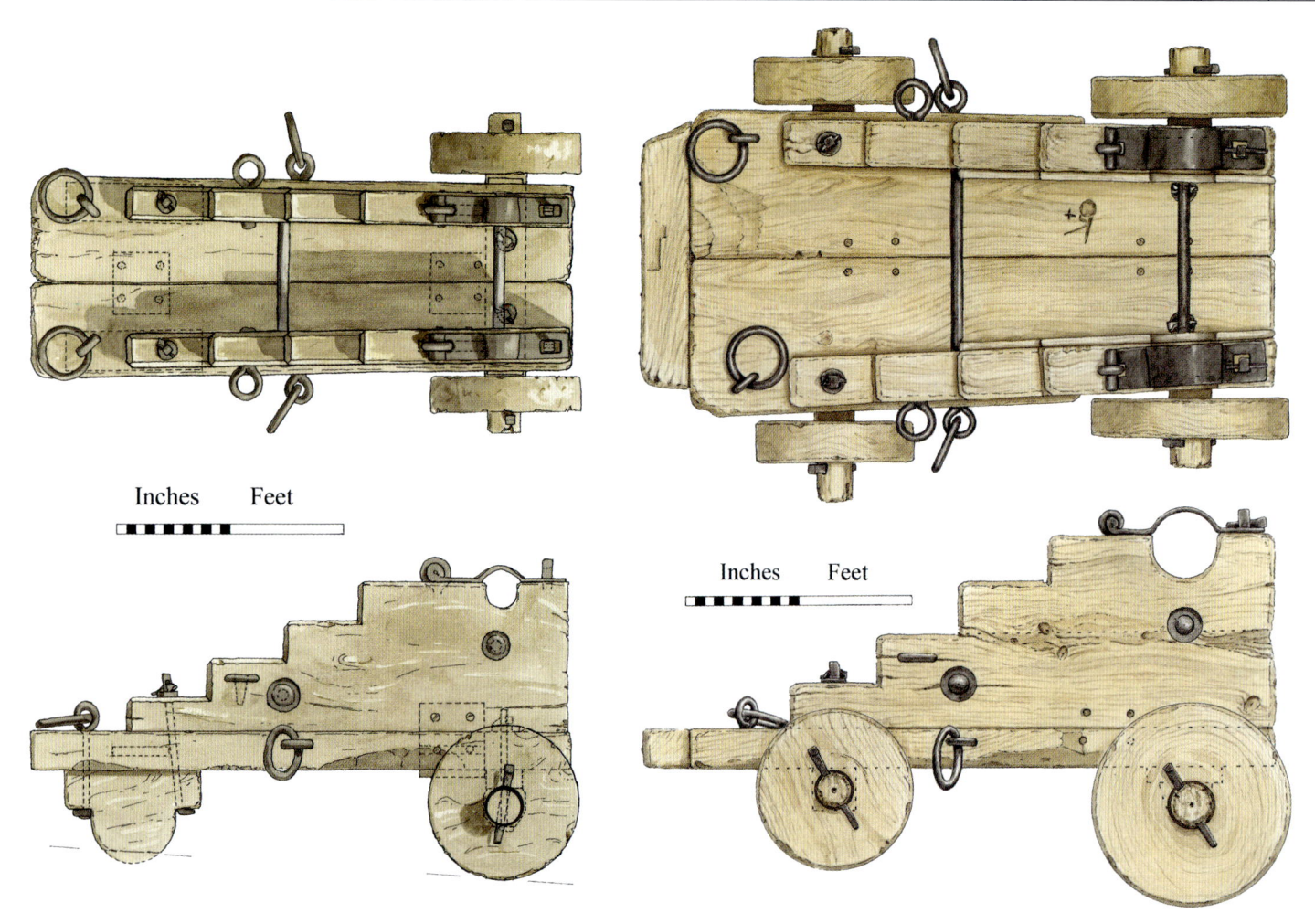

Below. *The saker carriage in the Curfew Tower, Windsor Castle. It is painted black, but gun carriages were generally left unpainted in sea service. Author.*

Below Right. *A 24lb gun carriage from the wreck of the* London, *1665. Although larger and earlier, it is very similar in construction to the Windsor Castle carriage. Author.*

Gunner's Stores

The size of the gunner's stores was dependent on the war or peace armament and the length of service for which his ship was expected to be commissioned. For fourth rates such as the *Tyger*, it varied from about 40 to 75 rounds of shot for each gun. A ship with a similar demi-culverin and saker main armament to that of the *Tyger* was the *Constant Warwick*. In 1681 she was sent on a six-month commission to Lisbon and on her return was ordered to Sheerness to be fitted for cruising in the Soundings.[931] Her gunner's stores were surveyed on 9 February 1682 by her captain, John Wood, and her gunner, Abraham Byam, as she rode on the Swale at Queenborough.[932] The survey followed a regulated procedure in which a printed form was filled in giving an account of the stores issued and used. The survey, which follows, has been edited to give the stores issued and the quantity of ammunition provided for each gun.

THE TYGER'S GUNS

The Windsor Castle saker and carriage, shown in the waist of a ship. Author.

THE MASTER SHIPWRIGHT'S SECRETS

Below. *A nine-foot-long demi-culverin dating from 1666, of the same weight and length as many of those belonging to the Tyger. It is preserved on the Cobb Harbour, Lyme Regis. Scale 1/16. Author.*

Below. *A seven-foot-long saker probably dating from the commonwealth period and similar to many guns on the upper deck of the Tyger. Julian Kingston collection. Author.*

Below. *A six-foot-six-inch-long saker dated to the middle of the 17th century and identified by the 'F' on the end of the trunnion as coming from the Finspong foundry in Sweden. Known as 'Finbankers', these weapons were used primarily by the Dutch, and captured guns are identified in the 1698 survey by their weights being recorded in pounds rather than by the English cwt, qtr and lbs. The Tyger's gun, 10113, is one of them. This example is preserved in the Winchelsea Museum. Author.*

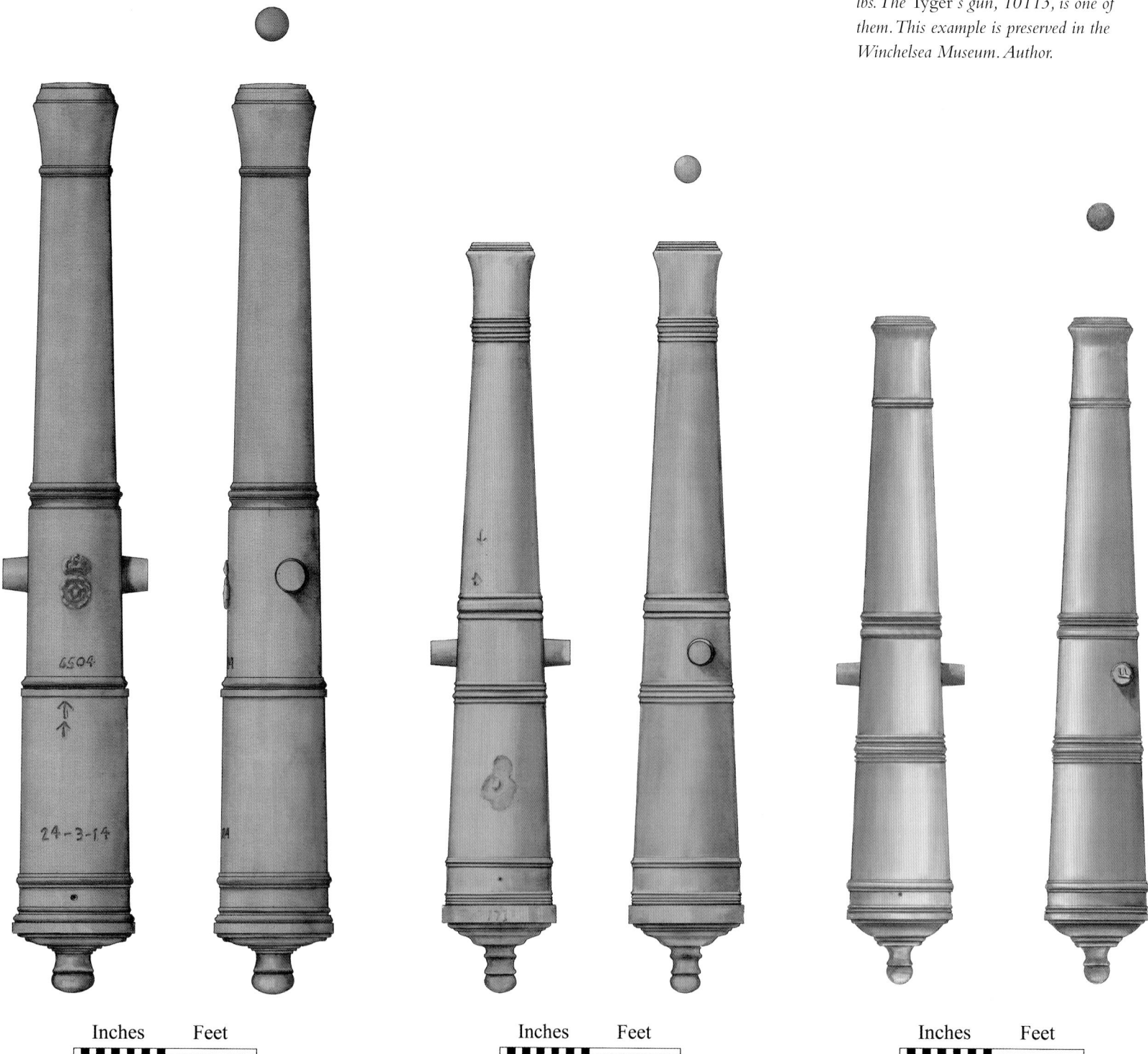

THE TYGER'S GUNS

Below. *This nine-foot-three-inch-long saker dating from the early 17th century still rests on its original naval carriage located in the Curfew Tower, Windsor Castle. It is of the same size and type as the Tyger's two saker chase guns, survey numbers 10121 and 10122. Author.*

Below. *A reconstruction of the Tyger's Rupertino saker cutt, number 10134, based on her recorded dimensions and other similar surviving guns. The design was based on guns which had been cut short due to cracking near the muzzle. They proved successful, and new guns were made to the same short length but included a muzzle swell. Author.*

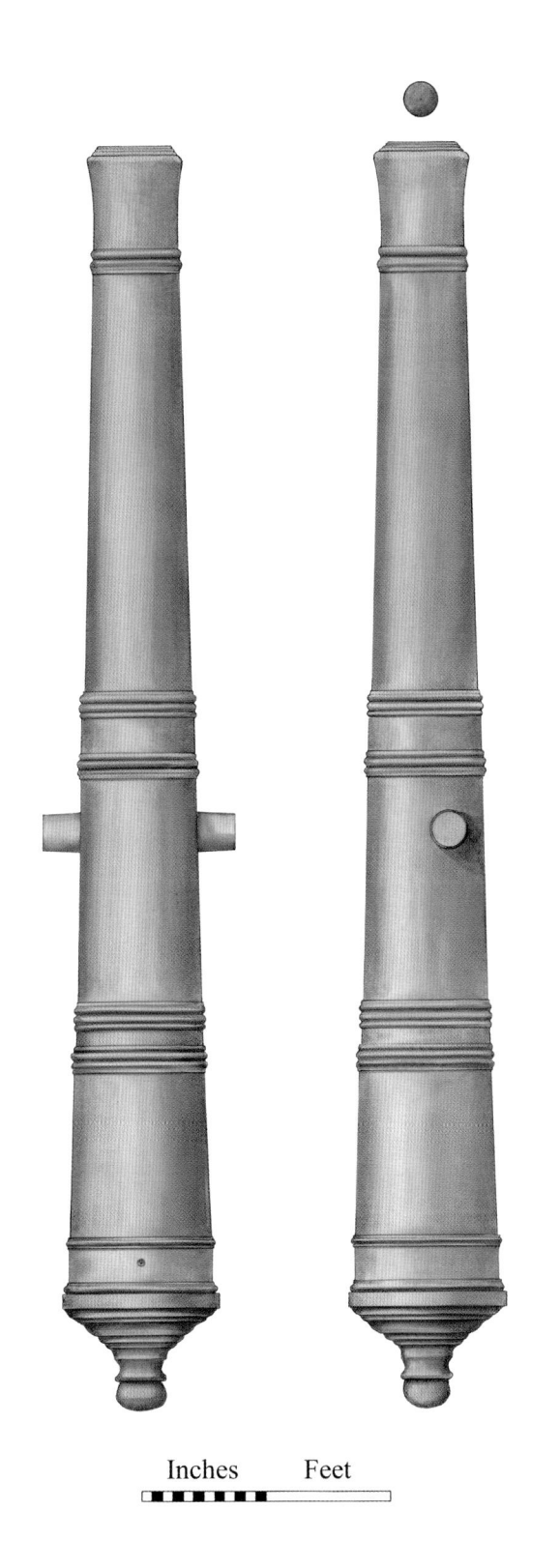

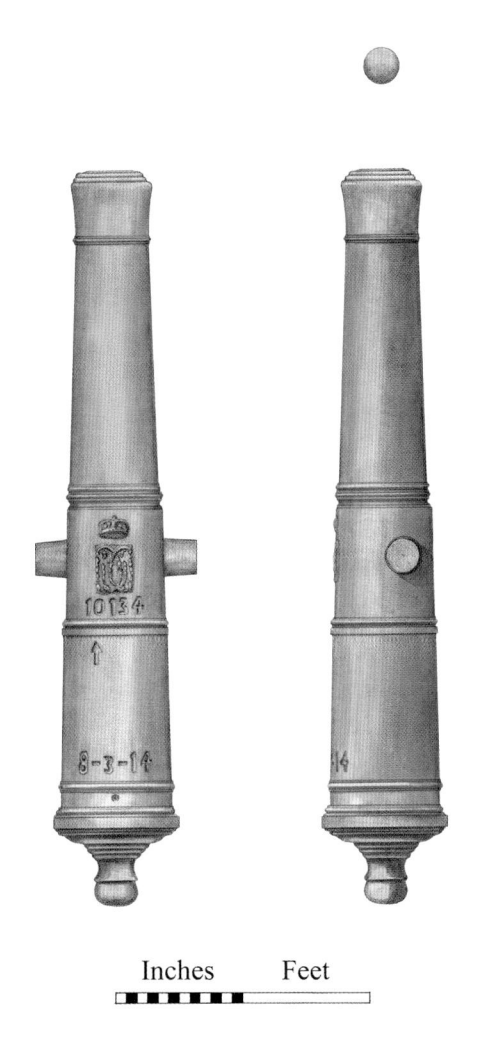

Gunner's stores of the *Constant Warwick*, 1682

Iron demi culverin gun	18	Pistol shot	1 cwt	Spare hoops	1 pair
Iron Saker gun	14	Aprons of lead	70	Canvas	200 ells
3 pounder gun	4	Crows of iron	36	Paper royal for cartridges	4 ream
Demi culverin ship carriage	18	Tackle hooks	15 pair	Oil	3 gallon
Saker ship carriage	14	Ladle hooks	20 pair	Tallow	1 cwt
3 pounder ship carriage	4	Lynch pins	15 pair	Starch	6 lb
Demi culverin (9lb) round shot. 40 per gun	720	Spikes	150	Needles	6 dozen
Saker (5 ¼ lb) round shot. 40 per gun	560	Forelocks	50 pair	Thread	6 lb
3 pounder round shot. 40 per gun	160	Sledges	1	Ordinary lanthorns	2
Demi culverin forged double headed shot	108	Great melt ladle	1	Dark lanthorns	1
Demi culverin cast double headed shot	42	Small melt ladle	2	Muscovia lights ordinary	2
Demi culverin tin case filled with musket shot	50	Nails 30d	200	Wad hooks	4
Saker tin case filled with musket shot	38	Nails 24d	200	Hand-crow levers	36
Boxes for tin case	7	Nails 20d	200	Rope sponges	36
Hand granadoes	30	Nails 10d	400	Powder horns	36
Fuze for hand granadoes	30	Nails 6d	500	Priming irons	6
Demi culverin ladles	4	Nails 2d	3000	Linstocks	6
Demi culverin sponges	6	Bed for carriage	36	Marlin	40 lb
Saker ladles	3	Coyne for carriage	72	Twine	6 lb
Saker sponges	4	Truck ordinary	3 pair	Wire	6 lb
3 pounder ladles	1	Truck extraordinary	2 pair	Hand screws	1
3 pounder sponges	2	Demi culverin axletrees	4	5 inch tarred rope	1 coil
Ladle staves	12	Saker & minion axletrees	3	3 inch tarred rope	2 coil
Demi culverin cases of wood for cartridges	24	Tampeons great	220	2 1/2 inch tarred rope	1 coil
Saker cases of wood for cartridges	18	Tampeons small	150	Breechings of tarred rope	36
3 pounder case of wood for cartridge	6	Pulleys great	20 pair	Tackles of tarred rope	72
Funnels of plate	1	Pulleys small	30 pair	Port tackles of tarred rope	36
Corn powder	79 barrels	Heads and rammers great	8 pair	Junk tarred rope	20 cwt
Match powder	5 cwt	Heads and rammers small	8 pair	Snaphance musquets	50
Three quarter pikes	18	Formers great	2	Musquetoons	2
Short pikes	18	Formers small	3	Musquets rod	18
Bills	4	Budge barrels	1	Bandeliers	50 collars
Hatchets	20	Tanned hides	6	Blunderbusses brass	1
Swords	20	Sheep skins	36	Blunderbusses iron	2
Hangers	30	Baskets	12	Pistols	8 pair
Musket shot	4 cwt	Port hooks	10		

CHAPTER 10

CONTEMPORARY SHIPBUILDING CONTRACTS UNVEILED

The surviving contracts made between the Navy and private shipbuilders are a wonderful source for studying and understanding ships of the second half of the 17th century. Ships built for the King in his own yards were ordered by warrant, but these give only a basic specification, as the master shipwrights were trusted to make ships of sufficient scantling, or size of timbers. As private shipbuilders were building at an agreed price and would benefit from using smaller scantling, a detailed contract was drawn up by the Navy Board giving the dimensions of all the timbers in the ship. Many fourth rates were trusted to approved private shipbuilders, and five of these surviving contracts dating from between 1649 and 1692 are published here. The ships vary in size, and one of them is a galley frigate. Together, they demonstrate the changes that took place in building practice over time as well as the growing sophistication and length of the contracts. The originals are very difficult to follow, being full of obscure wording which only the peculiarly dedicated student of 17th-century shipbuilding would try to follow. There are usually no headings or breaks, making it difficult to find a particular item. In one or two cases where a word was lost or was indecipherable, similar contracts were consulted to obtain the correct meaning. The contracts appear to contain one or two mistakes, probably as they are copies of originals, but unless these mistakes are obvious they have not been changed. The contracts have been transcribed, with headings inserted and notes added where necessary for further explanation. To make things as clear as possible, references have been added with a visual glossary. Some words confusingly changed their meaning over time: after the 1673 contract, sleepers were called footwaling and footwaling was called ceiling. In typical 17th-century lack of conformity, hanging knees were sometimes called up and down knees and clamps sometimes called risings.

Visual glossary for fourth-rate ships built between 1649 – 1692

1 Keel
2 False keel
3 Stem
4 False Stem
5 Sternpost
6 Inner false post
7 Rising wood
8 Sternpost knee
9 Half timbers
10 Main or wing transom
11 Transom
12 Half transom
13 Timber and room
14 Floor timber
15 Lower futtock or naval timber
16 Chock
17 Middle or second futtock
18 Upper futtock or gundeck timber
19 Toptimber
20 Hawse piece
21 Keelson
22 Limber board
23 Sleepers, called footwaling after 1673
24 Middleband
25 Gundeck clamp
26 Footwaling, called ceiling after 1673
27 Orlop beam
28 Orlop lodging knee
29 Orlop false beam
30 Orlop carling
31 Floor rider
32 Futtock rider
33 Breasthook
34 Keelson standard
35 Crotch
36 Fashion piece
37 Transom knee

38 Main mast saddle or step
39 Well
40 Shot locker
41 Chain pump
42 Gundeck beam
43 Gundeck lodging knee
44 Gundeck hanging knee
45 Pillar in hold
46 Helm port transom
47 Fore mast saddle or step
48 Mizzen mast saddle or step
49 Gundeck carling
50 Gundeck ledge
51 Gundeck waterway
52 Gundeck plank
53 After bitt pin
54 Fore bitt pin
55 Cross piece
56 Bitt pin standard
57 Main mast partner
58 Main capstan partner
59 Main capstan
60 Main capstan false partner
61 Gundeck head ledge
62 Gundeck hatch
63 Manger
64 Gundeck scupper
65 Gundeck spirketting
66 Rowing scuttle
67 Upperdeck clamp
68 Gundeck gunport
69 Gundeck standard
70 Gundeck ceiling
71 Gundeck pillars
72 Upperdeck beams
73 Upperdeck lodging knee
74 Upperdeck hanging knee
75 Upperdeck long carline
76 Upperdeck short carline

77 Upperdeck ledge
78 Upperdeck waterway
79 Upperdeck plank
80 Upperdeck head ledge
81 Upperdeck coaming
82 Jeer capstan partner
83 Upperdeck spirketting
84 Upperdeck string over port
85 Upperdeck gunport
86 Upperdeck ceiling
87 Jeer capstan
88 Upperdeck grating hatch
89 Upperdeck scupper
90 Quarterdeck and forecastle clamp
91 Great cabin rising
92 Forecastle beam
93 Forecastle hanging knee
94 Forecastle transom
95 Forecastle transom knee
96 Forecastle bulkhead
97 Cookroom
98 Quarterdeck beam
99 Quarterdeck hanging knee
100 Upperdeck standards
101 Steerage bulkhead
102 Quarterdeck plank
103 Quarterdeck spirketting
104 Quickwork
105 Roundhouse
106 Great cabin window transom
107 Fore mast partner
108 Mizzen mast partner
109 Topsail sheet bitt
110 Jeer bitt
111 Lower main wales
112 Strake between main wales
113 Strake above and below main wales
114 Plank to the bilge
115 Underwater plank
116 Plank below channel wales
117 Channel wales

118 Plank between channel wales
119 Quickwork above channel wale
120 Chainwales
121 Forefoot
122 Cathead
123 Knee of the head
124 Cheek of the head
125 Rail of the head
126 Trail board
127 Beast
128 Bracket of the head
129 Keelson of the head
130 Standard of the head
131 Cross piece of the head
132 Tack dead block
133 Gammoning mortice
134 Quarter galleries
135 Gripe
136 Gunwale
137 Planksheer
138 Fife rail
139 Lower counter
140 Chesstree
141 Rudder
142 Iron brace
143 Gudgeon
144 Pintel
145 Iron hoop and brace
146 Tiller
147 Gooseneck
148 Whipstaff
149 Elm cleat
150 Sweep
151 Belfry
152 Pissdale
153 Furnace
154 Limber hole
155 Channel spur
156 Iron clasp for davit
157 Garboard strake
158 Kevel
159 Cleat
160 Upperdeck mast partners

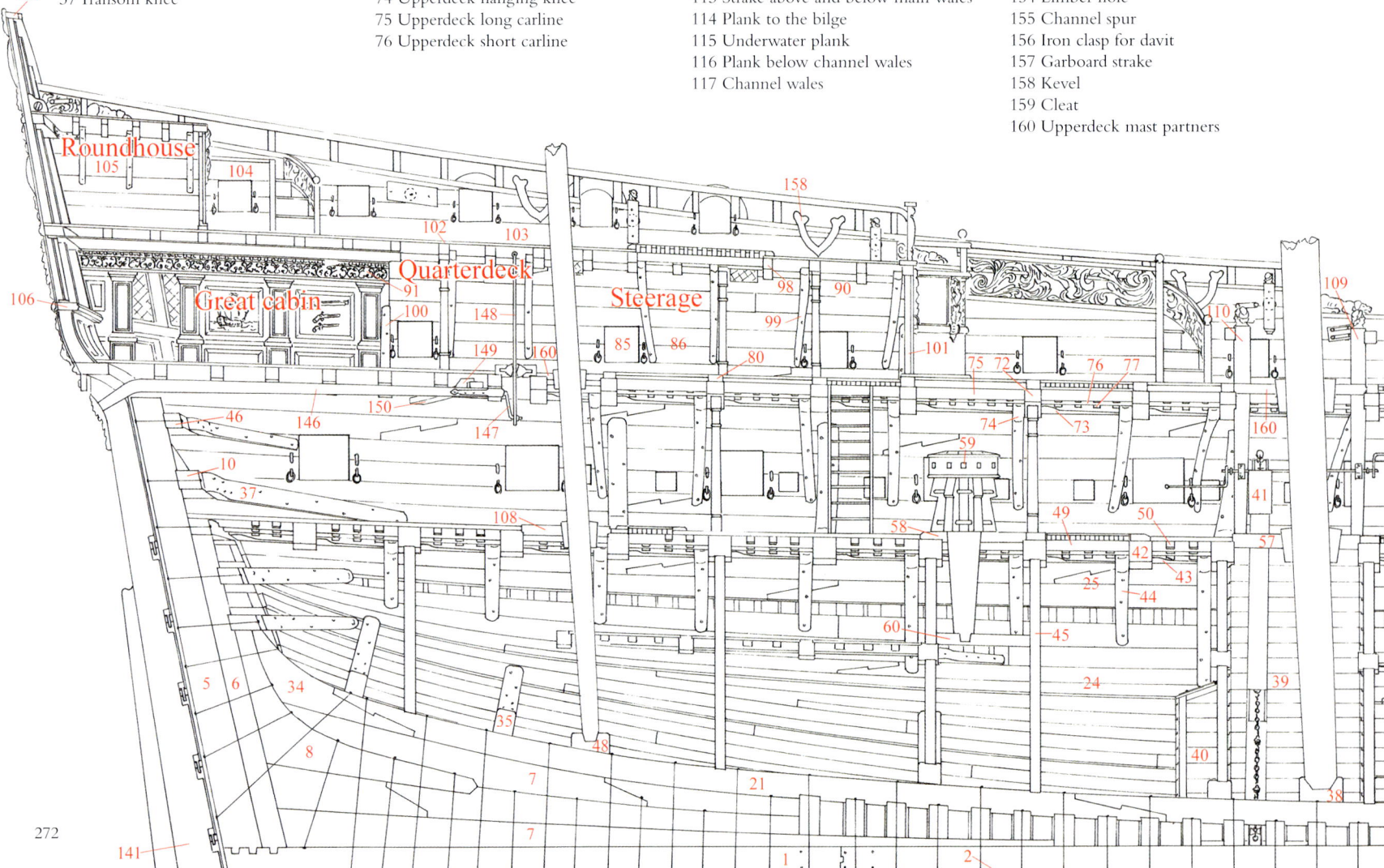

272

273

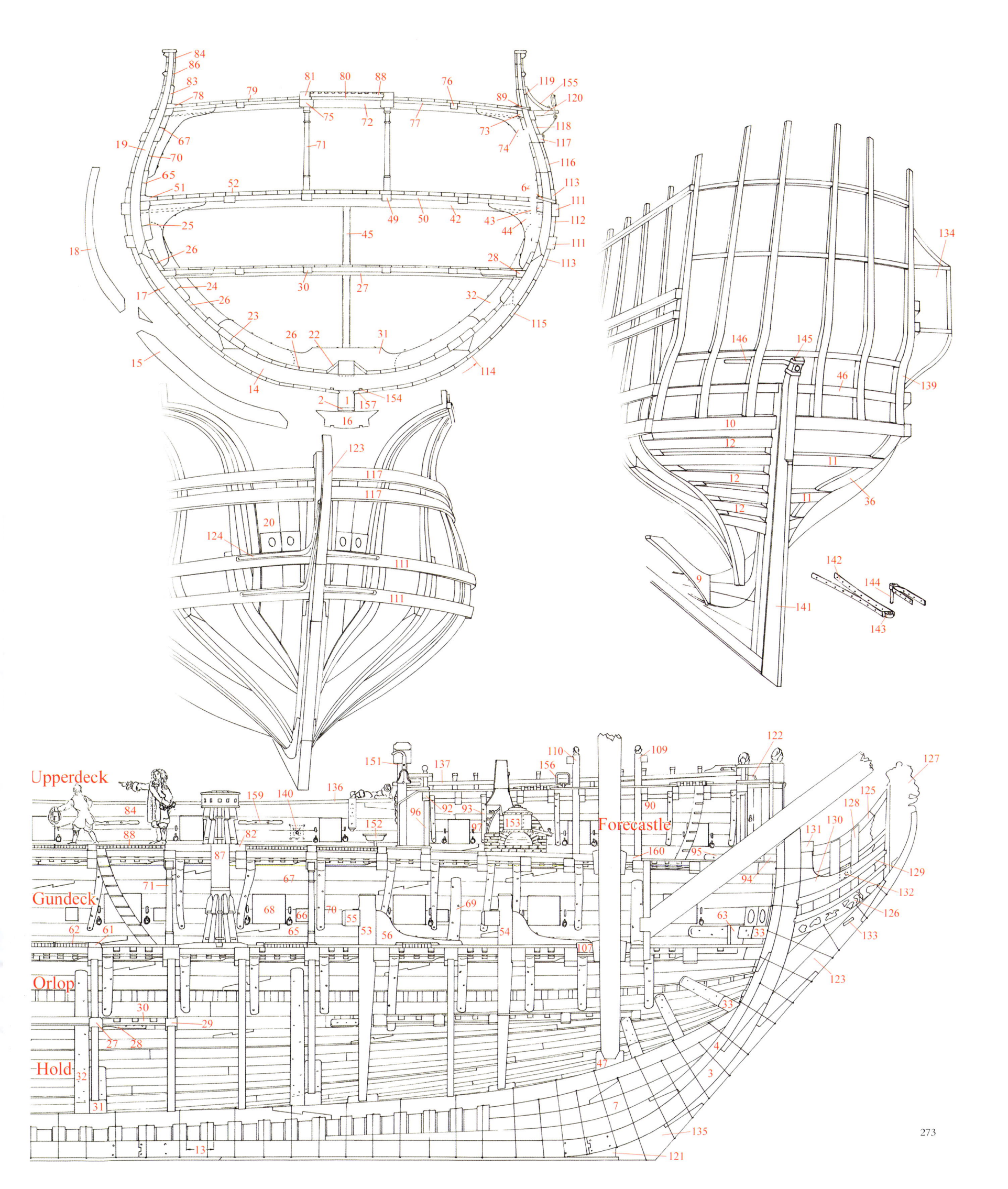

Stem
The stem(3) to be of good sound oak timber thirteen inches thwartships and fifteen inches fore and aft. The scarphs not to be less than four foot & bolted with six bolts by an inch auger. The false stem(4) to be twenty two inches broad and eight inches thick and one foot at least in the scarph well bolted with inch bolts into the stem.

Sternpost
The stern post to(5) be sixteen inches thwart and eighteen inches fore and aft in the head and two foot fore and aft upon the keel. The false post within(6) the stern post to be fifteen inches fore and aft and eighteen inches thwartships.

Rising wood
To bring on upon the keel two substantial pieces of rising wood(7) of sufficient length for the run of the ship and then to bring on a well grown substantial knee(8) upon the same, faid to the stern post. The up and down arm to reach the lower transom and the arm fore and aft to be twelve foot long at least. And to be bolted by bolts by an inch and eighth auger through the keel and through the post at every twenty inches distance and to fix other rising wood(7) upon the said knee of sufficient height and substance for the support of the half timbers(9) which are to be of good lengths and to be well let into the same and fastened. To bolt the upper rising wood(7) to the lower in the same manner and with the same size of bolts as aforementioned.

Transoms
The main transom(10) to be twelve inches and the lower transoms(11) ten inches thick and to lie within eighteen inches one from another and to fit the space between them on each side the post with a sufficient number of half transoms(12) of such substance as that their upper and lower edges may lie two inches at least clear of the whole transoms to give space for air.

Timber and room
The Ranou Nic Armorer*Floor timbers*
And the floor timbers (14)to be thirteen inches fore and aft and fourteen inches up and down upon the keel and of such length in the floor of the ship that their heads may lie eighteen inches at least above the bearing thereof by a straight line from the bottom of the keel and to be eleven inches (moulded) in and out at the wrongheads or ten inches wrought in the length of the floor and before and abaft the same to be nine inches wrought in and out and every other floor timber to be bolted through the keel with one bolt by an inch and eighth auger.

Lower futtocks
The lower futtocks(15) to fill the rooms full and workmanlike for the strength of the frame below and to be twelve foot long at least and to be chocked(16) down within twenty inches of the keel.

Middle futtocks(17)
These are not mentioned here but are in the specification for the ceiling.

Upper futtocks
The upper futtock(18) head to take the lower sill of the ports in the midships and the heel of the toptimbers, the lower edge of the gundeck clamps within board and wale without board to be ten inches fore and aft and eight inches in and out at the breadth and to stand an inch and half apart for air

Toptimbers
And the toptimbers(19)to be three inches in and out aloft. And the whole frame to be well chocked with good sound timber. *The chocks referred to are the triangular chocks at the head and heels of the timbers that allowed straighter timber to be used in making the futtocks.*

Hawse pieces
To place two substantial hawse pieces(20) on each side the stem to be twenty two inches broad each piece at least.

Keelson
The keelson(21) not to have more than four pieces and to run fore and aft and to overlaunch the scarphs of the keel to be fourteen inches up and down and fifteen inches broad scored down upon the timbers an inch and half at least. The scarphs to be four foot long at least and to be bolted through every other timber and through the keel by bolts by an inch and eighth auger.

Limber board
The limber board(22) to be two inches and a half thick and fourteen inches broad on each side the keelson.

Planking in hold
To bring on six strakes of foot waling(23) on each side at the wrongheads to be fifteen inches broad each of them, two to be six inches and two to be five inches and two four inches thick to run fore and aft and to be wrought narrow in due proportion each way the six inch stuff to lie upon the chocking of the floor heads and middle futtocks and the other strakes of five and four inches thick one of each to be brought on above and below the same. To bring on two strakes of middlebands(24) on each side fourteen inches broad to be five inches thick and two strakes of gundeck clamps(25) six inches thick and fourteen inches broad to be tabled one into the other and scarphed Flemish hook and butt three foot and six inches long at least to have an opening fore and aft below the clamps ten inches broad for airing the frame and all the rest of the ceiling in hold to be good oak plank(26) of three inches thick.

Orlop
To place five orlop beams(27) in hold twelve inches broad and ten inches deep, one abaft the mast, two at the main hatchway, one at the bulkheads of the cockpit abaft the other at the bulkheads of the boatswain's storeroom forward. To lie from the lower edge of the gundeck beams five feet, to be kneed with one good fore and aft knee(28) to be three foot long at each arm at least and bolted with five bolts by an inch and one eighth auger and to place other false beams(29) and carlings(30) in hold convenient for ye platforming and storerooms to be made thereon.

Riders
To fay and fasten three bends of floor(31) and lower futtock riders(32). The floor riders to be fifteen inches fore and aft eleven inches up and down upon the keel and eleven inches at the wrongheads. To be twenty foot long at least and bolted with seven bolts in each arm by an inch and eighth auger the futtock riders to be fourteen inches fore and aft and eleven inches thick and to be fifteen foot long each and bolted with nine bolts of the same size.

Breasthooks in hold
The bresthooks(33) forward to be eleven inches deep and to lie one under the hawses, one at the deck and three more in hold to be as long as possible but none less than seven foot in the arm. To be bolted with six bolts in each arm by an inch and eighth auger.

Crotches and standard
To end the keelson abaft with a good standard(34) well bolted into the post and keelson and to fay two crotches(35) in the breadroom eleven inches sided with arms each way five or six foot long and bolted with bolts by an inch and eighth auger, eighteen inches asunder to overlaunch the thick stuff and ceiling in hold over the fashion pieces(36) abaft and to meet upon the middle of the false stem afore and to fay all the breasthooks, crotches and knees upon the same in good workman like manner.

Transom knees
And to fay on each side one large knee(37) ten inches sided & ten foot long at ye main transom & knees of proper scantling at all ye rest of the transoms and bolted as aforesaid.

Mast saddles
To fix saddles(38, 47, 48) in their proper places for all the masts and of fit substances and fastenings.

Hold
And to complete build and part off with ordinary deals all the necessary platforms and store rooms in hold. To plaster with lime and hair and line with slit deals all bulkheads and places where powder comes. To line the breadroom with good dry deal. To build a well(39) with two inch oak plank and two convenient shot lockers(40) about the mast and to furnish all cabins and store rooms with doors, locks, shelves and other partitions in every respect as the same is done in others of their Majesties ships of ye like quality or shall hereafter be set off and directed as appointed to be done by the Surveyor of their Majesties Navy.

Pumps
To place one pair of chain pumps(41) in hold and to fix them with cisterns, dales and scuppers through the sides.

Upperdeck

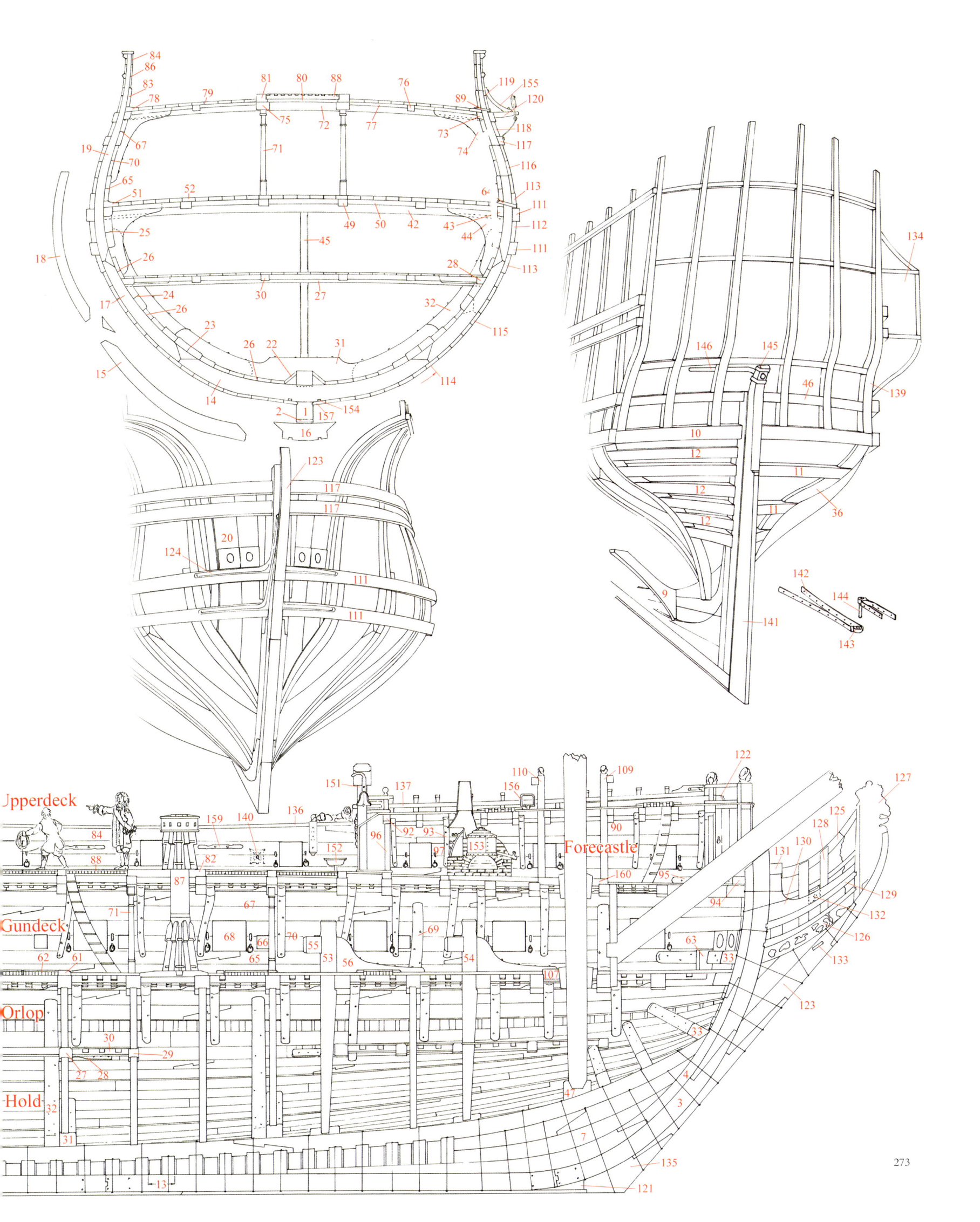

Gundeck

Orlop

Hold

Forecastle

273

The *Foresight* Contract, 1649

Building contract for the fourth-rate Foresight, built by Jonas Shish of Deptford and launched in 1650 (NMM SPB/8). The ship measured 121 feet 2 inches; length of gundeck: 102 feet 0 inches; keel length: 32 feet 0 inches breadth and 13 feet 0 inches depth in hold.

This indenture made the twenty fourth day of December 1649 between John Holland, Thomas Smith & Robert Thompson, Commissioners of the Navy for and on the behalf of the State of the one part and Jonas Shishe of Deptford in the County of Kent, Shipwright, on ye other Witnesseth that the said Mr Jonas Shish for the consideration hereafter in these predrafts mentioned and expressed: Doth covenant, promise grant & agree to and with the said John Holland, Thomas Smith & Robert Thomson, for & on the behalf of the State that he the said Jonas Shishe, his executors, administrators, servants or assignees shall & will at his & their own charges well & workmanlike erect & build off the stocks for the use of the State, one good, strong & substantial new ship or frigate of good, sound, well seasoned timber and plank of English oak, beech & elm.

Overall Dimensions
And that the said ship or frigate shall be of the dimensions hereafter mentioned & shall be erected & built in manner & form following (that is to say) The said ship or frigate shall contain in length by the Keel one hundred foot. Breadth thirty one foot without the plank. Depth under the beam eleven foot six inches by a right line. Breadth at transom eighteen foot. The Rake fore & aft to be twenty & three foot.

Keel
The said keel(1) to be fourteen inches square, to have long scarphes, tabled and to be well bolted with a bolt of one inch auger. The bottom to be sheathed(2) with a three inch plank.

Stem
To have a firm substantial stem(3) with a sufficient false stem(4) of 7 inches thick and twenty or two and twenty inches broad.

Sternpost
To have a substantial stern post(5) of eighteen or twenty inches deep with a long armed knee(8). To be well bolted with an inch & half quarter auger bolt fastening the same together. *As suggested here the knee probably fitted directly in the corner against the sternpost and keel. No mention is made of an inner false sternpost.*

Timber and room
The space of timber and room(13) to be two foot & two inches at the most.

Floor timbers
The floor timbers (14) of the ship to be fourteen inches ½ up & down upon the keel and eleven inches in & out at the wrongheads at ye bearing & ten afore & abaft. Naval timbers *(Sometimes known as the lower tier of futtocks.)*
The naval timbers(15) to fill the rooms and to have six feet scarph.

Middle futtocks(17) and toptimbers(19)
They are not mentioned here, although the Toptimbers are referred to later in the contract relating to payment.

Gundeck Timbers
(Sometimes known as the third or upper tier of futtocks. The dimension refers to the head of the upper futtocks and the heel of the toptimbers.)
The timbers upward at the gundeck(18) to be eight inches in & out & to have the like scarph and the rooms to be filled with timber.

Keelson
To have a substantial keelson(21) fourteen inches up and down and sixteen inches broad and to run fore and aft to be well bolted by an inch & inch quarter auger through every timber and through the stem.

Planking in hold
To put in six strakes of sleepers(23) in hold on each side the wrongheads of six inches thick & fifteen inches broad three of them to be 7 inches thick on the bearing amidships & to run fore and aft. Two strakes of middlebands(24) on each side of five inches thick and fifteen inches broad and to run fore and aft. Two strakes of clamps(25) each side fore & aft under the beams of ye gundeck of 8 inches thick & sixteen inches broad to be hooked and tabled one into the other for [to prevent] retching and all the rest of the footwaling(26) in hold to be of good three inch plank.

Riders
With a floor rider(38) for the step of ye main mast and pillars(45) in hold under the beams.

Gundeck beams
The beams(42) of the gundeck to be twelve or 13 inches up & down & 14 or 15 fore & aft and to lie one under each port & one betwixt and to be kneed with four knees to each beam (Vizt) two fore & aft(43) hooked into the beams and two up & down(44) and to be well bolted with four bolts to each knee of an inch auger.

Gundeck
And to have a double tire of carlings(49) on each side fore and aft six inches thick & nine inches broad and the ledges(50) to lie within eight inches of each other and lay the said deck with long plank(52) of two inches thick and by the side in the wake of the ordnance with two inch plank and to have waterways(51) of six inches thick and fourteen inches broad. To put out eight leaden scuppers(64) on each side the gundeck.

Storerooms
To make as many hatches in the hatchway(62) as shall be convenient of two inch plank with the hatchway abaft the mast for the stowing of provisions & hatchway to the Stewardroom and hatchway for the convenience of the Cookroom and for Boatswain and Gunner's store rooms & powder rooms. *There were probably three hatches in the main hatchway.*

Hawsepieces
To make a manger(63) on the same deck and to put in four hawse pieces(20) & to cut out four hawse holes.

Bitts
To place two pair of carrick bitts(53, 54) with cross pieces(55) and knees(56).

Standards
To bring on four standards(69) on each side the gundeck and to shovel(*sole*) them with plank and to bolt them with two or three bolts in each arm of an inch auger.

Gundeck spirket
To have one strake of spirket wale(65) on the same deck of four inch plank and one gunwale(65) scored into timbers fore & aft six inches thick to the lower edge of the ports. *The gunwale mentioned here is the upper piece of spirketting. It would have been scored down by about an inch into the gaps between the frame timbers to prevent retching – that is, the ends of the ship bending downward. After this contract of 1649, both pieces are referred to as spirketting.*

Gundeck gunports
To cut out twelve ports(68) on each side the same deck with two abaft and to make & hang them with hooks & hinges and to fit & drive all ringbolts, eyebolts, rings & staples.

Gundeck fittings
To place partners for ye main(57) & fore mast(107) partners(58) & a step(60) for the main capstan(59) with a step(48) for the mizzen mast and to have turned pillars(71) under the beams with staircase up into the quarterdeck and stairs & ladders to all conveniences. To bring on five breasthooks(33) besides the step(47) of the foremast with good long arms. To have three bolts in each arm and one through the stem by an inch & ½ quarter auger.

Transoms
To have as many transoms(10, 11) abaft below the ports as may lie within eighteen inches one of another and one transom(46) at the upper edge of the ports under the helm port with good substantial knees(37) with long arms with three bolts in each arm by inch auger.

Upperdeck clamps
To bring on risings(67) fore & aft of four inches thick & of oak sixteen inches broad under ye beams of the upperdeck and to ceil(70) up between decks with three inch plank fore & aft.

Upperdeck beams
The beams(72) of the said deck to be eight inches up and down and ten inches broad and to lie between eight & nine foot asunder excepting in the wake of the gunroom & in the hatchway where to have two carling knees & to be in height between the said decks between plank & plank six foot & four inches in ye midships and go flush fore and aft six of the said beams to be double kneed with four knees to each beam two fore and aft(73) the other up and down(74) and the rest to be kneed with two knees to each beam up & down(74) and to be well bolted with bolts of ¾ & ½ quarter auger. *The beams are further apart at the main hatchway and to strengthen the structure, lodging knees, called here carling knees, are fitted in the corners between the carling and the beam.*

Upper deck
To have one tier of long carlings(75) fore & aft with sufficient ledges(77). To lay the said deck(79) with good Sprutia deals & to make a quick waist with ports in ye wake of the guns in the quarter to lay with two inch plank. *Note: there is only one tier of upper-deck carlings each side at the hatchway in this ship.*

Upper-deck fittings
With a Spirketting(83) of two inch plank abaft the mast. To have coamings(81), head ledges(80) with a grating hatchway(88) before & abaft the mast & a grating forward for the vent of the smoke of the ordnance. To fit topsail sheet bitts(109) with catts(122), davit, clasp of iron(156). partners for the jeer capstan(82) & partners for the mizzen mast(108) & to put out as many scuppers(89) abaft upon that deck as shall be convenient.

Capstans
To make main(59) and jeer(87) capstan with capstan bars and iron pawls.

Quarterdeck and forecastle
To make a large quarterdeck & forecastle & roundhouse with a bulkhead & doors with two ports in the same and as many of the beams in the wake of the steerage, forecastle & roundhouse to be kneed as shall be convenient and in the wake of the cabin to have a rising(91) under the beams of six inches thick and the beams to be dovetailed & bolted into the same.

Upper-deck ports
To cut out four ports(85) on each side under the quarterdeck and two right aft and to make & hang port lids with hooks & hinges & ports in ye forecastle. To have a transom(106) abaft under the windows in the cabin & one in the wake of the ports and the same kneed & bolted.

Roundhouse
And to have as large a roundhouse as the work with convenience shall give leave with a bulkhead and a door to the same. To cut out three ports(85) on each side in the forecastle.

Hold
To make all the platforms in hold with bulkheads and partitions (Vizt) for the powder room & Gunner's storeroom, Sailroom, Boatswain's storeroom, Cookroom(97) with cabins & conveniences for lodgings, Carpenter's & Steward's storeroom, fish room & convenience for the Captain's provisions with Bread room and ceiling the same.

Without board planking
And without board the ship to be planked up from the keel to the chainwales with three inch plank(114, 115,116) excepting four strakes(112, 113) of four inch plank (Vizt) one below the wale, two between & one above the wale and to have two firm wales(111) of eleven inches up & down & 8 inches thick and chain wale(117) of five inches thick and ten inches broad to be inbowed and the work upward so high as the waist to be wrought upward with two inch plank(119) and the quarter with Sprutia deal or two inch plank.

Head
To have a fair head with a firm substantial knee(123) and cheeks(124), treble rails(125) with trail board(126), beast(127) & brackets(128), keelson(129) & cross pieces(131) & supporters under the catts(122).

Stern
To have a fair lower counter(139) with rails & brackets, stern, gallery with carved work & brackets, windows & casements into the cabin. To have a fair upright with a complete pair of Arms, cherubims heads, pilasters & terms. To have a fair pair of turrets(134) with windows into the cabin complete.

Rigging attachments
To have a pair of chesstrees(140) with fore & main chainwales(120) well bolted with chain bolts and chain plates.

Gripe
To have a ranch gripe(135) well bolted with dovetailes & a stirrup on the skeg well bolted.

Rudder
To make & hang on a complete rudder(141) with braces(142), gudgeons(143) & pintles(144) with a muzzle for the head & a tiller(146) thereto.

Rails
To rail(137) & gunwale(136) the said ship fore & aft with brackets & hancing pieces complete.

Masts and yards
To make a complete suite of masts with caps & crosstrees, yards, boltsprit, topgallant masts, flag staff & ancient staff. All the said masts & yards to be ready made fit for ship of that burthen.

Decoration
Likewise to do & perform all the carved work.

Details
And to finish & provide all the materials for the same likewise all the joiner's work finding deals, locks, hinges for doors, settlebeds & cabins with workmanship thereunto both within board & without. And to plane the ships all over underwater. All plumber's work lead & leaden scuppers etc. All glazier's work, slate glass for casements & scuttles for cabin windows etc. All painters' work for painting & gilding within board and without. And to do & perform whatsoever belongs to the Carpenter to do for the finishing & completing of the hull in like manner as is done & performed to the like frigate in the State's yard.

Materials and delivery
And the said Jonas Shishe for himself his executors & administrators doth covenant & undertake at his or their cost & charges to find & provide all manner of iron work, bolts, spikes, nails, bradds etc, likewise all timber, plank, boards, treenails White & black oakum, pitch, tar, rozin, hair, oil, brimstone & all other materials that shall be needful to be used or spent in or about the works & premises aforesaid for the complete finishing thereof. And in like manner to discharge & pay all manner of workmanship touching all and every part of the works herein expressed to be done and performed. And to finish complete & launch the said ship or frigate into the River of Thames to & for the use of the State as aforesaid on or before the last day of [July] now next ensuring the date of this present Indenture.

Quality
Provided that if at any time during building of the said ship or frigate herein contracted according to the dimensions, proportions & scantlings herein expressed & set forth or intended to be expressed or set forth there shall be found & discovered upon due survey to be made thereon by such person as shall be thereunto appointed any unsound & insufficient timber, plank or other materials or any insufficient workmanship and performance prejudicial to the state. That then after due notice thereof given in writing by the said Surveyor to the said Jonas Shishe or to the chief master workman under him there shall be an effectual and speedy reforming of all and every such default in stuff or workmanship. And that the said amendment or reformations shall be certified in writing by the said surveyor to ye said Commissioners of the Navy in the service of the State in this behalf.

Payment
And the said John Holland, Thomas Smith & Robert Thomson for & on the behalf of the State aforesaid covenate and grant to & with the said Jonas Shishe his executors, administrators and assignees by these predrafts that they the said John Holland, Thomas Smith & Robert Thomson or some of them for & on the behalf of the state shall according to the custom of the Office of the Navy sign and make out bills to the Treasurer of the Navy to be paid to the said Jonas Shishe his executors, administrators or assignees for the sum of six pounds and ten shillings per ton for every ton that said ship or frigate shall be of in burthen being measured & calculated according to the accustomed rule of Shipwright's Hall. The said payments to be made in manner & form following (that is to say) two third parts of the whole in money and one third part in timber to be valued as is

hereafter mentioned. And it is agreed that the first two thirds in money shall be paid at four payments as followeth (vizt) one fourth part at the ensealing of these predrafts, one other fourth part when all the lower futtocks are in & fastened, one other fourth part when all the tier of toptimbers are in, both the wales on and the lower deck laid, and the other fourth part or what is shall amount unto within ten days after launching & finishing of works respecting the hull, carving, joining or painting fit to be done after the launching, and for ye other third part of the whole agreed as above said to be paid in timber it is agreed that the same shall be paid in manner following (vizt) either by so many trees to be assigned by the said John Holland, Thomas Smith & Robert Thomson to the said Jonas Shishe or his assignees out of one or more of the late King's parks & chases by or before the first of March next. The said trees to be valued by two or more of us the Commissioners of the State's Navy, which valuation, if it be disliked by the said Jonas Shishe, then the said third part to be paid in money by or before the first of June next. Lastly for all such sum or sums of money that shall grow due by virtue of this contract the said Jonas Shishe is content to accept of ye Parliament for payment. In witness whereof the parties above named have interchangeably set their hands the day & year above written.

John Holland Tho:Smith Rob Thomson

Endorsement

We having overlooked in this covenant & amended what was thought fit & the builders having seen the dimensions & scantling etc; herin contained we conceive for the better expediting of the work you may proceed to agreement with them, who before they receive the papers are willing to conform to all such other particulars and articles for both sorts of ships as shall be propoprtionable to the dimensions, which we leave to your Worship's consideration & remain yours to be commanded:
John Taylor
Christopher Pett
Jonas Shish

On the back of the contract are rough drawings of the head and stern of a ship which is probably the Foresight.

The *St Patrick* Contract, 1665

Building contract for the fourth-rate St Patrick, built by Francis Bailye in Bristol and launched on 9 May 1666 (NA SP29/141). The ship measured 102 feet 0 inches keel length, 33 feet 10 inches breadth and 14 feet 6 inches depth in hold.

15 April 1665
Contract with Mr Bailey building a 4th rate
Bristol

This indenture made the fifteenth day of April in the year of our lord one thousand six hundred and sixty five between the Principal Officers and Commissioners of his Majesty's Navy for and on the behalf of his Majesty of the one part and Mr Francis Bailye Shipwright of Bristol of the other part witnesseth that the said Mr Francis Bailye for the consideration hereafter expressed does covenant, promise and grant to and with the said Principal Officers and Commissioners for and on the behalf of his Majesty that he the said Mr Francis Bailye his executors, administrators, servants and assignees shall and will at his and their own proper cost and charges well and workmanlike erect and build off the stocks for the use of the King, one good and substantial new ship or frigate of good and well seasoned timber and plank of English oak and elm.

Overall dimensions
And that the said ship or frigate shall contain in length by the keel one hundred feet. Breadth from outside to outside of the plank thirty two feet and six inches. Depth under the beams thirteen feet and six inches. Breadth at transom twenty one feet four inches. Rake forward twenty one feet. Rake aft five feet.

Keel
The keel(1) to be fifteen inches square in the midships to be sheathed with a three inch plank(2). To have long scarphs tabled in the keel and to be well bolted with a bolt of an inch auger.

Stem
To have a firm substantial stem(3) with a sufficient false stem(4) of eight inches thick and two feet four inches broad.

Sternpost
To have a substantial stern post(5) of two feet deep and another post within(6) it to be fastened by the lower transom with a long armed knee(8) to be well bolted with an inch and quarter auger bolt fastening the same together. *The knee was probably fitted on top of the rising wood in order to reach the lower transom.*

Timber and room
The space of timber and room(13) to be two feet and two inches.

Floor timbers
The floor timbers(14) of the said ship to be fifteen inches up and down upon the keel and eleven inches in and out at the wrongheads.

Naval timbers
(Sometimes known as the lower tier of futtocks)
The naval timbers(15) to fill the rooms being at least twelve inches fore and aft and to have at least six feet scarph.

Middle futtocks(17) and toptimbers(19)
They are not mentioned here, although the toptimbers are referred to later in the contract relating to payment.

Gundeck timbers
(Sometimes known as the third or upper tier of futtocks. The dimension refers to the head of the upper futtocks and the heel of the toptimbers.)
The timbers upward at the gundeck(18) to be nine inches in and out and to have the like scarph and the rooms to be filled with timber.

Keelson
To have a substantial keelson(21) of three pieces fourteen inches up and down and eighteen inches broad and to run fore and aft to be well bolted by an inch and quarter auger through every timber and through the stem.

Planking in hold
To put in six strakes of sleepers(23) in hold on each side the wrongheads of six inches thick four, and the other two of seven inches: to be also thick fore and aft and fifteen inches broad and to run fore and aft. To have two strakes of middlebands(24) on each side of five inches thick and fifteen inches broad and to run fore and aft. Two strakes of clamps(25) on each side fore and aft under the beams of the gundeck of six inches thick and fifteen inches broad to be hooked one into the other to prevent reaching and all the rest of the footwaling(26) in hold to be good three inch plank.

Riders
With four bends of riders in hold three of them to be placed before the mast and the other abaft to be eighteen inches broad and sixteen inches deep, the floor riders(31) and futtock riders(32) twelve inches square to have seven or eight bolts a piece in the floor riders and five in the futtock riders to be bolted with an inch and quarter bolt and to have seven foot scarph upwards and downwards.

Orlop
And to place four false beams(27) over them of a convenient height and to make platforms upon them for the strengthening of the ship and stowing cables thereon to be fourteen inches broad and twelve inches deep to have two knees and a standard at each end to be well bolted with five bolts of inch and 1/8 auger. *The false beams secured here with knees and standards are called orlop beams after this contract, while the earlier contract for the Foresight has no mention of them. The false beams mentioned in the 1692 contract of Norwich were light orlop beams without knees.*

Mainmast step
A rider(38) for the step of the mainmast.

Pillars in hold
And pillars(45) in hold under the beams.

Gundeck beams
The beams of the gundeck(42) to be fourteen inches broad and twelve up and down and to lie four feet asunder excepting in the hatchway which must be seven feet asunder and to be kneed with four knees to each beam Viz: two fore and aft(43) hooked into the beams and two up and down(44) and to be well bolted with four bolts to each knee of an inch and 1/8 auger.

Gundeck
And to have a double tire of carlings(49) on each side fore and aft six inches thick and nine inches broad and the ledges(50) to lie within eight inches one of another and lay to the said deck with four strakes besides the waterway of three inch plank fourteen inches broad and the rest of the said deck(52) with two inch planks the three inch plank and waterways to be treenailed and to have waterways(51) of six inches thick and fourteen inches broad. To put out eight leaden scuppers(64) on each side of the gundeck.

Storerooms
To make as many hatches in the hatchways(62) as shall be convenient of two inch plank with the hatchway abaft the mast for the stowing of provisions and hatchway to the Steward's room and for Boatswain and Gunner's storerooms and powder rooms.

Hawse pieces
To make a manger(63) on the same deck and to have two, four inch scuppers in it of lead and to put in four hawse pieces(20) and to cut out four hawse holes.

Bitts
To place two pair of carrick bitts(53, 54) with cross pieces(55) and knees(56).

Standards
To bring on six standards(69) on each side the gundeck and to shovel (sole) them with plank and to bolt them with three bolts on each arm of an inch and a half quarter auger.

Gundeck spirket
To have two strakes of spirket wales(65) on the same deck of four inch plank fore and aft four inches thick to the lower edge of the ports.

Gundeck gunports
To cut out twelve ports(68) on each side the same deck with two abaft and to make and hang them with hooks and hinges and to fit and drive all ringbolts, eyebolts, rings and staples.

Gundeck fittings
To place partners for the main(57) and foremast(107) partners ten inches thick and a step(60) for the main capstan with a step for the mizzen(48) mast upon the keelson. To have four inch plank upon the gundeck(52) between the bitts and the main partners. To raise the hatches(62) above the deck and bolt the planks into the beams and to have turned pillars(71) under the beams and under the pillars a four inch plank for the pillars to rest upon. With staircase up into the quarterdeck and stairs and ladders to all conveniences. To bring on five breasthooks(33) besides the step(47) of the foremast with good long arms to have three bolts in each arm and one through the stem by an inch and half quarter auger.

Transoms
To have as many transoms(10, 11) abaft below the ports as may lie within twelve inches one of another and one transom(46) at the upper edge of the ports under the helm port to take hold of the sternpost with good substantial knees(37) with long arms with three bolts in each arm by inch auger. *Half transoms are not mentioned and were probably not used.*

Upperdeck gunports
To make a whole tire of round and square ports(85) on the upperdeck and forecastle to be all square ports under the quarterdeck and forecastle and to garnish them with carved works.

Upperdeck clamps
To bring on risings(67) fore and aft of three inches thick and sixteen inches broad under the beams of the upperdeck and ceiled up between decks with three inch plank(70) fore and aft.

Upperdeck beams
The beams(72) of the said deck to be eight inches up and down and ten inches broad and to lie between five and six feet asunder excepting in the wake of the gunroom and in the hatchway where to have carling knees and to be in height between the said deck between plank and plank six feet and six inches in the midships and to go flush fore and aft all the said beams to be double kneed with four knees to each beam from the gunroom forward that is to say two fore and aft(73) and the other up and down(74) and the beams in the gunroom to be single kneed with two knees in each beam up and down and to be well bolted with two bolts in each arm with bolts of three quarters auger. *The beams are further apart at the main hatchway, and to strengthen the structure, lodging knees, called here carling knees, are fitted in the corners between the carling and the beam.*

Upper deck
To have one tier of long carlings(75) fore and aft with sufficient ledges(77). To lay the said deck with good Sprutia deals or dry two inch plank(79) with a waterway(78) of four inch plank seasoned and in the wake of the guns in the quarter to lay with two inch plank. *Note: there is only one tier of upper-deck carlings each side at the hatchway in this ship.*

Upper-deck fittings
With a Spirketting(83) of two inch plank fore and aft. To have coamings(81), head ledges(80) with a grating hatchway(88) before and abaft the mast and the grating forward for the vent of the smoke of the ordnance. To fit topsail sheet bitts(109) with catt(122), davit, clasp of iron(156). partners for the main(58) and jeer(82) capstans and partners for the mizzen(108) masts and to put out nine scuppers(89) on each side on the upperdeck.

Capstans
To make a main(59) and a jeer capstan(87) with capstan bars and iron pawls.

Quarterdeck and forecastle
To make a large quarterdeck and forecastle with two round bulkheads in each side of the said forecastle abaft for cabins and in the midships of the said bulkhead to place the cookroom(97) for roasting and boiling and to set all the bulkheads upon six inch plank eight inches broad and to be well fayed to the deck and laid with tar and hair. To have four ports in the same and as many of the beams(98) in the wake of the steerage bulkhead and the beams(92) in the forecastle to be all kneed(93, 99) (because of the anchors) as shall be convenient and in the wake of the cabin to have a rising(91) under the beams of six inches thick and the beams to be dovetailed and bolted into the same.

Upper-deck ports
To cut out four ports(85) on each side under the quarterdeck and two right aft and to make and hang portlids with hooks and hinges. To have a transom(106) abaft under the windows in the cabin and one in the wake of the ports and the same to be kneed and bolted.

Roundhouse
And to have as large a roundhouse as the works with conveniency will give leave with a bulkhead and a door to the same and cut out three ports on each side in the forecastle.

Hold
To make all the platforms in hold with bulkheads and partitions Viz: for the powder room and Gunner's storeroom, sailroom, Boatswain's storeroom with cabins and conveniencies for lodging, Carpenter's and Steward's storerooms, fish room and conveniencies for the Captain's provisions with a bread room and ceiling the same and to pay it with rozin being through dried with charcoal.

Without board planking
And without board. The ship to be planked up from the keel to the chainwales with three inch plank(114, 115, 116) excepting three strakes of four inch plank Viz: one below the wale(113), one between(112) and one above the wales(113) and to have two firm wales(111) of eleven inches up and down and seven inches thick and to have two wales(117) for the conveniency of the chain plates and bolts to go fore and aft both to be five inches thick and ten inches broad and the work upwards(119) so high as the waist from the upper wale to be wrought up with two inch plank and the quarter with Sprutia deal landed work or two inch plank.

Head
To have a fair head with a firm and substantial knee(123) and cheeks(124) treble rails(125) with trail board(126), beast(127) and brackets(128), keelson(129) and cross pieces(131), standards(130) and supporters under the catts(122).

Stern
To have a fair lower counter(139) with rails and brackets and open gallery with a house of office in it with carved works and brackets, windows and casements into the cabin. To have a fair upright and to put in it a complete pair of King's Arms, cherubims heads, pilasters and terms. To have a fair pair of turrets(134) with windows into the cabin complete.

Rigging attachments
To have a pair of chesstrees(140) with fore and main chainwales(120) well bolted with chain bolt and chain plates.

Gripe
To have a ranch gripe(135) well bolted with dovetailes and a stirrup on the skeg well bolted.

Rudder
To make and hang on a complete rudder(141) with braces(142), gudgeons(143) and pintles(144) with a muzzle for the head and a tiller thereto.

Rails
To rail and gunwale the said ship fore and aft and to put on with brackets and hancing pieces complete.

Masts and yards
To make a complete suite of masts suitable to such a frigate with caps and crosstrees, yards, boltsprit, topgallant masts, flag staff and ancient staff. All the said masts and yards to be ready made fit for ship of that burthen and to set them likewise good.

Decoration
And perform all the carved work.

Details
And to finish and provide all the materials for the same likewise all the joiner's work finding deals, locks, iron bars, hinges for storerooms, Steward rooms, doors, settlebeds and cabins with workmanship thereunto both within board and without. And to plane the ships all over underwater. All plumber's work lead and leaden scuppers and all glazier's work slate glass and casements and scuttles for cabin windows etc. All painters' work for painting and gilding within board and without. And to do and perform whatsoever belongs to the Carpenter to do for the finishing and completing the hull in like manner as is done and performed to the like ship and frigate built in the King's yard.

Materials and delivery
And the said Mr Francis for him his executors and administrators doth covenant and grant at his and their own cost and charges to find and provide all manner of iron work, bolts, spikes, nails and bradds etc, likewise all timber, plank, boards, treenails which are to be all mooted from the chainwale downward. White and black oakum, pitch, tar, rozin, hair, oil, brimstone and all other materials that shall be needful to be used or spent in or about the works and premises aforesaid for the complete finishing thereof and in like manner to discharge and pay all manner of workmanship touching all and every part of the works herein expressed to be done and performed and to finish complete and launch the said ship or frigate into the harbour of Frome River near Bristol to be delivered safe afloat to and for the use of the King as aforesaid on or before the last day of November next ensuring the date of this present Indenture.

Quality
Provided that if at any time during the building of the said ship or frigate herein contracted according to the dimensions, proportions and scantlings herein expressed and set forth or intended to be expressed and set forth there shall be found and discovered upon due survey to be made therein by such person as shall be thereunto appointed any unsound and unsufficient timber, plank or other materials or any insufficient workmanship and performance prejudicial to the King that then after due notice thereof given in writing by the said Surveyor unto the said Francis Bailye or to the chief master workman under him there shall be an effectual and speedy rectifying of all and every such default in stuff and workmanship and the said amending or reformations shall be certified in writing by the said surveyor to the said Principal Officers and Commissioners of the Navy in the service of the King in this behalf.

Payment
And the said Principal Officers and Commissioners of the Navy for and on the behalf of the King shall according to the custom of the office of the Navy sign and mark out bills to the Treasurer of the Navy to be paid to the said Francis Bailye his executors, administrators or assignees for the sum of six pounds six shillings for every ton that said ship or frigate shall be of in burthen being measured and calculated according to the accustomed rule of Shipwrights Hall the said payments to be made in manner and form following. That is to say one thousand pounds of lawful English money in hand which he acknowledges to have received and the like consideration in manner and form following that is to say four payments Viz: one fourth thereof to be paid when the lower futtocks are in and fastened, Another fourth when all the tiers of toptimbers are in place and both the wales are in, and another fourth part when all her decks are laid and the last fourth or what shall be behind and unpaid within thirty days after launching and complete finishing of all the work respecting her hull, carving, joining or painting etc fit to be done. Lastly for all such sum and sums of money that shall grow due by virtue of this contract the said Francis Bailye is content to accept of the King for payment. In witness whereof the parties above said to their part have interchangeably set their hands the day and year above written.

Henry Brouckner John Tippetts William Batten Francis Bailye Samual Pepys

Proposed Contract for Two Fourth Rates, 1673

Building contract for a proposed number of fourth rates to be built in Ireland

(TNA ADM106/3071).
18 May 1673
Contract with Sir Edward Spragge and Sir Nicholas Armorer for the building of 2 fourth rate ships in Ireland
28th May 73 Imprest £1000:00s:00d
4 March 73/4 Imprest taken in & Cancelled

This indenture made the eighteenth day of May in the year of our lord 1673 in pursuance of an order from his Royal Highness James Duke of York, Lord High Admiral of England and dated the sixth day of January 1672 between the Principal Officers and Commissioners of his Majesty's Navy for and on the behalf of his Majesty of the one part and Sir Edward Spragge and Sir Nicholas Armorer, Knights of the other part witnesseth that they the said Sir Edward Spragge and Sir Nicholas Armorer for the consideration hereafter expressed do covenant, promise and grant to and the said Principal Officers and Commissioners for and on the behalf of his Majesty that they the said Sir Edward Spragge and Sir Nicholas Armorer their executors, administrators, servants or assignees shall and will at their or some of their own proper costs and charges well and workmanlike erect and build off the stocks for the use of the King four good and substantial new ships or frigates of good and well seasoned timber and plank of Irish oak and elm.

Overall dimensions
And that each of the said ships or frigates shall contain in length by the keel one hundred and nine feet. Each in breadth from outside to outside of the plank thirty three feet each. Depth in hold from the top of the ceiling to the upper edge of the gundeck beam fifteen feet. Breadth at transom twenty one feet four inches. Rake forward of the stem fourteen feet at the harpin. Rake aft five feet to the main transom.

Keel
The keel(1) to be fifteen inches square in the midships to be sheathed with a three inch plank for a false keel(2). To have five foot scarphs tabled in the keel and to be well bolted with six bolts of an inch auger.

Stem
To have a firm substantial stem(3) of fifteen inches thwart ships and sixteen inches fore and aft with a sufficient false stem(4) of eight inches thick and two feet four inches broad with scarph four feet and six inches long, six bolts in each scarph.

Sternpost
To have a substantial stern post(5) of two feet deep and another post(6) within it to be fastened to the lower transom with a long armed knee(8) of six foot each arm to be well bolted with an inch and quarter auger bolt fastening the same together. *The knee was probably fitted on top of the rising wood in order to reach the lower transom.*

Timber and room
The space of timber and room(13) to be two feet and two inches.

Floor timbers
The floor timbers(14) of each of the said ships to be fifteen inches up and down upon the keel and eleven inches in and out at the wrongheads and to be thirteen inches fore and aft.

Naval timbers
(Sometimes known as the lower or first tier of futtocks.)
The naval timbers(15) to fill the rooms being at least twelve inches fore and aft and to have at least six feet and six inches scarph.

Middle futtocks(17)
(Sometimes known as the futtock or second tier of futtocks, they are not mentioned here but are referred to later in the contract relating to payment.)

Gundeck timbers
(Sometimes known as the third or upper tier of futtocks. The dimensions refer to the head of the upper futtocks and the heel of the toptimbers.)
The timbers upward at the gundeck(18) to be nine inches in and out and ten inches fore and aft to have the scarph of six foot and the rooms to be filled with timber.

Keelson
To have a substantial keelson(21) of three pieces fourteen inches up and down and eighteen inches broad and to run fore and aft to be well bolted with inch and quarter auger through every other timber and to bolt every other floor timber through the keel and through the stem.

Toptimbers
The toptimbers(19) to be in proportion unto the floor.

Planking in hold
To put in six strakes of sleepers(23) in hold on each side of the wrongheads four of six inches thick and the other two of seven inches thick and fifteen inches broad and to run fore and aft. To have two strakes of middlebands(24) on each side of six inches thick and fifteen inches broad and to run fore and aft, two strakes of clamps(25) on each side fore and aft under the beams of the gundeck of seven inches thick and sixteen inches broad each to be hooked one into the other to prevent reaching and all the rest of the footwaling(26) in hold to be good four inch plank. *After this contract of 1673 sleepers are referred to as footwaling and footwaling as ceiling.*

Riders
With four bends of riders in hold, three of them to be placed before the mast and the other abaft to be eighteen inches broad and sixteen inches deep, the floor riders(31) and futtock riders(32) twelve inches square, to have seven or eight bolts a piece in the floor riders and six in the futtock riders to be bolted with an inch and quarter bolt and to have seven feet scarph upwards and downwards.

Orlop
And to place six orlop beams(27) of a convenient height and to make platform upon them for the strengthening of the ship and stowing cables thereon to be fourteen inches broad and thirteen inches deep. To have two knees and a standard at each end or riders of thirteen feet long and two feet deep at the beam tailed into the beam and to be well bolted with eight bolts of inch and half quarter auger with one knee fore and aft to each end of every beam. *The visual glossary shows only one lodging knee(28) at each end of the orlop beam, which was later practice.*

Mainmast step
To have a rider or saddle(38) for the step of the mainmast of sufficient bigness.

Pillars in hold
And pillared(45) in hold under the beams of the gundeck and orlop seven inches square.

Gundeck Beams
The gundeck(42) to be fourteen inches broad and thirteen inches up and down and to be placed one beam under each port of the gundeck and one beam between each port of the gundeck excepting in the hatchway which must be seven feet asunder and to be kneed with four knees to each beam of nine inches thick Viz: two fore and aft(43) hooked into the beams and two up and down(44) and to be well bolted with five bolts to each knee of an inch and half quarter auger.

Gundeck
And to have a double tire of carlings(49) on each side fore and aft six inches thick and nine inches broad and the ledges(50) to lie within eight inches one of another and six inches broad five inches deep. The waterways(51) to be of seven inches in the chine in thickness and fourteen inches broad all the rest of the gundeck to be laid with good three inch plank(52) well seasoned and of good lengths. The three inch plank and waterways to be treenailed and spiked with two spikes in each beam and two treenails in each ledge. To put out eight leaden scuppers(64) on each side of the gundeck.

Hatchways
To make as many hatches(62) in the hatchways as shall be convenient of two inch plank with the hatchway abaft the mast for the stowing of provisions and hatchway to the Stewards room and for Boatswain's and Gunner's storerooms and powder rooms to be built of ordinary deals of such bigness and contrivance as equals any of his Majesty's ships of the like burthen.

Hawse pieces
To make a manger(63) on the same deck and to have two four inch scuppers in it of lead and to put in four hawse pieces(20) two feet six inches broad each and to cut out four hawse holes in them.

Bitts
To place two pair of carrick bitts seventeen and sixteen inches square the aftermost(53) and fourteen and thirteen inches square the foremost(54) pair with cross pieces(55) and knees(56).

Standards
To bring in six standards(69) on each side of the gundeck and to bolt them with three bolts in each arm of an inch and half quarter auger.

Breasthooks in hold
To have four breasthooks(33) in hold fourteen inches deep and fourteen foot long and seven bolts in each breasthook of an inch and half quarter auger.

Gundeck spirket
To have two strakes of spirket wales(65) on the same deck of five inch plank fore and aft to the lower edge of the port.

Gundeck gunports
To cut out eleven ports(68) on each side the same deck two foot ten inches broad and two foot eight inches deep with two abaft and to make and hang them with hooks and hinges and to fit and drive all ringbolts, eyebolts, rings and staples.

Gundeck fittings
To place partners for the main(57) and foremast(107) ten inches thick and a pillar for the main capstan(60) to be iron bound for the end of the spindle to stand in. And a step(48) for the mizzen mast upon the keelson. To have four inch plank(52) upon the gundeck between the bitts and the main partners. To raise the hatches over the deck and bolt the planks into the beams. And to have turned pillars(71) under the beams a four inch plank for the pillars to rest upon. With staircases up into the quarterdeck and stairs and ladders to all conveniences. To bring two breasthooks(33) between decks fourteen foot long fourteen inches deep, seven bolts in each hook of inch and half quarter auger.

Transoms
To have as many transoms(10, 11) abaft below the ports as may lie within twelve inches one of another and one transom(46) at the upper edge of the ports under the helm port to take hold of the sternpost with good substantial knees(37) to all the transoms with long arms with three bolts in each arm by inch auger. *Half transoms are not mentioned and were probably not used.*

Upper-deck gunports
To make a whole tire of square ports(85) on the upperdeck to be two foot six inches broad and two foot four inches deep, forecastle to be all square ports two foot broad twenty two inches deep under the quarterdeck and forecastle and to garnish them with carved work fore and aft oval manner.

Upper-deck clamps
To bring on risings(67) fore and aft of four inches thick and sixteen inches broad under the beams of the upperdeck and ceiled(70) up between decks with three inch plank fore and aft.

Upper-deck beams

The beams(72) of the said deck to be eight inches up and down and eleven inches broad and to lie between and under each port and not to exceed six feet asunder excepting in the wake of the gunroom and those beams to be two foot asunder and in the hatchway there to have carling knees and to be in height between the said deck between plank and plank six feet and seven inches in the midships. And the beam to round nine inches to go flush fore and aft all the said beams to be double kneed with four knees to each beam of six inches thick from the gunroom forward that is to say two fore and aft(73) and the other up and down(74) and the beams in the gunroom to be single kneed with two knees in each beam up and down and to be well bolted with two bolts in each arm at the beam and three bolts at the side with bolts of three quarters auger. *The beams are further apart at the main hatchway, but to strengthen the structure, lodging knees, called here carling knees, are fitted in the corners between the carlings and the beams.*

Upper deck

To have two tier of carlings on each side fore and aft one long carling(75) and the other short(76) with sufficient ledges(77). To lay the said deck(79) with good Sprutia deal of two inch and a half thick or dry two inch and half plank with a waterway(78) of four inch plank seasoned and in the wake of the guns in the quarter to lay with two inch oaken plank.

Upper-deck fittings

With a Spirketting(83) of three inch plank fore and aft. To have coamings(81), head ledges(80) with grating hatchway(88) before and abaft the mast and the grating forward for the vent of the smoke of the ordnance. To fit topsail sheet bitts(109) with catts(122), davit, clasp of iron(156), partners for the main(58) and jeer(82) capstans and partners for all the masts(160) and to put out nine scuppers(89) on each side on the upperdeck.

Capstans

To make a main(59) and a jeer(87) capstan with capstan bars and iron pawls.

Quarterdeck and forecastle

To make a large quarterdeck and a forecastle with two round bulkheads in each side of the said forecastle(96) and quarterdeck(101) abaft for cabins and in the midships of the said bulkhead to place the cookroom(97) for roasting and boiling and to set all the bulkheads upon six inch plank eight inches broad and to be well fayed to the deck and laid with tar and hair. To have two ports in the bulkhead of the forecastle and two in the bulkhead of the steerage and as many of the beams(98) in the wake of the steerage bulkhead and the beams in the forecastle(92) to be all kneed(93, 99) (because of the anchors) as shall be convenient and in the wake of the cabin to have a rising(91) under the beams of six inches thick and the beams to be dovetailed and bolted into the same.

Upper-deck ports

To cut out eleven ports(85) on the upperdeck between each port of the gundeck of the bigness afore mentioned and two ports right aft and to make and hang portlids with hooks and hinges. To have a transom(106) abaft under the windows in the cabin and one under the ports and the same to be kneed and bolted with five bolts in each knee.

Roundhouse

And to have as large a roundhouse as the works with conveniency will give leave with a bulkhead and a door to the same.

Hold

To make all platforms in hold with bulkheads and partitions Viz: for the powder room and Gunner's storeroom, sailroom, Boatswain's storeroom with cabins and conveniencies for lodgings, Carpenter's and Steward's storerooms, fish room and conveniencies for the Captain's provisions with a bread room to be sheathed with lead or tin plate it being through dried with charcoal (the leading, tinning or plating which breadroom is to be at his Majestie's charge).

Without board planking

And without board the ships to be planked(114, 115, 116) up from the keel to the chainwales with three inch plank excepting six strakes of four inch plank Viz: two below the wale(113) and two between(112) and two above the wales(113). And to have two firm wales(111) of thirteen inches up and down and seven inches thick and to have two wales(117) for the conveniency of the chain plates and bolts to go fore and aft both to be five inches thick and nine inches broad. To have one strake of three inch plank(118) between the chainwale wale ten inches broad and the work upwards so high as the waist from the upper wale to be wrought up with two inch plank(119) and the quarter with Sprutia deal of two inches thick.

Head

To have a fair head with a firm and substantial knee(123) and cheeks(124) triple railed(125) with trail board(126), beast(127) and brackets(128), keelson(129) and cross pieces(131), standards(130) and supporters under the catts(122).

Stern

To have a fair lower counter(139) with rails and brackets and open gallery with a house of office in it with carved work and brackets, windows and casements into the cabin. To have a fair upright and to put in it a complete pair of King's Arms, cherubims heads, pilasters and terms. To have a fair pair of turrets(134) with windows into the cabin complete.

Rigging attachments

To have a pair of chesstrees(140) fore, main(120) and mizzen chainwales well bolted, chain bolts and chain plates sufficient for number and size fixed complete.

Gripe

To have a rank gripe(135) well bolted and fastened with dovetailes and a stirrup on the skeg well bolted.

Rudder

To make and hang a complete rother(141) with six pair of braces(142), gudgeons(143) and pintles(144) and a muzzle for the rother head and a tiller thereto.

Rails

To rail and gunwale the said ships fore and aft.

Masts and yards

To find make and set a complete suite of masts and yards for each ship fit and suitable to such ship or frigate fitted with caps and crosstrees, flag, ensign and jackstaff.

Decoration

And likewise to do and perform all carved works and garnish them answerable to ships of the like bigness built in his Majesty's own yards.

Details

To find and provide all manner of materials requisite for building and complete finishing the said ships. Likewise to do and perform all joiner's works both within and without board finding deals, locks, iron bars, hinges for storerooms, Steward room's, settlebeds and cabins. And to plane the ships all underwater. To find all plumber's and mason's work and stuff with lead and leaden scuppers. To find do and perform all glazier's work of muscovia glass and to find and hang all casements and scuttles for cabin windows etc. To find and all painter's work and gilding both within board and without board and to do and perform whatsoever belongs to the Carpenter to do for the complete finishing the hulls in like manner as is done and performed to the like ships and frigates built in his Majesty's yards.

Materials and delivery

And the said Sir Edward Spragge and Sir Nicholas Armorer for themselves and the survivor of them their executors and administrators do covenant and grant at their own cost and charges to find and provide all manner of iron works of the best Spanish iron likewise to find and provide all timber, planks, boards and treenails (the treenails to be all mooted from the chainwale downward). They are also to find white and black oakum, pitch, tar, rozin, hair, oil, brimstone and all other materials that shall be needful to be used or spent in or about the works and premises aforesaid for the complete finishing thereof and in like manner to do discharge and pay all manner of workmanship touching all and every part of the works that are or ought to have been herein expressed to be done and performed and to finish, complete, launch and rigg the said four ships or frigates into the harbour of Waterford in the Kingdom of Ireland for the use of the King as aforesaid in manner following. That is to say two of the said ships to be completely finished, launched and rigged on or before the first day of August which shall be in the year of our Lord God 1674, one other on or before the first day of August 1675 then next ensuing and a fourth on or before the first day of August 1676. And they do further oblige themselves and the survivor of them their executors and administrators from time to time to transport at their own charge from London to Ireland what rigging, cables, sails, guns, Boatswain's, Gunner's and Carpenter's stores and all other stores or provisions which shall be provided by his Majesty for the use of the said four ships and each of them respectively before the several and respective days before limited and expressed for their launching and complete finishing at Ireland as aforesaid. The said stores being brought at his Majesty's charge to the side of such vessel or vessels as shall be provided by the said Sir

Edward Spragge and Sir Nicholas Armorer to receive the same. And further that they the said Sir Edward Spragge and Sir Nicholas Armorer shall and will deliver the said ships and every of them with all the guns and other stores aforesaid free of all charge to his Majesty (reasonable convoy excepted) as well relating to victuals and wages to mariners and all other charges or disbursements whatsoever into his Majesty's harbour at Portsmouth within three months after the launching of each of the said ships wind and weather permitting.

Quality

Provided that if at any time during the building of the said ships or frigates herein contracted for according to the dimensions, proportions and scantling herein expressed and set forth or intended to be expressed and set forth there shall be found and discovered upon due survey to be made thereon by such person or persons as shall be thereunto appointed by the Principal Officers and Commissioners of his Majesty's Navy any unsound and unsufficient timber, plank or other materials either before or after the same shall be wrought up into the ship or any insufficient workmanship and performance in any of the premises prejudicial to the King that then after due notice thereof given in writing by the said Surveyor or Surveyors unto the said Sir Edward Spragge and Sir Nicholas Armorer or to the chief master workman under them there shall be an effectual and present stop put to the employing of any such goods so excepted against and a rectifying of and reforming of all and every such default both in the stuff and workmanship so excepted against which said changings, amendments and reformations when duly complied with the said Surveyor or Surveyors shall from time to time certify to in writing to the said Principal Officers and Commissioners of the Navy to have been so complied with.

Payment

And the said Principal Officers and Commissioners of the Navy for and on the behalf of the King shall (according to the custom of the Office of the Navy) sign and make out bills to the Treasurer of the Navy to be paid to the said Sir Edward Spragge and Sir Nicholas Armorer or the survivor of them their executors, administrators or assignees for the sum of eight pounds six shillings and three pence for every ton that each of the said ships or frigates shall be of in burthen being measured and calculated according to the accustomed rule of Shipwrights Hall the said payments to be made in manner and form following. That is to say one thousand pounds of lawful money of England within twenty days after the signing of this contract and in and during the time that the two former of the said ships shall be in building the sum of five hundred pounds apiece more so soon as their floors shall be bolted. Five hundred pounds more for each when the second futtocks shall be all in the said ships. Five hundred pounds more for each when the orlop beams shall be all fast and footwaling in. Five hundred pounds more for each when their gundeck beams are all fast and kneed. Five hundred pounds more for each when their upperdeck beams are fast and kneed. Five hundred pounds more for each when their quarterdecks and forecastle beams are all fast and kneed and five hundred pounds more for each when the said two ships are completed and launched and the remainder which shall be due for those two ships when they shall be delivered as aforesaid at Portsmouth unto the Master Attendant for the use of his Majesty. And in and during the time that the third of the said ships shall be in building a bill shall be made out for payment of the sum of five hundred pounds on the last day of August which shall be in the year of our Lord God 1674 and from thenceforth the like bills for payment shall be made out in the same manner and form in all respects as for either of the two former ships is before expressed and in and during the time that the fourth or last ship herein contracted for shall be in building a bill shall be made out for payment of the sum of five hundred pounds on the last day of August which shall be in the year of our Lord God 1675 and from thenceforth the like bills for payment shall be made out in the same manner and form in all respects as shall be for either of the aforesaid ships. And lastly for all and singular such sum and sums of money that shall grow due by virtue of this contract the said Sir Edward Spragge and Sir Nicholas Armorer are content to accept of bills on the Treasurer of his Majesty's navy for payment. In witness whereof the parties aforesaid to these pnte indentures have interchangeably set their hands and sealed the day and year first above written.

Sealed and delivered by Sir Edward Spragge in the presence of
E Spragge Tho Billopp The Ranou Nic Armorer John Stanley
Sealed and delivered by Sir Nicolas Armorer
In the presence of Wm Hewer Tho Lewsley

The *Mary Galley* Contract, 1686

Building contract for the fourth-rate Mary Galley, *built by John Deane at Blackwall and launched in 1687 (TNA ADM106/3070). The ship measured 104 feet 0 inches keel length, 29 feet 6 inches breadth and 11 feet 0 inches depth in hold.*

31 July 1686

Contract between the Principal Officers and Commissioners of his Majesty's Navy and Mr John Deane for building a new ship or frigate answerable to the *James Galley* by 20 April next at the rate of £7:18:0 per ton.

Blackwall

This indenture made the one and thirtieth day of July in the year of our lord 1686 between the Principal Officers and Commissioners for managing the affairs of his Majesty's Navy for and on the behalf of his Majesty on the one part and Mr John Deane of the other part witnesseth that the said John Deane for the consideration hereafter expressed doth covenant, promise and grant to and with the said Principal Officers and Commissioners for and on the behalf of his Majesty that he the said John Deane his executors, administrators, servants or assignees shall and will at his and their own proper costs and charges well and workmanlike erect and build off the stocks for the use of the King one good and substantial new ship or frigate of good and well seasoned timber and plank.

Overall dimensions

And that the said ship or frigate shall contain in length by the keel one hundred and four feet, breadth from outside to outside of the plank twenty nine feet, depth in hold from the top of the ceiling to the upper edge of the gundeck eleven feet. Breadth of the transom fifteen feet. Rake forward of the stem seven feet at the harpin. Rake aft four feet to the main transom.

Keel

The keel(1) to be twelve inches square in the midships to be sheathed with a three inch plank for a false keel(2). To have three feet six inches scarph to be laid [tabled] in the keel and to be well bolted with six bolts of an inch auger.

Stem

To have a firm substantial stem(3) of twelve inches thwart ships and thirteen inches fore and aft with a sufficient false stem(4) of six inches thick and one foot six inches broad with a scarph of three foot two inches long, six bolts in each scarph of inch auger.

Sternpost

To have a substantial stern post(5) of two feet deep with a long armed knee(8) of four feet six inches long each arm to be well bolted with an inch and one eighth auger bolt fastening the same together. *The inner false post is not mentioned, although it was a standard fitting at this time. It is probable this contract is based on an old version, as the wording is the same as that in the 1649 contract.*

Timber and room

The space of timber and room(13) to be two feet and two inches.

Floor timbers

The floor timber (14)of the said ship to be ten inches fore and aft, up and down upon the keel twelve inches in and out at the wrongheads to be nine inches.

Lower futtocks

The futtocks(15) fore and aft nine inches except five floor timbers in the midships which are to be ten inches fore and aft and to have five foot scarph.

Middle futtocks(17)

They are not mentioned, but would be defined by the dimensions of the floor timbers and the upper futtocks. It is possible that this small ship had only two tiers of futtocks.

Upper futtocks

The timbers(18) upward at the orlop eight inches sided and six inches in and out and to have the scarph five feet.

Keelson
To have a substantial keelson(21) of five pieces twelve inches up and down and one foot one inch broad and to run fore and aft, to be well bolted with an inch and one eighth auger through every other timber and to bolt every other floor timber through the keel or through the stem.

Toptimbers
The toptimbers(19) to be in proportion unto the floor timber.

Planking in hold
To put in five strakes of foot waling(23), two of five inches thick and three of six inches thick fourteen inches broad on each side the wrongheads. To put two strakes of middlebands(24) on each side of four inches thick and fourteen inches broad to run fore and aft, and all the rest of the ceiling(26) in hold to be good three inch plank.

Riders
With three floor riders(31) in hold and three pair of futtock riders(32), two of them to be placed afore the mast and the other abaft, broad one foot three inches and deep twelve inches, the floor riders and futtock riders ten inches square, to have seven bolts a piece in the floor riders and six bolts in the futtock riders to be bolted with inch 1/8 auger and to have five feet scarph upwards and downwards.

Mainmast step
To have a saddle (38) for the step of the mainmast of sufficient bigness and pillars(45) in the hold under the beams of the lower deck six inches squared.

Lower-deck beams
The lower deck beams(42) to be twelve inches broad and twelve inches up and down. To have a rider and knee(43) at each end of each beam, the rider to be ten feet long and eight inches thick and thirteen inches deep at the beams and to be well bolted with six bolts to each rider of inch and 1/8 auger. The beams to lie under every port and to be no more than eight feet asunder except in the hatchway.

Breasthooks in hold
To have three breasthooks(33) in hold twelve inches deep and ten feet long and seven bolts in each breasthook of inch 1/8 auger.

Lower-deck spirket
To have two strakes of spirket wales(65) where the lower deck lies of six inches oak plank to be hooked one into the other fore and aft up to the lower edge of the scuttles for oars.

Rowing scuttles
To cut out eighteen scuttles(66) on each side of the same deck fifteen inches broad and twenty four inches deep. The sills of each of the said scuttles to be two feet and eight inches from the water at the ships full draught.

Partners
To place partners for the main(57) and foremast(107) five inches thick and a pillar(60) for the main capstan to be iron bound for the end of the spindle to stand in. And a step(48) for the mizzen mast upon the keelson. To have quarter deck stairs and ladders to all conveniences.

Between-deck bresthooks
To bring two breasthooks(33) between the decks of nine feet long each and twelve inches deep and seven bolts in each hook of inch auger.

Transoms
To have as many transoms(10, 11) abaft below the ports as may lie within twenty inches one of another and one transom(46) at the upper edge of the ports under the helm port to take hold of the sternpost with good substantial knees(37) to all the transoms with long arms with three bolts in each arm by an inch auger.

Upper-deck gunports
To make twelve square ports(85) on the upperdeck and to be twenty six inches broad and twenty four inches deep and garnish them with carved work fore and aft.

Upper-deck clamps
And to bring on risings(67) fore and aft of oak plank four inches thick and fourteen inches broad under the beams of the upperdeck and ceiled(70) up between decks with two inch plank fore and aft.

Upper-deck beams
The beams(72) of the said deck to be eight inches up and down and eight and a half inches broad and to lie one between and one under each port and not to exceed five feet asunder excepting in the wake of the gunroom and those beams to be four foot asunder and to be in height on the said lowerdeck between plank and plank seven feet in the midships, and the beams to round seven inches, the said deck to go flush fore and aft. All the said beams to be well kneed with one knee(74) to each end except two beams next to each mast and at the bitts(53, 54) which beams are to be double kneed at each end with knees of five inches and half thick from the gunroom forward and the beams in the gunroom to be kneed with two knees(74) up and down to each beam and to be well bolted with five bolts in each knee of inch auger. *This lightly built galley frigate was seven feet in height between decks, an average of more than six inches higher than in the previous contracts. This was to allow for the easier use of oars.*

Upper deck
To have one tier of carlings(75) on each side fore and aft with sufficient ledges(77). To lay the said deck with good well seasoned oak plank(79) at least three strakes within the waterways(78) and the rest with good Sprutia deal of two inches thick or dry plank of like size with waterways of three inch oak plank. *Note: there is only one tier of upper-deck carlings each side at the hatchway in this ship.*

Upper-deck fittings
With a Spirketting(83) of three inch oak plank fore and aft. To have coamings(81), head ledges(80) with gratings and hatchways(88) before and abaft the mast and gratings to vent the smoke of the ordnance. To fit topsail sheet bitts(109), catts(122), davit with a clasp of iron(156) to it, partners for the main(58) and jeer(82) capstans and partners for all the masts(160) and to put out eight leaden scuppers(89) on each side of the upperdeck.

Capstans
To make a main(59) and jeer(87) capstan and a set of capstan bars and iron pawls to each.

Hold
To make all platforms in the hold with bulkheads and partitions Viz: for the powder room and Gunner's storeroom, sailroom and Boatswain storeroom Carpenter's and Steward's storeroom, fishroom and conveniences for the Captain's provisions and bread room.

Without board planking
Without board, the ship to be planked(114, 115, 116) up from the keel to the chainwale with three inch good oak plank well seasoned excepting three strakes of four inch plank Viz: one below the wales(113) and one between(112) and above(113) the wales to be of good well seasoned English oak plank. And to have two firm wales(111) of ten inches up and down and seven inches thick and to have two wales(117) for the conveniency of chain plates and bolts to go fore and aft to be four inches thick and eight inches broad. To have one strake of three inch oak plank(118) between the chainwales nine inches broad and the work upward so high as the waist to be wrought up to the top of the side with two inch oak plank(119) and the quarters with Prutia deal inch and half thick.

Head
To have a fair head with a firm substantial knee(123) and cheeks(124) treble railed(125) with carved trail board(126), beast(127) and brackets(128) a keelson(129) and cross pieces(131) and standards(130), catts(122) and supporters under them.

Stern
To have a fair lower counter(139) with rails and carved brackets, windows and casements into the cabin. To have a fair upright and put into it a complete pair of King's Arms, or other ornaments with complete pilasters and terms. To have a fair pair of turrets(134) with carved works and windows into the cabins complete.

Rigging attachments
To have a pair of chesstrees(140) and main chainwales(120) well bolted and to place chain bolts and chain plates sufficient.

Gripe
To have a gripe(135) fixed complete well bolted and with dovetailes and stirrup upon the skeg well bolted.

Rudder
To make and hang a complete rudder(141) with five pair of braces(142) and gudgeons(143) and pintles(144) with a muzzle for the head and a tiller thereto.

CONTEMPORARY SHIPBUILDING CONTRACTS UNVEILED

Rails
To rail and gunwale the said ship fore and aft and to put on carved brackets and garnish hancing pieces complete.

Masts and yards
To find and make a complete suite of masts suitable to such a frigate with caps and crosstrees, yards, bowsprit, topgallant masts and ensign staff. All the said masts and yards to be ready made fit for ships of that burthen.

Decoration
To do and perform all the carved and painted work answerable to his Majesty's ship the James Galley.

Details
And to furnish and provide all the materials for the same likewise all joiner's work finding all deals, locks, iron bars, hinges for storerooms, Steward rooms, doors and cabins with workmanship thereunto both within board and without. And to plane the ship all underwater, all plumber's work, lead and leaden scuppers and all glazier's work the same to be done of slate glass both for casements and scuttles for cabin windows. And also to provide the casements and to do and perform whatsoever belongs to the Carpenter to do for the finishing and completing the hull in the like manner as is done and performed on his Majesty's ship the James Galley before named.

Materials and delivery
And the said John Deane for himself his executors and administrators doth covenant and grant at his and their proper costs and charges to find and provide all manner of iron work of the best Spanish iron, spikes, nails, bradds, likewise all timber, planks, boards, treenails which are to be all mooted from the chainwale downward. White and black oakum, pitch, tar, rozin, hair, oil, brimstone, masts and all other materials that shall be material to be used or spent in or about the work and premises aforesaid for the complete finishing thereof and in like manner to pay and discharge all manner of workmanship touching all and every part of the work herein expressed to be done and performed and to finish and complete and launch the said ship or frigate in the river of Thames and there deliver her safe afloat for the use of his Majesty as aforesaid by the nine and twentieth day of April next ensuing the date hereof.

Quality
And it is further agreed that if at any time during the building the said ship or frigate herein contracted for according to the dimensions, proportions and scantling herein expressed and set forth or intended to be expressed and set forth there shall be found upon due survey to be made thereon by such persons as shall be thereunto appointed, any unsound and insufficient timber, plank or other materials or any insufficient workmanship and performance prejudicial to the King that then after due notice thereof given in writing by the said Surveyor unto the said John Deane or the chief master workman under him there shall be an effectual and speedy reformtion of all and every such default in stuff or workmanship and the said amendment or reformation shall be certified in writing by the said Surveyor to the Principal Officers and Commissioners of the Navy for the service of the King in this behalf.

Payment
And the said Principal Officers and Commissioners of the Navy for and on the behalf of the King shall according to the custom of the Office of the Navy issue and make out bills to the Treasurer of the Navy to be paid unto the said John Deane his executors, administrators or assignees for the sum of seven pounds eighteen shillings for every ton that the said ship or frigate shall be in burthen being measured and calculated according to the accustomed rules of the Shipwrights Hall excepting as is hereafter excepted about the allowance for the rake. The said payments to be made out of such money as shall be appointed for that use in manner and form following. That is to say six hundred pounds of lawful money of England at signing the contract, six hundred pounds more at laying the keelson, six hundred pounds when the upper futtocks are up, five hundred pounds at laying the orlop beams, five hundred pounds when the upper deck is laid, three hundred pounds when ready to launch, the rest when launched and completely finished in the river of Thames for the use of his Majesty. It is further agreed that if the said ship shall be exceeded of her dimensions or scantlings contrary to what is herein before agreed that no satisfaction or allowance shall be made for such other work or increase of scantling unless the said increase of work and scantling shall have been made by order first given therein by the said Principal Officers and Commissioners of the Navy. It is also agreed that her rake fore and aft shall be allowed in proportion to the rake of the Kingfisher to her other dimensions. In witness whereof the parties aforesaid to these present indentures have interchangeably set their hands and seals the day and year above written.

Memorandum
It is further agreed before the signing and sealing hereof that the with instructions Mr Deane shall at his own cost and charges build two platforms for the convenient stowing of cables.

John Deane
Signed and sealed and delivered in
Vide The board letter 10th September 86 to
the presence of Cha GravesMr Furzer for increase of scantlings.
Jn Peeter Viz: 6 inches broader in midships
3 inches deeper in hold
3 inches lower between decks
Bill past for £3972 13th April 87
Samuel Pepys

The *Norwich* Contract, 1692

The contract for the fourth-rate Norwich (TNA ADM106/3071) is much more detailed than those written earlier.
When launched in 1693, the ship measured 123 feet 8 inches on the gundeck, 101 feet 6 inches keel length, 33 feet 10 inches broad and 13 feet 6½ inches depth in hold.

Norwich
Contract 5th August 1692 with Mr Robert and John Castle for building one new 4th rate for their Majesties.
To be launched in ten months.
£9:10s:0p per ton
This indenture made the fifth day of August in the year of our lord one thousand six hundred and ninety two between the Principal officers and Commissioners of their Majesty's Navy for and on behalf of their Majesties of the one part and Mr Robert & Mr John Castle of Deptford Shipwrights of the other part witnesseth that they the said Robert & John Castle for the considerations hereafter mentioned do covenant, bargain, promise and grant to and with the said Principal officers and Commissioners of their Majesties Navy that he the said Robert & John Castle their executors, administrators, servants and assignees shall and will at their own proper costs & charges well & workmanlike erect and build at their yard at Deptford for the use of their Majesties one good and substantial new ship or frigate to be of the Fourth Rate and wrought with good and well seasoned timber & plank of English oak and elm (excepting only East country sound white crown plank for the outsides under water)

Overall dimensions
And the said ship or frigate shall contain in length upon the gundeck between the rabbets of the main stem & main post one hundred and twenty three foot. The breadth to the outsides of the plank in the midships to be thirty three foot six inches. And not to be accounted more for the calculating the ships burthen. Depth in hold from the upper edge of the ceiling to the upper edge of the gundeck beam thirteen foot six inches. The rake of stern post to the upper edge of the wing transom to be four foot six inches. And the rake forward to the lower harpin to be left to the discretion of the builder saving that it shall not exceed the proportion of three fifth parts of the main breadth not be less than half thereof. And the length of the keel for the computation of the ships burthen shall be accounted one hundred foot and no more. The length of the wing transom to be twenty foot within the plank. The height between decks at the side to be six foot four inches and the depth of the waist from the deck to the upper edge of the plankshere to be four foot and eight inches. And the scantlings, scarphings, bolts, fastenings workmanship complete and entire finishing of the hull or body of the said ship shall be as followeth (Vizt)

Keel
The keel (1) to be of elm (not more than in three pieces and to [be] fourteen inches square in the midships with scarphs four foot four inches long at least and each scarph tabled and laid with tar and hair to be well bolted with six bolts by an inch auger. The false keel(2) to be three inches thick of elm laid with tar and hair and well fastened with treenails, spikes and staples.

283

Stem
The stem(3) to be of good sound oak timber thirteen inches thwartships and fifteen inches fore and aft. The scarphs not to be less than four foot & bolted with six bolts by an inch auger. The false stem(4) to be twenty two inches broad and eight inches thick and one foot at least in the scarph well bolted with inch bolts into the stem.

Sternpost
The stern post to(5) be sixteen inches thwart and eighteen inches fore and aft in the head and two foot fore and aft upon the keel. The false post within(6) the stern post to be fifteen inches fore and aft and eighteen inches thwartships.

Rising wood
To bring on upon the keel two substantial pieces of rising wood(7) of sufficient length for the run of the ship and then to bring on a well grown substantial knee(8) upon the same, faid to the stern post. The up and down arm to reach the lower transom and the arm fore and aft to be twelve foot long at least. And to be bolted by bolts by an inch and eighth auger through the keel and through the post at every twenty inches distance and to fix other rising wood(7) upon the said knee of sufficient height and substance for the support of the half timbers(9) which are to be of good lengths and to be well let into the same and fastened. To bolt the upper rising wood(7) to the lower in the same manner and with the same size of bolts as aforementioned.

Transoms
The main transom(10) to be twelve inches and the lower transoms(11) ten inches thick and to lie within eighteen inches one from another and to fit the space between them on each side the post with a sufficient number of half transoms(12) of such substance as that their upper and lower edges may lie two inches at least clear of the whole transoms to give space for air.

Timber and room
The Ranou Nic Armorer*Floor timbers*
And the floor timbers (14) to be thirteen inches fore and aft and fourteen inches up and down upon the keel and of such length in the floor of the ship that their heads may lie eighteen inches at least above the bearing thereof by a straight line from the bottom of the keel and to be eleven inches (moulded) in and out at the wrongheads or ten inches wrought in the length of the floor and before and abaft the same to be nine inches wrought in and out and every other floor timber to be bolted through the keel with one bolt by an inch and eighth auger.

Lower futtocks
The lower futtocks(15) to fill the rooms full and workmanlike for the strength of the frame below and to be twelve foot long at least and to be chocked(16) down within twenty inches of the keel.

Middle futtocks(17)
These are not mentioned here but are in the specification for the ceiling.

Upper futtocks
The upper futtock(18) head to take the lower sill of the ports in the midships and the heel of the toptimbers, the lower edge of the gundeck clamps within board and wale without board to be ten inches fore and aft and eight inches in and out at the breadth and to stand an inch and half apart for air

Toptimbers
And the toptimbers(19) to be three inches in and out aloft. And the whole frame to be well chocked with good sound timber. *The chocks referred to are the triangular chocks at the head and heels of the timbers that allowed straighter timber to be used in making the futtocks.*

Hawse pieces
To place two substantial hawse pieces(20) on each side the stem to be twenty two inches broad each piece at least.

Keelson
The keelson(21) not to have more than four pieces and to run fore and aft and to overlaunch the scarphs of the keel to be fourteen inches up and down and fifteen inches broad scored down upon the timbers an inch and half at least. The scarphs to be four foot long at least and to be bolted through every other timber and through the keel by bolts by an inch and eighth auger.

Limber board
The limber board(22) to be two inches and a half thick and fourteen inches broad on each side the keelson.

Planking in hold
To bring on six strakes of foot waling(23) on each side at the wrongheads to be fifteen inches broad each of them, two to be six inches and two to be five inches and two four inches thick to run fore and aft and to be wrought narrow in due proportion each way the six inch stuff to lie upon the chocking of the floor heads and middle futtocks and the other strakes of five and four inches thick one of each to be brought on above and below the same. To bring on two strakes of middlebands(24) on each side fourteen inches broad to be five inches thick and two strakes of gundeck clamps(25) six inches thick and fourteen inches broad to be tabled one into the other and scarphed Flemish hook and butt three foot and six inches long at least to have an opening fore and aft below the clamps ten inches broad for airing the frame and all the rest of the ceiling in hold to be good oak plank(26) of three inches thick.

Orlop
To place five orlop beams(27) in hold twelve inches broad and ten inches deep, one abaft the mast, two at the main hatchway, one at the bulkheads of the cockpit abaft the other at the bulkheads of the boatswain's storeroom forward. To lie from the lower edge of the gundeck beams five feet, to be kneed with one good fore and aft knee(28) to be three foot long at each arm at least and bolted with five bolts by an inch and one eighth auger and to place other false beams(29) and carlings(30) in hold convenient for ye platforming and storerooms to be made thereon.

Riders
To fay and fasten three bends of floor(31) and lower futtock riders(32). The floor riders to be fifteen inches fore and aft eleven inches up and down upon the keel and eleven inches at the wrongheads. To be twenty foot long at least and bolted with seven bolts in each arm by an inch and eighth auger the futtock riders to be fourteen inches fore and aft and eleven inches thick and to be fifteen foot long each and bolted with nine bolts of the same size.

Breasthooks in hold
The bresthooks(33) forward to be eleven inches deep and to lie one under the hawses, one at the deck and three more in hold to be as long as possible but none less than seven foot in the arm. To be bolted with six bolts in each arm by an inch and eighth auger.

Crotches and standard
To end the keelson abaft with a good standard(34) well bolted into the post and keelson and to fay two crotches(35) in the breadroom eleven inches sided with arms each way five or six foot long and bolted with bolts by an inch and eighth auger, eighteen inches asunder to overlaunch the thick stuff and ceiling in hold over the fashion pieces(36) abaft and to meet upon the middle of the false stem afore and to fay all the breasthooks, crotches and knees upon the same in good workman like manner.

Transom knees
And to fay on each side one large knee(37) ten inches sided & ten foot long at ye main transom & knees of proper scantling at all ye rest of the transoms and bolted as aforesaid.

Mast saddles
To fix saddles(38, 47, 48) in their proper places for all the masts and of fit substances and fastenings.

Hold
And to complete build and part off with ordinary deals all the necessary platforms and store rooms in hold. To plaster with lime and hair and line with slit deals all bulkheads and places where powder comes. To line the breadroom with good dry deal. To build a well(39) with two inch oak plank and two convenient shot lockers(40) about the mast and to furnish all cabins and store rooms with doors, locks, shelves and other partitions in every respect as the same is done in others of their Majesties ships of ye like quality or shall hereafter be set off and directed as appointed to be done by the Surveyor of their Majesties Navy.

Pumps
To place one pair of chain pumps(41) in hold and to fix them with cisterns, dales and scuppers through the sides.

Gundeck beams

The beams of the gundeck(42) to be fourteen inches fore and aft and twelve inches deep to lie one under each port and another between them fore and aft saving in the wake of the main hatchway where they must lie seven foot asunder. Every beam of the deck shall be tailed into the clamps and double kneed at each arm by one lodging(43) and one hanging(44) knee eight inches and a half sided and the hanging knees of the gundeck in the wake of the orlop beams shall be of length sufficient to reach below the said beams and to be bolted one to another with one fore and aft bolt by an inch auger and all the rest of the knees not to be less than four foot and six inches in the shortest arm and to be bolted with bolts driven after an inch auger at every fifteen inches space that each arm shall be in length or with the number of bolts proportionally.

Pillars in hold

To place under each beam in hold and upon the keelson one pillar(45) to be seven inches square.

Helm port and main transom and knees

To place a transom under the helm port(46) to be nine inches thick and of sufficient breadth to take hold of the stern post to be kneed at each arm with a knee(37) sided eight inches the fore and aft arm to be ten foot long and the short arm five foot at least. And one other pair of knees to be nine inches sided to be ten foot long in the fore and aft arm and five foot at least in the shorter to be brought on upon the spirketting against the main transom. Each pair of knees to be bolted with bolts of an inch auger at every eighteen inches space each arm is in length or with the number of bolts proportionally as aforesaid.

Gundeck

The carlings(49) of the gundeck to be eight inches broad and seven inches deep. To have two tier on each side fore and aft. The ledges(50) to be five inches broad and four and half deep and to lie within nine inches one of another. And the waterways(51) to be five inches thick and fourteen broad and the flat of the deck to be laid with oak plank(52) full three inches wrought, treenailed and spiked with two spikes in each arm.

Bitts

To place two pair of substantial bitt pins the aftermost pair(53) to be sixteen inches and the foremost pair(54) to be fifteen inches square in the head with cross pieces(55) to each pair of the same dimensions. To step them in hold and to place them abaft the beams to be stepped with a score of an inch deep into the same and bolted with two bolts in each of an inch auger. To score the cross pieces into the bitts about two inches and to brace them together with four pair of substantial iron hooks and eyes to fix two pair of good standards(56) upon four substantial carlings placed between beam and beam in the wake of them the fore and aft arm to be as long as the bitt pins are in distance and bolt them with five bolts in each standard by an inch and eighth auger.

Gundeck partners

To place partners for the main(57) and fore(107) masts eight inches thick and for the main capstan(58) six inches thick and to cut out holes there for the mast.

Main capstan

To make and set one main capstan(59) drum fashioned to be twenty three inches diameter in the barrel with a suit of good ashen bars, chains and pins and iron pawls, sole and iron hoop in the heel and ribs and hoop in the partners. To place a pair of false partners(60) under the beams of three inches thick to keep them from rising.

Gundeck fittings

To make all the hatchways and scuttles that are convenient and needful to pass and repass to the several apartments in hold. To raise them with head ledges(61) two inches above the deck at least. The hatches(62) to the main hatchway to be in as many parts as convenient to be made with two inch oak plank and fitted with the needful hatchings and eyes capable to hold the weight of them. To place rails and banisters about the Stewards room hatchway for the more convenient going up and down to the same. To build a manger(63) forward on with three inch plank of sufficient compass about the hawses. To put out four scuppers, (64) two of four inches (at ten pounds to the foot square) and two of three inches diameter (made of cast lead at eight pounds to the foot square) The like three inch scuppers for the deck fore and aft as many as are sufficient.

Gundeck spirket

The spirketting(65) to be of oak four inches thick and fourteen inches broad and tabled in the meeting edges two inches one into ye other and scarphed hook and butt Flemish fashion at least four foot long.

Upper-deck clamps

To bring on two strakes of upper deck clamps(67) under the upperdeck beams four inches thick and twelve inches broad, or to be single clamps so much in depth as the work of the ports will admit of, tabled and wrought with a Flemish hook and butt four foot long at least.

Gundeck gunports

To space and cut out eleven ports(68) of equal distance one from another from the first to the last port on each side to be two feet four inches fore and aft & two foot one inch up and down.

Standards

To place five pair of good standards(69) upon shovels (soles) of three inch plank to be twelve inches sided the up and down arm to take hold of the upper clamps, two pair to stand a little abaft the main mast against the main chains and two other pair in the wake of the fore chains and another in the bulkhead of the gunroom to be bolted with bolts of an inch and eighth auger.

Gundeck ceiling

And to birth up the side within board in the wake of ye ports with good three inch oak plank(70). To drive all the needful ring and eye bolts by ye ports and all the ring and swivel bolts upon the flat of the deck for handling the guns and for stoppers and iron hooks into the beams for the port ropes.

Turned pillars

To place two tier of turned pillars(71) in such places as are convenient under the beams fore and aft six inches square.

Upper-deck beams

The upper deck beams(72) to be ten inches broad and nine inches deep and to be placed over every beam of the lower deck except in the wake of the cookroom(97) where they shall lie two foot asunder and in the wake of the great cabin where they shall be eight inches broad and seven inches deep and to lie two foot asunder. And the two beams of the upper deck over the main capstan to be placed at convenient distance for the taking up and down of the same, all the beams to be double kneed at each arm with one lodging(73) and one hanging knee(74) seven inches sided where the distance will permit it and where not one hanging knee only and every hanging knee fore and aft to take hold four inches upon the spirketting at least. To have one good spike through the end thereof into the same and to be well bolted with five bolts in each by seven eighths of an inch auger. To have a tire of long fir carlings(75) fore and aft next the hatchways twelve inches broad and eight inches deep set down upon the beams and the short carlings(76) next the side to be five inches deep and seven broad. The ledges(77) to be three inches deep and four broad and to lie eight inches asunder. The waterway(78) to be four inch oak plank and under the bulkheads and wake of the guns on each side fore and aft and one strake next the coaming carling where iron work is driven to be all two inch oak plank(79) the rest of the plank of the flat of the deck to be oak plank or Prussia deals of like thickness. *No separate long carling and coaming are mentioned in this contract as they are in the earlier versions. It says the long carling is set down upon the beams so that the upper side could act as the coaming. It is also referred to here as the 'coming carling'. This arrangement would help stop retching, where the ship's ends bend downward.*

Upper-deck partners

The partners of the mast(160) and Jeer capstan(82) to be six inches thick.

Upper-deck spirketting

The spirketting(83) to be fourteen inches broad and three inches thick and the string(84) over the ports in the waist to be four inch oak plank and eight or nine inches broad.

Upper-deck gunports

To cut out on each side eleven ports(85) to be two foot fore and aft & one foot ten inches up and down and birth up in the wake of the ports fore and aft with two inch Prussia deals(86). To drive all the needful ring and eye bolts at the ports and upon the flat of ye deck for the service of the guns.

Upper-deck capstans

To make & set one drumhead Jeer capstan(87) to be twenty inches diameter in the barrel to fit it with a set of new pins and chains and pawls of iron, a sole and hoop in the heel and partners with a good set of ashen bars.

Upper-deck hatches
To make all the needful grating hatches(88) fore and aft to be three inches in the clear between ledge and batten.

Upper-deck scuppers
And to put out a convenient number of cast lead scuppers(89) three inches diameter.

Clamps of the quarterdeck and forecastle
The clamps(90) or risings of the quarterdeck and forecastle to be three inches thick and fifteen inches broad of oak and the rising(91) in the great cabin into which the beams shall be tailed and bolted to be of elm fourteen inches broad and six inches thick at the upper edge and two inches at the lower edge thereof. To be well bolted through and through the side by bolts of three quarters of an inch auger at every twenty inches distance and the same to be well clenched within board.

Forecastle beams
The foremost beam(92) of the forecastle to be eight inches fore and aft and six inches up and down, the rest to be six inches broad and five inches up and down and to lie two foot asunder and every other beam to be kneed with one hanging knee(93) to be four inches and half sided, the arm to be four foot and a half long at least and to be well bolted with five bolts by a three quarters inch auger.

Forecastle
To place a transom(94) under the ports in the fore peak six inches deep and eight inches in and out to be kneed(95) at each end with a fore and aft knee sided to five inches & bolted with five bolts of a three quarter inch auger. To place stanchions in the bulkhead(96) of the forecastle six inches broad and four inches thick and to birth up the same with two inch Prussia deals rabbeted one edge upon another. To part off a space for a cookroom(97) and enclose it with three inch oak plank and to line the same with double white plates fit for the bricklayers work putting a thread of sish tar and hair in the seams of the deck and ledding the same in the wake thereof.

Quarterdeck beams
The quarterdeck beams(98) to be six inches broad and five deep to lie two foot asunder, every other beam to be kneed with one hanging knee(99) at each end to be four inches and a half sided and four foot and a half long in the lower arm. To be well bolted with five bolts by a three quarters inch auger.

Quarterdeck
To place a bittacle about the mizzen mast. To part off the steeridge and great cabin by two bulkheads and to place three pair of small standards upon the upper deck one pair in the aftermost bulkhead(96) of the forecastle another in the bulkhead of the steerage(101) and the other in the bulkhead of the great cabin(100) to be six inches sided and to be bolted with six bolts by a three quarters inch auger. The waterway upon the forecastle and quarterdeck to be two inches and a half oak plank and the rest upon the flat thereof to be two inches Prussia deal. The spirketting(103) to be oak plank of two inches thick and the rest of the quickwork(104) to be birthed up with inch and half Prussia deals.

Roundhouse
To build a roundhouse abaft upon the quarterdeck for the Lieutenant and Master

Great cabin
& under the tier of lights baft in the great cabin to prick in a transom(106) between the timbers to be six inches deep and twelve inches fore and aft and to knee the same with a substantial long armed iron knee well bolted and the like transom and ye same fastenings above the lights for the ending and fastening the plank of the quarterdeck.

Finishing within board
To place topsail sheet bitts(109) and jeer bitts(110) and cross pieces about each mast. To make all ladders, gangways, gratings, ranges(159), kevills (158) etc and to complete the ship entirely within board. *This contract is the first to mention jeer bitts.*

Wales
The lower wales(111) without board to be twelve inches broad and seven inches in and out. To have a strake(112) of fourteen inches broad and four inches thick between them and one other strake(113) above them and one other below them of the same breadth and thickness.

Outboard planking and channels
And to work out to the wrongheads in the wake of the bilge with four inch oak plank(114) the rest underwater to be birthed up with full three inch oak plank(115) when tis wrought upon the ships side and from the strake above the wales to the channel wales & between them to be three inch oak plank(116) and between to be three inch oak plank(118). The channel wales(117) to be ten inches broad and four inch and a half thick wrought and to chock the timbers with oak in the wake of the main and fore chain bolts the gunnel strake to be twelve inches broad and four inches thick and the quickwork(119) from the channels to the top of the side to be wrought up with two inch Prussia deal the main and fore channels(120) to be four inch and a half thick and of sufficient lengths and breadths. The mizzen channels of three inches thick well bolted with a three quarter inch auger bolts. To place a pair of catts(122) forward to be twelve inches square at the head. To strengthen the ships frame where the chainwales fit, chocks are to be fitted in the gaps between the toptimbers.

Port lids
To make and hang all the ports on each side upon each deck and in each bulkhead with the same size of plank ye side is wrought with to line the same with elm board well nailed.

Head
To build a fair head with a firm and substantial knee(123) and cheeks(124), treble rails(125), trail board(126), beast(127), brackets(128), keelson(129) and standards(130) and to finish the same with gratings, stools of easement for the sailors with a dead block(132) carved for the tack between the rails and with mortise(133) in the knee for the gammoning, washboards under the cheeks.

Galleries and rails
To build a fair pair of galleries(134) and to finish the same and stern abaft with high and handsome carved works. To gunwale(136) rail and planksheer(137) ye ship fore and aft and to finish the same abaft with a handsome fife rail(138) and with a breast rail at the fore part of ye quarterdeck.

Steering
To make and hang a substantial rother(141) with strong well wrought iron braces(142), gudgeons(143) and pintels(144) to be placed three foot distance one from another. To strap the head with iron hoops and braces(145) and to fix a tiller(146) therein with a gooseneck(147) and whipstaff(148) an elm cleat(149) well bolted to swing upon a sweep(150) of three inch oak plank bolted under the beams in the gunroom for that purpose.

Lights
To fit the ship with lights sash fashion to be glazed with stone ground glass twelve by ten inches square.

Caulking and graving
To caulk the ship all over within board and without and in every seam of five inch plank to drive six double threads of ochum and two of hair in every seam of four inch five double threads of ochum and two of hair in every seam of three inch four double threads of ochum and one of hair in every seam of two inch two double threads of ochum only and to grave the ship under water with white stuff but if rozin can't be had then with black and to tallow upon it three or four strakes under water and the sides from the wales upward to be paid with rozin and tallow.

Conditions
An whereof several disputes do frequently arise between the said Principal officers and Commissioners and their contractors touching the complete and exact finishing of the ships they build for their Majesties service by claiming an overcharge for works they pretend are not within the reach of their contracts whereby is brought on their Majesties service exceeding disappointments in that the said contractors do leave many light and small matters undone that are more of charge in the delay and waste of time at the ships fitting to sea than of weight or expense in themselves and that this practice also doth render the first estimate of such works (which are prepared and presented to the Right Honourables the Lords of their Majesties Treasury for the usual supplies) sometimes very uncertain. It is therefore further concluded and agreed by and between the said Principal officers and Commissioners and the said Robert & John Castle their executors, administrators, servants and assignees to build, fit and completely finish every work matter, contrivance, convenience or accommodation (to be fixed to the hull) fit either for the company that sail in her or requisite for the conduct and guidance of the ship in the sea in such manner and in such quality as ships of the like kind are finished

with upon their first going from any of their Majesties ports into sea service excepting in such works as shall hereafter be excepted (Vizt) They the said Robert & John Castle shall perform all the Shipwrights, Joyners, Caulkers, Labourers, Sawyers Plasterers, Bricklayers, Smiths, Platterers, Glaziers, Plumbers, Carvers and Painters works of what kind so ever which are proper to be made done and fixed to ye hull of ye ship And in such order quality, quantity scantling, number and sound workmanship as shall be directed to be done and performed either by the said Principal officers themselves or their overseers toward the completing of every needful works at once and particularly within the painters work to gild with gold the Lyon of the head and a pair of King's arms in the stern and also at his charge to prove the tightness of his own work by the customary watering of ships after their first caulking. And it is hereupon further agreed to except and reserve unto their Majesties proper charge over and above what is expressed or intended to be expressed or set forth in this present contract or what is or ought to be contained in the general or particular works aforementioned these other works following namely the contractor shall not make or provide the suite of masts nor coats of canvas about them nor the pumps and pump cases, nor wheels, chains, axeltrees and winches belonging to them nor provide or set the furnaces nor provide iron work or other materials for setting of them nor make or provide bucklers to the hawses or half ports nor sheath the ship nor make alterations of any works within board with respect to accommodation if once done and approved by the said Principal officers & Commissioners or their overseers. These and no other works are excepted in this contract but everything else whatsoever is needful and useful to be done in any of their Majesties ships of this quality in order to the full completing them with all requisites either comprehended in or under any of the officers relating to the artificers before named or comprehended to be done by and with any materials whatever saving such as are just before excepted shall be caused, directed and done by the said Robert Castle & John Castle their servants and assignees according to the disposition and direction of the said Principal officers and Commissioners or their assignees without delay. And the said Robert & John Castle their executors, administrators, servants and assignees do grant at his and their own proper costs and charges to find and provide all materials of timber and plank, iron of the best Spanish or what is equal in goodness to the same pitch, tar, ochum, tallow and treenails to be of the growth of Sussex well seasoned or equal in quality to them and all other materials whatsoever necessary and fitting to be had and expended in & about the premises aforesaid & he doth also oblige himself that all ye said materials so used & expended shall be good, sound seasoned and well conditioned in every particular according to their respective kinds and in like manner to pay and discharge all manner of workmanship touching all and every part of the works herein expressed or ought to have been expressed and to launch and deliver the said ship afloat for their Majesties use in the River of Thames to such officers as shall be appointed to receive her within ten months from ye day hereof. And it is further agreed that the said Principal officers and Commissioners of their Majesties Navy shall have liberty to appoint such person or persons as they shall see fit to overseer and survey every part of the works before mentioned and if at any time during the building of the said ship herein mentioned according to the dimensions, scantlings and conditions herein expressed there shall be found and discovered by the said surveyor or surveyors any unsound or insufficient materials or workmanship contrary to the meaning of this contract that then upon due notice of all or any such defect in stuff or workmanship given in writing by the said surveyor or surveyors to the said Robert & John Castle or their master workman under them there shall be an effectual and speedy amendment and reforming of all and every such default in stuff or workmanship and the said amendments shall be certified in writing by the surveyor or surveyors to the said Principal officers and Commissioners of the Navy for their Majesties service in that behalf and the said Principal officers and Commissioners for and on behalf of their Majesties shall according to the custom of the office make out and sign bills to the Treasurer of the Navy to be paid to the said Robert & John Castle their executors or assignees after the rate of nine pounds & ten shillings (of lawful money of England) per ton for every ton the said ship shall measure with the dimensions and limitations before exprest and no otherwise. And the tonnage to be cast according to the accustomed rule of Shipwrights Hall. And it is also agreed that in case the said ship shall be exceeded in dimensions or scantlings contrary to what is therein set forth and declared no allowance shall be made for any such over work or increase of scantling unless ye same hath been made by order first given in writing by the said Principal officers and Commissioners of the navy. And it is agreed upon the considerations aforesaid that a bill of imprest shall be made out for the sum of one thousand pounds upon the signing of this contract, another for the sum of one thousand pounds more when the keel, stem and post is raised, floor across and keelson bolted, another for the sum of one thousand pounds more when all the futtocks are in another for the sum of one thousand & two hundred pounds more when the gundeck beams and toptimbers are in and wales about and the ship planked within board and without up to the same, another for the sum of one thousand & two hundred pounds more when the beams of the upperdeck, quarterdeck and forecastle are all in and decks laid and the sides of the ship planked up within and without and perfect. Bill for the remainder which shall be due for the said ship after she shall be launched and delivered safe afloat to such officers as shall be appointed to receive her and a certificate of the performance of the whole work according to the tenor of this contract made and given by such persons as shall be appointed by the said Commissioners thereunto which bills are to be assigned for payment in manner following (Vizt) The first for ready money the rest in course according to the register and custom of this office; and it is also agreed for the further encouragement of the dispatch of this ship that the contractor for his second payment aforementioned shall have a bill made out & registered at ye signing hereof to take its course of payment & ye rest to be made out in order as certificates shall be produced for the same & paid in course as is before specified .

Sealed & delivered by Mr Rob:t Castle
In the presence of   Rob:t Castell
Jn: Peters   John: Castell
Edw'd Gerard
Sealed & delivered by Mr John
Castle in ye presence of Jos'h Griffin
Jn: Peters

APPENDIX 1

THE MEDWAY WARRANT

Warrant to Sheerness dockyard for building the fourth-rate Medway dated 17 March 1690 from the collection of Arnold Kriegstein, California. The ship measured 145 feet 3 inches on the gundeck, 38 feet broad and 15 feet 8 inches depth in hold.

By ye Principal Officers & Commissioners of their Majesties Navy:

Whereas ye Right Honourable ye Commissioners of ye Admiralty by their order of yesterdays date have been pleased to direct a Fourth rate ship, one of ye twenty seven ships appointed by ye late Act of Parliament to be built for their Majesties to be forthwith gone in hand with & to be built at their Majesties yard at Sheerness. These are in pursuance thereof to require and direct you to set up a fourth rate in your yard causing as good a progress as you can to be made in ye hewing of ye frame before you lay her keel, it having been observed in building some ships heretofore that ye keel has been laid so early as that it has proved defective before ye ship could be finished; ye said ship to be of ye dimensions as follows. Viz:

Length of ye Gundeck
One hundred forty five foot
Extreme Breadth in ye Midship
Thirty seven & a half foot
Depth in Hold
Fifteen two thirds foot

And in burthen about Nine hundred tunns; & to carry sixty four guns according to ye said Act.

In ye building of which ship you are strictly enjoined to observe ye above said dimensions & not to vary from them as you will answer ye contrary at your perils as also to use all ye dispatch & good husbandry therein that possibly may be.

And whereas there is money appropriated by ye Parliament on purpose for building ye said Twenty seven ships, which money is not upon great penalties to be diverted or applied to any other use or service as you will find by ye said Act: Reference thereunto being had: it is therefore absolutely necessary that all such stores as shall be received on ye account should be kept apart from any other, & that your books both for wages & stores & all other things relating to ye said ships should be kept distinct in all respects as was done in ye time of ye building ye XXX ships formerly built by Act of Parliament: whereof you are required to make room for such stores as shall be provided on ye account & according to ye respective duties of your places to be very exact & punctual in keeping distinct accounts of all matters relating to ye said ships in ye same manner as was done for ye aforesaid XXX ships, taking care in all bills for stores & other services relating thereto, to express in a longer charter your ordered indentures that ye said stores where received or services done for or by ye account of ye said Twenty seven ships appointed by Act of Parliament to be built for their Majesties; but in regard ye money appropriated for ye said ships does not yet come in nor no stores are as yet provided for them we are by virtue of ye same order from ye Right Honourables ye Commissioners of ye Admiralty to give you leave to make use of such materials now in stores provided for yet other services of ye Navy as may be spared from those services for ye building of this ship upon a due valuation & appraisement first made thereof to the end ye Money be charged to her account and made good out of such materials as shall be provided for her of ye monies appointed by ye Act of Parliament for building ye 27 New ships or paid out of that money & lastly you are respectively required according to ye duties of your places to be so punctual in keeping all ye accounts relating to ye building and equipping of the said ship distinct and separate from all other accounts whatever as that you may be able at any time when called to it, to give a particular account of ye charge thereof & for so doing this shall be your warrant, Dated at ye Navy Office this 17th March 1690

J Tippetts   R Beach   Cha Sergison
Tho Willshaw

To ye Master Attendant, Master Shipwright, Storekeeper, Clerk of ye Check & Survey of their Majesties yard at Sheerness

APPENDIX 2

THE MORDAUNT SURVEY

The survey for the *Mordaunt*, built in 1681 (TNA ADM1/3552 p937-945), dates from 28 September 1682, when she was being considered for purchase into the Navy. It consists of three parts, the first being the important dimensions, the second an estimate of her worth, including the boatswain's and carpenter's stores, and the third an account of the dimensions of the gunports.

P937
Rt Honourables

In pursuance of Your Honourables late orders unto this Board of the 19th of this instant September, We have caused a fresh survey to be taken of the ship *Mordaunt* by two of the Elder Brethren of Trinity House, two of the Members of the Company of Shipwrights and the Master Shipwright and Master Attendant of his Majesties yard at Deptford. And do herewith present unto your Honourables a copie of the valuation which the said gentlemen have given under their hands of the said ship's hull, masts, yards, boats, ground tackle, rigging and other things belonging to her amounting in the whole to the sum of five thousand eight hundred forty two pounds eight shillings and six pence. Your Honourables will also receive enclosed a paper containing the dimensions of the said ship and her masts, yards and boats. We humbly remain

Your Honourables obedient servants
Navy Office
30th September 1682
Richard Haddock John Narbrough James Sothern

P939
The dimensions of the Ship *Mordaunt* with the number of ports and cabins and the dimensions of her masts, yards, boats etc.

Tuns		Feet	Inches
567 2/94	Length by the keel	101	9
	Breadth from outside of the plank	32	4½
	Rake afore	18	0
	Rake abaft	6	2
	Depth in hold from plank to plank	13	0
	Height between decks from plank to plank	6	2
	Length on the gundeck from the rabbet of the stem to the rabbet of the post	121	6
	Height in the steerage	6	4
	Height in the forecastle	5	9
	Height in the Roundhouse	5	9
		Number	
Double cabins on the gundeck		3	
Single cabins on the gundeck		7	
Single cabins on the transom in the gunroom		2	

Single cabins in the steerage	4
Single cabins in the forecastle	4
Single cabins under the main gangways	2
Single cabins under the gangways of the poop	2
Two cabins in the roundhouse for the Captain★ & Master and a stateroom abaft it	
The ship produceth in burthen according to the custom of Shipwrights Hall	$565\ ^{78}/_{94}$ of a tun $567\ ^{13}/_{94}$

No platforms nor beams in the hold, there is a bread room that will stow 2400 of bread and two store rooms for the Boatswain and Carpenter and all gangways aloft completed according to a man of war of the same burthen.

Dimensions of the boats	**Length aloft Feet Inches**	**Breadth Feet Inches**
Shallop	32 00	09 08
Barge	31 00	06 00
Pinnace	25 00	05 06

P940

The dimensions of the masts and yards

	Length in yards	**Diameter in inches**
Main mast	27	22
Main top mast	$16^2/_3$	$13¼$
Main topgallant mast	$8^2/_3$	6
Fore mast	24	20
Fore top mast	14	12
Fore topgallant mast	6	5
Mizon mast	23	$14½$
Mizon top mast	$8^2/_3$	7
Bowsprit	16	21
Spritsail top mast	$4^1/_6$	$5½$
Main yard	24	$16¾$
Main topsail yard	14	10
Main topgallant yard	$7^1/_3$	$5¾$
Fore yard	21	$15½$
Fore topsail yard	13	$9½$
Fore topgallant yard	$5^1/_3$	$5¼$
Mizon yard	21	11
Mizon topsail yard	$8^1/_3$	6
Crojack yard	13	7
Spritsail yard	$14½$	11
Spritsail topsail yard	8	$5¾$

★*Possibly a mistake; 'Captain' may be 'Chaplain'*

	Feet	Inches
Length of the forecastle	19	00
Length of the waist from bulkhead to bulkhead	61	00
Length of the steerage	21	00
Length of the great cabin	19	00
Length of the half deck	27	08
Length of the poop	18	08
Number of ports in the forecastle		4
Number of ports in the waist		12
Number of ports in the steerage		4
Number of ports in the great cabin		4
Number of ports on the half deck		6
Number of ports in the roundhouse		4
Number of ports on the gundeck		22
Number of ports on the gundeck right aft		4
	Feet	Inches
The main hatchway in length	9	4
The main hatchway in breadth	5	4

P941

28th September 1682

An estimate of the value of the hull, masts yards, boats, ground tackle, rigging, sails, boatswains and carpenters stores etc of the ship *Mordaunt* as followeth, Vizt.

Materials	Quantity			Value		
				£	s	d
For the hull, masts, yards and sheathing with longboat, pinnace and galliot with oars to them.				4669	5	0
	cwt	qtr	lbs		150	16	0
Anchors of	26	2	21	one			
	24	2	14	one			
	23	2	0	one			
	11	2	21	one			
	6	0	14	one			
	1	2	14	one			
Cables of	16 inch			one new	64	7	0
	16 inch			one half worn	32	3	6
	15½ inch			one half worn	29	14	0
	11 inch			one eighth worn	28	7	6
Cablets of 6 inch				two half worn	9	7	0
Hawsers of 6½ inch				8 fathoms	0	13	6
Coils of	3 inch			50 fathoms	1	10	0
	2½ inch			40 fathoms	0	14	0
	2 inch			88 fathoms	0	16	6

	1½ inch	two 12 fathoms	1	10	0
	¾ inch	one coil	0	5	6
Junck of 12½ inch		33 fathoms	3	0	0
Tarred lines		Four	0	3	6
Tarred marlin and housing		Twenty pounds	0	6	8
Platts		Thirteen	3	18	0
Blacking		Twelve barrels	0	3	0
Brushes small scrubbing		One	0	1	0
Cloth awning of canvas one suite		150 yards	3	15	0
			5000	16	8

P942

Materials	Quantity				Value		
					£	s	d
Colours ensign ¾ worn	one				0	5	0
Colours ensign silk ¼ worn	one				10	0	0
Colours flag silk ¼ worn	one						
Colours vane silk 9 yards	two						
Kersey toparm & waistcloth	suite 130 yards est				18	0	0
Tarpaulins	suite 1/2 worn				2	0	0
Compass with wood boxes	one				0	5	0
Copper covers for furnaces	two						
		Cwt	Qtr	Lbs	29	14	0
Copper funnels	one	0	0	19			
Copper furnaces fitted with cocks	two	3	0	21			
Copper hoods	one	0	0	20			
Glasses half hour	two				0	1	4
Iron bilboes with iron shackles	one pair				1	0	0
Iron booms	four				1	2	6
Iron top chains	two				4	10	0
Iron creepers	two				1	2	6
Iron crows	two				1	2	6
Iron topmast fidds	two				1	3	6
Iron fire fork	one				0	2	6
Iron fire shovel	one				0	1	6
Iron tongs	one pair				0	2	6
Iron boats grapnel of 60 lb	one				1	10	0
Iron boats grapnel of 50 lb	one						
Iron furnace doors	one				0	3	6
Iron grommets and staples	24 grommets 1 staple				0	3	6
Iron hooks can	Two pair				0	3	6
Iron hooks fish	one				0	3	6
Iron hooks flesh	one				0	1	0

Iron hooks tackle	nineteen		0	15	0
Iron hanging locks	three		0	3	0
Iron marlin spikes	eight		0	10	0
Iron mauls for topmasts	two		0	11	0
Iron Scrapers	thirty nine		2	8	0
Iron vane spindles	six		0	6	0
Iron small thimbles	thirty		0	12	6
Iron large thimbles	six		0	4	6
Lead scupper and copper saw pan trimmed	one each		0	12	0
Poop lanthorns	three		18	0	0
Top lanthorn	one				
			96	19	10

P943

Materials	Quantity	Value		
		£	s	d
Hand leads	two	0	2	6
Sails		313	0	0
Spritsail courses	two			
Topsails	one			
Fore courses	two			
Fore topsails	two			
Fore topgallant sail	one			
Main courses	two			
Main topsails	two			
Main topgallant sail	one			
Mizon courses	two			
Mizon topsail	one			
Main staysail	one			
Mizon staysail	one			
Main topsail staysail	one			
Fore topsail staysail	one			
Main studding sails	two			
Topsail studding sails	two			
Boat sails	seven			
Grindstone unfixed	one	0	3	0
Bittacle	one	0	10	0
Blocks		9	3	0
Catt iron bound	two			
Spar	sixty six			
Top	two			
Top tackle	two			

Top tackle iron bound	two			
Vyoll	one			
Winding tackle double	one			
Winding tackle treble	one			
Cann buoy	one	0	10	0
Splicing fidd	one	0	1	0
Scoops	two	0	3	0
Serving mallets	seven	0	3	6
Shovels steel shod	fourteen	0	14	0
Spanish tables	three	1	10	0
Trucks for shrouds small	thirty two	0	5	0
Trucks long	twelve	0	4	0
Trucks for vane spindles	three	0	0	6
Trucks for studding sail booms	ten	0	4	6
Her standing & running rigging all complete	In tuns est	300	0	0
Her blocks & strapping as is now overhead		70	0	0

Carpenters Stores

Iron stancions	Twenty four	3	15	0
		700	9	0

P944

Materials	Quantity	Value		
		£	s	d
Hatch bars	one suite	1	10	0
Grating bars with bolts	one suite	4	10	0
Pitch pot and ladle	one each	0	11	0
Logger heat	one	0	7	0
Two handed saw	one	0	5	0
Spare tiller & anchor stock	one each	5	10	0
Fidd and bolster	one each	0	3	6

		Cwt	Qtr	Lbs			
Lead weights for ballast for the barge	two	2	0	11	1	10	0
Chain pumps fixed	two				18	2	6
Hand pumps fixed	two				8	15	0
Half ports fitted with canvas	ten				2	10	0
Grindstone	one				0	9	0
					44	3	0
					700	9	0
					96	19	10
					5000	16	8
					5842	08	6

The total is: five thousand eight hundred forty two pounds eight shillings and six pence
John Hill   Jonas Shish
Henry Lowe   Abraham Greaves
Thomas Willshawe   John Shish

A true Copy
James Sothern

P945

	Feet	Inches
Height of the ports from the upper deck	01	06
Fore and aft	02	03
Up and down	01	11
Height of the ports from the gun deck	02	04
Fore and aft	02	10
Up and down	02	05

ENDNOTES

[1] Pepys, Diary, 22 July 1664
[2] Pepys, Naval Minutes, NRS 200
[3] Further Correspondence of Samuel Pepys, p.26
[4] Pepys Library, PL1490, p.145, printed in NRS, Catalogue of Pepysian MSS, vol. 1, pp.76–77
[5] William Sutherland, *Ship-builders Assistant*, 1711, p.76
[6] N.M.M. SPB 50, Essay by William Sutherland
[7] Richard Endsor, *Restoration Warship*, 2009, pp.21–23
[8] Anon., *The Secret History of Deptford*, 1717, pp.6–7
[9] Frank Fox, *Great Ships*, 1980
[10] Anon., *The Secret History of Deptford*, 1717, p.7
[11] *Diary of John Evelyn*, 13 May 1680
[12] C. Knight, *The Mariner's Mirror*, vol. 18, p.411
[13] NA, ADM1/3554, p.317
[14] J. D. Davies, *Pepys's Navy*, 2008, pp.184–185
[15] TNA ADM/1737, p.142
[16] TNA ADM2/1737, p.218
[17] Extracted from Pepys Library PL1337
[18] TNA ADM7/633
[19] Ed. J. R. Tanner, *Samuel Pepys's Naval Minutes*, 1925, p.115
[20] J. D. Davies, *Kings of the Sea*, p.69
[21] John Ehrman, *The Navy in the War of William III*, p.14
[22] Frank Fox, *Great Ships*, pp.73–74
[23] TNA ADM1/3546, p.85
[24] CSPDom Add 1660–70, p.163, printed in Catalogue of Pepysian MSS, NRS, vol. II, p.9
[25] TNA ADM2/1746, 13 June 1673
[26] TNA ADM1/3545, p.475
[27] TNA ADM1/3545, pp.479–480
[28] Catalogue of Pepysian MSS, NRS, vol. II, p.9
[29] TNA ADM1/3545, p.480
[30] TNA ADM1/3545, pp.121–122
[31] TNA ADM1/3545, p.123
[32] TNA ADM1/3545, p.343
[33] Catalogue of Pepysian MSS, NRS, vol. II, p.97
[34] TNA ADM1/3545, p.180
[35] Anon., *The Secret History of Deptford*, 1717, p.7
[36] J. D. Davies, *Gentlemen and Tarpaulins*, 1991
[37] TNA ADM106/333, f338
[38] TNA ADM106/333, f312
[39] TNA ADM106/30, 3 December 1674
[40] Ed. J. R. Tanner, *Samuel Pepys's Naval Minutes*, 1925, p.244
[41] TNA ADM106/30, 3 December 1674
[42] TNA ADM106/30, 14 December 1674
[43] TNA ADM106/30, 3, 14, 18 December 1674; ADM1/3546 p.483–484; ADM106/31 27 March 1675
[44] TNA ADM106/384 f214
[45] TNA ADM106/3119, p.225
[46] TNA ADM1/3563, pp.493–497
[47] TNA ADM106/298, f437
[48] TNA ADM106/380, f232; for further details of the *Hampshire* repairs, see *The Mariner's Mirror*, vol. 91, 2005, pp.67–81
[49] TNA ADM52/5
[50] TNA ADM1/3560, p.975
[51] TNA ADM52/5
[52] David J. Hepper, *British Warship Losses in the Age of Sail, 1650–1859*, p.8
[53] Richard Endsor, *Restoration Warship*, 2009, pp.188–189
[54] David J. Hepper *British Warship Losses in the Age of Sail, 1650–1859*, p.4
[55] TNA ADM 106/291, pp.217, 234
[56] TNA ADM1/3545, pp.645, 679
[57] TNA ADM2/1737, p.138
[58] TNA ADM106/406, f9
[59] *The London Gazette*, no 2697, 14 September 1691
[60] TNA ADM 106/2507, p.84
[61] TNA ADM91/1, p.1107
[62] TNA ADM7/169
[63] TNA ADM106/37, 30 August 1677
[64] TNA ADM106/3118, p.299
[65] TNA ADM106/43
[66] TNA ADM 106/3119, p.172
[67] TNA ADM1/3563, p.563
[68] TNA ADM106/372, f262; ADM106/371, f154
[69] TNA ADM106/372, f264, f331
[70] TNA ADM106/373, f191
[71] TNA ADM106/373, f331; ADM106/368, f357
[72] TNA ADM106/30 December 1674
[73] Richard Endsor, *Restoration Warship*, 2009, pp.126–140
[74] TNA ADM106/2507, no 6
[75] TNA ADM106/293, f223
[76] TNA ADM106/2507, no 41
[77] TNA ADM106/2507, no 32
[78] TNA ADM1/3575, pp.267, 373
[79] TNA ADM95/14, pp.237–243
[80] A full account is given in *The Mariner's Mirror*, vol. 94, no. 3, August 2008
[81] TNA ADM106/3118, p.27
[82] NMM AND/31
[83] Frank Fox, *Great Ships*, p.143
[84] TNA ADM18/60, p.75
[85] Pepys Library PL1339
[86] TNA ADM18/60, p.37
[87] TNA ADM106/3118, p.140
[88] Captain John Smith, *A Sea Grammar*, pp.15–16
[89] TNA ADM106/3118, p.333
[90] TNA ADM8/1
[91] TNA ADM106/3119, p.109
[92] TNA ADM8/1
[93] TNA ADM106/358, p.279
[94] TNA ADM106/358, p.298
[95] TNA ADM106/358, p.306
[96] TNA ADM106/362, p.600
[97] TNA ADM106/358, p.326
[98] TNA ADM1/3552, pp.337, 361
[99] TNA ADM106/358, p.420
[100] TNA ADM106/358, p.392
[101] TNA ADM106/516, 18 October 1698
[102] TNA ADM8/1
[103] TNA ADM7/168
[104] TNA ADM52/121, p.4
[105] TNA ADM51/4398, p.3
[106] TNA ADM106/372, p.282, 284
[107] TNA ADM8/1
[108] TNA ADM106/372, p.337
[109] TNA ADM106/372, p.339
[110] TNA ADM106/372, p.288
[111] TNA ADM106/372, p.343
[112] TNA ADM106/3566, pp.68–69
[113] TNA ADM106/372, p.341
[114] TNA ADM106/372, p.292
[115] TNA ADM106/372, p.294
[116] TNA ADM106/372, p.297
[117] TNA ADM106/378, p.356
[118] TNA ADM7/168, 4 June 1685
[119] TNA ADM106/378, p.271
[120] TNA ADM106/378, p.379
[121] TNA ADM106/378, p.364
[122] TNA ADM106/378, p.381
[123] *Hoppus's Measurer*, 1790, intro, p.xl
[124] TNA ADM106/378, p.387
[125] TNA ADM106/378, p.331
[126] TNA ADM106/378, p.391
[127] TNA ADM106/378, p.393
[128] Samuel Pepys, *Memoires Relating to the State of the Royal Navy*, 1690, p.193
[129] Bodleian Library, Oxford, Rawlinson MSS, C429
[130] Philip Aubrey, *The Defeat of James Stuart's Armada*, 1692, p.23
[131] TNA ADM51/4398, p.2
[132] TNA ADM51/4398, p.2
[133] TNA ADM1/3563, p.293
[134] TNA ADM51/4398, p.1
[135] TNA ADM106/397, p.514
[136] TNA ADM106/397, p.514
[137] TNA ADM1/3562, p.795
[138] TNA ADM1/3562, p.823
[139] TNA ADM1/3562, p.905
[140] TNA ADM95/14, p.1
[141] TNA ADM51/4398, p.1

ENDNOTES

142 TNA ADM1/3564, p.319
143 TNA ADM1/3564, p.357
144 TNA ADM1/3564, p.539
145 TNA ADM106/3290; ADM1/3564, p.543
146 TNA ADM52/121, p.5
147 TNA ADM1/3566, p.836
148 TNA ADM52/121, p.5
149 TNA ADM8/3
150 Josiah Burchett, *Memoirs of Transactions at Sea During the War with France*, 1703, p.193
151 TNA ADM52/121 6 TNA ADM51/4398, p.6
152 TNA ADM1/3571
153 TNA ADM8/3 ADM51/4398 6 ADM52/121, p.6
154 Letter from Captain Archibald Hamilton to Navy Board, 26 July 1694
155 ADM52/121, p.6
156 TNA ADM1/3575, p.189
157 TNA ADM1/3575 249, p.277
158 TNA ADM51/4398, p.9
159 TNA ADM1/3578, p.835
160 TNA ADM51/4398, p.9
161 TNA ADM95/14, p.105
162 TNA ADM8/5
163 TNA ADM1/3583
164 TNA ADM180/20
165 TNA ADM106/3541, part1
166 William Keltridge, *His Book*, p.20; NMM AND/31
167 *The Diary of Henry Teonge*, published by George Routledge, 1927, p.197
168 NAADM106/3566, p.44
169 TNA ADM106/3541, part1
170 TNA ADM106/3068
171 TNA ADM1/3567, p.691
172 TNA ADM1/3560, p.717
173 TNA ADM180/20
174 TNA ADM106/3541
175 Ed. J. R. Tanner, *Samuel Pepys's Naval Minutes*, 1925, pp.17, 241, 371
176 Robert Gardiner, *The Sailing Frigate*, 2012, p.8
177 Thanks to Frank Fox, private correspondence, p.201
178 TNA ADM1/3560, p.391
179 *Diary of John Evelyn*, 1 October 1661
180 Richard Endsor, 'The Van de Velde Paintings for the Royal Yacht Charlotte, 1677', *The Mariner's Mirror*, 2008, vol. 94, no 3, pp.264–275
181 Catalogue of Pepysian MSS, NRS, vol. IV, pp.519–520
182 Catalogue of Pepysian MSS, NRS, vol. IV, pp.641–642
183 TNA ADM106/41, 3 March 1679
184 TNA ADM2/1748, 27 March 167
185 TNA ADM 106/44, 22 December 1679
186 Isle of Wight History Centre, 17th-century catamaran
187 Bodleian Library, Oxford, Rawlinson MSS, A178, f264
188 Samuel Pepys, *Letters and Second Diary*, pp.150–15
189 Richard Endsor, *Restoration Warship*, 2009, pp.72–73
190 Pepys Library, PL1339
191 TNA ADM106/2507, no 51, June 1678
192 Print by Phillips, c.1680; Edmund Dummer drawing, PL2934; section through a first rate, painting NMM; Blaise Ollivier drawing, *18th Century Shipbuilding: Remarks on the Navies of the English and Dutch*, 1737; details of the *Lenox*, PL1339; artefacts from wrecks of the *Stirling Castle* and the *Dartmouth*, Thomas Blanckley, *A Naval Expositor*, William Keltridge, *His Book*, NMM AND/31 and Edward Battine, *Methods of Building Ships of War*, NMM SPB/28
193 William Keltridge, *His Book*, 1675, p.219
194 Edmund Dummer, drawing, c.1680, Magdalene College, Cambridge
195 TNA ADM106/3118
196 TNA ADM106/33, 25 January 1676
197 TNA ADM106/35, 31 January 1677
198 TNA ADM106/35, 16 November 1676
199 TNA ADM106/35, 20 November 1676
200 TNA ADM 106/317, f106
201 TNA ADM106/35, 11 December 1676
202 NS ADM106/320, p.226
203 Ed. David Roberts, *18th Century Shipbuilding: Remarks on the Navies of the English and Dutch*, 1737, remarks of Blaise Ollivier, p.158
204 NMM, no 538
205 TNA ADM1/3547, p.785
206 TNA ADM3/276, p.155
207 TNA ADM106/35
208 TNA ADM106/352, f117
209 TNA ADM1/3547, p.929
210 TNA ADM106/3118, p.198
211 TNA ADM3/276, 20 January 1677
212 TNA ADM3/276, 27 January 1677
213 TNA ADM106/3119, p.190
214 TNA ADM20/29
215 TNA ADM106/379, f407
216 Ed. David Roberts, *18th Century Shipbuilding: Remarks on the Navies of the English and Dutch*, 1737, remarks of Blaise Ollivier, p.158
217 Brian Lavery, *Arming and Fitting of English Ships of War*, 1987, p.197
218 Richard Endsor, 'The Loss of *Stirling Castle*, 1703', *The Mariner's Mirror*, 2004, vol. 90, pp.92–102
219 Brian Lavery, *Arming and Fitting of English Ships of War*, 1987, p.37
220 TNA ADM106/3118, p.201
221 TNA ADM106/3119, p.167
222 TNA ADM106/3119, p.172
223 Museum of London exhibit
224 *The Mariner's Mirror*, 1969, vol. 55, p.115
225 For a fuller history, see Thomas Oertling, *Ships' Bilge Pumps*, 1996, Texas A & M
226 Thomas Oertling, *Ships' Bilge Pumps*, 1996, Texas A & M, p.56
227 TNA ADM106/345, p.312
228 Pepys Library, PL1339
229 Reconstructed from building list of the *Lenox* of 1678, PL1339; draught of the *Resolution*, 1708, NMM ZAZ6228; model of the *Royal Oak*, 1713; Kriegstein Collection, contemporary prints of section through a first rate and an account of the sizes of pumps, William Sutherland, *Ship-building Unvail'd*, 1717, part II, pp.238–241
230 Catalogue of Pepysian MSS, NRS, vol. II, p.66
231 Catalogue of Pepysian MSS, NRS, vol. II, p.145
232 TNA ADM3/276, 16 September 1676; Catalogue of Pepysian MSS, NRS, vol. IV, p.351
233 Catalogue of Pepysian MSS, NRS, vol. IV, p.485
234 TNA ADM106/35, 16 November 1676
235 TNA ADM106/341, f365
236 TNA ADM106/36, 20 April 1677
237 TNA ADM3/276, 30 June 1677
238 TNA ADM2/1741, p.249
239 TNA ADM1/3555, p.855
240 TNA ADM1/3555, p.881
241 TNA ADM1/3555, p.897
242 TNA ADM1/3556, p.101
243 TNA ADM1/3556, p.137
244 TNA ADM1/3556, p.121
245 TNA ADM106/3542, part II, 15 January 1687
246 Samuel Pepys, *Letters and the Second Diary*, ed. R. G. Howarth, pp.175–176
247 Bodleian Library, Oxford, A179, f100, f138
248 TNA ADM1/3556, p.429
249 TNA ADM1/3556, p.435
250 TNA ADM1/3556, p.429
251 TNA ADM1/3557, p.105
252 TNA ADM106/698, 19 May 1688
253 TNA ADM1/3557, p.603
254 TNA ADM1/3571, p.722
255 TNA ADM1/3565, pp.161–163
256 TNA ADM1/3574, p.1093
257 TNA ADM106/497, f263
258 TNA ADM1/3580, p.641
259 Pepys Library, PL1339
260 Thomas Blanckley, *A Naval Expositor*, 1750
261 For further developments of the pump until the end of the 17th century, see Dan Pascoe, *International Journal of Nautical Archaeology*, 2014, pp.145–159
262 TNA ADM2/1748, f75
263 Richard Endsor, *The Mariner's Mirror*, August 2008, p.268
264 TNA ADM2/1748, f75
265 TNA ADM20/29
266 Blaise Ollivier, published in *18th Century Shipbuilding: Remarks on the Navies of the English and Dutch*, 1737, tr. David Roberts, pp.54, 147–147; Thomas Blanckley, *A Naval Expositor*, p.87
267 TNA ADM106/3069, 21 January 1690
268 Blaise Ollivier, published in *18th Century Shipbuilding: Remarks on the Navies of the English and Dutch*, 1737, tr. David Roberts, pp.54–147; Thomas Blanckley, *A Naval Expositor*, p.87
269 Brian Lavery, *The Ship of the Line*, vol. II, p.41
270 *Further Correspondence of Samuel Pepys*, pp.264–269
271 TNA ADM106/2888, p.1
272 Brian Lavery, *Arming and Fitting of English Ships of War*, 1987, p.61
273 Catalogue of Pepysian MSS, NRS, vol. II, p.184
274 TNA ADM106/31, 12 May 1675
275 TNA ADM106/31, 19 June 1675
276 TNA ADM106/3118, p.193
277 TNA ADM106/3118, p.224
278 TNA ADM106/3118, p.79
279 TNA ADM106/3118, p.301
280 TNA ADM106/3118, pp.26, 314
281 TNA ADM106/37, 5 November 1677; TNA ADM3/276, 20 October 1677
282 Anon., *The Secret History of Deptford*, 1717, p.7
283 TNA ADM39, 9 April 1678
284 TNA ADM106/337, f185
285 TNA ADM1/3553, p.57
286 TNA ADM106/37
287 TNA ADM1/3555, p.429
288 TNA ADM1/3581, p.751
289 (TNA ADM1/3583, 22 March 1698
290 Catalogue of Pepysian MSS, NRS, vol. IV, p.642
291 TNA ADM1/3572, p.635
292 Edmund Bushnell, *The Compleat Ship-Wright*, 4th ed., 1678, pp.46–48
293 TNA ADM1/3572, p.635
294 NMM image x.13500, reproduced in David Lyon, *Navy Sailing List*, 1993, p.119
295 TNA ADM3/278, 17 March 1681
296 *The London Gazette*, no 2026, 16 April 1685
297 TNA ADM2/1741, p.224
298 TNA ADM2/1749; TNA ADM106/45, 3 February 1680
299 Thomas Miller, *The Compleat Modellist*, 2nd ed., 1664, p.6
300 R. C. Anderson, *Seventeenth Century Rigging*, 1955, pp.4–5
301 R. C. Anderson, *Seventeenth Century Rigging*, 1955, p.85
302 *Samuel Pepys's Naval Minutes*, NRS, p.16
303 J. D. Davies, *Kings of the Sea*, p.71
304 TNA ADM106/3119, p.187
305 *Diary of Samuel Pepys*, vol. 9, ed. Robert Latham and William Matthews, p.26
306 TNA ADM91/1, f280v
307 Thomas Miller, *The Compleat Modellist*, 1664, p.6
308 Author's collection, *Dimensions of Old Ships*
309 John Franklin, *Navy Board Ship Models*, 1989, p.50, Conway's Ship Modelling
310 William Sutherland, *Ship-builders Assistant*, 1711, p.25
311 Thomas Miller, *The Compleat Modellist*, 2nd ed., 1664, p.1
312 Anthony Deane, *Deane's Doctrine*, 1670, Pepys's Library, PL2910 p39, reproduced by Brian Lavery, Conway Maritime Press, 1981, p.52
313 Edmund Bushnell, *The Compleat Ship-Wright*, 1678, p.5
314 TNA ADM7/827, p.88v, printed by Salisbury and Anderson, SNR, Occasional Publication no 6
315 *18th Century Shipbuilding: Remarks on the Navies of the English and Dutch*, 1737, tr. David Roberts, p.36
316 TNA ADM91/1, f280v
317 R. C. Anderson, *The Mariner's Mirror*, 1919, vol. 5, no 4
318 Catalogue of Pepysian MSS, NRS, vol. IV, p.468
319 TNA ADM 18/63, 7 March 1678
320 TNA ADM 106/40, 11 September 1678
321 Pepysian MSS, *Admiralty Letters*, pp.ix, 213, 218, printed in Catalogue of Pepysian MSS, NRS, vol. I, p.246
322 R. C. Anderson, SNR, Occasional Publication no 5, Notes
323 Catalogue of Pepysian MSS, NRS, vol. IV, p.430
324 William Sutherland, *Ship-builders Assistant*, 1711, p.76
325 NMM Catalogue of Ship Models, 1670–1671, part 1, no 1670–1
326 Richard Ollard, *Man of War*, p.157; Anne Doedens and Jan Souter, *De Ramp Van Vlieland en Tesrschelling*, 1666, pp.223–224
327 TNA ADM2/1737
328 John Charnock, *Biographica Navalis*, vol I, pp.62, 335–336; Rif Winfield, *British Warships 1603–1714*, p.92
329 ADM52/112, part1
330 TNA ADM8/1, 6 July 1674
331 ADM8/1, 6 July 1674
332 TNA ADM106/3118, pp.75–77
333 TNA ADM 106/298, f399; a fuller account from Daniel Furzer

[334] NMM AND/31, p.28
[335] TNA ADM106/3118; ADM106/3119
[336] Catalogue of Pepysian MSS, NRS, vol. IV, p.18
[337] Catalogue of Pepysian MSS, NRS, vol. IV, p.28
[338] TNA ADM42/486
[339] Catalogue of Pepysian MSS, NRS, vol I, p.150
[340] Catalogue of Pepysian MSS, NRS, vol I, p.43–47
[341] Bodleian Library, Oxford, Rawlinson, A175, ff284–285
[342] Catalogue of Pepysian MSS, NRS, vol IV, p.80
[343] TNA ADM106/3118
[344] Catalogue of Pepysian MSS, NRS, vol. IV, p.100
[345] Catalogue of Pepysian MSS, NRS, vol. IV, pp.103–104
[346] Catalogue of Pepysian MSS, NRS, vol. III, p.50
[347] TNA ADM106/32, 7 August 1675
[348] TNA ADM106/3118, pp.39–40
[349] TNA ADM106/310, f308
[350] TNA ADM18/60, f165
[351] TNA ADM106/310, f316
[352] TNA ADM106/3118, p.77
[353] Catalogue of Pepysian MSS, NRS, vol. IV, pp.223–224
[354] Catalogue of Pepysian MSS, NRS, vol. IV, p.228
[355] Catalogue of Pepysian MSS, NRS, vol. IV, p.265
[356] TNA ADM1/3547, p.61
[357] TNA ADM1/3548, p.277
[358] Catalogue of Pepysian MSS, NRS, vol. IV, p.280
[359] Catalogue of Pepysian MSS, NRS, vol. IV, p.288
[360] TNA ADM106/3118, pp.143–145
[361] Calendar of Treasury Book, vol.V, p.77
[362] TNA ADM1/3548, p.275
[363] TNA ADM106/317, f16
[364] TNA ADM106/323, f172
[365] TNA ADM106/323, f174
[366] Catalogue of Pepysian MSS, NRS, vol. III, p.380
[367] Catalogue of Pepysian MSS, NRS, vol. IV, p.409
[368] Ed. Anchitell Grey, *Debates of the House of Commons*, 1763, vol. IV, p225
[369] Calendar of Treasury Book, vol.V, p.606
[370] TNA ADM2/1748, p.45; TNA ADM106/36, 14 May 1677
[371] TNA ADM1/3547, p.1167
[372] TNA ADM106/3118, p.225
[373] TNA ADM106/323, f206
[374] Act of Parliament for Building Thirty Ships of War, 15 February 1677, p.146
[375] TNA ADM106/323, f220
[376] TNA ADM106/323, f232
[377] Richard Endsor, *Restoration Warship*, 2009, p.35
[378] TNA ADM106/3118, p.247
[379] TNA ADM106/323, f310
[380] TNA ADM106/323, f101
[381] Pepys Library, PL2266 f28, published by Brian Lavery, *The Ship of the Line*, vol. 1, p.196
[382] Pepys Library, PL1339
[383] TNA ADM7/168, f90v
[384] TNA ADM106/323, f220; ADM106/355, f305, f361
[385] TNA ADM1/3554, p.311
[386] J. D. Davies, *Kings of the Sea*, p.223
[387] TNA ADM20/21
[388] M. A. Faraday, Camden Society
[389] Richard Endsor, *Restoration Warship*, 2009, p.24
[390] TNA ADM106/40
[391] TNA ADM106/39
[392] TNA ADM106/45
[393] TNA ADM106/355, f351
[394] TNA ADM106/39
[395] TNA ADM106/49
[396] TNA ADM106/44, 15 November 1679
[397] TNA ADM106/50
[398] Catalogue of Pepysian MSS, NRS, vol. IV, pp.332, 351
[399] TNA ADM3/276
[400] Catalogue of Pepysian MSS, NRS, vol. IV, p.69
[401] Catalogue of Pepysian MSS, vol. IV, NRS, pp.97–98
[402] Catalogue of Pepysian MSS, NRS, vol. IV, p.216
[403] TNA ADM106/34, 25 July 1676
[404] TNA ADM8/1, f79
[405] TNA ADM 8/1, f81
[406] TNA ADM3/276, p.10, 22 June 1677
[407] TNA ADM3/276, p.119
[408] TNA ADM18/60, f111
[409] TNA ADM8/1, f85v
[410] Riff Winfield, *British Warships 1603–1714*, p.117
[411] TNA ADM8/1, f124v
[412] TNA ADM8/1, ff140–150v
[413] TNA ADM2/1752
[414] TNA ADM2/1749, 15 July 1679
[415] TNA ADM2/1752, 20 July 1679; ADM3/277, 3 August 1679
[416] TNA ADM2/1752, 18 July 1679
[417] TNA ADM3/277, 10 July 1680
[418] TNA ADM3/277, 3 August 1680
[419] TNA ADM106/347, f73
[420] TNA ADM3/277, 5 October 1680
[421] TNA ADM106/47
[422] TNA ADM106/358, ff434–435
[423] TNA ADM2/1753, p.203
[424] TNA ADM2/1750, 16 May 1681
[425] TNA ADM3/278, 21 May 1681
[426] TNA ADM106/357, f346
[427] TNA ADM3/277, 1 and 29 April 1680
[428] Bodleian Library, Oxford, Rawlinson MSS, A175, f317
[429] Catalogue of Pepysian MSS, NRS, vol. IV, p.383
[430] *Samuel Pepys's Naval Minutes*, NRS, p.201
[431] Catalogue of Pepysian MSS, NRS, vol. I, p.227
[432] *Samuel Pepys's Naval Minutes*, NRS, p.242; WO 55/333, p.110
[433] J. D. Davies, *Kings of the Sea*, p.139
[434] TNA ADM1/3545, 12 November 1673
[435] TNA ADM1/3545, p.827
[436] TNA ADM2/1737, p.129
[437] TNA ADM1/3546, p.865
[438] TNA ADM2/1737
[439] TNA ADM2/1737, p.320
[440] TNA ADM106/32, 14 August 1675
[441] TNA ADM18/63, p.119
[442] TNA ADM1/3546, p.1115
[443] www.talkingpointsmemo.com/edblog/artifacts-of-our-past, accessed 20 August 2014
[444] ADM106/33, 25 January and 24 March 1676
[445] TNA ADM106/37
[446] TNA ADM1/3548, p.5
[447] TNA ADM2/1748
[448] TNA ADM1/3571, pp.225–231
[449] Pepys Library, PL2991
[450] NMM DAR0001, reproduced in Riff Winfield, *British Warships 1603–1714*, p.86
[451] *Samuel Pepys's Naval Minutes*, NRS, p.244
[452] TNA ADM8/1 f28-f37v
[453] SNR, Occasional Publication no 5, p.23; R. C. Anderson, *The Mariner's Mirror*, 1963, vol. 49, no 4
[454] J. D. Davies, *Kings of the Sea*, p.73
[455] TNA ADM1/3563, p.293
[456] TNA ADM3/276, p.13
[457] TNA ADM7/168, p.44; TNA ADM106/33, 8 January 1676
[458] TNA ADM3/276, f21
[459] TNA ADM18/60, p.209
[460] TNA ADM7/168, p.44
[461] BM, Add, MSS, 22183, f26, part printed in *The Mariner's Mirror*, 1958, vol. 44, no 3, p.252
[462] BM, Add, MSS, 22183, f22
[463] TNA ADM3/276, p.23
[464] TNA ADM7/168, p.46
[465] TNA ADM106/33
[466] TNA ADM1/3547, pp.321–325
[467] TNA ADM1/3567, pp.965–969
[468] TNA ADM1/3547, p.325
[469] TNA ADM3/276, p.111
[470] TNA ADM106/34; ADM 1/3547, p.619
[471] TNA ADM106/3541, part 2; TNA ADM108/20, ADM8/1
[472] Pepys Library, PL2879, pp.111–122, reproduced in Frank Fox, *Great Ships*, pp.193 and TNA, WO55/1762
[473] TNA ADM106/3070
[474] TNA ADM1/3547, 27 September 1676
[475] TNA ADM51/4226
[476] TNA ADM106/3118, p.193
[477] TNA ADM51/4226
[478] TNA ADM106/355, p.323
[479] TNA ADM1/3554, p.897
[480] TNA ADM106/380, p.30
[481] TNA ADM106/3069, 1 March 1693, 19 February 1689
[482] TNA ADM106/367, p.49
[483] TNA ADM95/14, p.43
[484] TNA ADM106/355, p.156
[485] TNA ADM106/351, pp.414, 422
[486] TNA ADM106/3542, part 2, 12 August 1686
[487] TNA ADM106/344, p.78
[488] ADM106/3070, contract for the *Mary Galley*
[489] Mainwaring, NRS, vol. 2, p.243
[490] TNA ADM106/3118, pp.269, 270
[491] NMM DAR0001, reproduced in Riff Winfield, *British Warships 1603–1714*, p.86
[492] *The Mariner's Mirror*, vol. 52, no 1, pp.86–87
[493] TNA ADM106/35, 11 December 1676
[494] B. S. Ingram, *Three Sea Journals of Stuart Times*, p.117
[495] NMM DAR0001, reproduced in Riff Winfield, *British Warships 1603–1714*, p.86 and Thomas Blanckley, *A Naval Expositor*, p.112
[496] TNA ADM1/3547, pp.597–598
[497] TNA ADM3/276, 9 November 1676
[498] TNA ADM106/35, 11 December 1676; ADM7/639, p.1; ADM7/640, p.1
[499] TNA ADM3/276, p.147
[500] TNA ADM106/35, 11–28 December 1676
[501] TNA ADM106/322, p.108
[502] TNA ADM8/1, pp.79v–81
[503] Catalogue of Pepysian MSS,, NRS, vol. III, pp.357, 363
[504] TNA ADM106/322, p.1
[505] TNA ADM106/322, p.5
[506] TNA ADM106/323, p.432
[507] TNA ADM106/322, p.11
[508] B. S. Ingram, *Three Sea Journals of Stuart Times*, p.116
[509] TNA ADM51/4142
[510] TNA ADM51/4142
[511] B. S. Ingram, *Three Sea Journals of Stuart Times*, p.117
[512] TNA ADM51/4142
[513] TNA ADM51/4226; ADM106/326, p.349
[514] TNA ADM3/276, 3 November 1677; ADM2/1748, p.121
[515] TNA ADM7/168, 15 November 1677
[516] TNA ADM106/37, 19 November 1677
[517] TNA ADM106/38, 10 October 1677
[518] TNA ADM3/276, 15 December 1677; ADM2/1748, 21 December 1677
[519] TNA ADM3/276, 27 December 1677
[520] TNA ADM106/3541, part2
[521] Published in John Franklin, *Navy Board Ship Models*, 1989, Conway's Ship Modelling, p.132
[522] TNA ADM2/1748, 24 April 1678; ADM106/38
[523] J. D. Davies, *Pepys's Navy*, 2008, p.174
[524] J. D. Davies, *Pepys's Navy*, 2008, p.27
[525] NRS, *Samuel Pepys's Naval Minutes*, 1925, p.71
[526] TNA ADM2/1748
[527] TNA ADM106/31, 23 June 1675
[528] TNA ADM106/355, p.253
[529] Bodleian Library, Oxford, Rawlinson MSS, A185, p.325
[530] TNA ADM106/41, 3 March 1679
[531] Bodleian Library, Oxford, Rawlinson MSS, A185, f325
[532] Pepys's Diary, 13 June 1664
[533] Pepys Library, Magdalene College, Cambridge and Rawlinson collection in the Bodleian Library, Oxford
[534] Pepys Library, Magdalene College, Cambridge, PL2820
[535] Edward Battine, *Methods of Building Ships of War*, 1684, PL977
[536] Pepys Library, Magdalene College, Cambridge, PL2934
[537] Bodleian Library, Oxford, Rawlinson MSS, A174, ff282–283
[538] Bodleian Library, Oxford, Rawlinson MSS, A341, ff160–162v
[539] Pepys Library, Magdalene College, Cambridge, PL2910, ed. Brian Lavery, Conway Maritime Press, 1981
[540] Pepys Library, Magdalene College, Cambridge, PL2910, ed. Brian Lavery, Conway Maritime Press, 1981, p.21
[541] William Sutherland, *The Ship-builders Assistant*, London, 1711, p.77
[542] William Sutherland, *The Ship-builders Assistant*, London, 1711, p.159
[543] Anon., *Treatise on Shipbuilding*, c.1625, 93v, TNA ADM7/827
[544] Thomas Miller, *The Compleat Modellist*, 1664
[545] Bodleian Library, Oxford, Rawlinson MSS, A185, f325
[546] TNA ADM106/3071, contract for building four fourth rates in Ireland, 1673
[547] Anthony Deane, *Deane's Doctrine*, 1670, Pepys Library, Magdalene College, Cambridge, PL2910, pp.65–66; Edward Battine,

547 *Methods of Building Ships of War*, 1684, NMM SPB/28, p.3; and William Keltridge, *His Book*, 1675, NMM AND/31, pp.268, 111
548 William Keltridge, *His Book*, 1675, NMM AND/31, p.111
549 Mungo Murray, *A Treatise on Ship-building and Navigation*, 1754, p.145
550 TNA ADM106/3070
551 *Methods of Building Ships of War*, 1684, NMM SPB/28, p.2
552 Anthony Deane, *Deane's Doctrine*, 1670, Pepys Library, Magdalene College, Cambridge, PL2910, p.63
553 NRS, *Samuel Pepys's Naval Minutes*, p.158
554 *The Compleat Ship-wright*, 1664, reproduced in Scholars' Facsimiles, vol. 481, New York, 1993, p.10
555 Thomas Miller, *The Compleat Modellist*, 2nd ed., 1664, p.4
556 Bodleian Library, Oxford, Rawlinson MSS, A175, f328
557 Anon., *Treatise on Shipbuilding*, c.1625, TNA ADM7/827, p.93v
558 Fragments of *Ancient English Shipwrightry*, Pepys Library, Magdalene College, Cambridge, PL2820, p.33
559 Fragments of *Ancient English Shipwrightry*, Pepys Library, Magdalene College, Cambridge, PL2820, p.29
560 Anthony Deane, *Deane's Doctrine*, 1670, Pepys Library, Magdalene College, Cambridge, PL2910, p.62
561 *Diary of John Evelyn*, 28 January 1682
562 William Keltridge, *His Book*, 1684, NMM AND/31, pp.281–284
563 William Sutherland, *Ship-builders Assistant*, London, 1711, p.76
564 William Sutherland, *Ship-builders Assistant*, London, 1711, p.58
565 William Sutherland, *Ship-builders Assistant*, London, 1711, p.82
566 William Sutherland, *Ship-builders Assistant*, London, 1711, p.60
567 William Sutherland, *Ship-builders Assistant*, London, 1711, p.4
568 Anon., *Treatise on Shipbuilding*, c.1625, 93v, TNA ADM7/827 pp.89v–90r
569 Anon., *Treatise on Shipbuilding*, c.1625, 93v, TNA ADM7/827 p.90r
570 Anon., *Treatise on Shipbuilding*, c.1625, 93v, TNA ADM7/827, p.92v
571 Anon., *Treatise on Shipbuilding*, c.1625, 93v, TNA ADM7/827, pp.92v–93r
572 Anon., *Treatise on Shipbuilding*, c.1625, 93v, TNA ADM7/827, p.93r
573 Anon., *Treatise on Shipbuilding*, c.1625, 93v, TNA ADM7/827, pp.93r–93v
574 *Marine Architecture*, incorporating Edmund Bushnell's *The Compleat Ship-Wright*, London, 1739, p.43
575 TNA ADM3/276, 18 November 1677
576 TNA ADM106/37, 27 August 1677
577 Frank Fox, *Great Ships*, p.174
578 TNA ADM42/486; ADM42/1601
579 TNA ADM106/323, f79
580 TNA ADM2/1750, p.133
581 NRS, *Samuel Pepys's Naval Minutes*, p.226
582 Pepys Library, Magdalene College, Cambridge, PL2910, ed. Brian Lavery, Conway Maritime Press, 1981, p.71
583 NMM AND/31, William Keltridge, *His Book*, p.285
584 SCM 1927-0837_0001 and 0002
585 *The Mariner's Mirror*, 1912, vol. 2, pp.164–166
586 TNA ADM2/1750
587 TNA ADM3/278
588 TNA ADM3/278
589 TNA ADM2/1750
590 TNA ADM1/3552, pp.937–945
591 TNA ADM3/278, 30 September 1682
592 TNA ADM3/278
593 TNA ADM106/367, f506
594 TNA ADM106/367, f512
595 TNA ADM6/425
596 TNA ADM3/278
597 TNA ADM2/1750, p.364
598 NMM Catalogue of Ship Models, part 1, 1681-1
599 Model of the *Lizard*, 1697, Pitt Rivers Museum, Oxford
600 John Franklin, *Navy Board Ship Models*, 1989, Conway's Ship Modelling, pp.95–97
601 Richard Endsor, *Restoration Warship*, 2009, p.31
602 I am indebted to Simon Stephens, Curator of Models at the National Maritime Museum, for his help and co-operation
603 TNA ADM1/3552, pp.937–945
604 NMM 598
605 TNA ADM2/1751; ADM1/3553, p.975
606 TNA WO55/1762
607 TNA ADM7/168, p.73
608 TNA ADM7/168, p.69
609 TNA ADM2/1751, p.70
610 TNA ADM106/370, f354
611 TNA ADM106/372, f377
612 TNA ADM106/370, f358
613 TNA ADM106/370, f362
614 TNA ADM106/370, f366
615 TNA ADM6/426
616 John Charnock, *Biographica Navalis*, 1794, vol. 1, p.338
617 TNA ADM106/370, f370
618 TNA ADM106/371, f534
619 TNA ADM106/370, f384
620 William Sutherland, *Ship-builders Assistant*, 1711, p.109
621 TNA ADM106/370, f482
622 TNA ADM106/371, f546
623 TNA ADM106/371, f510
624 TNA ADM49/25
625 ADM8/1
626 TNA ADM106/370, f482
627 TNA ADM106/371, f536
628 TNA ADM2/1741, p.18
629 TNA ADM106/370, f488
630 TNA ADM106/370, f396
631 TNA ADM106/371, f540
632 TNA ADM106/371, f542
633 TNA ADM106/371, f544
634 TNA ADM106/371, f548
635 TNA ADM106/369, f29
636 David Hepper, *British Warship Losses*, 1994, p.16
637 TNA ADM2/1750
638 TNA ADM3/278, p.138
639 TNA ADM106/355, f299
640 TNA ADM3/278
641 TNA ADM 106/359, f528
642 TNA ADM2/1750, p.182
643 TNA ADM 7/168, p.26
644 TNA ADM3/278
645 TNA ADM2/1750
646 TNA ADM 106/376, f193
647 TNA ADM1/3566, p.535
648 (TNA ADM7/168
649 TNA ADM106/3556
650 TNA ADM2/1741, p.8
651 Richard Endsor, *Restoration Warship*, 2009, pp.100–114
652 TNA ADM106/381, f269
653 TNA ADM106/381, f267
654 TNA ADM106/3542
655 TNA ADM106/382, f146; ADM106/383, f413
656 TNA ADM106/3566, p.66
657 ADM1/3556, p.295
658 *Journal of Edward Gregory*, 3 May 1687, author's collection
659 *Journal of Edward Gregory*, 4 May 1687, author's collection
660 TNA WO55/1762
661 Essay by William Sutherland, NMM SPB50
662 *The Mariner's Mirror*, 1970, vol. 56, no 2, p.154
663 Science Museum Catalogue no 40B
664 John Franklin, *Navy Board Ship Models*, 1989, Conway's Ship Modelling, pp.150–153
665 Frank Fox, personal communication, 2012
666 Science Museum, 1927-0837, 0001 and 0002
667 Richard Endsor, *Restoration Warship*, 2009, p.46
668 NRS, Catalogue of Pepysian MSS, vol. IV, p.607
669 TNA ADM180/20
670 TNA ADM106/355, f253
671 TNA ADM106/370, f342
672 TNA ADM180/20
673 TNA ADM106/370, f342
674 Frank Fox, *Great Ships*, p.176
675 TNA ADM1/3552, pp.937–945
676 NMM VV598
677 TNA ADM106/384, f214
678 TNA ADM106/355, f273
679 TNA ADM51/3953
680 NMM Catalogue of Ship Models, part 1, 1681-1
681 TNA ADM106/3119, p.7
682 TNA ADM7/169
683 TNA ADM106/3119, pp.3–8
684 ASV Senato comunicate del C.X 1619-28-1, ex S.M.LXXXV-17 and 22
685 William Sutherland, *The Ship-builders Assistant*, 1711, p.76
686 TNA ADM1/3554, p.311
687 TNA ADM18/64, p.28
688 *The Mariner's Mirror*, vol. 41, p.48
689 TNA ADM106/3071
690 TNA ADM106/3542, part 1
691 TNA ADM1/3567, pp.965–969
692 NMM 1221
693 TNA ADM106/367, p.49
694 William Sutherland, *The Ship-builders Assistant*, 1711, p.26 and early draughts
695 John Franklin, *Navy Board Ship Models*, 1989, Conway's Ship Modelling, p.8–11
696 BL Stowe 428, f42, reproduced in Brian Lavery, *The Ship of the Line*, vol. 1, p.194
697 TNA ADM106/3071
698 TNA ADM8/2 and 3
699 Anthony Deane, *Deane's Doctrine*, 1670, Pepys Library, Magdalene College, Cambridge, PL2910, p.73
700 TNA ADM106/357, f251
701 NMM AND/31, p.115
702 NMM MAS/KLT1-6
703 NMM SPB/28, p.7
704 TNA ADM106/3118, pp.44, 69
705 TNA ADM106/3118, p.94
706 William Keltridge, *His Book*, p.32
707 TNA ADM106/344, p.148
708 TNA ADM106/3119
709 William Sutherland, *The Ship-builders Assistant*, 1711, p.77
710 *A Treatise on Ship-building and Navigation*, Mungo Murray, 1754, p.145
711 William Sutherland, *The Ship-builders Assistant*, 1711, p.77
712 *A Treatise on Ship-building and Navigation*, Mungo Murray, 1754, p.145
713 *A Treatise on Ship-building and Navigation*, Mungo Murray, 1754, p.166
714 TNA ADM42/486
715 William Sutherland, *The Ship-builders Assistant*, 1711, pp.77, 82
716 Anon, *The Shipbuilder's Repository*, 1788, pp.370–373
717 Anon, *The Shipbuilder's Repository*, 1788, p.374
718 *A Treatise on Ship-building and Navigation*, Mungo Murray, 1754, p.166 and plate VII
719 John Franklin, *Navy Board Ship Models*, 1989, Conway's Ship Modelling, p.150
720 Anon., *The Shipbuilder's Repository*, 1788, pp.375–376
721 Anon., *The Shipbuilder's Repository*, 1788, pp.376–377
722 William Sutherland, *The Ship-builders Assistant*, 1711, p.82
723 Anon., *The Shipbuilder's Repository*, 1788, p.377
724 Anon., *The Shipbuilder's Repository*, 1788, p.376
725 Anon., *The Shipbuilder's Repository*, 1788, pp.379–380
726 Anon., *The Shipbuilder's Repository*, 1788, p.370
727 William Sutherland, *The Ship-builders Assistant*, 1711, p. 77; Anon, *The Shipbuilder's Repository*, 1788, p.381
728 TNA ADM106/355, p.299
729 *Hoppus's Measurer*, 1790, intro, p.xl
730 TNA ADM2/1752, p.146
731 TNA ADM106/3538, part2
732 Richard Endsor, *Shipwright*, 2013, pp.48–53
733 TNA ADM95/14, p.87
734 William Sutherland, *The Ship-builders Assistant*, 1711, p.260
735 Damian Goodburn, *Woodworking of the Mary Rose*, p.75
736 William Sutherland, *The Ship-builders Assistant*, 1711, p.26
737 TNA ADM106/323; ADM106/329; ADM106/3538
738 William Sutherland, *The Ship-builders Assistant*, 1711, pp.26, 70
739 TNA ADM106/3538
740 William Sutherland, *The Ship-builders Assistant*, 1711, p.46
741 William Sutherland, *The Ship-builders Assistant*, 1711, p.49
742 William Sutherland, *The Ship-builders Assistant*, 1711, p.47
743 *The Mariner's Mirror*, vol. 84, no 2, p.179
744 John Franklin, *Navy Board Ship Models*, 1989, Conway's Ship Modelling, p.150
745 William Sutherland, *Ship-building Unvail'd*, 1717, p.58
746 William Sutherland, *Ship-building Unvail'd*, 1717, p.124
747 *Philosophical Transactions of the Royal Society of London*, vol. 32, 1 January 1722
748 TNA ADM106/3542; ed. David Roberts, *18th Century Shipbuilding: Remarks on the Navies of the English and Dutch, 1737*, p.71; William Sutherland, *Ship-building Unvail'd*, 1717, pp.88, 54, 184
749 William Sutherland, *Ship-building Unvail'd*, 1717, part 1, p.126

[750] TNA ADM106/373 267
[751] TNA ADM106/349 339
[752] William Sutherland, *Ship-building Unvail'd*, 1717, p.127
[753] William Sutherland, *Ship-building Unvail'd*, 1717, p.77
[754] TNA ADM106/364, p.115
[755] Richard Endsor, *Restoration Warship*, p.114
[756] TNA ADM106/351, p.321
[757] TNA ADM20/31, p.799
[758] TNA ADM106/354, p.256; TNA ADM49/24, p.57
[759] Richard Endsor, *Restoration Warship*, p.73
[760] TNA ADM1/3551, p.849
[761] TNA ADM106/355, p.225
[762] TNA ADM106/355, p.229
[763] TNA ADM106/351, p.292
[764] Catalogue of Pepysian MSS, NRS, vol. I, p.189
[765] TNA ADM1/3545, p.179
[766] TNA ADM49/123
[767] TNA ADM106/317, p.202
[768] TNA ADM106/377, p.541
[769] TNA ADM106/372, p.565
[770] TNA ADM1/3548, p.291
[771] TNA ADM106/337, p.35
[772] TNA ADM1/3575, p.149
[773] TNA ADM106/353, p.375
[774] TNA ADM106/350, p.676
[775] TNA ADM106/2507, p.93
[776] TNA ADN106/352, p.372
[777] NMM AND/31; William Keltridge, *His Book*, p.227
[778] TNA ADM106/370, p.321; ADM1/3554, p.37
[779] TNA ADM2/1752, p.264
[780] TNA ADM016/3069
[781] TNA ADM106/352, p.473
[782] TNA ADM106/355, p.303
[783] TNA ADM106/357, p.234
[784] TNA ADM106/3069
[785] TNA ADM106/3538
[786] TNA ADM106/355, p.335
[787] TNA ADM3/278
[788] TNA ADM2/1750, p.27; ADM3/278, p.43; ADM106/48
[789] TNA ADM2/1750; ADM3/278, p.45; ADM106/48
[790] TNA ADM3/278, p.51
[791] TNA ADM106/355, p.241
[792] TNA ADM106/355, p.247
[793] TNA ADM3/278, 10 May 1681
[794] TNA ADM106/355, f253
[795] TNA WO55/1762, Account of the Navy, 16 March 1688
[796] TNA ADM106/3641, part 1
[797] TNA ADM3/278, 21 June 1681
[798] TNA ADM3/278, 25 June 1681
[799] TNA ADM106/42, 11 September 1679
[800] TNA ADM106/49; ADM3/278; ADM1/3552, p.155
[801] TNA ADM106/355, p.271
[802] TNA ADM106/358, p.18
[803] TNA ADM106/49; ADM2/1750, p.66
[804] TNA ADM106/355, p.275
[805] TNA ADM106/3520, p.2
[806] TNA ADM1/3554, p.247
[807] TNA ADM1/3554, pp.311–312
[808] TNA ADM18/63, p.154
[809] TNA ADM106/355, p.361
[810] TNA ADM106/355, p.275
[811] TNA ADM106/355, p.359
[812] ADM106/355, p.205
[813] TNA ADM106/355, p.363
[814] TNA ADM1/3552, p.167
[815] Thanks to Frank Fox for the method of illustrating the masting and rigging layout
[816] NMM AND/31
[817] NMM AND/31
[818] BM ADD MS9303
[819] Richard Endsor, *Restoration Warship*, p.87
[820] TNA ADM106/355, p.273
[821] TNA ADM106/353, p.466
[822] TNA ADM51/3953
[823] William Sutherland Shipbuilding Unveiled, p.23
[824] William Keltridge *His Book*, p.217
[825] TNA ADM106/355, pp.115, 374
[826] TNA ADM51/3953
[827] TNA ADM106/353, p.466
[828] TNA ADM106/355, p.281
[829] John Charnock, *Biographica Navalis*, vol. 2, p.88
[830] Charles Dalton, *English Army Lists and Commission Registers, 1661–1714*, vol. 1, pp.197, 265, 288
[831] TNA ADM106/347, pp.125, 127
[832] TNA ADM3/277
[833] HMC Finch MSS II, p.167, reprinted in J. D. Davies, *Kings of the Sea*, pp.133–134
[834] TNA ADM3/278
[835] TNA ADM106/49; ADM3/278
[836] TNA ADM33/119
[837] TNA ADM106/353, p.464
[838] TNA ADM33/119
[839] TNA ADM6/425
[840] TNA ADM106/49
[841] TNA ADM3/278
[842] TNA ADM33/119
[843] Edwin Chappell, *The Tangier Papers of Samuel Pepys*, p.173
[844] TNA ADM106/353, p.468
[845] TNA ADM51/3953
[846] TNA ADM2/1726
[847] TNA ADM106/355, pp.376, 378
[848] TNA ADM51/3953
[849] TNA ADM106/353, p.462
[850] TNA ADM3/278
[851] TNA ADM51/3953
[852] TNA ADM2/1753
[853] TNA ADM2/1753
[854] Van de Velde painting of the *Tyger* at Berkeley Castle
[855] TNA ADM106/355, p.160
[856] Luttrell I, p.117, reprinted in J. D. Davies, *Kings of the Sea*, p.134
[857] TNA ADM51/3953
[858] David Hepper, *British Warship Losses*, p.12
[859] TNA ADM51/3953
[860] TNA ADM2/1753
[861] TNA ADM2/1750
[862] TNA ADM18/63, p.282
[863] TNA ADM51/3953
[864] J. D. Davies, *Gentlemen and Tarpaulins*, pp.186–187
[865] TNA ADM106/358, p.473
[866] TNA ADM106/353, p.471
[867] TNA ADM2/1753, p.266
[868] TNA ADM51/3953
[869] TNA ADM51/3953
[870] J. D. Davies, *Kings of the Sea*, pp.133–134
[871] J. D. Davies, *Kings of the Sea*, p.134
[872] TNA ADM33/119
[873] TNA ADM2/1751, p.57
[874] TNA WO49/85, p.19. Thanks to Charles Trollope for supplying this information
[875] Adrian Caruanna, *History of Sea Ordnance*, vol. 1, p.56
[876] TNA WO47/5, p.121
[877] British Museum Add 9302, reproduced in Frank Fox, *Great Ships*, p.185
[878] TNA ADM8/1, p.2
[879] TNA WO47/5, p.97v
[880] TNA ADM7/677, p.1
[881] TNA ADM7/827, p.76; ADM49/123
[882] TNA ADM2/1748, pp.131–140
[883] TNA ADM8/1, p.90v
[884] Richard Endsor, *Restoration Warship*, p.151
[885] Catalogue of Pepysian MSS, NRS, vol. 1, p.233
[886] Richard Endsor and Frank Fox, *The Great Ordnance Survey of 1698*,
[887] TNA WO47/2
[888] Frank Fox, private correspondence
[889] TNA ADM2/1750, p.27; WO47/9
[890] TNA ADM2/1750, 15 March 1681
[891] NMM, Michael Robinson Catalogue, p.595
[892] TNA ADM106/355, p.253
[893] TNA ADM1/3552, p.151
[894] TNA ADM106/3541, part 1; ADM8/1, p.177
[895] TNA ADM7/827, p.82
[896] Pepys Library, PL2879, pp. 111–122, reproduced in Frank Fox, *Great Ships*, p.193, The demi-culverins are erroneously called drakes in the Pepys list
[897] TNA ADM 3/278, 2 July 1681
[898] TNA WO51/24, p.32
[899] TNA WO51/24, p.24
[900] TNA WO51/24, p.136; TNA WO51/24, p.39
[901] TNA WO51/24, p.136
[902] TNA WO51/24, p.121
[903] TNA WO55/1736, p.171v
[904] TNA ADM2/1753; ADM3/278, 23 August 1681
[905] TNA ADM3/278, 27 August 1681
[906] TNA ADM2/1753, 27 August 1681
[907] TNA WO51/24, p.39 and 7 September 1681
[908] TNA NA51/24, 7 September 1681
[909] TNA WO51/24, 25 October 1681
[910] TNA WO51/24, 6 September 1681
[911] TNA WO51/24, 9 September 1681
[912] TNA ADM51/3953
[913] TNA WO51/18, pp.71v, 227r, reproduced in Sarah Barter Bailey, *Prince Rupert's Patent Guns*, p.127
[914] TNA ADM1/3559, p.911
[915] TNA ADM1/3559, p.385
[916] TNA ADM106/373, p.70
[917] TNA WO55/1762
[918] NMM ADM/L/T/116
[919] TNA ADM8/2, 1 July–1 October 1689
[920] TNA ADM7/827, p.85
[921] TNA ADM51/4369, part5; ADM52/112, part9; NMM ADM/L/T/121, 28 October 1695
[922] TNA WO55/1736, reproduced by Richard Endsor and Frank Fox in *The Great Ordnance Survey of 1698*
[923] TNA WO51/18, 71v, pp.226r–227r, reproduced in Sarah Barter Bailey, *Prince Rupert's Patent Guns*, pp.105–106, 127
[924] TNA ADM7/827, p.85
[925] TNA ADM106/46, 13 May 1680
[926] Royal Collection
[927] TNA WO55/1717
[928] TNA WO50/13, p.19
[929] Richard Endsor, *Restoration Warship*, pp.160–161
[930] TNA WO51/9, p.146
[931] TNA ADM8/1
[932] TNA WO55/1718

INDEX

accidents 20–23, 60
Admiralty
  and decoration 46
  and disabled sailors 79
  galleasses 86–88
  and galley frigates 89, 97–98
  and guns 258–259
  and innovations 47–48, 49, 57–58, 60, 63
  and maintenance and repairs 25, 30, 32, 35
  and *Margaret Galley* 86
  and *Mordaunt* 123
  new board formed after Oates scare 99, 214
  and plans for rebuilding fleet after Dutch Wars 67, 68–73, **71**
  and *St Albans* 147
  shipbuilding administration 15, 146
  and *Tyger* 208, 235, 236, 237, 252, 256, 260, 261, 262
  and wood and timber supplies 145
  and yachts 44–45
Adventure 82, 98, 165
Advice 20, **21**
Algiers 80–83, **83**
anchors **30**, 56, 239–246, **239**, 291
Anderson, R. C. 62, 123, 148–149
Anne 21–23, 55, **84–85**, 95
apprentices 13
Assistance 39–40, **40**, **41**, 66

Bagwell, Elizabeth 12
Bagwell, Owen 198
Bagwell, William 12, 35, 111
ballast 27, 147, 164, 237, 239
Bantry Bay, Battle of (1689) 33
Barbary States conflict 80–97, **81**, **84–85**
Barfleur, Battle of (1692) 37
Battine, Edward 103, 108–109, 192
Beachy Head, Battle of (1690) 33–35
Beckford, Susanna 232
Berkeley, Captain Charles, 2nd Baron Berkeley of Stratton 7, 174, 246–257, **248**, 261, 262
Berkeley, Admiral John, 3rd Baron Berkeley of Stratton 7, 257
Berkeley Castle 7
Berkeley family 7, 246
Betts, Isaac 11, 235
bevelling 201, **202**, **210**
bitts **59**, 123, 150, **152–154**, 274, 279, 285
boat types **238**, 257, 290
boatswains 25, 28, 31, 32, 79, 96, 247
bows **140–141**, **142**
bowsprits 60, **61**, 147, 240, 290
Boyne 149, **149**
breaking up 41, 73–76, **74**, **76–77**, 78
bresthooks 37, 282, 284
Brisbane, John 252, 255, 256
Brisco, John 247, 256

Bristol 58, 189, **192**
Brouncker, Lord 15
bucklers **33**
bulkheads
  contracts for 275, 277, 280, 282, 284, 285, 286
  other ships 123, 125, 126, **138**, **139**, 150, **163**, 235
  Tyger 164–165, 174, **176–178**, 198, **227**, 235
bumpkins 60, **61**
buoys **145**
Bushnell, Edmund 59, 62, 109, 115, 118

cabins and store rooms 25, 231–235, **234**
  contracts for 274, 275, 277, 280, 282, 283, 284, 286, 289–290
Canning, Captain 48, 96–97
capstans 50, **51**, **152–154**, **183–185**, 213
  contracts for 275, 277, 280, 282, 285
careening 22
carlings 232
Castle, Robert 37, 40, 283
catamarans *see* double-bottomed vessels
Catherine of Braganza, Queen 16
caulking 286
Centurion 235
Charles II **11**, **100–101**
  and Berkeley 246, 256
  and catamarans 45
  and decoration 46
  and disabled sailors 79
  falls out with new Admiralty Board 99
  and frigates 42, 88–89, 92, 96, 97–98
  and innovations 42, 47–49, 52–55, 56–58, 60, 63, 150
  interest in shipbuilding and ships 8, 10, 15, 41
  and other ships 16–19, 27, 123, 144, 145, 146–147
  and plans for rebuilding fleet after Dutch Wars 67–73, **71**, 76
  Popish Plot 99, 214
  and *Tyger* 99, 174, 196–197, 235–236, 237
  Tyger visit 7, **9**, **250**, 252, **252–253**, **254–255**
  and warrants 78
  and yachts 44–45
Charles 44–45, 55, 99, 149, **149**
Charles Galley **90–91**, 95
  accommodation 233
  in action 80, 81, 96–98
  building and design 92–94, 108
  guns 99
  ranges 47, 49
  structural problems 189
Charlotte 27, 44, **45**, 56
Chatham
  finishing work at 28, 33

repairs at 33, 35, 39
  shipbuilding at 146, 148–149
chocks 147, **195**, 203, **215**, **230**, 284, 286
clamps **232**, 274, 277, 279, 282, 285
compasses **143**, 144–145
Constant Warwick 41, 42, 68, 82, 236, 266–270
contracts 8, 15, 271–287
cookroom 42–43, 46–47, **48**, 125, 126
  contracts for 274, 275, 277, 280, 285, 286
corsair ships 82–85
cross pillars 29, 185
crosspawls 213
crotches **37**, 284
Cutty Sark 79

Danby, Thomas Osborne, Earl of 68–70, **68**, 73, 74
davits **187**
Deane, Sir Anthony **16**
  and corsair ships 83–85
  Doctrine of Naval Architecture 104, 108–109, 110, **110**, 121, 192
  and galley frigates 92, 94, 98
  and innovations 46, 56–57, 58
  and keel length 62
  and Pepys 11
  and *Saudadoes* 12, 15–16, 18
  shipbuilding 15, 18, 256, 265
  survey of fleet at end of Dutch Wars 67
Deane, Anthony Junior, 88, 89–93
decks
  contracts for 274–275, 276–277, 279–280, 282, 285–286
  height between 108, 126, 151
  other ships 126, **135–137**, **159–161**
  Tyger 151, **219–222**, 231, **263**
decoration 46, **47**, 126, **226**, 231, 235
  contracts for 275, 278, 280, 283
Deptford **78**, 79
  Charles II's relationship with 10
  double dock **72–73**
  Great Storehouse **197**
  hulk at **246**
  master shipwrights 12, 13
  oar making at 94
  other ships from 122, 145–148, 214, 274, 283
  pump trials 55
  rebuilds, repairs and refits at 37, 39–40, 68
  St Nicholas Church **104**
  ship launches 99, **100–101**, 236
  shipbuilding trades employed at 13
  shipwrights discharged 214
  survey of ships at 65
  and *Tyger* 64–66, 68–79, 196–197
Deptford 148–149, 164
desalination 60

301

Diamond 24, 25, 50, **51**, 144–145
discipline 251
double-bottomed vessels (catamarans) 45–46, **46**
Dragon 25, 65
draughts and plans
 calculating and drawing curves 8, 106, **107**, 108–121, **110–121**, 124–125, **125**, 165, 198, **199**
 fourth-rate ship **166–167, 168–169, 170–171**
 Mordaunt 124–126, **125, 131–134**
 rising lines of breadth 165
 St Albans 150
 Shish's *Dimensions* on 99, **102–103**, 104–109, 116–121, 151, 164
 Tyger 151–192, **179–185**, 198
Duchess 44, 99, 197
Dummer, Edmund
 Draught of the Body 103
 drawings by **30, 170–171**
 and galley frigates 86–88
 plan of Deptford 198
 recording lines 124
 and ship dimensions 60–61, 62
 Venice visit 165
 and *Woolwich* 35, 37
Dunkirk 260
Duteil, Sir Jean Baptiste 85

Eagle 23, 24, 65, 67, 68
Elizabeth 23, 60
engines 59–60, **59**
English Channel **249**
Evelyn, John 12, 44, **44**, 110
Ewbank, Colonel 47, 48–49
explosions 20–23

Fagge, Thomas 192, 194, 201
fashion pieces 206, **208, 213, 228**
figureheads 46, 123, 174
Finch, Lord Daniel 236, 246
finishing work 28, 235–236
fire buckets 60, **61**
fishes 29
flats *see* midship flats
floor timbers **149**
 assembling 209, **209, 210, 212, 213, 215, 229**
 bevelling 201, **202**
 contracts for 274, 276, 279, 281, 282, 284
 in general 82, 191
 moulds 198
forefeet 63
Foresight
 accident avoided 23
 accommodation 233–234
 contract 274–276
 frame 192
 overview 65
 sheathing 58
 space and room 188
 and *Tyger* 235

forward rising wood moulds 206, **206**
frames
 arrangement **148–149**, 149
 making 208–209, **208–213**
 Tyger 174–192, **176–178, 194, 195**, 208–209, **208–213, 215, 228–230, 232**
frigates, first 42–43
furniture **233, 234**
furring *see* girdling and furring
Furzer, Daniel 29, 35, 192
futtocks
 assembling 209, **213, 215, 229**
 bevelling 201, **202**
 contracts for 274, 276, 279, 281, 284
 and models **148**, 149, **149**, 164
 moulds 203–206, **203–207**
 third-rate ship **194**
 Tyger **178, 182**, 189–192, **195**

galleasses 86–88
galley frigates 86–99, **89, 90–91, 92–93**
 see also Charles Galley; *James Galley*
galleys 85–88, **86–87, 88**
Garland 95, 188–189
girdling and furring 18–20, **19**
Glorious Revolution (1688) 33
Gloucester 255
Godfrey, Captain William 251, 255
Golden Marigold 97, 98
Goodman, Jasper 82–83
Gourdon, Sir Robert 55–56
graving 246, 286
Greenwich 41, 231
gudgeons 21
Guernsey 65, 73
gun carriages 265, **266, 267**
gundecks 126, 151, **220–222, 231, 263**
 contracts for 274, 276, 277, 279, 285
gunports
 contracts for 274, 277, 279, 282, 285
 other ships 29, 126, **140–141**, 150, **192**, 291, 295
 in Shish's *Dimensions* 108
 Tyger **176–178**, 188–189, **220–222, 231, 263**
gunpowder 20–23, 235
guns
 Charles Galley 99
 gun establishment of 1677 258–260
 gunner's stores 266–270
 manning **259**, 265
 Tyger **188**, 236, 256, 258–270, **268–269**
 warships 18

Haddock, Sir Richard 74–75, 236, 239, 255, 289
Hampshire **19**, 20, **47**, 97, **150**, 164
Hampton Court 56, 99, 197, 208, 211
Happy Return 32, 256
Harding, Fisher 35, 37, 55, 93, 147, 149–150
hatches and hatchways 279, 286
hawse pieces 56, 123, **229, 230**
 contracts for 274, 277, 279, 284

heads **139, 227**, 231
 contracts for 275, 277, 280, 282, 286
hearths 29, 42, 47–49, **48**
Helby, Joseph 28, 231
Henrietta 45, 58, 252
Herbert, Arthur 81, 82
Hestor 21, 22
La Hogue, Battle of (1692) 37
holds 275, 277, 280, 282, 284
Holmes 68
Hooke, Robert 15
Hudson, Henry 246, 247
hull lines 164–165

ice 24
irons **21**

James II
 deposed 33
 and *Gloucester* accident 255
 and innovations 55, 56–57, 63
 naval appointments made by 85
 and Oates 99
 and ships 11, 44, 45, 88–89, 148
James Galley **72–73, 90–91, 94**
 in action 80, 81, 96–98
 building and design 92–94
 launch 73, **92–93**, 94
 oars 95
 ranges 47, 48–49
 sheathing 58, 94
Jersey 52, 85, 233
Les Jeux (renamed *Play Prize*) 42, **43**
Johnson, Henry 25, 39, 92
Jull, Stephen 257

Katherine 18, 96
keels
 contracts for 274, 276, 278, 281, 283
 length 60–63, **62**, 150–151, 289
 making 208–209, **212, 213**
keelsons 209, 230, 274, 276, 279, 282, 284
Keltridge, William 108, 110, 121, 125, 164, 165, 192
Killigrew, Admiral Henry 144–145
Kingfisher 49, 82
knees **29**, 56–57, **57, 232**, 284, 285

Lawrence, Joseph 11, 31–33, 42–43
Leadman, John 28, 231
leaks 23–24
ledges **232**
Lee, Robert 11, 35, 146, 237
Lenox **196**
 building and design 28, 54, 56, 147, 208, 211
 launch 99, **100–101**, 197
 name 44
 replica project **79**, 197
Leopard 235
Lewsley, Thomas 144, 231
Lodgingham, Robert 55, 56

INDEX

London 12, 20, 237
longboats **238**, 257
Lymehouse 35–37
Lyon **34**, 58

maintenance 25–28
Margaret Galley 85–86, **86–87**
Mary 252, 262
Mary Galley 93, 108, 151, 281–283
Mary Rose **259**
master shipwrights
   duties 13–15
   rewards for building new ships 237
masts and yards
   contracts for 275, 276, 278, 279, 280, 282, 283, 284
   other ships 144, 147, 290
   pilfering 14
   Tyger **175**, 237, 240, 241, **242**
Medway 288
midship flats **107**, 108, 117, 126, 150, 164
Miller, Thomas 62, 104, 109
   drawings by **244**
mooring 24, **24**, **30**
Mordaunt **124**, **140–141**
   cabins 232
   draught 124–125, **125**, **131–134**
   establishment 126
   hull lines 164
   models **122**, 123–126, **127–142**
   in the Navy 142–145
   overview 122–145
   stores and equipment 143–145, **143**, **145**
   survey 123, 142–143, 289–295
   views of design features **127–130**, **135–142**, **150**
Moreland, Sir Samuel 50, 51–52, **54**, 55–56
Morgan, Henry 79, 247
moulds 198–206, **199–208**
Murray, Mungo 108, 198
   drawings by **201**

Narborough, Sir John 58, 80–81, 83, 97, 289
Navy Board
   and accommodation 233–234
   and contracts 271
   and discharge of staff 232
   and galleasses 86–88
   and galley frigates 89, 92, 96, 97–98
   nd ice 24
   and innovations 48, 49, 55–56, 57–58, 59, 60
   and maintenance and repairs 25, 35, 39
   and other ships 123, 143, 147
   plans for rebuilding fleet after Dutch Wars 67, 70, 73
   rewards given to master shipwrights 237
   and shipbuilding administration 10
   and *Tyger* 66, 68, 74–75, 236, 237, 261
   and yachts 44, 45
Newcastle 60, 108, 234–235
Nonsuch 42, 97
Norwich 188, 192, 283–287

oar scuttles 188–189, **192**, **193**, 282
oars and rowing 94–96, **95**
Oates, Titus 98–99
Old James 67, 76
Ordnance Board 258–259, 261, 262
orlops **183–185**, **234**, 235, 275, 279, 284
Oxford 20, 82, 99, 233, 235

paddle wheels 42, 59–60, **59**
paintwork 235
pensions 79
Pepys, Samuel **11**
   and catamarans 46
   on Charles II and ships 42, 44, 45
   and disabled sailors 79
   and discipline 251
   favourite shipwrights 11–12, 13
   and guns 258–260
   and innovations 52, 56–57, 58, 60–61, 63
   and Moreland 55
   and naval administration 10–11
   and other ships 16, 18, 20, 23, 33, 83–85, 92, 98
   and plans for rebuilding fleet after Dutch Wars 67, 68, 70, 74
   proposal to repair whole fleet 31
   resigns 99
   shipbuilding documents collection 103–104
   on shipwrights and mathematics 121
   and Shish's *Dimensions* treatise 99, 103, 104
   and Tangier expedition 31
   and *Tyger* 7, 8, 262
   and warrants 78
   and women 12
Pett, Peter 42
Pett, Peter II 64
Pett, Sir Phineas 11, 48–49, 60, 116
Pett, Phineas II 18–20, 64
Pett, Phineas III 27, 44, 63, 92–93, 98
Pett, Mrs (wife of above) 98–99
Pett, Mr (master shipwright) 47
Petty, Sir William 15, 45–46, **45**
Phoenix 23, 57–58, 144, 233
pinnaces 56, **238**, 257, 290
pintles **21**, 275, 278, 280, 282
planking 214–231, **215**, **230**
   contracts for 274, 275, 276, 277, 279, 280, 282, 284, 286
plans *see* draughts and plans
Play Prize see Les Jeux
Plymouth 20, 23, 67, 80
Popish Plot 99, 214
Portsmouth 20, 25, 28–30, 33, 38, 95
Portsmouth 39, 55, 65
Portsmouth, Louise de Kérouaille, Duchess of **44**
Powis, John 257
Prince 18–20
Princess 192
pumps 51–56, **51–54**, 146–147, 284

quickwork 29

ranges, iron 47–49, **48**
Read, Herbert 150
repairs and refits overview 18–20, 28–39
Reserve 23, 50, 88
ribbands 202, 209, **209**, **212**, **213**
riders 37, 274, 276, 279, 282, 284
rigging *see* sails and rigging
rising and narrowing lines *see* draughts and plans
rising timbers **215**, **229**, 284
rot 25, 28
roundhouses **138**, 275, 277, 280, 286
Rounsevall, John 247
rowing *see* oars and rowing
Royal James 56, 57, 121
Royal Katherine 15
Royal Navy
   care for disabled sailors 79
   condition at end of Dutch Wars 67–73, 74
Royal Oak 12
Royal Society 15, 45–46
Royal Sovereign 24, 56, 149, **149**
rudders **21**, 275, 278, 280, 282, 286
Rupert, Prince 85, 89

sails and rigging
   contracts for 275, 278, 280, 282
   maintenance 27
   other ships 293–294
   Tyger 239, **240–245**
St Albans 122, 145–151, **152–163**, 164
St David 192, 246
St Patrick 192, 276–278
Sapphire 95, 256, 265
Saudadoes 12, 15–18, **17**
scuppers 29, **29**, 231
Sedgemoor 55, 148–149, 164
sheathing 28, 30, 42, 57–58
Sheeres, Sir Henry 8, 46, 98, 109
Sheerness **149**
   accidents 21–22, 24
   history 12–13
   shipbuilding and repairs at 31–33, 35, 37, 288
ship weight, calculating 121
shipbuilding
   contracts 8, 15, 271–287
   costs for fourth-rate ship 146
   government control 10, 15
   visual glossary **272–273**
   warrants 288
ships
   alterations 15–18
   launching **236**, 237
   ordering new 15
shipwrights
   percentage at Deptford 13
   see also master shipwrights
Shish, John **232**
   background, character and skills 12–13
   and *Centurion* 235
   death 147

Dimensions treatise 99, **102–103**, 104–109, 116–121, 151, 164
  on explosion of *Anne* 21–22
  and innovations 48–49, 60
  and *Mordaunt* 123, 143, 144, 145–147, 295
  overview 8
  and paintwork 235
  Pepys's attitude to 11, 12
  and *St Albans* 122, 145–147
  and *Tyger* 65, 68, 73–75, 99, 151, 164, 165, 197, 198, 209, 211, 231, 236, 237, 239, 261
Shish, Jonas
  background and character 11, 12, 13
  and discharge of shipwrights 214
  models 149, **149**
  and *Mordaunt* 295
  rebuilding *Saudadoes* 16–18
  and ship repairs 68
  shipbuilding 20, 235, 274
  and *Tyger* 65–66, 99
Shish, Thomas
  and innovations 60
  positions held 13
  shipbuilding 99, 231, 235
  and *Tyger* 65
  and *Woolwich* 30, 31
Shovell, Captain Cloudesley 55, 265
sirmarks **182**, **199**, **202**, **203**, **204**, **209**
slavery 82, 85–88, **89**, 96
Smith, Thomas 52–55
Solby, Edward 256, 257
Solebay, Battle of (1672) 57, 64
Sovereign 67, 237
space and room 174–188, **195**
Sparks, Cuthbert 79, 247
Spence, Robert 149, 150
square tucks **26**, 37
standards **27**
  contracts for 274, 276, 277, 279, 280, 284, 285, 286
  iron 56
steering wheels 49, **50**
stems 62, 63, 150, 165, 209, **212**, **213**
  contracts for 274, 276, 278, 281, 284
sternposts 111, 209, **213**, **228**
  contracts for 274, 276, 278, 281, 284
sterns
  contracts for 275, 278, 280, 282
  other ships **127**, 150, **155**, **168–169**
  Tyger **190–191**, **226–228**, 231
Stigant, William 116–117
Stirling Castle 49, **50**, **233**
store rooms *see* cabins and store rooms
stores 31, 32, 143, 145, 251–252, 266–270, 294
Sutherland, William
  on draughts 104
  on frames and planking **212**, **215**
  on keels 62, **62**, 63
  on master shipwrights 12
  and moulds 198
  on rising and narrowing lines 111, **112**, **199**
  on *St Albans* 148

The Ship-builders Assistant **62**, 104, 111, **112**, **121**, 198, **199**, **212**, **215**
Ship-building Unvail'd 50, 54, **62**
sweeps 108, 198

tallowing 29
Tangier 31, 85–86
taxation 70, 74
Taylor, James 39, 276
Thompson, Albion 246–247, 256, 262
Tippetts, Sir John
  discharges shipwright 214
  and galley frigates 92
  and other ships 19, 25
  Pepys on 11
  and plans for rebuilding fleet after Dutch Wars 67, 73–74
  and sheathing 58
  and *Tyger* 68, 74–75
toptimbers
  assembling and arrangement **148**, 209, **210**, **229**
  bevelling 201, **202**, 210
  contracts for 279, 282, 284
  moulds 206, **207**
  other ships 82, 192
  Tyger 174–188, **178**, **182**, 192, **195**, 198, 206, 209, **210**, 229
transoms 150, 206, **213**, **228**
  contracts for 274, 277, 279, 282, 284, 285
treenails 230–231, **230**
Tyger
  in action 42, 256–257, 262
  appearance and decoration 165–74, **226**, 231, 235
  broadside **223–225**
  bulkheads 164–165, 174, **176–178**, 198, **227**, 235
  cabins and store rooms 231–235, **234**
  decks 151, **219–222**, 231, **263**
  draught 151–192, **179–185**, 198
  draught of similar ships 122–151
  drawings and paintings **175**, **187**, **188**, **190–191**, **247**, **249**, **250**
  fitting out 239–246
  frame 174–192, **176–178**, **195**, 208–209, **208–213**, **215**, **228–230**, **232**
  gunports and scuttles **176–178**, 188–189, **193**, **220–222**, 231, **263**
  guns **188**, 236, 256, 258–270, **268–269**
  hull lines 164–165
  King's visit 7, **9**, **250**, 252, **252–253**, **254–255**
  launch 237
  manning 76–79, 246–251, 257, 265
  masts and yards **175**, 237, 240, 241, **242**
  models **148**
  paintwork and finishing work 235–236
  previous life and appearance 64, **69**, **72–73**
  rebuild plans 99
  rebuilding 196–236
  rebuilding sequence **211**

runs aground 255
sails and rigging 239, **240–245**
sections **150**, **183–186**
setting sail 251
ship's boats **238**
size and dimensions 150
stern **190–191**, **226–228**, 231
stores 251–252
surveys and disassembly 65–79, **70–71**, **74**, **76–77**
topside **216–218**

Velde, Willem van der the Elder
  Mordaunt drawings **124**, 126, **140–141**
  Navy salary 174
  other artwork by **26–27**, 37, **38–39**, 44, 49, **95**
  studio 260
  Tyger artwork 165–174, 189, **247**, **249**, **254–255**, 260
Velde, Willem van der the Younger
  artwork by **17**, **43**, 44, **88**, **92–93**, **94**
  studio 260
Victory 41

wales 126, **140–141**, **142**, 174, **228**, **232**, 286
Walker, Isaac 235
warrant officers 76–79, **76–77**, 247
Warren, Sir William 231, 237
Wheeler, Captain Francis 257
Willshaw, Captain Thomas 88, 288, 295
Winchester 192
Windsor Castle 265, **267**, **269**
Wood, Captain John 266
wood and timber
  carpenter's stores 31, 32, 294
  contracts for 274–275, 276–277, 279–280, 281–282, 284–286
  for planking 214–231, **215**, **230**
  and repairs 25, 32
  for *St Albans* 145–146
  supplies 13–15, **14**
  for *Tyger* 208, 231
  working frames 208–209, **209–213**, **215**, **228**
  see also floor timbers; toptimbers
Woolwich
  master shipwrights 11, 13
  repairs and refits at 30–31, 35, 37, 38, 39, 42–43
  safety of anchorage 24, 30–31
  shipbuilding at 28, 92, 146, 148
Woolwich **26–27**, 28–39, **38–39**

yachts 44–45, **45**
York, Duke of *see* James II

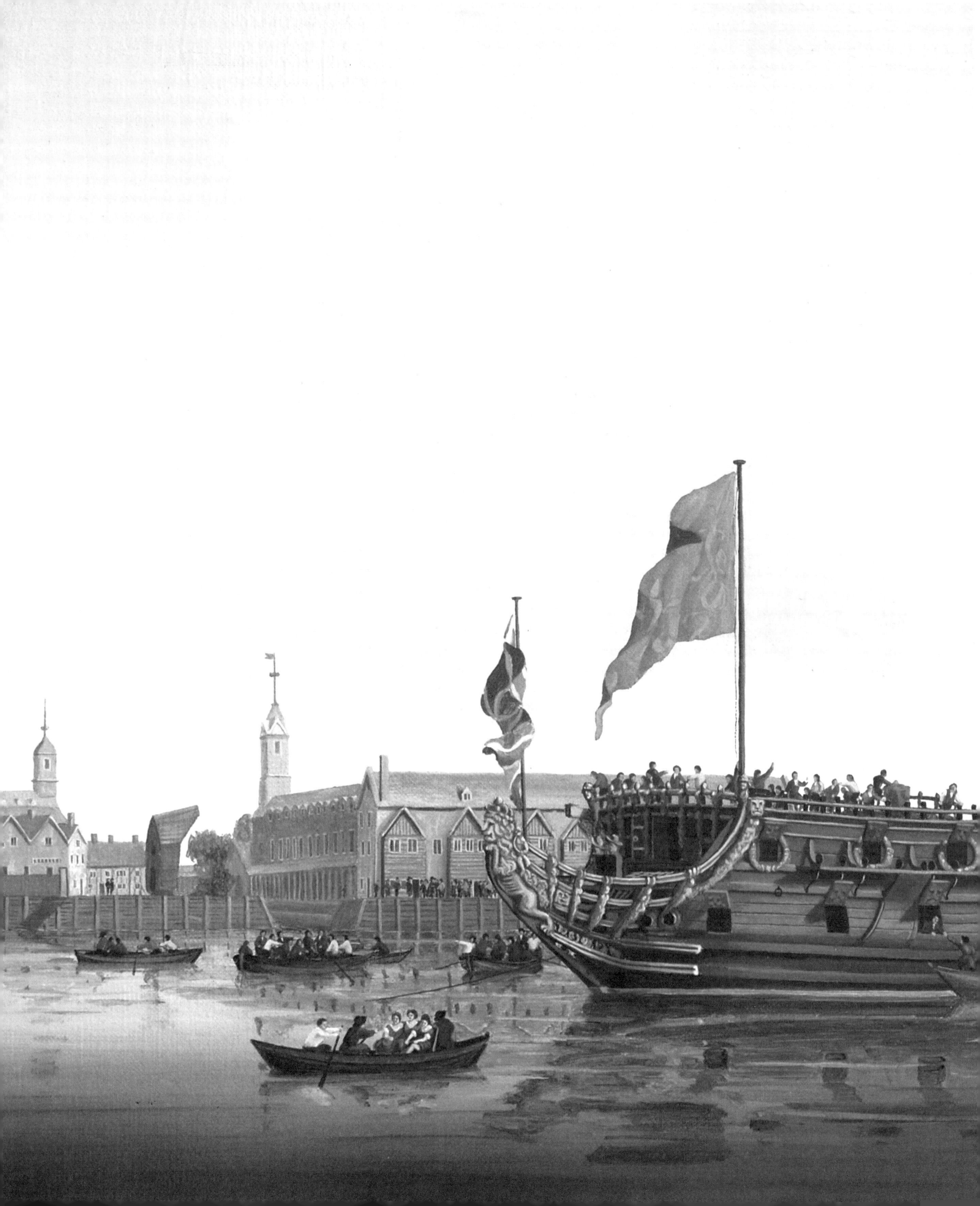